Molecular Catalysis

Volume 1

Series Editors

Piet W. N. M. van Leeuwen, Institut National des Sciences Appliquées de Toulouse, LPCNO, Toulouse, France

Carmen Claver, Departament de Química Física i Inorgànica, Universitat Rovira i Virgili, Tarragona, Spain

Nicholas Turner, School of Chemistry, University of Manchester, Manchester, UK

This book series publishes monographs and edited books on all areas of molecular catalysis, including heterogeneous catalysis, nanocatalysis, biocatalysis, and homogeneous catalysis. The series also explores the interfaces between these areas. The individual volumes may discuss new developments in catalytic conversions, new catalysts, addressing existing reactions and new reactions regarded as desirable from a societal viewpoint. The focus on molecular insight requires an appropriate attention for synthesis of catalytic materials, their characterization by all spectroscopic and other means available, and theoretical studies of materials and reaction mechanisms, provided the topic is strongly interwoven with catalysis. Thus the series covers topics of interest to a wide range of academic and industrial chemists, and biochemists.

More information about this series at http://www.springer.com/series/15831

Piet W. N. M. van Leeuwen · Carmen Claver
Editors

Recent Advances
in Nanoparticle Catalysis

 Springer

Editors
Piet W. N. M. van Leeuwen
LPCNO
Institut National des Sciences
Appliquées de Toulouse
Toulouse, France

Carmen Claver
Departament de Química Física i Inorgànica
Universitat Rovira i Virgili
Tarragona, Spain

ISSN 2522-5081   ISSN 2522-509X   (electronic)
Molecular Catalysis
ISBN 978-3-030-45825-6   ISBN 978-3-030-45823-2   (eBook)
https://doi.org/10.1007/978-3-030-45823-2

This Springer imprint is published by the registered company Springer Nature Switzerland AG
The registered company address is: Gewerbestrasse 11, 6330 Cham, Switzerland

Preface

In the last two decades we have witnessed a spectacular growth of nanoparticle catalysis and clearly in the new series of Molecular Catalysis this topic is of great interest. A look at the numbers shows that "spectacular growth" is even an understatement. Thirty years ago, the number of hits for "nanoparticle" and "catalysis" found in Scifinder or other search machines is zero, which does not mean that it did not exist at all, but it was not documented as such. Twenty years ago, barely one publication a week appeared on this topic, but today the number amounts to almost 200 publications per week, and an accumulated total of almost 90,000 indicates what size of research area we are dealing with. No way, as in many other fields today, we can put together comprehensive volumes on nanoparticle catalysis. The area is also extremely diverse, in materials, their preparation, their use, and the catalytic reactions explored. The number of catalytic reactions explored is much smaller than the number of materials developed, but the diversity is so enormous that general lessons are hard to be deduced from the results. Thus, from the one million of catalysis results on nanoparticle catalysis it will be difficult to develop general theories, not in the least because we do not report our results in a format suited for later machine learning. The area outnumbers homogeneous catalysis, while comparison with heterogeneous catalysis is less clear due to overlap. Compared to both homogeneous and heterogeneous catalysis new catalytic activity is found for nanosized particles; no need to explain this here. Furthermore, the number of metal atoms as potential participants as active centers is larger in nanoparticles than in particles of micrometer size. It is often argued that an advantage of MNPs over molecular homogeneous catalysts is their ease of separation, but we doubt that, and MNPs for industrial use need to be immobilized in most reactions. In the last decade the two classic areas of homogeneous and heterogeneous have found that MNP catalysis brings the two closer to one another; in particular ligand modification and support influences have led to an increase of understanding in catalysis in a broad sense.

This book provides the reader a rapid overview of the latest developments in this field of catalysis with chapters representing the relevant activities. A broad area has been covered describing not only the established topics in depth, but also several new, recent topics have been included, some of which may be new to the interested researcher. The topics have been selected in such a way that readers of related disciplines will find a quick introduction to catalysis with nanoparticles and those readers familiar with the field will find an update of recent advances and a few future developments. In the first chapters the emphasis is on new materials with exceptional properties in catalysis such as very small clusters, single atom layers, and precise clusters. For all new catalytic systems the authors dwell upon the synthesis of the new particles, but also present the new avenues for catalysis. Several chapters report on new support materials that have a strong influence on the catalytic outcome. The last chapters focus on catalytic reactions in which selectivity plays a major role and they highlight the effect of ligands and supports on the MNP catalyst activity and selectivity. Catalysis with nanoparticles is exponentially expanding and receives a lot of attention as many exciting findings are being reported. Control over their synthesis, characterization, stabilization and use offers potentially a rich area with possibilities that far exceed those of classic heterogeneous and homogeneous catalysis.

We are very grateful to the authors who have contributed to this volume and who have shown in only 12 chapters the diversity covered in the field of nanoparticle catalysis.

Toulouse, France   Piet W. N. M. van Leeuwen
Tarragona, Spain   Carmen Claver
December 2019

Contents

Contributors

J. M. Asensio LPCNO, INSA, CNRS, Université de Toulouse, Toulouse, France

M. Rosa Axet LCC-CNRS, Université de Toulouse, INPT, UPS, Toulouse, France

D. Bouzouita LPCNO, INSA, CNRS, Université de Toulouse, Toulouse, France

Marc Camats Laboratoire Hétérochimie Fondamentale et Appliquée (UMR 5069), Université de Toulouse, CNRS, Toulouse, France

Israel Cano Departamento de Física Aplicada, Facultad de Ciencias, Universidad de Cantabria, Santander, Spain

Kun Cao Huazhong University of Science and Technology, Wuhan, People's Republic of China

B. Chaudret LPCNO, INSA, CNRS, Université de Toulouse, Toulouse, France

Rong Chen Huazhong University of Science and Technology, Wuhan, People's Republic of China

Jorge A. Delgado Centre Tecnològic de la Química, Tarragona, Spain

Amarajothi Dhakshinamoorthy School of Chemistry, Madurai Kamaraj University, Madurai, India

Jairton Dupont Institute of Chemistry, Universidade Federal do Rio Grande do Sul, UFRGS, Porto Alegre, RS, Brazil

Hermenegildo Garcia Departamento de Quimica, Instituto Universitario de Tecnologia Quimica (CSIC-UPV), Universitat Politecnica de Valencia, Valencia, Spain

Cyril Godard Departament de Química Física i Inorgánica, Universitat Rovira i Virgili, Tarragona, Spain

Montserrat Gómez Laboratoire Hétérochimie Fondamentale et Appliquée (UMR 5069), Université de Toulouse, CNRS, Toulouse, France

Rajenahally V. Jagadeesh Synergy Between Homogeneous and Heterogeneous Catalysis, Leibniz-Institut für Katalyse e. V, Rostock, Germany

Rongchao Jin Department of Chemistry, Carnegie Mellon University, Pittsburgh, PA, USA

Antonio Leyva–Pérez Instituto de Tecnología Química (UPV–CSIC), Universitat Politècnica de València–Consejo Superior de Investigaciones Científicas, València, Spain

Site Li Department of Chemistry, Carnegie Mellon University, Pittsburgh, PA, USA

Xiao Liu Huazhong University of Science and Technology, Wuhan, People's Republic of China

Luis M. Martínez-Prieto Instituto de Tecnología Química (ITQ), Universitat Politècnica de València-Consejo Superior de Investigaciones Científicas (UPV-CSIC), Valencia, Spain

Yuanyuan Min LCC-CNRS, Université de Toulouse, INPT, UPS, Toulouse, France

Judit Oliver–Meseguer Instituto de Tecnología Química (UPV–CSIC), Universitat Politècnica de València–Consejo Superior de Investigaciones Científicas, València, Spain

A. Palazzolo SCBM, CEA, Université Paris Saclay, Gif-sur-Yvette, France

G. Pieters SCBM, CEA, Université Paris Saclay, Gif-sur-Yvette, France

Daniel Pla Laboratoire Hétérochimie Fondamentale et Appliquée (UMR 5069), Université de Toulouse, CNRS, Toulouse, France

Muhammad I. Qadir Institute of Chemistry, Universidade Federal do Rio Grande do Sul, UFRGS, Porto Alegre, RS, Brazil

David Rivillo Institute of Chemistry, Universidade Federal do Rio Grande do Sul, UFRGS, Porto Alegre, RS, Brazil

Philippe Serp LCC-CNRS, Université de Toulouse, INPT, UPS, Toulouse, France

Bin Shan Huazhong University of Science and Technology, Wuhan, People's Republic of China

Nathália M. Simon Institute of Chemistry, Universidade Federal do Rio Grande do Sul, UFRGS, Porto Alegre, RS, Brazil

S. Tricard LPCNO, INSA, CNRS, Université de Toulouse, Toulouse, France

Piet W. N. M. van Leeuwen LPCNO, Institut National des Sciences Appliquées-Toulouse, Toulouse, France

Chapter 1
Ligand-Free Sub-Nanometer Metal Clusters in Catalysis

Judit Oliver–Meseguer and Antonio Leyva–Pérez

Abstract Recent advances in the synthesis and characterization of ligand-free sub-nanometer metal clusters, either in solution or supported on solids, have enabled their rational use as catalysts in new reactions. These clusters expose all their metal atoms to outer molecules without the potential steric/electronic interferences of ligands, while, at the same time, show defined molecular orbitals to ultimately control the catalytic reaction. Therefore, these clusters somehow combine the advantages of single atom and metal nanoparticles for molecular activation and catalysis. This chapter aims at describing how to prepare, better characterize and apply, on a realistic scale, ligand-free sub-nanometer metal clusters for catalytic processes (many of them of industrial interest) during the last ten years.

Keywords Sub-nanometer metal clusters · Catalysis · Ligand-free · Industrial reactions

1.1 Introduction

1.1.1 History and Nanotechnology Projection

The prefix *nano* gives the name to a discipline that is considered the scientific sign of our generation, the nanotechnology, although this is not a novelty for humanity. The book written by the philosopher and medical doctor Francisci Antonii in 1618, *Aurea–Auro Potabile*, is considered as the first book on colloidal Au. The reputation of soluble Au until the Middle Ages was to disclose fabulous curative powers for various diseases, such as heart and venereal problems, dysentery, epilepsy and tumors and for diagnosis of syphilis. Also, this book includes considerable information on the formation of colloidal Au solutions and their medical uses, including successful practical cases (Fig. 1.1), whereas the extraction of Au started in the 5th millennium

J. Oliver–Meseguer · A. Leyva–Pérez (✉)
Instituto de Tecnología Química (UPV–CSIC), Universitat Politècnica de València–Consejo Superior de Investigaciones Científicas, Avda. de Los Naranjos s/n, 46022 València, Spain
e-mail: anleyva@itq.upv.es

© Springer Nature Switzerland AG 2020
P. W. N. M. van Leeuwen and C. Claver (eds.), *Recent Advances in Nanoparticle Catalysis*,
Molecular Catalysis 1, https://doi.org/10.1007/978-3-030-45823-2_1

Fig. 1.1 Extraction of the Panacea aurea, sive, Tractatus duo de ipsius auro potabili, from Francisci Antonii

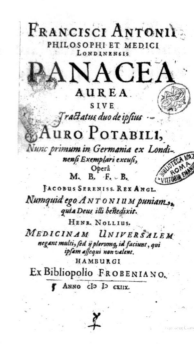

B.C. near Varna (Bulgaria) and reached 10 tons per year in Egypt around 1200–1300 B.C., when the marvelous statue of Touthankamon was constructed, it is probable that "soluble" Au appeared around the 5th or 4th century B.C. in Egypt and China [11].

Thus, the discovery of nanoparticles was not a novelty in modern times, the merit was to discover that they were here. In 1857, Faraday used for the first time the concept nanoparticle [16], and we had to wait one century for the development of the nanotechnology after the improvement of the instruments that give access to the nanometric scale, like the scanning tunneling microscope (STM), promoted by Gerd Binning and Heinrich Rörher in 1982 (Nobel Prizes awarded in 1986) and the atomic force microscope (AFM), developed also by Binning with Christoph Gerber and Calbin Quate in 1987. Two years later, Don Eigler and Erhard Schweizer wrote the letters IBM (where they worked) manipulating 35 Xenon atoms on a Ni surface, showing the possibility to move atoms from a surface using the tip of the STM (Fig. 1.2).

1.1.2 Differences Between Nanoparticles and Sub-Nanometer Clusters

A nanoparticle is every particle with dimensions between 1 and 10^3 nm (Fig. 1.3). The chemical properties of these nanoparticles depend not only on the type of constituent

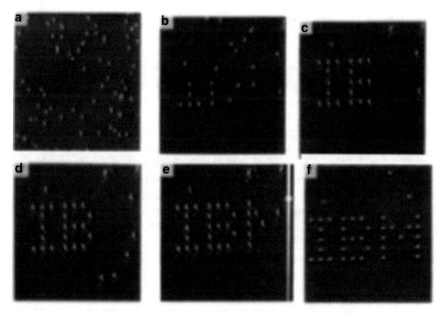

Fig. 1.2 A sequence of STM images taken during the construction of a patterned array of xenon atoms on a nickel (1 1 0) surface. The atomic structure of the nickel surface is not resolved. The ⟨1 1 0⟩ direction of the surface runs vertically. **a** The surface after xenon dosing, **b–f** various stages during the construction of the word IBM. Each letter is 50 (Å) from top to bottom

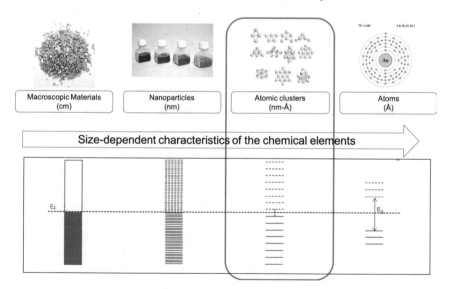

Fig. 1.3 Scheme of the size-dependent characteristics of metallic chemical elements, here illustrated for gold

atoms but also on the relationship surface/volume, the size and shape and the oxidation state, among other parameters. In the case of metallic nanoparticles, the most relevant characteristic is the plasmon. Bulk plasmons are quantum waves from electrons that are produced when the electrons are perturbed with an incident beam, changing the equilibrium position and vibrating in a given characteristic frequency. In other words, the plasmon is a collective oscillation of the conductive electrons that are shining by an appropriate wavelength.

Metal clusters are extremely small particles with a diameter less than 1 nm. Because of that, they are located between the bulk and atomic states of the corresponding metal. Due to the possibility to observe finite-size effects on the physical properties of metal clusters and to understand their microscopic origins, they have attracted physicists over the last four decades. Development of new experimental and theoretical methods has led to a discovery of a variety of remarkable size-specific phenomena and physicochemical properties. During this development, the community has come to be convinced that metal clusters are promising functional units of novel materials and has tried to develop cluster-based materials using "small is different" and "every atom counts" claims [18, 67].

1.1.3 Jellium Model

The characteristic features of small metal clusters, like the electronic shell structure and band gap, can be well understood in terms of quantum motion of the delocalized valence electrons and positively charged ionic core. This concept, known as the Jellium model for neutral metal clusters, can also be applied to charged systems [58].

This model was developed initially for clusters in the gas phase, and the cluster is replaced by an electronic structure in layers that consists of spherical Jellium (UEG—Uniform Electron Gas or HEG—Homogeneous Electron Gas) positively charged and surrounded by electrons. It is considered that the electrons move in a medium field potential occupying, according to the Aufbau principle, energy levels. This model represents a good approximation since it preserves most of the physicochemical characteristics of the clusters. The total energy, as a function of the cluster size, can be calculated by the approximation that represents the covered electronic levels and corresponds with the most stable clusters that possess the "magic numbers."

Density functional theory calculations to determine the geometric structures of the metal clusters are often used to confirm the experimental data and approximations used by the Jellium model. In general, these studies demonstrate also that there is an oscillation in the stability and the electronic properties of the clusters in function of the atom numbers, neutral clusters with odd number of atoms being more stable than ionic clusters with even atoms.

For all this, the Jellium model provides a good approximation to the electronic behavior of the clusters, describing the dependency of the photoemission energy with the number of atoms in the cluster and following Eq. 1.1:

$$E = \frac{E_F}{N^{1/3}} \tag{1.1}$$

where N is the number of the atoms in the cluster, E_F is the Fermi energy of the metal (tabulated) and E is the band gap energy, which can be approximated to the emission band in the photoluminescence experiments. With this formula in hand, it is possible to calculate the number of atoms of an unknown cluster from a simple photoluminescence spectrum.

1.2   Synthesis and Characterization of Ligand-Free Metal Clusters

When Faraday treated (gold chloride) with phosphorus to generate particles, in what he called "activated Au," he had the intuition to propose that the red-colored Au was in the form of very small particles and that the color may vary as a function of the size, something later found to be correct. Faraday's aggregates were in turn Au nanoparticles, which are still prepared today from a similar reduction procedure. Although speculative, it is possible that Faraday obtained clusters in his more diluted experiments since Au clusters are able to persist in nanomolar aqueous solutions.

There are two paths to prepare metal clusters (Fig. 1.4). In the bottom–up way, the metal clusters are prepared starting from small aggregate or single atoms, like metal complexes or salts. On the contrary, in the bottom–down path, the starting materials are bigger nanoparticles from which one can leach small aggregates of the metal.

The synthesis of large-scale metal clusters with atomic precision has experienced great advances during the last years; [32] however, most of the methods still rely on extremely expensive techniques that only recently start to be replaced by affordable wet procedures, with potential industrial application.

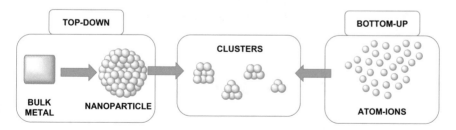

Fig. 1.4 Top–down and bottom–up approaches for the synthesis of atomic metal clusters

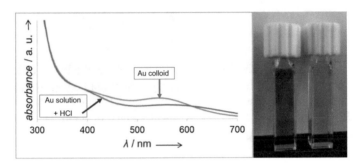

Fig. 1.5 UV/Vis spectra of the Au colloid before and after the addition of HCl and corresponding photograph (Fig. adapted from Ref. [53]. Copyright © 2013 by John Wiley & Sons, Inc.)

1.2.1 Synthesis of Metal Clusters in Solution

A first synthetic method to prepare sub-nanometric metal clusters in solution, without the aid of ligands, consists in the electrochemical etching of metallic plates in high-diluted solutions (10^{-6} M) with typical yields less than 5%. More recently, chemical methods relying on the wet reduction of a metal salt or complex with soft reduction agents (amides solvents, H_2,...) have been described [2, 7, 49, 50, 52, 53, 54, 56, 58, 59, 65]. Specifically, Au, Pd, Ag, Rh, Cu and Pt clusters between 3 and 15 atoms can be formed in reductive conditions starting from the corresponding salts and complexes. Both the electrochemical and the wet chemical methods must be carried out in very diluted solutions to avoid later agglomeration of the clusters. These diluted clusters are very useful in reactions where the amount needed to perform the catalysis is low (a few part per million, ppm, or even part per billion, ppb, amounts).

Following a top–down method, it is possible to prepare Au clusters starting from bigger nanoparticles (5–10 nm) in acid media, preferentially HCl solutions [53]. The red–purple previous aqueous colloidal solutions turn transparent after acid treatment (Fig. 1.5). UV/Vis measurements confirmed the formation of small atom Au clusters after the addition of HCl and the complete absence of the original plasmon band at approximately 550 nm.

1.2.2 Synthesis of Supported Metal Clusters

Metal clusters are formed from suitable precursors supported on solids. As it occurs in solution, these supported metal clusters can be prepared either by bottom–up or top–down approaches [3].

Following bottom–up approaches (Fig. 1.6), different solid supports have been employed, including polymers, inorganic oxides, metal-organic frameworks (MOFs), mesoporous carbon and zeolites. For instance, the bio-compatible ethylene–vinyl alcohol (EVOH) copolymer encapsulates and mildly reduces either Au, Pd, Pt or Cu

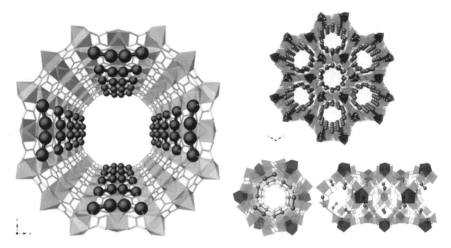

Fig. 1.6 Pd$_4$ (left) and Pt$_2$ clusters in MOFs. Metal ions of the whole net have been depicted as yellow, light and dark blue forms and ligands as gray sticks (adapted from Refs. [45 and 60]. Copyrights © 2018 by John Wiley & Sons, Inc.)

metal salts inside the polymer matrix, to give sub-nanometer clusters [17, 51, 54]. Another example of bottom–up approach is the synthesis and determination of Au clusters supported on nanoceria (nCeO$_2$). nCeO$_2$ has a high number of vacancies to stabilize catalytically active Au species, and the amount and nature of these Au species depend on the metal precursor and the reducing treatment employed. Hydrogenation at 200 °C of nCeO$_2$ properly impregnated with HAuCl$_4$ gives up to 15% of cationic few-atom Au clusters. For nanoporous materials, it is possible to achieve the multigram-scale chemical synthesis of up to 8 wt% sub-nanometer clusters of Pd$_4$ or Pt$_2$ on different MOFs. The synthesis of this kind of materials involves three steps: First, a robust and water-stable 3D MOF is prepared; [25, 44] then, Pd^{2+} or Pt^{4+} cations are incorporated either by anchoring the metal salt in thioether arms or by cationic exchange, and finally, the Pd$_4$ or Pt$_2$ units are obtained by reduction with NaBH$_4$. Complementary, nanoporous zeolites can also be used to synthesize well-dispersed Pt and Pd di and trinuclear clusters, with an exquisite control of the Lewis acidity [61]. Also, mesoporous carbon is a good support to stabilize monodispersed zero-valent Pt$_{5-12}$ clusters prepared from reduction of Pt–thiolate complexes, inaccessible by chemical methods [31]. In order to obtain a precise size of the metal clusters, soft-landing techniques are used in combination with gas-phase cluster ion sources and mass spectrometry. This approach is particularly effective for investigations of small nanoclusters (less than 20 atoms), where the rapid evolution of the atomic and electronic structure makes it essential to have precise control over cluster size. Cluster deposition allows for independent control of cluster size, coverage and stoichiometry (e.g., the metal-to-oxygen ratio in an oxide and oxide cluster) and can be used to deposit the clusters on nearly any substrate without constraints of nucleation and growth [69].

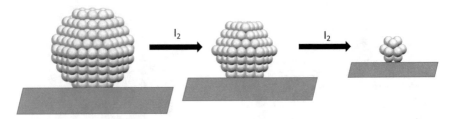

Fig. 1.7 Disaggregation of supported metal nanoparticles to clusters with iodine

Following top–down approaches (Fig. 1.7), redox active halogen species can be used to dislodge metal nanoparticles into sub-nanometer clusters. For instance, I_2 solutions broke Au NPs supported on charcoal, Al_2O_3, TiO_2 and ZnO, into sub-nanometric Au clusters and isolated atoms [50, 62, 63].

1.2.3 Metal Cluster Characterization Techniques

1.2.3.1 UV–Vis and Fluorescence

Ultraviolet-visible (UV–vis) spectrophotometric analysis is probably the easiest and fastest characterization method to determine the presence or absence of metal clusters in solution and also when supported on solids (diffuse reflectance mode). The appearance of absorption bands in the UV–A and blue visible regions (250–400 nm) can be indicative of the presence of metal clusters, which have defined valence orbitals, in clear contrast with metal nanoparticles which present plasmon bands in a different and non-interfering area, around 550 nm (Fig. 1.8). Based on the Jellium model presented above, the number of atoms of the clusters, N, can be calculated by the simple Eq. 1 ($E_F = 5.32$ eV for bulk Au) and Eg the HOMO–LUMO energy band gap, respectively (HOMO 1/4 highest occupied molecular orbital, LUMO 1/4

Fig. 1.8 Experimental absorption UV/Vis and the corresponding emission (inset) spectra for the AuIPrCl-catalyzed hydration of 1–octyne just after the induction time, where the formation of Au clusters (is completed) [53]. Copyright © 2013 by John Wiley & Sons, Inc.

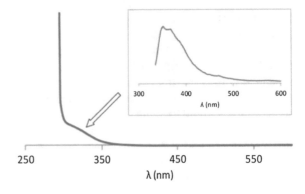

lowest unoccupied molecular orbital), which is obtained from the UV–vis spectrum. Thus, the absorption wavelength is directly correlated with atomicity and vice versa. Complementary emission (fluorescence) analysis circumvents the possible mask or interference by other absorbing species in UV–vis that could be present in the analyte, such as organic molecules and other metal compounds, since the clusters have the particularity of behaving as potent quantum dots. Thus, irradiation of the clusters in their corresponding absorbing wavelengths gives clear emission bands, which does not occur with most of organic compounds and metal precursors, including nanoparticles (see Fig. 1.8).

1.2.3.2   Mass Spectrometry

Routine and high-resolution electrospray ionization–mass spectrometry with quadrupole detectors are commonly employed to determine the empirical formula of metal clusters with ligands in solution, and matrix-assisted laser desorption/ionization time-of-flight (MALDI–TOF) spectrometers are used for solid samples. Following this, these mass spectrometry techniques can be employed not only to determine the mass of ligand-free metal clusters but also for selection, separation, isolation and deposition of individual clusters on solid surfaces [69]. Most metals show a unique isotopic pattern which unveils the atomicity of the cluster. Even for monoisotopic metals, such as the case of Au with a practical single isotope at 197 Da., mass spectrometry is useful since metal clusters appear beyond the minimum detectable mass of the instrumentation, typically 400 Da., which is a clear advantage with respect to lighter metals (Fig. 1.9) [55].

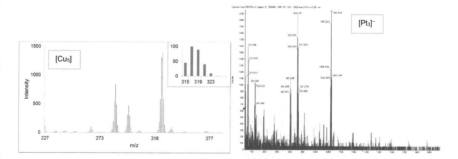

Fig. 1.9 Left: ESI–QTOF spectrum of small (Cu5) clusters in ethanol solution in negative ion mode with the simulation of the relative intensity peaks. Electrospray ionization/mass spectrometry with a quadrupole time-of-flight (ESI–QTOF) measurements of the samples. Right: ESI–QTOF measurements of Pt samples taken at 60 °C with 0.005 mol% of Karstedt's catalyst (reprinted with permission from Refs. [11 and 16]. Copyright © 2015 by American Chemical Society and Copyright © 2019 John Wiley & Sons, Inc., respectively)

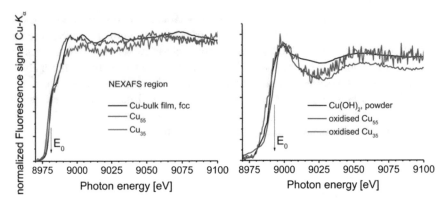

Fig. 1.10 NEXAFS region of the pristine (top) and oxidized (bottom) Cu_n clusters (n = 35, 55) in comparison with the respective bulk sample, Cu and $Cu(OH)_2$ (Fig. from [57]. Copyright © by The Royal Society of Chemistry 2016)

1.2.3.3 Synchrotron Techniques: EXAFS and XANES

Synchrotron techniques give information about the metal bonding and oxidation state and have been used for the determination of metal clusters on solids. However, determination of ligand-free metal clusters in solution with these techniques proved difficult since the highly diluted (typically micro- or nanomolar) conditions of the metal clusters are far below the detection limits of the instrumentation. For this reason, only few examples are available [57] (Fig. 1.10).

1.2.3.4 High Resolution Transmission Electron Microscopy (HR–TEM)

Transition metal elements are heavy, having very good response to the electron beam in EM techniques, particularly in unscattered, dark field mode. Thus, HR–TEM and high angle annular dark field–scanning transmission electron microscopy (HAADF–STEM) are routinely applied to the determination of supported metal nanoparticles and also clusters near or below the nanometer regime. For ultra-small metal clusters, aberration corrected HR–TEM must be employed. The solid support plays a role during visualization, and solids containing excessively heavy atoms or organic substances are more difficult to evaluate, since the former hide the metal atoms and the latter burn under the strong electron beam and spoil the microscopy detector. However, careful measurements have allowed the visualization of Au_5 clusters on carbon nanotubes and EVOH polymers. Indeed, HAADF–STEM has been employed to follow the evolution of supported Au atoms into Au clusters during reaction [10, 39] (Fig. 1.11).

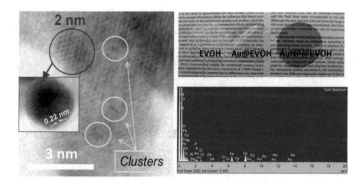

Fig. 1.11 Left: Aberration-corrected HR–TEM micrograph of Au@EVOH; the circles in yellow indicate Au clusters, the circle in red indicates a 2 nm Au NP, and the inset shows the interplanar crystallographic distance for Au. Top right: Photograph of neat EVOH (left, colorless), Au@EVOH (middle, yellow) and the material synthesized with carvacrol, which leads to plasmonic Au nanoparticles (right, red). Bottom right: A representative EDX spectrum of the area indicated for clusters, showing the presence of Au (Fig. from Ref. [50]. Copyright © by The Royal Society of Chemistry 2017)

1.2.3.5   Single Crystal X-Ray Diffraction

Perhaps, the more powerful technique for the structural determination of metal clusters is single crystal X-ray diffraction (SC–XRD). For ligand-stabilized clusters, this technique has been routinely employed, and for instance, the SC–XRD structure of a quasi-linear Pd_4 cluster stabilized and protected by polyarene ligands has been reported (Fig. 1.12) [47, 48]. However, this compound does not find application in catalysis since ligand exchange with the reactants triggers decomposition of the cluster even below 0 °C. In principle, decomposition may not occur in a ligand-free cluster suitably accommodated within a solid support, thus enabling heterogeneous catalysis by metal clusters. Indeed, the same quasi-linear Pd_4 cluster could be obtained in a robust and crystalline MOF structure with ability to incorporate and reduce metal cations [20]. The use of this type of MOFs has recently opened not only the possibility to prepare ligand-free few-atom sub-nanometer clusters but also to characterize them by SC–XRD and do catalysis with them. The MOF acts as a convenient crystalline matrix to host the ligand-free metal clusters and to obtain suitable monocrystals for diffraction, as it has been achieved not only for Pd but also for Pt clusters [45, 60].

1.2.3.6   Other Techniques

Dynamic light scattering (DLS) and zeta potential measurements

These techniques have been long employed to determine the size (DLS) and charge (zeta potential) of colloidal Au nanoparticles with the same instrumentation. Recent

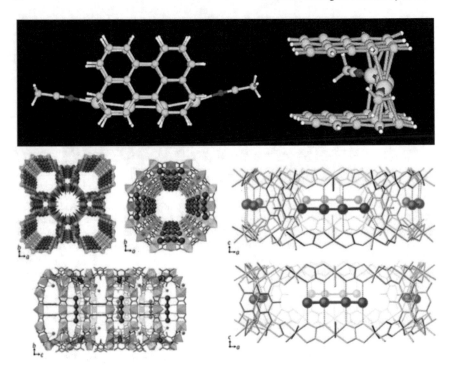

Fig. 1.12 Top: Single-crystal X-ray diffraction of a Pd$_4$ cluster stabilized by perylene ligands (Pd atoms are represented as yellow spheres) and within a MOF structure (Pd atoms are represented as purple spheres (Fig. extracted from Ref. [48]. Copyright © 2003 by American Chemical Society) Bottom: Perspective view along the b crystallographic axis of two portions (left) of the crystal structure of the MOF enclosing the two crystallographic not equivalent ligand-stabilized quasi-linear [Pd$_4$] clusters and DFT optimized structures (right). (Fig. adapted from Ref. [20] Copyright © 2017 by Nature Publishing Group)

advances of the technique allow the determination of the size and charge in soluble sub-nanometer metal clusters, down to a limiting resolution of 0.4 nm. Generally, polar solvents such as water, alcohols or amide solvents (*N, N*–dimethyl formamide, DMF, or *N*–methyl pyrrolidone, NMP) must be employed. Indeed, cationic Au$_{3-7}$, Pd$_{3-4}$ and Pt$_{3-5}$ clusters have been satisfactorily determined with this technique [17, 37, 49, 53]. The somehow related diffusion-ordered spectroscopy (DOSY) technique, which seeks to separate NMR signals of the clusters according to their diffusion coefficients, is also a valuable tool for organic-stabilized Au clusters but not for ligand-free metal clusters since a signal of the organic ligand must be monitored in the experiments [9, 21].

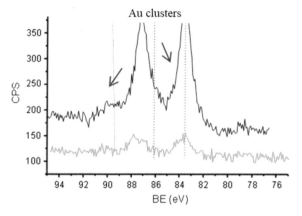

Fig. 1.13 XPS spectrum of a sample of Au–nCeO$_2$ after hydrogenation of 1 g of gold chloride on nanoceria at 200–300 °C under a flow of 100 ml per min of N$_2$:H$_2$ (10:1) (black line). The percentage of cationic Au in the reduced sample is 15% after deconvolution. For comparison, a sample of Au–nCeO$_2$ with a stronger reduction treatment—1 g of AuCl on ceria with 5 ml of phenylethanol at 160°C for 1 h (green line) —is also presented. The percentage of cationic Au decreases significantly to 2% (Fig. from Ref. [50]. Copyright © by The Royal Society of Chemistry 2017)

X-ray photoelectron spectroscopy (XPS)

XPS is a surface-sensitive and quantitative technique that allows one to determine the oxidation state of supported metal clusters. Either barely anchored to the solid surface or free in solution, metal clusters have been characterized with this technique, provided that the measurement is carried out with a relatively mild power in the X-ray beams, otherwise the metal clusters may further aggregate under operating conditions [50, 54]. Figure 1.13 shows a representative example of Au clusters.

Reaction test

The ester-assisted hydration of alkynes is a reaction exclusively catalyzed by Au clusters of 3–7 atoms (Au$_{3-7}$), and it has been developed as an analytical tool to unambiguously quantify sub-nanometer Au clusters and differentiate them from salts and nanoparticles (Fig. 1.14). For this, the reaction test was first validated with different samples of well-characterized sub-nanometer Au clusters on nanoceria (Au–nCeO$_2$) and then applied to a series of new solids containing sub-nanometer Au clusters [50]. This reaction test also allows for a very rapid quantification of the Au clusters in solution or on solids (after leaching them in situ under the test reaction conditions) without the requirements of any instrumental characterization but only a quantitative method to measure the evolution of the organic reaction, such as gas chromatography (GC) or nuclear magnetic resonance (NMR), using the initial rate of the reaction as a quantitative and linear parameter respect to the amount of Au$_{3-7}$ clusters in solution. (The reaction will be treated in detail in Sect. 1.3.2.1).

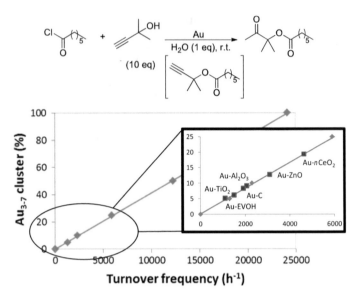

Fig. 1.14 Calibration line obtained for neat Au$_{3-7}$ clusters respect to their initial turnover frequencies and quantification of sub-nanometric Au clusters in the different Au-supported solids prepared here (inset), according to the test (Fig. from Ref. [50]. Copyright © by The Royal Society of Chemistry 2017)

1.3   Catalysis Using Metal Clusters

The most obvious benefit of a nanoscale catalyst is that it maximizes the surface area of the material and thereby increases the number of accessible sites for external reagents. For catalysis, exposing all metal atoms to outer reagents minimizes catalyst loading and save significant costs when using scarce noble metals (e.g., Au, Pd and Pt), which are nonetheless widely used in heterogeneous catalysis. With the development of modern methodologies to engineer the physical size, morphology and termination of nanoscale materials, the promise of exploiting the unique chemical properties of very small nanoparticles (< 5 nm) has now become feasible. No better example is the unique activity of small Au nanoparticles (< 2 nm), which are able to catalyze a wide range of oxidation reactions, whereas bulk Au is essentially inert, a singular property we are all familiar with in the macroscopic world [69]. Despite the unarguable success of nanocatalysis, recent advances in controlled atomic aggregation have stimulated enormous research interest for chemists to seek atomically precise particles; thus, cluster science is expected to bring a third upsurge of research interest due not only to the fascinating properties of clusters in solution but also to solid-supported clusters for catalysis. With the developments in instrumentation and technology, precise information regarding the breaking or formation of chemical bonds will become more clearly accessible. While the chemistry of solid-supported clusters and monolayer protected clusters (MPC) has become a topic of substantial current interest, information on structural and electronic properties is often corroborated only by free gas-phase clusters. The stability of MPCs is often understood within the conceptual framework of superatoms validated in the gas phase on the basis of electronic shells.

In this respect, we summarize here the research advances in the catalytic activity of ligand-free metal clusters, either supported or not [14, 34, 38, 66, 70]. For that, we divide the reactions considering the new bond formed (carbon–carbon, carbon–heteroatom or heteroatom–heteroatom) and also in hydrogenation reactions. Oxidation reactions have been covered previously, and we do not discuss those here [29, 30].

1.3.1 Carbon–Carbon Bond–Forming Reactions

1.3.1.1 Heck, Suzuki and Sonogashira Couplings

Pd-catalyzed cross-coupling reactions are recognized as fundamental transformations in synthetic chemistry [71]. The general mechanism for these reactions (also under ligand-free conditions) involves an oxidative addition–reductive elimination cycle over a Pd(0) species generated in situ. Despite considerable effort during last three decades, the exact nature of the Pd^0 catalytic species remains a matter of debate [13]. It was recently found that Pd_{3-4} clusters are formed from either Pd salts, complexes and nanoparticles in N–methylpyrrolidine (NMP) under heating conditions and are responsible for the catalytic activity during the Heck, Sonogashira, Stille and Suzuki coupling reactions of different iodo– and bromo–derivatives. The ligand-free Pd_{3-4} clusters can be stored in aqueous solution to be used on demand and catalyze, for example, the Heck reaction under industrially viable conditions in high yields and with unprecedented turnover frequencies in some cases (Fig. 1.15). Despite this high activity, the activation of chloro–derivatives was not possible with this catalytic system, which highlights the still important role of ligands for particular molecule activations.

The use of metals in the form of sub-nanometer clusters can lead to a completely unexpected catalytic behavior for a given metal, and the Heck reaction is an illustrative example, recently uncovered [17]. As commented above, Pd clusters seem to be very active species for the Heck reaction which, however, is not surprising in terms of metal nature since Pd is the most common metal catalyst for this type of C–C coupling reaction. Indeed, the metal above Pd in the group X of the Periodic Table, Ni, is also active for this type of couplings but, in striking contrast, the metal below in the group X, Pt, is barely reported as a catalyst for C–C cross-coupling reactions. The lack of catalytic activity of Pt for these coupling reactions has been traditionally ascribed to the difficulties associated to single Pt atoms (in organometallic complexes) to efficiently shift between two electron oxidation states and thus promote the oxidative addition and the reductive elimination steps necessary to perform the coupling, without rapidly aggregating and losing activity. However, the

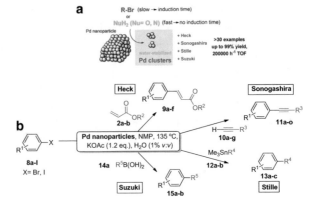

X	R¹	R²/R³/R⁴/R⁵	Product	Pd (ppm)	TOF₀ [a]	Yield (%)[b]
I	p-COMe 8a	R²= nBu 2a	9a	3	2.3·10⁵	99
	H 8b	R²= nBu 2a	9b	30	1.2·10⁴	99
		R³= nOct 10a	11a	300	2.0·10³	73
		R³= Ph 10b	11b	300	7.2·10³	83
		R⁵= Ph 14a	15a	30	3.2·10³	68
	p-Br 8c	R²= nBu 2a	9c	300	3.6·10³	99
		R³= nOct 10a	11c	300	3.1·10³	99
	p-OMe 8d	R²= nBu 2a	9d	300	3.2·10³	99
		R²= 2-Ethylhex 2b	9e	3	8.8·10⁴	99
Br	p-COH 8e	R³= nOct 10a	11d	3	2.0·10⁵	99
		R³= Ph 10b	11e	300	8.4·10³	99
	o-COH 8f	R⁴= nBu 12a	13a	30	8.2·10³	91
	p-COMe 8g	R²= nBu 2a	9a	300	3.1·10⁴	99
		R³= Ph 10b	11f	300	1.8·10⁴	99
		R³= nOct 10a	11g	300	2.1·10³	99
		R³= nOct 10a	11g	0	-	<5
		R⁴= nBu 12a	13b	3	2.0·10⁵	70[c]
		R⁴= nBu 12a	13b	0	-	<5
		R⁴= Ph 12b	13c	3	7.9·10⁴	44
		R⁵= Ph 14a	15b	0.3	1.0·10⁵	61[c]
	H 8h	R²= nBu 2a	9b	300	8.8·10²	64
		R³= nOct 10a	11a	300	5.2·10³	99
		R³= m-tol 10c	11h	300	6.6·10³	99
		R³= o-tol 10d	11i	300	1.2·10³	96
		R³= p-anisole 10e	11j	300	2.4·10³	99
		R³= o-anisole 10f	11k	300	2.0·10³	94
		R³= nDodec 10g	11l	300	8.0·10²	97
		R⁵= Ph 14a	15a	30	7.6·10³	87
		R⁵= Ph 14a	15a	0	-	<5
	p-Me 8i	R³= nOct 10a	11m	300	2.0·10³	83
	o,p-OMe-Napth 8j	R³= nOct 10a	11n	300	4.9·10³	99
	o,o,p-Isoprop 8k	R²= 2-Ethylhex 2b	9f	300	1.1·10⁴	35[c]
	p-OMe 8l	R²= nBu 2a	9d	300	4.7·10²	40
		R²= nBu 2a	9d	0	-	<5
		R³= Ph 10b	11j	300	2.8·10²	50

Fig. 1.15 Reaction scheme for C–C coupling reactions catalyzed by Pd clusters formed in situ and catalytic results **a** TOF₀ is calculated on the basis of total amount of palladium and indicated as h⁻¹. **b** Product yields after 24 h. **c** Reaction temperature: 160 °C (Fig. from Ref. [5]. Copyright © 2014 by John Wiley & Sons, Inc.)

recent finding that Pt single atoms and clusters can be formed in solution has enabled the Pt-catalyzed Heck reaction of iodo– and bromo–derivatives [17]. Experimental studies in combination with DFT calculations strongly support the feasibility of the coupling mechanism steps on the Pt atoms, thus representing one example where the formation of ligand-free metal clusters unveils a somehow hidden catalytic behavior for a particular metal.

Recent theoretical studies on the electronic properties and activation energies for the three steps of the Suzuki cross-coupling reaction have shown that Pd/Ni bimetallic clusters supported on defected graphene can be excellent catalysts [72]. In fact, reducing the size of the clusters from Ni_{13} to Ni_4 enhances the activity because of the increased negative charge Pd. Bimetallic Pd/Ni clusters were found to offer even lower activation energies for all three steps of the Suzuki reaction because of charge donation from the Ni atoms to the Pd atoms making the bimetallic clusters a highly active catalyst.

1.3.1.2 Buchner Reaction

The ring-opening cyclopropanation of benzenes with alpha-diazoesters (Fig. 1.16) is known as the Buchner reaction. This reaction, despite its uniqueness to form otherwise very difficult to prepare cycloheptatrienes, giving access to a plethora of advanced organic intermediates, has found little use in industrial organic synthesis since the only efficient catalysts for the transformation are based on extremely expensive, soluble and unrecoverable Rh_2 salts [1].

Mixed-valence Pd_4 (0, I) clusters supported within a MOF have shown a high catalytic activity and selectivity, comparable to Rh_2 salts, for the Buchner reaction, enabling the performance of the reaction in continuous flow. The Pd_4 clusters have the appropriate electron density to activate the diazo-compound and generate the required carbene to open the benzene derivative. It is possible that, in line with

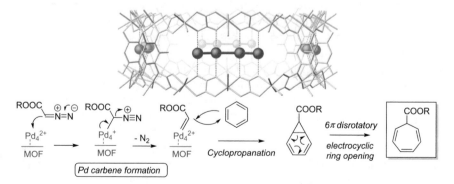

Fig. 1.16 Top: Pd_4 Cluster structure optimized by DFT calculations, on the basis of the real SC–XRD structure. Bottom: Proposed mechanism for the Pd4-catalyzed intermolecular Buchner reaction (Fig. from Ref. [20]. Copyright © 2017 by Nature Publishing Group)

the easy coordination of arene derivatives to this particular type of quasi-linear Pd_4 cluster, [46] the benzene molecule is also activated by the Pd_4 cluster to promote the final coupling. This recoverable Pd_4–MOF solid catalyst, with an estimated price 10–100 times lower than the Rh catalysts, constitutes a paradigmatic shift in the Buchner reaction, opening new avenues in the use of this reaction at larger scales under flow conditions.

1.3.1.3   Homocoupling of Alkynes

The homocoupling of alkynes is a classical organic reaction performed during years with Cu salts, in the Glaser, [24] Eglinton, [15] or Hay [27] versions. However, no significant reports on the use of the other two group XI metals, Ag and Au, as catalysts for the reaction have been shown, despite the electronic resemblance of these metals. When one examines the different accepted mechanisms for the Cu-catalyzed homocoupling of alkynes, it can be seen that a key step during the coupling is the oxidation of Cu(I) to Cu(III), a step intrinsically difficult for Ag but not so much for Au. Indeed, a second key feature during the homocoupling reaction is the plausible dimerization of the Cu atoms through coordinated alkyne bounds. Following this mechanistic rationale, it has been recently found that mixed-valence Au(I, III) clusters with ligands are very active species for the homocoupling of alkynes [36]. The particular structure of the cluster sterically discriminates between linear carbon chain alkynes with 10 or 12 atoms during the oxidative homocoupling of alkynes: the former is fully reactive, and the latter is practically unreactive. A distal size selectivity occurs by the impossibility of trans-metalating two long alkyl chains in an A-framed, mixed-valence di–Au (I, III) acetylide complex, as shown in Fig. 1.17. The reductive elimination of two alkyne molecules from a single Au(III) atom occurs extremely fast, in < 1 min at $-78°$ C (turnover frequency 40.016 s^{-1}). Notice that the extremely high catalytic activity of Au(III) and the stability of mixed Au(I, III) clusters for the homocoupling of alkynes is somewhat related to the relativistic effects present in the gold because is a lateheavy metal and absent in Cu and Ag [35]. The subtle steric and electronic discrimination of alkynes by this Au-catalyzed system allows the heterocoupling of two different alkynes in equimolecular amounts regardless of the nature of the terminal triple bond (Fig. 1.17).

It is not necessary to ligate the Au atoms through ligands to promote the homocoupling of alkynes since very small Au nanoparticles catalyze the aerobic coupling of alkynes [5]. For the latter, O_2 is dissociated as catalyzed by the air-tolerant Au nanoparticles, without significant oxidation of the metal. In contrast, ligand-free soluble sub-nanometer metal clusters do not dissociate O_2 in the presence of the alkyne and, thus, do not catalyze the aerobic coupling, [6] which makes sense considering the higher affinity of cationic Au atoms for alkynes than for O_2. These results exemplify the dramatic catalytic differences that can be found for a given metal in different aggregation forms, in this case Au, and when different ligands and supports are employed.

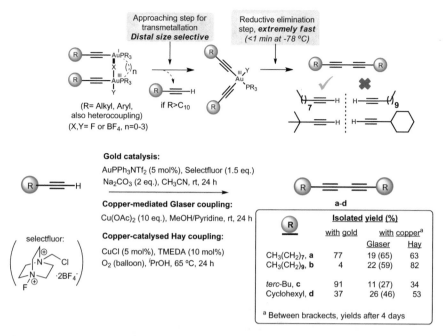

Fig. 1.17 Schematic mechanism of the Au-catalyzed oxidative homocoupling of terminal alkynes in solution. Notice the different reactivity of alkynes a–d found under the Au-catalyzed conditions reported here and typical copper-catalyzed conditions (Glaser or Hay conditions). Isolated yields are the average of two runs (Fig. from Ref. [36]. Copyright © 2015 by Nature Publishing Group)

1.3.1.4 Hashmi Phenol Synthesis

The rearrangement of alkynylfurans to prepare phenol derivatives has been reported by Hashmi et al. using Au salts and complexes [26]. Upon performing the reaction under conditions similar to those reported albeit with a lower amount of $AuCl_3$ (0.4 mol%), the corresponding product was smoothly formed after an induction time, as evidenced by ^1H NMR spectroscopy at different times (Fig. 1.18) [53]. MALDI–TOF measurements showed that the reaction started only after the formation of Au clusters comprising 3–4 atoms. These Au clusters can be formed also starting from AuCl and also from supported Au clusters on CeO_2 or leached from Au nanoparticles on TiO_2, with the assistance of a Brönsted acid.

1.3.1.5 Conia–ene Reaction

The Conia–ene reaction is an early example of Au-catalyzed carbon–carbon bond formation. In a pioneering work, Toste et al. showed that the Au(I) complex $AuPPh_3OTf$ (3 mol%) catalyzes the intramolecular coupling of electron-deficient

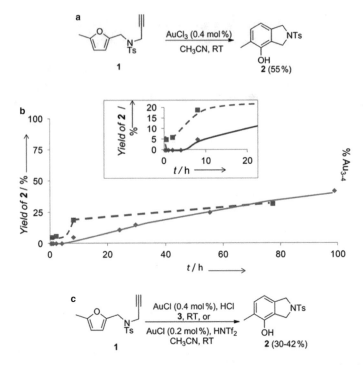

Fig. 1.18 a Phenol synthesis catalyzed by AuCl$_3$ (0.4 mol%). **b** Yield time plot for the phenol synthesis showed in scheme A above (**c**) and percentage of 3–4 atom Au clusters in solution according to MALDI–TOF MS measurements (**a**). The inset shows a magnification of the induction period when clusters formed. **c** Scheme for phenol synthesis by using AuCl as a catalyst in sub-molar amounts under two different reaction conditions (Fig. from Ref. [53]. Copyright © 2013 by John Wiley & Sons, Inc.)

α carbons to carbonyl groups with alkynes, whereas AuCl$_3$ was a very unselective catalyst, and only 30% of the cyclopentane product was obtained with a higher loading (10 mol%) of the Au salt after complete conversion [33]. Upon lowering the amount of AuCl and AuCl$_3$ to 0.02 mol%, no conversion was observed (Fig. 1.19) [53], but under acidic conditions (0.2 mol% of HOTf), the corresponding Au$_{3-6}$ clusters are formed to catalyze the reaction with similar results as AuPPh$_3$OTf.

1.3.2 Carbon–Heteroatom Bond-Forming Reactions

1.3.2.1 (Ester-Assisted) Hydration of Alkynes (C–O)

This reaction is the first example of catalysis by Au clusters formed in solution, as far as we know [49]. It can also be employed as a catalytic test to prove the presence of Au$_{3-5}$ clusters since the reaction only proceeds when the Au clusters are present in

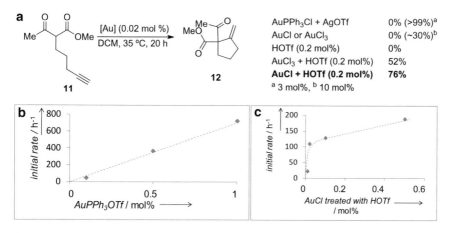

Fig. 1.19 a Results of the Au-catalyzed Conia–ene reaction of alkyne with different Au catalysts. **b** Initial rate dependence on the concentration of the initial AuPPh3OTf complex catalyst. **c** Initial rate dependence on the concentration of initial AuCl treated with HOTf. Lines are a guide for the eye (Fig. from Ref. [53]. Copyright © 2013 by John Wiley & Sons, Inc.)

the reaction media, even starting from part per billion amounts of Au salts, yielding a TON value as high as ~10^7 in some cases (Fig. 1.20). Notice that this value was substantially greater than any value reported for a non-enzymatic catalyst at room temperature at that time [4].

A systematic study of the cluster formation in different solvents and acids revealed that a variety of very small Au clusters, from 3 to 10 atoms, were formed under reaction conditions, with a size distribution controlled by the nature of both the solvent and the acid, as determined by UV–vis spectroscopy and mass spectrometry. However, Au_5 and Au_8 clusters stabilized on the dendrimer polyamineamide–ethanol, (PAMAM–OH) were synthesized independently and used as a catalyst, and the results showed that Au_5–PAMAM did indeed catalyze the ester-assisted hydration of alkynes, whereas Au_8–PAMAM showed a much lower activity, with no induction time in both cases. These results confirm the sensitivity of the reaction to the number of Au atoms in the cluster, showcasing perhaps one of the first examples of this class. It must be noted that the catalytic efficiency of the Au clusters when supported on the dendrimer was at least two orders of magnitude lower than that of those formed in situ in the propargyl alcohol, reflecting the inhibiting effect of ligands during the catalysis with few-atom soluble clusters.

1.3.2.2 Bromination of Aromatics and Alkynes (C–Br)

Together with the ester-assisted hydration of alkynes, the bromination of aromatics was one of the first examples of catalysis by Au clusters (Fig. 1.21) [49]. In this case, the formation of Au_7–Au_9 clusters occurred in the presence of the solvent

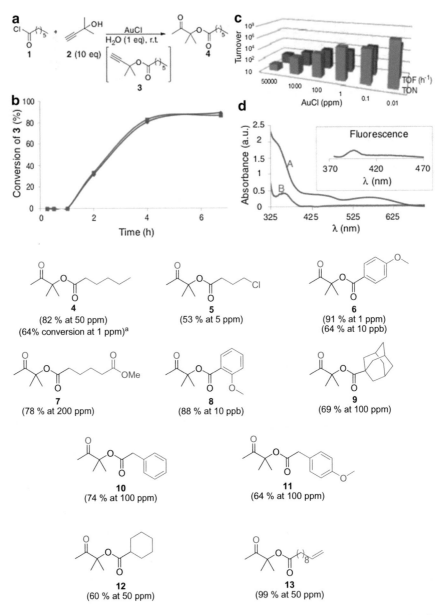

Fig. 1.20 **a** Ester-assisted hydration of in situ formed alkyne reaction. **b** Plot time conversion for AuCl (squares) and HAuCl₄ (diamonds) at 100 ppm, after correction with the blank experiment. **c** Turnover number (TON) and turnover frequency (TOF) for different amounts of AuCl, calculated as moles of 4 formed per mole of AuCl at final conversion (TON) and as the initial reaction rate after the induction time per mole of AuCl (TOF). The final yield of 4 is > 90% in all cases, except for 0.01 ppm, where it is ~ 50%. **d** Absorption measurements (a.u., arbitrary units) for the hydration of 3 containing the Au active species along the induction time (a) and when the reaction proceeds (b) with the corresponding fluorescence (inset, irradiated at 349 nm). Bottom, reaction scope (isolated yields and ppm of Au in parentheses (Fig. from Ref. [49]. Copyright © 2012 by AAAS)

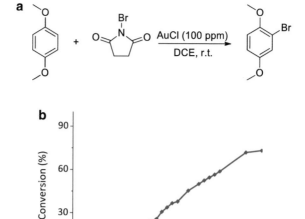

Fig. 1.21 a Studied reaction scheme. **b** Plot time conversion for the bromination of arene (Fig adapted from Ref. [49]. Copyright © 2012 by AAAS)

1,2–dichloroethane (DCE). In contrast to the ester-assisted hydration of alkynes, the reaction performed with Au_5–PAMAM and Au_8–PAMAM showed higher activity for the latter ones, in line with the Au_{7-9} free clusters.

The ω-bromination of terminal alkynes catalyzed by Au complexes in solution is also a good example for C–Br bond formation catalyzed by Au clusters. Figure 1.22 shows that upon performing the reaction in the presence of $AuPtBu_3NTf_2$ (Tf = trifluoromethanesulfonyl) and by systematically decreasing the amount of Au from 2 to 0.1 mol%, a reaction induction time appeared [53]. Excellent yields of the bromated compound were obtained in all cases. Monitoring the reaction by ^{31}P NMR spectroscopy showed the progressive degradation of the Au complex during the reaction, whereas ^{19}F NMR spectroscopy showed the concomitant formation of free triflimidic acid ($HNTf_2$). According to these spectroscopic measurements, the degradation of $AuPtBu_3NTf_2$ provides the two elements needed for the formation and stabilization of the Au clusters: ligand-free Au species and a strong Brönsted acid.

Figure 1.23 shows that Au clusters are also formed from colloidal Au nanoparticles (10 ± 2.5) nm solution in HCl media and ω-bromination of phenylacetylene. Thus, this reaction illustrates how bottom–up (from Au salts) and top–down (from Au nanoparticles) synthetic approaches to metal clusters can equally work for a particular reaction. Given that a mixture of Au clusters from 3 to 10 atoms was formed in situ during reaction, it could be possible to carry out at the same time reactions catalyzed by different clusters. Indeed, an excess of N–bromosuccinimide (NBS) together with water present in the reaction medium allowed the one-pot Au-catalyzed hydration of a bromoalkyne to give $\alpha,\alpha,$'–dibromoketone in a single step in reasonable yield

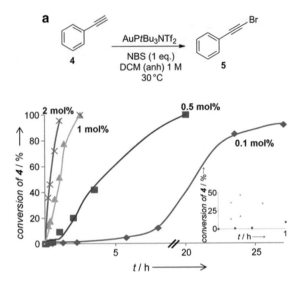

Fig. 1.22 Conversion time plot for the ω–bromination of phenylacetylene [4] with different amounts of the Au(I) AuPtBu₃NTf₂ complex catalyst. The inset maximizes the initial time (Fig. from Ref. [53]. Copyright © 2013 by John Wiley & Sons, Inc.)

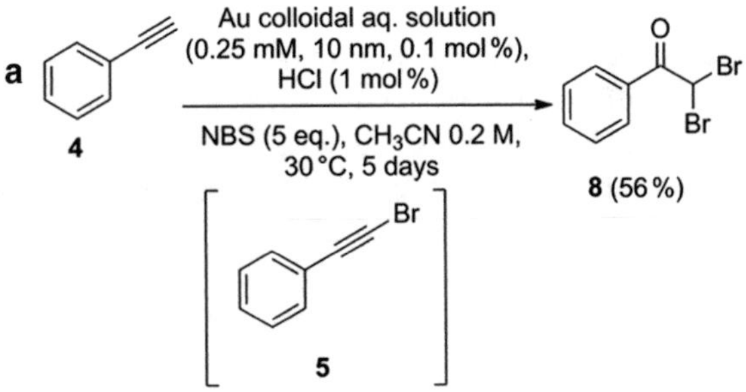

Fig. 1.23 a Bromination–hydration cascade with (10 ± 2.5) nm Au colloidal aqueous solution (0.25 mm, 0.1 mol%) treated with concentrated HCl (1 mol%, Fig. from Ref. [53]. Copyright © 2013 by John Wiley & Sons, Inc.)

after two consecutive Au-catalyzed reactions, one involving σ C–H activation and the other involving π activation of the alkyne.

1.3.2.3 Hydrosilylation of Alkynes (C–Si)

The hydrosilylation of alkynes gives three different possible vinylsilanes depending on the catalyst used (Fig. 1.24). Extensive research has been performed to obtain and functionalize the β-products (anti-Markovnikov addition), which can be formed

Fig. 1.24 Possible products in the hydrosilylation of terminal alkynes (top) and reaction scope for the Pt₃ clusters-catalyzed α– hydrosilylation: 0.005 mol% (50 ppm) of Kardstedt's catalyst (with respect to the alkyne) at 110 °C in toluene (0.5 M), GC yields, isolated yields of α isomer between parenthesis, and α/β ratio of the mixture, before isolation and calculated by GC–MS and ¹H NMR, between brackets, **a** 5–gram scale, **b** NMR yield, **c** Ratio (αα)/(αβ)/(ββ) = 43/47/10 (bottom, Fig. adapted from Ref. [59]. Copyright © 2017 by John Wiley & Sons, Inc.)

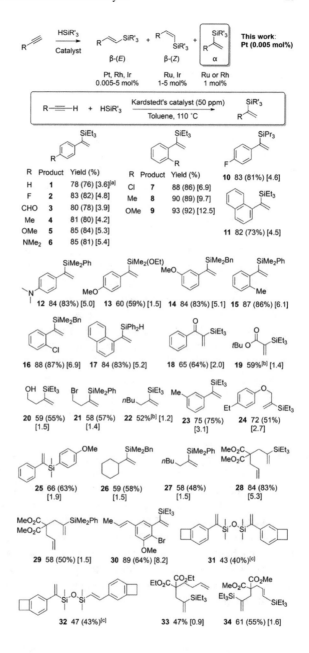

in high yield and selectivity with ppm amounts of Pt catalyst or with other metal catalysts including Rh and Ir for the β–(E) product (the most thermodynamically stable) and Ru and Ir for the β–(Z)–vinylsilane. In order to get the α-vinylsilane, Pt_3 clusters were made to catalyze the Markovnikov hydrosilylation of terminal alkynes to afford a wide variety of new α-vinylsilanes in good isolated yields using low amounts of catalyst [59]. These clusters can be formed in situ with < 100 ppm of simple Pt compounds or with externally added Pt "Chini" clusters [41].

In order to prove that the Pt clusters were the active catalysts and not the precursors, UV–Vis absorption spectroscopy measurements were performed during reaction at 110 °C, in which the α-isomer is formed. These measurements showed the appearance of new bands at approximately 300 nm, and the fluorescence spectrum confirmed the expected emission band for Pt_{3-4} (ca. 360 nm according to the Jellium model) when the α product was predominant. In contrast, only the plasmonic band of Pt nanoparticles was observed under β-favored reaction conditions, that is, reaction temperatures <70° C, without any absorption or emission band corresponding to sub-nanometer clusters. These results supported the formation of Pt_{3-4} clusters during the hydrosilylation reaction at 110 °C with 0.005 mol% of Kardstedt's catalyst, without nanoparticle formation, and indicate that Pt clusters are the species needed for the α isomer formation. These results were confirmed also by Electro-Spray–Ionization Quadrupole Time-Of-Flight Mass Spectrometry (ESI–QTOF) and zeta potential measurements. Moreover, the Chini clusters [8] $[NEt_4]_2[Pt_3(CO)_6]_3$, $Na_2[Pt_3(CO)_6]_5$ and $Na_2[Pt_3(CO)_6]_{10}$ showed the production of α-vinylsilanes under classical (anti-Markovnikov) reaction conditions. The absence of induction time at the beginning in the reaction and the presence of the Pt_3 UV–VIS signals during the reaction confirmed that the Pt_3 clusters are the active catalysts for obtaining the α-isomer.

Considering the reactivity trends of the Pt_3 clusters, we can discuss the origin of the inverse selectivity toward the hydrosilylation with respect to the traditional Pt compounds. The Chalk–Harrod and a modified Chalk–Harrod mechanism are presented in Fig. 1.25, which is the most accepted mechanism for hydrosilylation of alkynes and alkenes catalyzed by Pt [42, 64]. Once the active catalyst is formed, the mechanism consists of four steps: (1) alkyne coordination to the Pt; (2) silane oxidative addition; (3) migratory insertion in Chalk–Harrod and silylplatination for modified Chalk–Harrod and (4) reductive elimination. Based on this mechanism, the key step that determines the regioselectivity of the reaction is the migratory insertion, which proceeds through hydroplatination for the former and silylplatination for the latter mechanism.

1.3.2.4 Andrussow Reaction (Cyanide Synthesis, C–N)

Hydrocyanic acid HCN is a very versatile molecule produced in multi-ton amounts using the Andrussow process. This process consists of an endothermic reaction between methane (CH_4) or carbon monoxide (CO) with ammonia (NH_3) at temperatures higher than 500 °C using Pt–Rh *gauze* catalysts in flow conditions to obtain

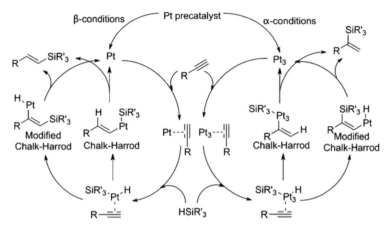

Fig. 1.25 Possible pathways for the hydrosilylation of alkynes to obtain $\beta-$ (left) and $\alpha-$alkenyl-silanes (right) through both Chalk–Harrod and modified Chalk–Harrod mechanism (Fig. from Ref. [59]. Copyright © 2017 by John Wiley & Sons, Inc.)

hydrocyanic acid. Usually, the yields obtained for different Pt precursors are less than 10% with poor selectivity.

Pt_2^0 clusters homogeneously distributed and densely packed within the channels of a metal-organic framework were prepared (Fig. 1.26), unambiguously characterized by different techniques including SC–XRD, and employed as catalysts to perform the reaction of CO and NH_3 at nearly room temperature with a high TOF ($TOF_0 = 1260\ h^{-1}$). When monoatomic Pt^{2+} inside the same MOF was used as a catalyst, the TON decreased until 56. Moreover, no induction time was observed for Pt_2^0 clusters, confirming they are the active catalysts [45].

The reaction product was not the expected HCN but NH_4CN, a compound only stable below 40 °C but thermodynamically more accessible than HCN. Thus, the extremely high catalytic activity of the supported Pt_2^0 clusters do not only permit to activate CO and NH_3 at much lower temperature but also to bypass the reaction manifold to a more exothermic reaction, thus facilitating the gas conversion and shifting the chemical equilibrium to the right. These results with the hybrid material are significant from an economic and environmental viewpoint.

1.3.2.5 Goldberg Coupling (C–N)

The direct coupling of aryl halides with amides is known as the Goldberg coupling, and it has been recently found that sub-nanometer Cu clusters formed by endogenous reduction of Cu salts and Cu nanoparticles in heating amide solvents are active and selective catalysts for this C−N bond-forming reaction (Fig. 1.27). The Cu_{2-7} clusters were formed from different Cu salts to catalyze not only the Goldberg but also related C–P, C–O and even C–C couplings (Sonogashira reaction). Sub-nanometer

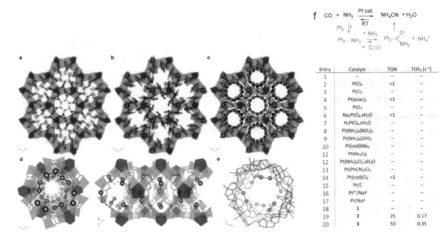

Entry	Catalyst	TON	TOF$_0$ (s^{-1})
1	–	–	–
2	PtCl$_4$	<1	–
3	PtCl$_2$	–	–
4	Pt(acac)$_2$	<1	–
5	PtO$_2$	–	–
6	Na$_2$PtCl$_6$·xH$_2$O	<1	–
7	H$_2$PtCl$_6$·xH$_2$O	–	–
8	Pt(NH$_3$)$_4$(NO$_3$)$_2$	–	–
9	Pt(NH$_3$)$_4$(OH)$_2$	–	–
10	Pt(cod)Me$_2$	–	–
11	PtMe$_3$Cp	–	–
12	Pt(NH$_3$)$_4$Cl$_2$·xH$_2$O	–	–
13	Pt(PhCN)$_2$Cl$_2$	–	–
14	Pt(cod)Cl$_2$	<1	–
15	Pt/C	–	–
16	Pt^{2+}/NaY	–	–
17	Pt/NaY	–	–
18	1	–	–
19	2	25	0.17
20	3	53	0.35

Fig. 1.26 Crystal structures of the MOF without metal (**a**, 1), with Pt$_2^+$ (**b**, 2) and with the Pt$_2$ clusters (**c**, 3). **d**. Perspective views, in detail, of a channel of 3 in the ab (left) and bc (right) planes. Copper and calcium atoms from the network are represented by cyan and blue polyhedra, respectively, whereas organic ligands are depicted as gray sticks. Yellow and purple spheres represent S and Pt atoms, respectively. Dashed lines represent the Pt···S interactions. **e**. Fragment of a channel of 3 emphasizing the interactions with the network. Pt–S and Pt...O interactions are represented by yellow and blue dashed lines, respectively. **f**. Pt-catalyzed synthesis of HCN at room temperature: Equation shows the reaction conditions: CO (4 bar, 1.5 mmol), NH$_3$ (2 bar, 0.75 mmol) and 0.0075 mmol Pt and a plausible reaction mechanism. Table includes the set of Pt catalysts tested under these reaction conditions (Fig. adapted from Ref. [45]. Copyright © 2018 by John Wiley & Sons, Inc.)

Cu clusters generated within a polymeric film show also activity for the C–N coupling and are a good example of metal cluster storage with full stability for months.

1.3.3 Heteroatom–Heteroatom Bond-Forming Reactions

1.3.3.1 Hydrosilylation of Alcohols

Alcohol hydrosilylated compounds are traditionally used as intermediates in organic synthesis. *O*–Silylation often competes so successfully with *C*–silylation that protective groups must be used to avoid *O*–silylation; however, only *O*–silylated compounds are formed for Pt catalysts used in low concentrations. Specifically for allyl alcohols, it was found that no *O*–silylation was obtained at 50 mM Pt catalyst concentration, and ~3 and 7% O–silylation occurred when [Pt] is 20 and 10 mM, respectively [75]. Furthermore, much more forcing conditions were required at the lower catalyst concentrations. These results suggest a change in the nature of the active catalyst species. Moreover, during the catalyst preparation, Pt$_2$ clusters were detected.

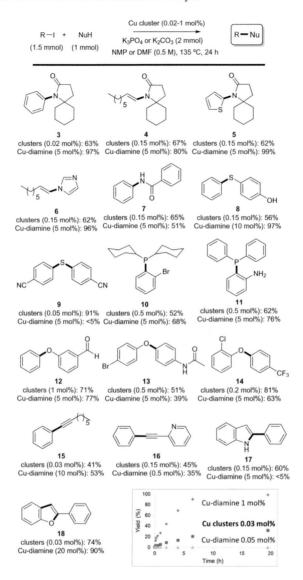

Fig. 1.27 Scope of the Cu cluster-catalyzed cross-coupling reactions. Isolated yields. The new bonds are in bold. For comparison, the values with Cu–diamine catalysts are also given, reaction conditions: substrates (1 mmol), CuI (5–20 mol%), N,N–methyl ethylenediamine (10–40 mol%), K$_3$PO$_4$ (2 mmol), anhydrous toluene or dioxane (0.25 M), nitrogen atmosphere, 110 °C, 24 h. The inset shows the kinetics for the Goldberg coupling between the iodobenzene and the amide under typical reaction conditions for diamine-assisted coupling and the conditions reported (Fig. from Ref. [54]. Copyright © 2015 by American Chemical Society)

1.3.3.2 Homocoupling of Thiols

The homocoupling of thiols is another clear example of catalysis by Au clusters. In this case, single Au atoms supported on functionalized carbon nanotubes only show activity in the oxidation of thiophenol with O_2 when they aggregate under reaction conditions into Au clusters of low atomicity (see induction period in Fig. 1.28) [5, 10]. The Au_{5-10} are extremely active for the reaction, with TOFs in the order of 10^5 h^{-1}, comparable to the activity of sulfhydryl oxidase enzymes. When clusters grow into nanoparticles of diameter ≥ 1 nm, catalyst activity drops to zero. Theoretical calculations show that only Au clusters of low atomicity are able to simultaneously adsorb and activate thiophenol and O_2 and that the strong Au–S interaction in 1 nm Au nanoparticles leads to the formation of very stable RS–Au–SR units that prevent further reaction (Fig. 1.28 right). The combination of activation of both reactants and facile product desorption makes Au clusters excellent catalysts.

In turn, the aerobic thiol coupling is a reaction mechanistically similar to alkyne coupling, where H_2O is generated as a by-product after O_2 dissociation and coupling on the Au atoms, which, in principle, should not occur on Au nanoparticles since thiols are recurrently used as ligands to generate and stabilize Au nanoparticles [18, 76]. Thus, it is not surprising that the reaction only occurs on few-atom Au

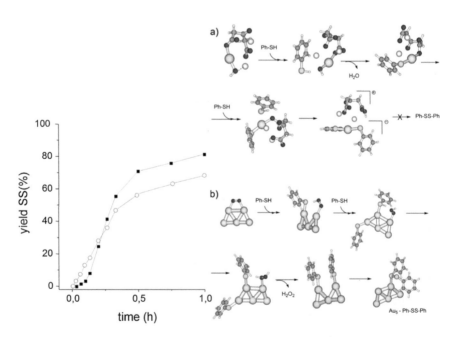

Fig. 1.28 Left: Yield to disulfide with reaction time over Au atoms (black squares) and over Au clusters (empty circles). Au nanoparticles are inactive. Right: Structures involved in the mechanism of thiol oxidation catalyzed by **a** AuI species and **b** Au_5 cluster. Au, S, C, O and H atoms are yellow, orange, red and white, respectively (Fig. adapted from Ref. [5]. Copyright © 2014 American Chemical Society)

clusters since, in contrast to alkynes, the strength of the coordination bond with thiols decreases as the Au particle decreases in size. The isolated Au atoms are not the true catalyst of the homocoupling but just precursors that evolve during reaction to few-atom Au clusters, which are indeed the catalytically active species (see Fig. 1.28).

1.3.4 Hydrogenation Reactions

The hydrogenation of unsaturated bonds is perhaps one of the older transformations with catalytic metals. Typically performed with finely divided metal powders and modernly with metal nanoparticles, the hydrogenation is widely accepted to proceed by H_2 dissociation and spillover of the H atoms on the metal and/or support surface. For this reason, the use of supported single metal atoms or clusters was not considered until recently. Indeed, it has been found that not only H_2 dissociation [40] but also H_2 formation [43] occurs very efficiently over supported isolated metal atoms and clusters [12, 19, 22, 74]. Two representative reaction examples for catalytic clusters follow, although much more can be found in the literature [28, 68].

1.3.4.1 CO$_2$ Methanation

The hydrogenation of CO_2 to methane (Sabatier reaction) is one of the oldest metal-catalyzed hydrogenation reactions, which is now revisited due to the urgent need of converting CO_2 to useful chemicals and, concomitantly, alleviating its global warming effect [23]. For that, it is convenient to design metal catalytic systems that operate at low temperatures (< 250 °C). It has been recently reported that Pt_2^0 clusters inside a MOF catalyze the reaction at temperatures below 150 °C, and comparison between isolated Pt atoms inside the same MOF and reference catalysts shows that the Pt cluster outperforms the rest of materials tested under the low-temperature conditions, including the industrial catalyst $Ru–Al_2O_3$ (Fig. 1.29) [45]. The lack of activity of Pt^{2+} discards these sites as catalytic active species and indicates that Pt_2^0 is the catalytic sites for the hydrogenation of CO_2.

1.3.4.2 Olefin Hydrogenation

Olefin hydrogenation is another typical example of a metal-catalyzed reaction with industrial application [73]. The Pt_2^0 clusters supported in the MOF also efficiently catalyze the hydrogenation of ethylene under industrial reaction conditions, in flow, at much lower temperature (60 °C) than current industrial processes with nanoparticles (200–400 °C) and with a sustained TOF of 250 h^{-1} (Fig. 1.30) [45]. Moreover, other $< C_6$ alkenes such as propylene, 1, 3–butadiene and 1–hexene, among others reacted similarly well. Also, when isomerically pure E–3–hexene was hydrogenated with

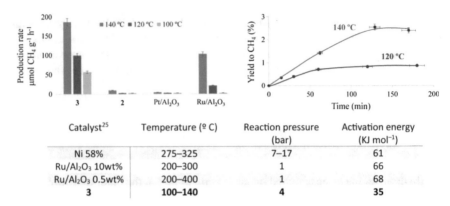

Catalyst[25]	Temperature (º C)	Reaction pressure (bar)	Activation energy (KJ mol⁻¹)
Ni 58%	275–325	7–17	61
Ru/Al₂O₃ 10wt%	200–300	1	66
Ru/Al₂O₃ 0.5wt%	200–400	1	68
3	100–140	4	35

Fig. 1.29 Pt-catalyzed methanation of CO_2. Reaction conditions: 7 ml CO2 (1 atm, 0.28 mmol), 7 ml N$_2$ (internal standard, 1 atm, 0.28 mmol), 7 ml H2 (4 atm, 1.12 mmol), MOF catalyst with Pt_2^+ (2) and Pt_2^0 (3) (8 wt%, 20 mg, 0.008 mmol metal) or M/Al₂O₃ (5 wt%, 32 mg, 0.008 mmol metal), 100–140 °C, 6 h. Table includes reaction values for reference literature catalysts and catalyst **3** (Fig. adapted from Ref. [45]. Copyright © 2018 by John Wiley & Sons, Inc.)

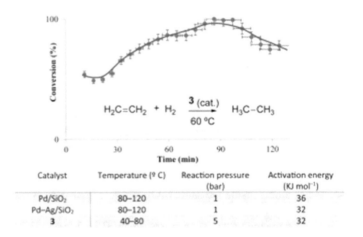

Catalyst	Temperature (º C)	Reaction pressure (bar)	Activation energy (KJ mol⁻¹)
Pd/SiO₂	80–120	1	36
Pd–Ag/SiO₂	80–120	1	32
3	40–80	5	32

Fig. 1.30 Pt-catalyzed hydrogenation of ethylene: Equation shows the reaction conditions: 2 ml/min C$_2$H$_4$, 6 ml/min H$_2$, atmospheric pressure, 60 °C, 50 mg of Pt_2^0–MOF (**3**, 0.010 mmol Pt). Table includes reaction values for reference literature catalysts and catalyst **3** (Fig. adapted from Ref. [45]. Copyright © 2018 by John Wiley & Sons, Inc.)

Pt_2^0 as a catalyst, the corresponding isomerized intermediate Z–3–hexene was found during reaction, which supports a Langmuir–Hinshelwood type mechanism.

1.4 Conclusions

Ligand-free sub-nanometer metal clusters with a precise number of metal atoms are extremely active and selective catalysts for a variety of organic reactions which include carbon–carbon, carbon–heteroatom and heteroatom–heteroatom bond-forming reactions and hydrogenation reactions, among others.

The synthesis of the tiny metal clusters is simple and can be performed by bottom–up (from metal salts and complexes) and top–down (from nanoparticles) approaches. In solution, mild reducing agents such as amide solvents and alcohols or dislodging agents such as Brönsted acids (HCl, HOTf) are employed, but often, these external agents are not necessary, and the same organic reagents trigger and organize the formation of the catalytically active metal clusters during reaction, provided that the metal is sufficiently diluted to avoid further agglomeration. These soluble metal clusters can be stored either in amide or alcohol solutions or in solids (polymeric films, inorganic oxides) to be used on demand for different organic reactions. For the synthesis of the solid-supported, ligand-free sub-nanometer clusters, much stronger reducing agents, such as $NaBH_4$ or H_2, can be employed since the strong interaction between the sub-nanometer cluster and the support avoids further agglomeration. These solid supported clusters are generally more homogeneous in atomicity, spatial distribution and size than those formed in solution and do not require dilution with loadings up to 8 wt% in a MOF.

Modern characterization techniques, including SC–XRD and aberration-corrected TEM, have sufficient technological ability to determine the exact number of atoms, oxidation state and topological distribution of the metal cluster, in other words, a complete structural and electronic information. With this in hand, and also in combination with well-advanced theoretical calculations for cluster chemistry, researchers should now be able to predict the catalytic behavior for a given metal cluster in different reactions and ultimately design the metal cluster needed for a target (and perhaps new) reaction.

References

1. Anciaux AJ, Demonceau A, Noels AF et al (1981) Transition-metal-catalyzed reactions of diazo compounds. 2. Addition to aromatic molecules: catalysis of Buchner's synthesis of cycloheptatrienes. J Org Chem 46:873–876. https://doi.org/10.1021/jo00318a010
2. Bayram E, Linehan JC, Fulton JL et al (2011) Is It Homogeneous or Heterogeneous Catalysis Derived from [RhCp*Cl2]2? In Operando XAFS, Kinetic, and Crucial Kinetic Poisoning Evidence for Subnanometer Rh4 Cluster-Based Benzene Hydrogenation Catalysis. J Am Chem Soc 133:18889–18902. https://doi.org/10.1021/ja2073438
3. Bittner AM, Wu XC, Balci S, et al (2005) Bottom-up synthesis and top-down organisation of semiconductor and metal clusters on surfaces. Eur. J. Inorg. Chem. 3717–3728. https://doi.org/10.1002/ejic.200500388
4. Bogdanović B, Spliethoff B, Wilke G (1980) Dimerization of propylene with catalysts exhibiting activities like highly-active enzymes. Angew Chemie Int Ed English 19:622–623. https://doi.org/10.1002/anie.198006221

5. Boronat M, Laursen S, Leyva-Perez A et al (2014) Partially oxidized gold nanoparticles: a catalytic base-free system for the aerobic homocoupling of alkynes. J Catal 315:6–14. https://doi.org/10.1016/j.jcat.2014.04.003

6. Boronat M, Leyva-Perez A, Corma A (2014) Theoretical and experimental insights into the origin of the catalytic activity of subnanometric gold clusters: attempts to predict reactivity with clusters and nanoparticles of gold. Acc Chem Res 47:834–844. https://doi.org/10.1021/ar400068w

7. Buceta D, Busto N, Barone G et al (2015) Ag2 and Ag3 Clusters: synthesis, characterization, and interaction with DNA. Angew Chemie Int Ed 54:7612–7616. https://doi.org/10.1002/anie.201502917

8. Calabrese JC, Dahl LF, Chini P et al (1974) Synthesis and structural characterization of platinum carbonyl cluster dianions bis, tris, tetrakis, or pentakis(tri-μ2-carbonyl-tricarbonyltriplatinum)(2-). New series of inorganic oligomers. J Am Chem Soc 96:2614–2616. https://doi.org/10.1021/ja00815a050

9. Carenco S, Leyva-Perez A, Concepcion P et al (2012) Nickel phosphide nanocatalysts for the chemoselective hydrogenation of alkynes. Nano Today 7:21–28. https://doi.org/10.1016/j.nantod.2011.12.003

10. Corma A, Concepción P, Boronat M et al (2013) Exceptional oxidation activity with size-controlled supported gold clusters of low atomicity. Nat Chem 5:775

11. Daniel M-C, Astruc D (2004) Gold nanoparticles: assembly, supramolecular chemistry, quantum-size-related properties, and applications toward biology, catalysis, and nanotechnology. Chem Rev 104:293–346. https://doi.org/10.1021/cr030698+

12. Ding K, Gulec A, Johnson AM (2015) Identification of active sites in CO oxidation and water-gas shift over supported Pt catalysts. Science (80);350:189 LP–192. https://doi.org/10.1126/science.aac6368

13. Durand J, Teuma E, Gómez M (2008) An overview of palladium nanocatalysts: Surface and molecular reactivity. Eur. J. Inorg. Chem. 3577–3586. https://doi.org/10.1002/ejic.200800569

14. Dyson PJ (2004) Catalysis by low oxidation state transition metal (carbonyl) clusters. Coord Chem Rev 248:2443–2458. https://doi.org/10.1016/j.ccr.2004.04.002

15. Eglinton G, Galbraith AR (1959) 182. Macrocyclic acetylenic compounds. Part I. Cyclotetradeca-1 :3-diyne and related compounds. J. Chem. Soc. 889–896. https://doi.org/10.1039/jr9590000889

16. Faraday M (1857) Experimental relations of gold {and other Metals) to Light. B y. Philos Trans 147:145

17. Fernandez E, Rivero-Crespo MA, Dominguez I et al (2019) Base-controlled heck, suzuki, and sonogashira reactions catalyzed by ligand-free platinum or palladium single atom and sub-nanometer clusters. J Am Chem Soc 141:1928–1940. https://doi.org/10.1021/jacs.8b07884

18. Ferrando R, Jellinek J, Johnston RL (2008) Nanoalloys: from theory to applications of alloy clusters and nanoparticles. Chem Rev 108:845–910. https://doi.org/10.1021/cr040090g

19. Flytzani-Stephanopoulos M, Gates BC (2012) Atomically dispersed supported metal catalysts. Annu Rev Chem Biomol Eng 3:545–574. https://doi.org/10.1146/annurev-chembioeng-062011-080939

20. Fortea-Perez FR, Mon M, Ferrando-Soria J et al (2017) The MOF-driven synthesis of supported palladium clusters with catalytic activity for carbene-mediated chemistry. Nat Mater 16:760–766. https://doi.org/10.1038/nmat4910

21. Frogneux X, Pesesse A, Delacroix S et al (2019) Radical-initiated dismutation of hydrosiloxanes by catalytic potassium-graphite. ChemCatChem. https://doi.org/10.1002/cctc.201900172

22. Fu Q, Saltsburg H, Flytzani-Stephanopoulos M (2003) Active nonmetallic Au and Pt species on ceria-based water-gas shift catalysts. Science (80);301:935 LP–938. https://doi.org/10.1126/science.1085721

23. Ghaib K, Nitz K, Ben-Fares F-Z (2016) Chemical methanation of CO2: a review. Chem Bio Eng Rev 3:266–275. https://doi.org/10.1002/cben.201600022

24. Glaser C (1870) Untersuchungen über einige derivate der zimmtsäure. Justus Liebigs Ann Chem 154:137–171. https://doi.org/10.1002/jlac.18701540202

25. Grancha T, Ferrando-Soria J, Zhou H-C et al (2015) Postsynthetic improvement of the physical properties in a metal-organic framework through a single crystal to single crystal transmetalation. Angew Chemie, Int Ed 54:6521–6525. https://doi.org/10.1002/anie.201501691
26. Hashmi ASK, Frost TM, Bats JW (2000) Highly selective gold-catalyzed arene synthesis. J Am Chem Soc 122:11553–11554. https://doi.org/10.1021/ja005570d
27. Hay AS (1962) Oxidative coupling of acetylenes. ii1. J Org Chem 27:3320–3321. https://doi.org/10.1021/jo01056a511
28. Hernández E, Bertin V, Soto J et al (2018) Catalytic reduction of nitrous oxide by the low-symmetry pt8 cluster. J Phys Chem A 122:2209–2220. https://doi.org/10.1021/acs.jpca.7b11055
29. Herzing AA, Kiely CJ, Carley AF, et al (2008) Identification of active gold nanoclusters on iron oxide supports for CO oxidation. Science (80) 321:1331 LP–1335. https://doi.org/10.1126/science.1159639
30. Ikuno T, Zheng J, Vjunov A et al (2017) Methane oxidation to methanol catalyzed by cu-oxo clusters stabilized in nu-1000 metal-organic framework. J Am Chem Soc 139:10294–10301. https://doi.org/10.1021/jacs.7b02936
31. Imaoka T, Akanuma Y, Haruta N et al (2017) Platinum clusters with precise numbers of atoms for preparative-scale catalysis. Nat Commun 8:688. https://doi.org/10.1038/s41467-017-00800-4
32. Imaoka T, Yamamoto K (2019) Wet-chemical strategy for atom-precise metal cluster catalysts. Bull Chem Soc Jpn 92:941–948. https://doi.org/10.1246/bcsj.20190008
33. Kennedy-Smith JJ, Staben ST, Toste FD (2004) Gold(I)-catalyzed conia-ene reaction of β-ketoesters with alkynes. J Am Chem Soc 126:4526–4527. https://doi.org/10.1021/ja049487s
34. Leyva-Perez A (2017) Sub-nanometre metal clusters for catalytic carbon-carbon and carbon-heteroatom cross-coupling reactions. Dalt Trans 46:15987–15990. https://doi.org/10.1039/C7DT03203J
35. Leyva-Perez A, Corma A (2012) Similarities and differences between the "relativistic" triad gold, platinum, and mercury in catalysis. Angew Chemie, Int Ed 51:614–635. https://doi.org/10.1002/anie.201101726
36. Leyva-Perez A, Domenech-Carbo A, Corma A (2015) Unique distal size selectivity with a digold catalyst during alkyne homocoupling. Nat Commun 6:6703. https://doi.org/10.1038/ncomms7703
37. Leyva-Perez A, Oliver-Meseguer J, Rubio-Marques P, Corma A (2013) Water-stabilized three- and four-atom palladium clusters as highly active catalytic species in ligand-free C-C cross-coupling reactions. Angew Chemie, Int Ed 52:11554–11559. https://doi.org/10.1002/anie.201303188
38. Liu L, Corma A (2018) Metal catalysts for heterogeneous catalysis: from single atoms to nanoclusters and nanoparticles. Chem Rev (Washington, DC, United States) 118:4981–5079. https://doi.org/10.1021/acs.chemrev.7b00776
39. Liu L, Zakharov DN, Arenal R et al (2018) Evolution and stabilization of subnanometric metal species in confined space by in situ TEM. Nat Commun 9:1–10. https://doi.org/10.1038/s41467-018-03012-6
40. Liu P, Zhao Y, Qin R, et al (2016) Photochemical route for synthesizing atomically dispersed palladium catalysts. Science (80) 352:797 LP–800. https://doi.org/10.1126/science.aaf5251
41. Longoni G, Chini P (1976) Synthesis and chemical characterization of platinum carbonyl dianions [Pt3(CO)6]n2- (n = .apprx.10,6,5,4,3,2,1). A new series of inorganic oligomers. J Am Chem Soc 98:7225–7231. https://doi.org/10.1021/ja00439a020
42. Marciniec B (2008) Hydrosilylation of unsaturated carbon—heteroatom bonds
43. Marcinkowski MD, Liu J, Murphy CJ et al (2017) Selective formic acid dehydrogenation on Pt-Cu single-atom alloys. ACS Catal 7:413–420. https://doi.org/10.1021/acscatal.6b02772
44. Mon M, Ferrando-Soria J, Grancha T et al (2016) Selective gold recovery and catalysis in a highly flexible methionine-decorated metal-organic framework. J Am Chem Soc 138:7864–7867. https://doi.org/10.1021/jacs.6b04635

45. Mon M, Rivero-Crespo MA, Ferrando-Soria J et al (2018) synthesis of densely packaged, ultrasmall Pt02 clusters within a thioether-functionalized mof: catalytic activity in industrial reactions at low temperature. Angew Chemie, Int Ed 57:6186–6191. https://doi.org/10.1002/anie.201801957

46. Murahashi T, Fujimoto M, Oka M (2006) discrete sandwich compounds of monolayer palladium sheets. Science (80) 313:1104 LP–1107. https://doi.org/10.1126/science.1125245

47. Murahashi T, Kato N, Uemura T, Kurosawa H (2007) Rearrangement of a Pd4 skeleton from a 1D chain to a 2D sheet on the face of a perylene or fluoranthene ligand caused by exchange of the binder molecule. Angew Chemie Int Ed 46:3509–3512. https://doi.org/10.1002/anie.200700340

48. Murahashi T, Uemura T, Kurosawa H (2003) Perylene – Tetrapalladium Sandwich Complexes. J Am Chem Soc 125:8436–8437. https://doi.org/10.1021/ja0358246

49. Oliver-Meseguer J, Cabrero-Antonino JR, Dominguez I (2012) Small gold clusters formed in solution give reaction turnover numbers of 107 at room temperature. Sci (Washington, DC, United States) 338:1452–1455. https://doi.org/10.1126/science.1227813

50. Oliver-Meseguer J, Dominguez I, Gavara R (2017) The wet synthesis and quantification of ligand-free sub-nanometric Au clusters in solid matrices. Chem Commun (Cambridge, United Kingdom) 53:1116–1119. https://doi.org/10.1039/C6CC09119A

51. Oliver-Meseguer J, Dominguez I, Gavara R et al (2017) Disassembling Metal Nanocrystallites into Sub-nanometric Clusters and Low-faceted Nanoparticles for Multisite Catalytic Reactions. Chem Cat Chem 9:1429–1435. https://doi.org/10.1002/cctc.201700037

52. Oliver-Meseguer J, Leyva-Perez A, Al-Resayes SI, Corma A (2013) Formation and stability of 3–5 atom gold clusters from gold complexes during the catalytic reaction: dependence on ligands and counteranions. Chem Commun (Cambridge, United Kingdom) 49:7782–7784. https://doi.org/10.1039/c3cc44104k

53. Oliver-Meseguer J, Leyva-Perez A, Corma A (2013) Very small (3–6 atoms) gold cluster catalyzed carbon-carbon and carbon-heteroatom bond-forming reactions in solution. Chem Cat Chem 5:3509–3515. https://doi.org/10.1002/cctc.201300695

54. Oliver-Meseguer J, Liu L, Garcia-Garcia S et al (2015) Stabilized naked sub-nanometric cu clusters within a polymeric film catalyze C-N, C-C, C-O, C-S, and C-P Bond-Forming Reactions. J Am Chem Soc 137:3894–3900. https://doi.org/10.1021/jacs.5b03889

55. Panyala RN, Pena-Mendez ME, Havel J (2009) Gold and nano-gold in medicine: overview, toxicology and perspectives. J Appl Biomed 7:75–91

56. Pastoriza-Santos I, Liz-Marzan LM (1999) Formation and stabilization of silver nanoparticles through reduction by N, N-dimethylformamide. Langmuir 15:948–951. https://doi.org/10.1021/LA980984U

57. Peredkov S, Peters S, Al-Hada M et al (2016) Structural investigation of supported Cun clusters under vacuum and ambient air conditions using EXAFS spectroscopy. Catal Sci Technol 6:6942–6952. https://doi.org/10.1039/C6CY00436A

58. Polozkov RG, Ivanov VK, Verkhovtsev AV (2013) New applications of the jellium model for the study of atomic clusters. J. Phys. Conf. Ser. 438. https://doi.org/10.1088/1742-6596/438/1/012009

59. Rivero-Crespo MA, Leyva-Pérez A, Corma A (2017) A ligand-free Pt 3 cluster catalyzes the markovnikov hydrosilylation of alkynes with up to 10 6 turnover frequencies. Chem - A Eur J 23:1702–1708. https://doi.org/10.1002/chem.201605520

60. Rivero-Crespo MA, Mon M, Ferrando-Soria J et al (2018) Confined Pt11 + water clusters in a MOF catalyze the low-temperature water-gas shift reaction with both CO_2 oxygen atoms coming from water. Angew Chemie, Int Ed 57:17094–17099. https://doi.org/10.1002/anie.201810251

61. Rubio-Marques P, Rivero-Crespo MA, Leyva-Perez A, Corma A (2015) Well-defined noble metal single sites in zeolites as an alternative to catalysis by insoluble metal salts. J Am Chem Soc 137:11832–11837. https://doi.org/10.1021/jacs.5b07304

62. Sa J, Frances S, Taylor R (2012) Redispersion of gold supported on oxides

63. Sá J, Goguet A, Taylor SFR et al (2011) Influence of methyl halide treatment on gold nanoparticles supported on activated carbon. Angew Chem Int Ed Engl 50:8912–8916. https://doi.org/10.1002/anie.201102066

64. Sakaki S, Mizoe N, Sugimoto M (1998) theoretical study of platinum(0)-catalyzed hydrosilylation of ethylene. chalk − harrod mechanism or modified chalk − harrod mechanism. Organometallics 17:2510–2523. https://doi.org/10.1021/om980190a

65. Scarabelli L, Coronado-Puchau M, Giner-Casares JJ et al (2014) monodisperse gold nanotriangles: size control, large-scale self-assembly, and performance in surface-enhanced raman scattering. ACS Nano 8:5833–5842. https://doi.org/10.1021/nn500727w

66. Serna P, Gates BC (2014) Molecular metal catalysts on supports: organometallic chemistry meets surface science. Acc Chem Res 47:2612–2620. https://doi.org/10.1021/ar500170k

67. Sharma S, Kurashige W, Niihori Y, Negishi Y (2016) Nanocluster Science. Elsevier Inc

68. Tian S, Fu Q, Chen W et al (2018) Carbon nitride supported Fe2 cluster catalysts with superior performance for alkene epoxidation. Nat Commun 9:2353. https://doi.org/10.1038/s41467-018-04845-x

69. Vajda S, White MG (2015) Catalysis applications of size-selected cluster deposition. ACS Catal 5:7152–7176. https://doi.org/10.1021/acscatal.5b01816

70. Wang N, Sun Q, Yu J (2019) Ultrasmall metal nanoparticles confined within crystalline nanoporous materials: a fascinating class of nanocatalysts. Adv Mater 31:1–23. https://doi.org/10.1002/adma.201803966

71. Wu X-F, Anbarasan P, Neumann H, Beller M (2010) From noble metal to nobel prize: palladium-catalyzed coupling reactions as key methods in organic synthesis. Angew Chemie, Int Ed 49:9047–9050. https://doi.org/10.1002/anie.201006374

72. Yang Y, Reber AC, Gilliland SE et al (2018) Donor/acceptor concepts for developing efficient suzuki cross-coupling catalysts using graphene-supported Ni, Cu, Fe, Pd, and Bimetallic Pd/Ni Clusters. J Phys Chem C 122:25396–25403. https://doi.org/10.1021/acs.jpcc.8b07538

73. Zea H, Lester K, Datye AK et al (2005) The influence of Pd-Ag catalyst restructuring on the activation energy for ethylene hydrogenation in ethylene-acetylene mixtures. Appl Catal A Gen 282:237–245. https://doi.org/10.1016/j.apcata.2004.12.026

74. Zhai Y, Pierre D, Si R (2010) Alkali-stabilized Pt-OH<em><sub>x</sub></em> species catalyze low-temperature water-gas shift reactions. Science (80) 329:1633 LP–1636. https://doi.org/10.1126/science.1192449

75. Zhang C, Laine RM (2000) Hydrosilylation of allyl alcohol with [HSiMe2OSiO1.5]8: Octa (3-hydroxypropyldimethylsiloxy) octasilsesquioxane and its octamethacrylate derivative as potential precursors to hybrid nanocomposites. J Am Chem Soc 122:6979–6988. https://doi.org/10.1021/ja000318r

76. Zitoun D, Respaud M, Fromen M-C et al (2002) Magnetic enhancement in nanoscale corh particles. Phys Rev Lett 89:37203. https://doi.org/10.1103/PhysRevLett.89.037203

Chapter 2
Atomically Precise Nanoclusters as Electrocatalysts

Site Li and Rongchao Jin

Abstract This chapter summaries recent advances in electrocatalytic application of atomically precise metal nanoclusters (NCs). Metal nanoclusters with determined structures can serve as new model catalysts for electrochemical catalytic study at the atomic level and offer insights into the underlying mechanisms. In recent years, electrocatalysis by metal nanoclusters has been reported and shows promise in several important reactions, including oxygen reduction reaction, water splitting, and CO_2 reduction reaction. By tuning the structure/ligand of the metal nanoclusters, it is possible to achieve catalytic property modification at the atomic level. Overall, the new material of atomically precise metal nanoclusters holds great promise in precise control of catalytic properties and investigation of the fundamental catalytic mechanism at the atomic level.

Keywords Metal nanocluster · Atomic precision · X-ray structure · Electrocatalysis · Doping

2.1 Introduction

2.1.1 Atomically Precise Metal NCs

Atomically precise metal nanoclusters have attracted broad interest due to the crystal structure availability and unique properties in optical and catalysis applications [1–3]. Compared with the traditional plasmonic metal nanoparticles, the ultra-small NCs (<3 nm) show quantized electronic structures because of the quantum confinement effect [4]. As a result, a single atom change can significantly alter the properties of NCs. The UV-vis spectrum can be used as the "fingerprints" of NCs [5] because it shows certain distinct peaks for each size of NCs, rather than similar plasmonic peaks for regular nanoparticles. Similarly, the catalytic properties of various metal NCs can be totally different because of the major change in surface and electronic structure of

S. Li · R. Jin (✉)
Department of Chemistry, Carnegie Mellon University, Pittsburgh, PA 15213, USA
e-mail: rongchao@andrew.cmu.edu

© Springer Nature Switzerland AG 2020
P. W. N. M. van Leeuwen and C. Claver (eds.), *Recent Advances in Nanoparticle Catalysis*, Molecular Catalysis 1, https://doi.org/10.1007/978-3-030-45823-2_2

NCs with subtle difference in atom numbers or size. Therefore, it is of great interest to correlate the structure and catalytic application using NCs as model catalysts [6]. Indeed, it is possible to build up a library of NCs structure–catalytic properties since the number of reported NCs is large enough. This library might offer great insights into the interpretation of catalytic process and reaction mechanism, and further offer some guidelines in the future design and synthesis of new NCs.

2.1.2   Electrochemical Catalysis with Atomically Precise Metal NCs

The global energy crisis and pollution issues have driven scientists to investigate the alternatives of fossil fuels. One of the strategies is using the secondary energy (such as solar energy and wind energy) derived electricity to split water for producing H_2 as a clean energy source [7]. In the water splitting system, hydrogen is produced at the cathode through hydrogen evolution reaction (HER), and oxygen is formed at the anode through oxygen evolution reaction (OER). Currently, Pt group metals are proved to be the most effective catalysts in HER, while Ir/Ru materials are successful in OER. However, the high cost of noble metals has motivated scientists to study alternative catalysts for these reactions.

On the other hand, the obtained hydrogen from HER and other fuels, such as methanol, can be utilized in the fuel cell system. The fuel cell is an electrochemical device to efficiently transform chemical energy of fuel without combustion [8]. Due to its high efficiency and environmentally friendly properties, fuel cells have been used in vehicles [9]. Currently, the disadvantage of fuel cell technique is the oxygen reduction reaction (ORR) in cathode electrode. It is believed that the ORR is the rate limiting reaction with very sluggish kinetics because of multi-electron transfer during the reaction [10]. Similar to the HER, ORR also favors Pt as the catalyst. Therefore, alternative catalysts are yet to be found to reduce the high cost of noble metals.

Another popular electrochemical catalytic reaction is the CO_2 electrochemical reduction reaction (CO_2RR). In the past decades, the global warming has been considered as a serious issue caused by massive CO_2 emission. To relieve the climate change pressure, one of the solutions is to utilize CO_2 as a resource to produce industrial chemicals and fuels [11, 12]. In the CO_2RR, catalytic materials are required to overcome the intrinsic inertness of CO_2 molecules [13]. Among the catalyst candidates, Au and Ag have been extensively studied because of their high selectivity toward CO [14]. Besides, Cu is also a good catalyst because of its versatility to form various carbon hydrates and low price.

To design better catalytic materials, it is essential to understand the mechanism behind these catalytic reactions. Thus, it is of great importance to find a system to correlate the structure and catalytic properties. Previously, several strategies, including size control and morphology control, have been used to investigate the relationships between structure and properties [15, 16]. For example, Zhu et al. [17] and Seoin et al. [18] reported the active site probing with Au catalysts of different morphologies from the view of experimental and computational modeling, respectively. Despite the well-designed experiment, the non-atomically monodispersed size of traditional nanomaterials significantly weakens the connection between the structure and properties.

In the past decades, the synthesis strategy of atomically precise metal NCs has been extensively investigated and a number of sizes of NCs between tens and hundreds of atoms (equivalent diameters ranging from sub-nanometer to ~2.2 nm) have been reported [1]. For applications as electrochemical catalytic materials, such NCs have several distinctive features such as high surface area and unique surface structure [19]. Besides, the atomic precision and crystal structure availability make metal NCs a perfect system to bridge the structure and properties.

In this chapter, several works about metal NCs as electrochemical catalysts are introduced with a focus on the atomic size effect, morphology effect, doping effect, and charge effect. The computational techniques used in the catalytic mechanism study are also summarized.

2.2 Synthesis and Structure Determination of Atomically Precise Metal NCs

2.2.1 Synthesis of Metal NCs

Here, we illustrate the size-focusing synthesis and structure determination using atomically precise $Au_{25}(SR)_{18}$ NCs as an example. Larger NCs such as $Au_{133}(SR)_{52}$ and $Au_{279}(SR)_{84}$ can also be synthesized by the size-focusing method [20, 21]. In the size-focusing method, a mixture of NCs with a controlled size distribution is first prepared by carefully controlling the ratio of gold precursor and reduction agent as well as other synthetic conditions. Then, the NCs mixture is subjected to size-focusing under harsh conditions, under which the unstable NCs decompose or convert to more stable ones. Eventually, only the most stable NCs can survive the size-focusing process [22].

In the case of Au_{25} [23, 24], the Au(III) salt is initially reduced to Au(I) by thiols at 0 °C in the first step. The as-obtained Au(I)-SR complex is then reduced by adding a $NaBH_4$ aqueous solution. Polydisperse NCs protected by thiolate are obtained after the reduction process. During the following size-focusing process, it can be observed from the evolution of the optical absorption spectra that the monodispersed Au_{25} NCs gradually become dominant, as shown in Fig. 2.1a. The mass spectrum also illustrates the molecular purity of Au_{25} (Fig. 2.1b).

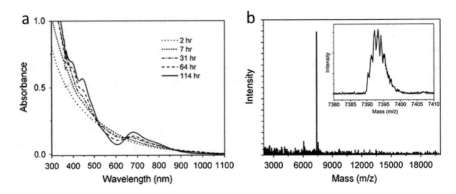

Fig. 2.1 **a** Evolution of the UV-vis spectra of the reaction product in the Au$_{25}$ synthesis, **b** mass spectrometry analysis of Au$_{25}$. Adapted from Ref. [24]. Copyright 2009 Royal Society of Chemistry

2.2.2   Structure Determination of Metal NCs

The crystal structures of NCs can be determined by X-ray crystallography. In the case of Au$_{25}$, the structure comprises a Au$_{13}$ icosahedral core and a Au$_{12}$(SR)$_{18}$ shell [23]. The Au$_{12}$ shell can be dissected into six dimeric –S–Au–S–Au–S– staple motifs (Fig. 2.2). Due to the atomic precision and the determined crystal structure of metal NCs, it is of great interest to use NCs as catalysts for catalytic mechanism study. Especially for electrochemical catalysis, it is challenging to capture the intermediates during the catalytic process and thus very little is known about the mechanism. However, the well-defined structure of nanocluster catalysts can now facilitate the computational modeling, thus providing opportunities to reveal the mechanism behind electrochemical catalysis.

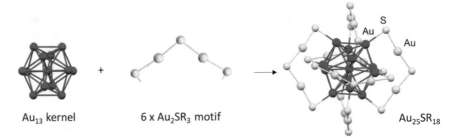

Au$_{13}$ kernel   6 x Au$_2$SR$_3$ motif   Au$_{25}$SR$_{18}$

Fig. 2.2 Dissection of the Au$_{25}$ cluster into Au$_{13}$ kernel and six surface motifs. Adapted from Ref. [5]. Copyright 2012 American Chemical Society

2.3 Hydrogen Evolution Reaction with Metal NC Catalysts

The hydrogen evolution reaction (HER) occurs at the cathode when an external voltage is applied. The reaction can be described in three steps:
Volmer reaction:

$$H^+ + M + e^- \rightarrow MH_{ads} \quad \text{(acidic)} \tag{2.1}$$

$$H_2O + M + e^- \rightarrow MH_{ads} + OH^- \quad \text{(alkaline)} \tag{2.2}$$

In the Volmer reaction, hydrogen absorbs on the catalytic material to form a MH_{ads} intermediate, followed by a Heyrovsky reaction or Tafel reaction. In the Heyrovsky reaction, the dihydrogen is formed through an electrochemical desorption:

$$MH_{ads} + H^+ + e^- \rightarrow M + H_2 \quad \text{(acidic)} \tag{2.3}$$

$$MH_{ads} + H_2O + e^- \rightarrow M + OH^- + H_2 \quad \text{(alkaline)} \tag{2.4}$$

The hydrogen can also undergo a chemical desorption process through the Tafel reaction:

$$2\,MH_{ads} \rightarrow 2\,M + H_2 \tag{2.5}$$

The binding energy of hydrogen to the catalyst is the key factor of HER activity. Among the HER catalysts, noble metals such as Pt and Pd have moderate binding energy with hydrogen, thus showing excellent HER activities [25]. However, the high cost and stability issue of Pt catalysts motivate the scientists to find alternative materials for HER.

Here, we summarize the doping effects and synergetic effects of Au NCs in HER. We also introduce how the computational technique is used in these cases to explain differences in catalytic activity and verify proposed mechanisms. These works offer some insights into the catalytic reactions, which is expected to further pave the way for future design of catalyst materials.

2.3.1 Pt or Pd Doped Au_{25} NCs in HER

The study of Au NCs as HER catalysts was reported by Kyuju et al. in 2017 [26]. A molecular-like Pt_1Au_{24} nanocluster was prepared and used as catalysts in homogeneous HER. Following this work, the same group further studied the Pd_1Au_{24}, Pd_2Au_{36} and Pt_2Au_{36} nanoclusters [27]. In the case of Au_{25} nanocluster, the doping atom (Pd or Pt) exclusively replaces the central gold atom in the nanocluster. The overall structure of nanoclusters remains unchanged after doping, while the mass spectrum and optical absorption spectrum obviously changed (Fig. 2.3). In the mechanistic study of nanoclusters as HER catalysts, Voltammetry was used to study the electron transfer properties (Fig. 2.4). The redox potentials are drastically changed after Pt doping. Also, compared with Au_{25}, Pt_1Au_{24} has more positive onset potential and higher current. Combining the voltammetry and linear weep voltammograms (LSVs) with different concentration of trifluoroacetic acid (TFA), it can be seen that the $[PtAu_{24}]^{1-/2-}$ peak at 1.10 V drastically rises with increasing TFA concentration, indicating that the $[PtAu_{24}]^{2-}$ is the major contributor for enhanced HER activity. Also, the production rate with the Pt_1Au_{24} catalyst is significantly higher than that of the commercial Pt/C catalyst. Additionally, the charge-state-dependent catalytic activity results show that the catalytic currents at potentials negative to the $[PtAu_{24}]^{1-/2-}$ exhibit a linear correlation with $[PtAu_{24}]$ and $[TFA]^{1/2}$, corresponding

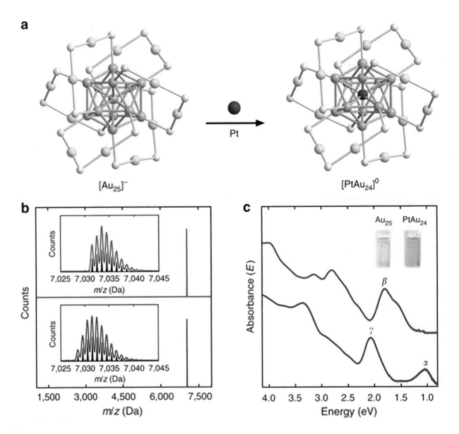

Fig. 2.3 **a** Structures of Au_{25} and $PtAu_{24}$ NCs (golden, Au atoms of the kernel; olive, Au atoms of the shell; gray, sulfur), **b** Mass spectrometry analysis of Au_{25} (red) and Pt_1Au_{24} (blue) NCs, **c** UV-vis-NIR absorption spectra of Au_{25} (red) and Pt_1Au_{24} (blue) NCs. Adapted with permission from Ref. [26]. Copyright 2017 Springer Nature

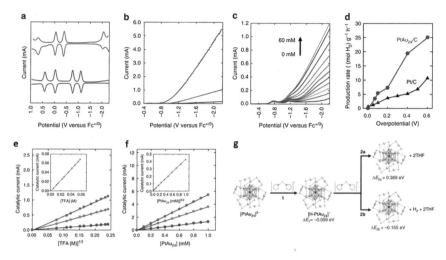

Fig. 2.4 **a** Square-wave voltammetry (SWV) of Au_{25} (red) and $PtAu_{24}$ (blue) NCs, **b** LSV of Au_{25} (red) and $PtAu_{24}$ (blue) NCs, **c** LSVs of $PtAu_{24}$ in THF in the presence of 0. 4. 8, 12, 17, 21, 27, 34, 45, 55, and 60 mM of trifluoroacetic acid (TFA), **d** H_2 production rates per mass of metals in catalyst at various overpotentials on $PtAu_{24}$ (blue) and Pt/C electrodes (red), **e** Dependence of the catalytic current Ic on the concentration of TFA in the presence of $PtAu_{24}$ (I mM), **f** Dependence of the catalytic current Ic on the concentration of $PtAu_{24}$ in TFA (1 M) solution at -1.3 V (blue), -1.8 V (green) and -2.2 V (purple), insets show dependence of the Ic on the concentration of **e** TFA and **f** $PtAu_{24}$ at -1.0 V, **g** calculated reaction energies for HER on $PtAu_{24}$. Adapted with permission from Ref. [26]. Copyright 2017 Springer Nature

to the Volmer–Heyrovsky mechanism [28, 29]. Meanwhile, the currents exhibit a linear correlation with [TFA] and $[PtAu]^{3/2}$ at -1.0 V where $[PtAu_{24}]^{1-}$ is dominant, corresponding to the Volmer–Tafel mechanism [28]. These results are reasonable considering the charge state of Pt_1Au_{24} at different potential. To be specific, the dominant $[PtAu_{24}]^{2-}$ at negative potential adsorbs a proton to form $[H–PtAu_{24}]^{1-}$ with negative charge, which is prone to react with another proton to evolve H_2. On the other hand, the dominant $[PtAu_{24}]^{1-}$ at -1.0 V will form $[H–PtAu_{24}]^0$ after adsorbing a proton. Therefore, the Tafel pathway is preferred at this potential.

DFT calculations are also utilized to explain the enhanced catalytic activity of Pt_1Au_{24} (Fig. 2.4g). The results show that the adsorption of the first proton on Pt_1Au_{24} is thermodynamically neutral, while the second proton adsorption is endothermic. This result is consistent with the charge-state-dependent catalytic activity. On the other hand, the geometry optimization shows that the proton is prone to bind with the Pt atom in the icosahedral center. Therefore, the stronger H–Pt interaction with respect to H–Au is a key factor for the enhanced HER activity. Using the same strategies, the group also studied the HER activities of Pd_1Au_{24}, Pt_2Au_{36} and Pd_2Au_{36} [27]. These works proved the versatility of metal NCs in catalytic mechanism study. Especially, the introduction of SWV and DFT calculation makes the metal NCs a perfect tool to correlate the structure and properties.

2.3.2 Boosting HER Activity with Au NCs/MoS₂ Composite

Besides the doping effects, metal NCs are also reported to show strong synergetic effects when loaded on other materials. In 2017, the synergetic effect was reported by Zhao et al. [30]. In this work, $Au_{25}(SR)_{18}$ and $Au_{25}(SePh)_{18}$ NCs are loaded on the MoS_2 ultra-thin nanosheets. These two NCs have a similar kernel structure but different protecting ligands. The TEM images and XPS spectra indicate the successful loading of Au NCs on the surface of MoS_2 (Fig. 2.5). The Mo 3d XPS of composites exhibits negative shifting compared with MoS_2, while the Au 4f spectra of composites show obvious positive shifting compared with Au_{25} NCs. The XPS results clearly indicate that the electron density transfers from Au_{25} to MoS_2. In the HER activity test, the thiolate-protected Au_{25} NCs exhibit a more positive onset potential and higher current density compared with MoS_2 nanosheets. On the other hand, the benzeneselenolate-protected Au_{25} NCs loaded MoS_2 nanosheets show similar synergetic effects. However, the enhancement is less obvious compared with the thiolate-protected Au_{25} NCs. To explain the different enhancement in HER catalytic activity, Au 4f XPS spectra of $Au_{25}(SR)_{18}/MoS_2$, $Au_{25}(SePh)_{18}/MoS_2$ and pure MoS_2 are obtained. The $Au_{25}(SR)_{18}/MoS_2$ composites exhibit more positive shifting compared with the $Au_{25}(SePh)_{18}/MoS_2$ composites, indicating stronger electron density transfer effects of thiol protected Au_{25}. Therefore, the electron interaction between MoS_2 nanosheets and Au NCs is a key factor for the HER activity. Based on these results, the authors proposed a dual interfacial effect, where the core/ligand interface of Au NCs and the MoS_2/Au NCs interface are both important in the HER catalytic activity.

Du et al. also studied the synergetic effects between Au_2Pd_6 NCs and MoS_2 [31]. In this work, DFT calculation is used to investigate the origin of the enhanced HER activity. The DFT calculation shows the $_\Delta G$ for a proton adsorbed at Au_2Pd_6/MoS_2 composite is more negative than that of MoS_2, indicating better HER activity of the composites. Besides, it is found that in the Au_2Pd_6 composites, both Au atoms and S atoms have appropriate $_\Delta G$ for proton adsorption. In contrast, only the Au–Pd bridge site has proper $_\Delta G$ in Au_2Pd_6 NCs, and no proper active site for proton adsorption can be found in defect-free MoS_2. Therefore, the significant increase in active sites in the composites is a key factor for boosted HER activity. Meanwhile, the DOS analysis also explained the enhanced activity of composites. The Au_2Pd_6 composites have a defect state near the Fermi level. This unique defect state narrows the band gap, leading to a better electronic conductivity.

2.4 Oxygen Evolution Reaction with Metal NCs

Oxygen evolution reaction (OER) is the other half reaction in water splitting and is indeed critical. Oxygen can be formed through several proton/electron-coupled steps in OER. The reaction can be described as follows:

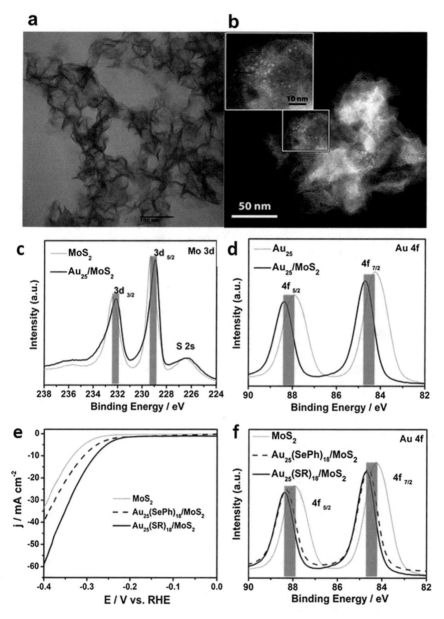

Fig. 2.5 **a** TEM image of the MoS$_2$ nanosheets, **b** HAADF-STEM image of the Au$_{25}$/MoS$_2$ composite, **c** high-resolution Mo 3d XPS spectra of MoS$_2$ nanosheet and Au$_{25}$/MoS$_2$ composite, **d** high-resolution Au 4f XPS spectra of Au$_{25}$ NCs and Au$_{25}$/MoS$_2$ composite, **e** HER LSV curves, **f** high-resolution Au 4f XPS spectra of MoS$_2$, Au$_{25}$(SePh)$_{18}$/MoS$_2$ and Au$_{25}$(SR)$_{18}$/MoS$_2$ composite. Adapted from Ref. [30]. Copyright 2017 Wiley-VCH

$$2 H_2O \rightarrow 4 H^+ + O_2 + 4 e^- \quad \text{(acidic)} \qquad (2.6)$$

$$4 OH^- \rightarrow 2 H_2O + O_2 + 4 e^- \quad \text{(alkaline)} \qquad (2.7)$$

In OER, the formation of oxygen requires a four-electron transfer, and the reaction kinetically favors single electron transfer at each step [32]. Therefore, catalysts are required to overcome the energy barrier and lower the high overpotential in the sluggish OER [33, 34]. To reduce the cost of catalyst materials, cheaper and efficient alternative catalytic materials are extensively studied to replace the current Ir-based materials. In previous reports, anchoring a small amount of gold onto cobalt-based materials can enhance the OER activity [35, 36]. However, the mechanism for the improvement was not well understood due to the variability and complicacy of the gold-loaded composites. In this section, the synergetic effects between Au NCs and $CoSe_2$ nanosheets are introduced. This unique composite may provide valuable insights into the mechanistic study by taking advantage of the precise atomic structures of Au NCs [1].

2.4.1 Au_n NCs Promote OER at the Nanocluster/$CoSe_2$ Interface

The OER performance of metal NCs was first reported by Zhao et al. in 2017 [37]. In this work, composites of Au_{25} and ultra-thin $CoSe_2$ nanosheets were synthesized and tested as OER catalysts. TEM images clearly show the ultra-thin nanosheet structure of $CoSe_2$ (Fig. 2.6a–c). In the high-angle annular dark field scanning transmission electron microscopy (HAADF-STEM), it can be observed that Au nanoclusters are homogeneously dispersed on the surface of $CoSe_2$ nanosheets.

In the electrochemical test (Fig. 2.6d–f), the $Au_{25}/CoSe_2$ composites show much smaller onset potential (1.406 V vs. RHE) and higher current density than $CoSe_2$ nanosheets and Au_{25}-loaded carbon. At 1.68 V, $Au_{25}/CoSe_2$ composites achieve a current density of 11.78 mA cm^{-2}, which is 2.4 times that of $CoSe_2$ nanosheets (4.92 mA cm^{-2}) and 20.7 times that of Au_{25}-loaded carbon (0.57 mA cm^{-2}). Also, the composites exhibit higher current density and smaller overpotential than commercial Pt/C catalysts. In the stability test, the polarization curve and UV-vis spectra of the $Au_{25}/CoSe_2$ composites exhibit the same features before and after 1000 cycles, indicating excellent stability of the composites as OER catalysts (Fig. 2.6g).

The XPS and Raman analysis of $CoSe_2$ and composites were conducted to explain the enhanced OER activity of $Au_{25}/CoSe_2$ composites (Fig. 2.7). The binding energy of Co 2p in the composites shows a ~1 eV decrease compared with $CoSe_2$, indicating electronic interaction between the Au_{25} and $CoSe_2$ nanosheet. Also, the Raman peak at ca. 657 cm^{-1} exhibits a shift toward higher wavenumber, suggesting the electronic interaction. It is believed that such an electronic interaction is a key factor that stabilizes the hydroperoxyl intermediates and optimizes interaction between $CoSe_2$ and oxygen.

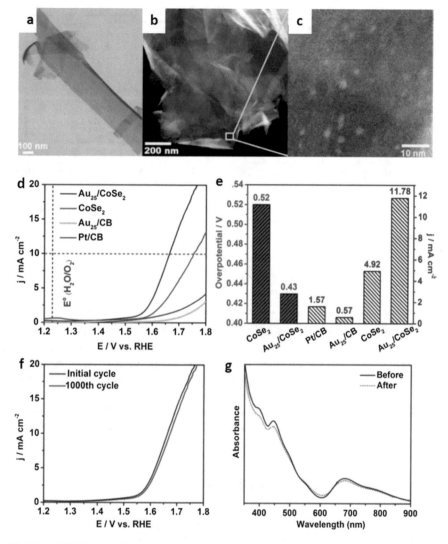

Fig. 2.6 a TEM image of $CoSe_2$ nanosheets, **b, c** HAADF-STEM images of $Au_{25}/CoSe_2$ composite, **d** OER polarization curves of $Au_{25}/CoSe_2$, $CoSe_2$, Pt/C and Au_{25}/C, **e** overpotential at the current density of 10 mA cm^{-2}, and the current density at the overpotential of 0.45 V for $Au_{25}/CoSe_2$, $CoSe_2$, Pt/C and Au_{25}/C catalysts, **f** stability test of $Au_{25}/CoSe_2$, HER LSV curves, **g** UV-vis spectra of Au_{25} NCs before and after the stability test. Adapted from Ref. [37]. Copyright 2017 American Chemical Society

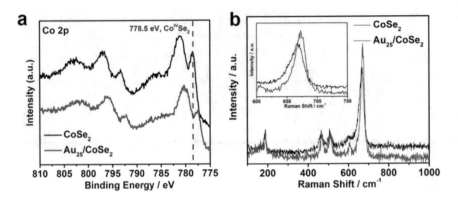

Fig. 2.7 **a** High-resolution Co 2p XPS spectra of $CoSe_2$ and $Au_{25}/CoSe_2$ composites, **b** Raman spectra of $CoSe_2$ and $Au_{25}/CoSe_2$ composites. Adapted from Ref. [37]. Copyright 2017 American Chemical Society

2.4.2 Au_n NC Size Effect in OER

The size of gold nanoclusters is also important for the catalytic activity. To study the potential size dependence of Au_n NCs for OER, Zhao et al. compared gold nanoclusters of $Au_{10}(SPh\text{-}^tBu)_{10}$, $Au_{25}(SR)_{18}$, $Au_{144}(SR)_{60}$ and $Au_{333}(SR)_{79}$, with the latter three being protected by the same phenylethanethiolate ligand. These NCs were loaded onto $CoSe_2$ (all at 2.0 wt%, denoted as $Au_n/CoSe_2$). The OER polarization curves show a moderate increase of OER activity with an increase in cluster size. The $Au_{333}/CoSe_2$ catalyst possesses the smallest overpotential (~0.41 V for 10 mA cm^{-2}) and the largest current density (15.44 mA cm^{-2} at the overpotential of 0.45 V).

2.5 Oxygen Reduction Reaction with Au NCs

The ORR is the rate-determining step of the fuel cell system because of its sluggish kinetics [38, 39]. In both acidic and alkaline electrolyte, different mechanisms have been documented for ORR. These processes can be described as follows:

$$\text{Acidic:} \quad O_2 + 4\,H^+ + 4\,e^- \rightarrow 2\,H_2O \tag{2.8}$$

$$O_2 + 2\,H^+ + 2\,e^- \rightarrow H_2O_2 \tag{2.9}$$

$$H_2O_2 + 2\,H + 2\,e^- \rightarrow 2\,H_2O \tag{2.10}$$

$$\text{Alkaline:} \quad O_2 + H_2O + 4\,e^- \rightarrow 4\,OH^- \tag{2.11}$$

$$O_2 + H_2O + 2\,e^- \rightarrow HO_2^- + OH^- \qquad (2.12)$$

$$HO_2{-} + H_2O + 2\,e^- \rightarrow 3\,OH^- \qquad (2.13)$$

From the equations, it can be seen that two possible pathways can be observed in ORR in both electrolytes. One of them is the direct $4e^-$ pathway where oxygen is reduced to H_2O in acidic electrolytes or OH^- in alkaline electrolytes. The other pathway is the $2e^-$ mechanism where H_2O_2 or HO_2^- is first formed before the sequent reduction to H_2O or OH^- with another $2e^-$ transfer. It is believed that the commercial Pt/C electrode favors the direct $4e^-$ pathways. However, the complicated surface structure of Pt/C catalysts makes it challenging to figure out the reaction occurring in the catalytic process [40–42]. To understand the ORR mechanism, several noble metal nanoparticle-based catalytic materials with either direct $4e^-$ or $2e^-$ pathways have been extensively studied [43–49]. Among these catalysts, Au has shown several unique properties in ORR. In 2007, Zhang et al. found that Pt catalysts can be stabilized against dissolution by modification with Au NCs [50]. In their electrochemical study, the Au NCs-modified Pt catalysts exhibit ultra-high stability where the polarization curve remains unchanged after 30,000 cycles. Later, Yin et al. reported in 2012 the Au NCs/graphene hybrids for high-performance ORR [51]. The hybrid catalytic materials exhibit high current density and excellent stability comparable to that of the commercial Pt/C catalysts. Yet, no study about combining the atomically precise Au NCs with DFT calculations is reported. The precise structure and ultra-small size make the Au NCs a perfect system to study the size effect in ORR. In this section, the reports on the size effects of atomically precise Au NCs are introduced.

2.5.1 Nanocluster-Derived Ultra-Small Nanoparticles for ORR

The size effect has been extensively studied for Au nanoparticles in the past decades [52, 53]. However, reports are rare for ultra-small Au nanoparticles (i.e., core diameter <2 nm) for ORR. In 2016, Wang et al. reported porous carbon-supported ultra-small nanoparticles as ORR catalysts using thiolate-capped Au_{25}, Au_{38} and Au_{144} NCs as precursors [54]. The average diameters of Au nanoparticles were estimated to be 3.7 $\pm$ 0.9 nm for Au_{25}-derived catalyst (AuPC-1), 4.9 $\pm$ 1.1 nm for Au_{38}-derived one (AuPC-2), and 5.8 $\pm$ 1.25 nm for Au_{144}-derived one (AuPC-3), as shown in Fig. 2.8. All the Au nanoparticles are larger than the sizes of the original nanoclusters because of aggregation of clusters during the calcination.

In the electrochemical test (Fig. 2.9), it is found that the AuPC-1 sample exhibits a peak current density similar to that of commercial Pt/C (0.57 mA cm^{-2}). The rotation ring and disk electrode (RRDE) measurements show that the onset potential is 0.95, 0.91, and 0.89 V for AuPC-1, AuPC-2, and AuPC-3, respectively. Also, the diffusion-limited current density of AuPC-1 (3.61 mA cm^{-2}) is obviously higher than

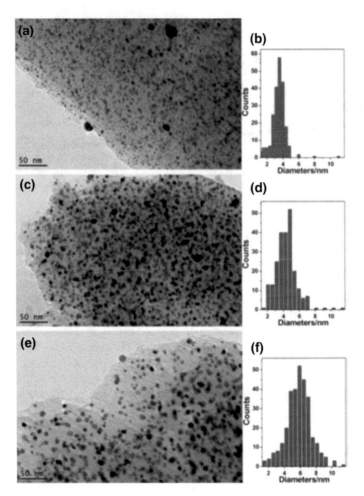

Fig. 2.8 TEM images of **a** AuPC-1, **c** AuPC-2, and **e** AuPC-3, with the corresponding size distribution in panels **b**, **d**, and **e**. Adapted with permission from Ref. [54]. Copyright 2016 American Chemical Society

that of AuPC-2 (3.21 mA cm^{-1}) and AuPC-3 (3.16 mA cm^{-2}). It is noted that the ORR activity of AuPC-1 is similar to that of the commercial Pt/C catalysts (0.95 V for onset potential and 4.98 mA cm^{-2} for limiting current density). Taking all the results together, one can find that the ORR activity increases with the decreasing Au nanoparticle size. Especially, the smallest AuPC-1 sample shows a comparable activity with the commercial Pt/C electrode.

In the Tafel plots (Fig. 2.9c), it can be seen that the specific activity increases with the decrease of AuPC nanoparticle size. At 0.8 V, the current density increases in the order of AuPC-3 (0.16 mA cm^{-2}) < AuPC-2 (0.195 mA cm^{-2}) < AuPC-1 (0.612 mA cm^{-2}) < commercial Pt/C (0.615 mA cm^{-2}). Additionally, similar

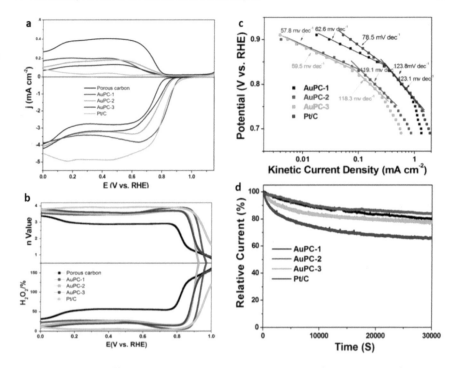

Fig. 2.9 a CV curves, **b** the ORR polarization curves, **c** Tafel plots, **d** Chronoamperometric profiles of AuPC-1, AuPC-2, AuPC-3 and Pt/C catalysts. Adapted with permission from Ref. [54]. Copyright 2016 American Chemical Society

features of the Tafel plots can be observed for all the catalysts. Two clear linear regions are displayed at low and high overpotentials. In the low overpotential region, the slopes of these catalysts are all close to 60 mV dec^{-1}, indicating that a pseudo-two-electron reaction might be the rate-determining step. However, the similar slopes of approximately 120 mV dec^{-1} for the four catalysts suggest that the rate-determining step is probably the first electron transfer to oxygen molecules. Also, the stability test of AuPC and commercial Pt/C catalysts indicate superior stability of Au nanoparticles in ORR. The relative current of AuPC-1, AuPC-2, and AuPC-3 shows a loss of 19.2%, 15.6%, and 22.7%, respectively, after 8 h. While for commercial Pt/C catalysts 35% loss of current is observed. The authors ascribe the superior performance of nanoclusters-derived ultra-small nanoparticles to the low-coordination surface Au atoms of small-sized nanoparticles and the synergetic effects between carbon and Au.

In summary, the small-sized Au particles are beneficial for the activation of oxygen, thus increasing the catalytic activity of ORR.

2.5.2 Size Effect of Au NCs in ORR

Au NCs were first reported for ORR by Chen et al. in 2009 [55]. Four Au NCs with different sizes: Au_{11}, $Au_{\sim25}$, $Au_{\sim55}$, and $Au_{\sim140}$ are synthesized. (On a note, we found that the UV-vis spectrum of "Au_{11}" [55] instead resembles that of the Au_{25} rod cluster [1]). In the electrochemical test, it was found that the "Au_{11}" exhibits the highest limiting current density and smallest overpotential in ORR. Overall, the catalytic activity decreases as the cluster size increases. However, the crystal structures of Au NCs were not obtained at that time. Therefore, the assignment of precise atom numbers was preliminary. Later, Jones et al. reported in 2018 a series of t-butylthiolate protected Au NCs with increasing sizes for ORR [56]. These four nanoclusters, namely Au_{23}, Au_{30}, Au_{46}, and Au_{65}, show distinct UV-vis spectra as shown in Fig. 2.10, and different colors were observed for these four nanoclusters. The same t-butylthiolate ligand and different core sizes make it ideal to compare the atomically precise size effect with this series of Au NCs.

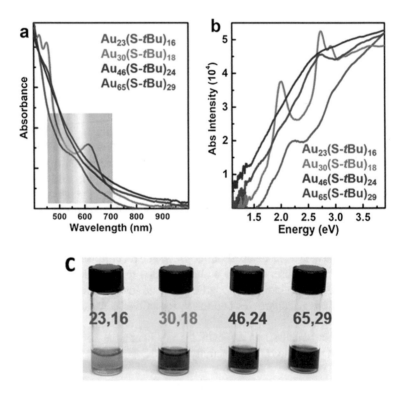

Fig. 2.10 **a** UV-vis spectra, **b** photon energy plot (eV), and **c** photograph of Au_{23}, Au_{30}, Au_{46} and Au_{65}. Adapted with permission from Ref. [56]. Copyright 2018 American Chemical Society

Table 2.1 Experimental parameters for Au NC-catalyzed ORR (n: the number of transferred electrons)

Sample	$\eta(V)$ at $j = -1$ mA cm^{-2}	n	HO$^-$ (%)
Au$_{23}$/SWNT	0.68	2.1	53
Au$_{30}$/SWNT	0.25	2.5	63
Au$_{46}$/SWNT	0.24	2.0	50
Au$_{65}$/SWNT	0.08	3.2	80

Data from Ref. [56]

In the electrochemical test (Table 2.1), the Au$_{65}$ exhibits a transfer of 3.2 electrons, which is higher than that of other NCs (approximately 2 electrons). Also, the potential at -1 mA cm^{-2} shows a trend of Au$_{65}$ < Au$_{46}$ < Au$_{30}$ < Au$_{23}$, indicating that the ORR catalytic activity increases as the nanocluster size grows. Therefore, it can be concluded that larger NCs can facilitate the ORR with smaller overpotential, higher diffusion-limiting current and higher selectivity toward OH$^-$ production.

2.5.3 Charge-State-Dependent ORR Activity of Au$_{25}$ NCs

In 2007, Negishi et al. reported the charge state of Au$_{25}$ NCs can be tuned between $-1, 0$ and $+1$ [57]. This unique property provides an ideal model to study the charge-state effect of Au NCs in electrochemical catalysis [58–60]. Later in 2014, Lu et al. synthesized these atomically precise Au$_{25}$ NCs protected by dodecanethiolate with different charge states ($-1, 0$ and $+1$) for ORR [61]. The UV-vis spectra clearly show the different features of the as-prepared NCs. In addition, the Au 4f$_{7/2}$ binding energy shows a positive shift when the charge state becomes more positive, further indicating the different charge state of Au$_{25}$ NCs (Fig. 2.11).

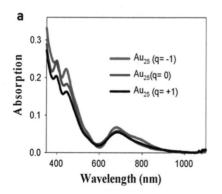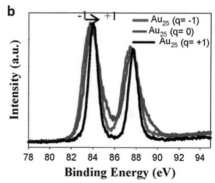

Fig. 2.11 **a** UV-vis spectra and **b** XPS spectra of Au$_{25}$ NCs with different charge states. Adapted with permission from Ref. [61]. Copyright 2014 Royal Society of Chemistry

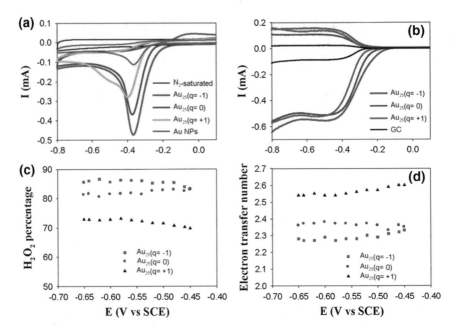

Fig. 2.12 **a** CV of the ORR on Au nanoparticles and Au$_{25}$ NCs with different charge states, **b** RRDE voltammograms recorded on glass carbon electrode and the Au$_{25}$ NCs, **c** selectivity of the H$_2$O$_2$, **d** the electron transfer number as a function of the applied potentials. Adapted with permission from Ref. [61]. Copyright 2014 Royal Society of Chemistry

The electrochemical test results show that the Au$_{25}{}^-$ shows a more positive onset potential and higher diffusion-limiting current density compared with Au$_{25}^0$ and Au$_{25}{}^+$ (Fig. 2.12). Also, the H$_2$O$_2$ production percentages show a trend of Au$_{25}{}^-$ (86%) > Au$_{25}^0$ (82) > Au$_{25}{}^+$ (72%), indicating that the two-electron pathway is dominant with the Au$_{25}$ nanoclusters. Thus, the Au$_{25}{}^-$ can be used as a promising catalyst for H$_2$O$_2$ production in ORR. The combined experimental results and previous DFT calculations suggest that charging the cluster can increase the chemical activity with respect to O$_2$ [62]; the authors proposed that the strong charge-state effects on H$_2$O$_2$ production can be attributed to electron transfer from the anionic Au$_{25}$ core into the LUMO (π^*) of O$_2$, activating the O$_2$ molecule and generating peroxo-like species.

2.6 CO$_2$ Reduction Reaction with Metal NCs

CO$_2$ reduction reaction (CO$_2$RR) has been extensively investigated in order to remediate the global climate change issues during the past decades. As a multiproton and multi-electron process, the CO$_2$ RR is a complicated process with several products produced at various voltages as shown in Table 2.2 [63]. Especially, the formation of CO$_2{}^-$ key intermediate consumes a large amount of energy. On the other hand, the

Table 2.2 Reduction potentials (vs. RHE) of various products in CO_2 reduction reactions

Product	Reaction	E^0 [V vs. RHE]
CO	$CO_2 + 2\,e^- + 2\,H^+ \rightarrow CO + H_2O$	−0.11
HCOOH	$CO_2 + 2\,e^- + 2\,H^+ \rightarrow HCOOH$	−0.25
HCOH	$CO_2 + 4\,e^- + 4\,H^+ \rightarrow HCOH + H_2O$	−0.07
CH_3OH	$CO_2 + 6\,e^- + 6\,H^+ \rightarrow CH_3OH + H_2O$	0.02
CH_4	$CO_2 + 8\,e^- + 8\,H^+ \rightarrow CH_4 + 2\,H_2O$	0.17
C_2H_4	$2\,CO_2 + 12\,e^- + 12\,H^+ \rightarrow C_2H_4 + 4\,H_2O$	0.06
CO_2^-	$CO_2 + e^- \rightarrow CO_2^-$	−1.5
H_2	$2\,H^+ + 2\,e^- \rightarrow H_2$	0.0

Data adapted from Ref. [63]

competing HER also hinders the efficient CO_2RR in aqueous solutions. Therefore, highly efficient catalysts are critically required to lower the energy barrier in CO_2RR [64–66].

Among the catalytic materials, Au has been extensively studied due to its high selectivity toward CO formation [14]. On the other hand, Cu is also attractive because of its versatility in forming various hydrocarbon products [67]. In this section, we summarize the Au and Cu NCs as catalysts for CO_2RR. The application of atomically precise NCs offers an opportunity for correlating the structure and catalytic properties, hence providing insights into the mechanism and also fundamental rules for future design of advanced catalytic materials for CO_2RR.

2.6.1 Au₂₅ for CO₂RR

In 2012, Kauffman et al. first reported atomically precise Au_{25} NCs as catalysts for CO_2RR [68]. The electrochemical results show that Au_{25} have much higher activity in CO_2RR than Au nanoparticles and bulk Au as shown in Fig. 2.13. To be detailed, the Au_{25} exhibits higher current density in LSV and higher CO formation rate than Au nanoparticles and bulk Au.

To explain the superior activity of the Au_{25} nanocluster, the same group used DFT calculations to obtain the free energy diagram of the CO_2RR process (Fig. 2.14) [69]. They proposed that partial ligand removal would occur in order to expose the active sites for CO_2 adsorption. The free energy diagrams of both fully ligand-protected Au_{25} and singly dethiolated Au_{25} cluster were obtained. In the energy diagrams, it can be seen that the most endergonic step for both cases is the *COOH formation. The U_{onset} for fully ligand-protected Au_{25} is −2.04 V, much larger than that of singly dethiolated Au_{25} cluster (−0.34 V) and experimentally value (−0.193 V), indicating that their proposal of ligand removal is correct. Therefore, they concluded that the cluster can facilitate the reduction of CO_2 by partial removal of the thiolate ligand.

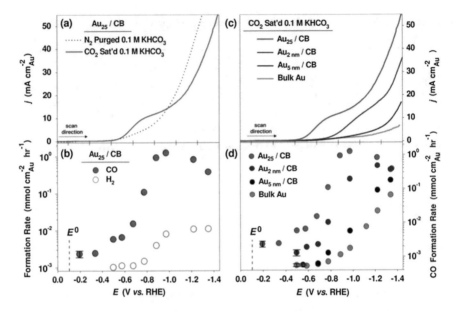

Fig. 2.13 **a** LSV of Au$_{25}$/CB, **b** potential-dependent H$_2$ and CO formation rates for Au$_{25}$/CB, **c** LSV of various Au catalysts in quiescent CO$_2$ saturated 0.1 M KHCO$_3$, **d** potential-dependent CO formation rates for the various Au catalysts. Adapted from Ref. [68]. Copyright 2018 American Chemical Society

The exposed Au site can reduce the free energy of the COOH intermediate formation, thus lowering the overpotential for CO formation. This group also reported the long-term stability of Au$_{25}$ NCs in CO$_2$RR. The results show that Au$_{25}$ can catalyze the CO$_2$RR for 6 days with steady production rate of 745 ± 59 L/(g$_{Au}$ h) and CO selectivity of 86 ± 5%, indicating the exceptional stability of Au$_{25}$ NCs in CO$_2$RR [70].

2.6.2 Atomic-Level Morphology Effects in CO$_2$RR

Previously, Au nanomaterials with different morphology have been extensively studied in testing the catalytic activity of facet, edge, and corner. Despite some interesting results, the nonavailable atomic-level structure of these nanomaterials made it difficult to connect the structure and the catalytic properties. To find more solid evidence of morphology effects of Au catalysts in CO$_2$RR, Zhao et al. prepared atomically precise Au$_{25}$ nanosphere and nanorod and tested their electrochemical performance as CO$_2$RR catalysts [71]. These two NCs exhibit distinct features in UV-vis spectra, corresponding to their spectroscopic fingerprints (Fig. 2.15). The Au$_{25}$ nanosphere comprises an icosahedral Au$_{13}$ core protected by six dimeric surface staples (–SR–Au–SR–Au–SR–), showing a spherical morphology; while the Au$_{25}$

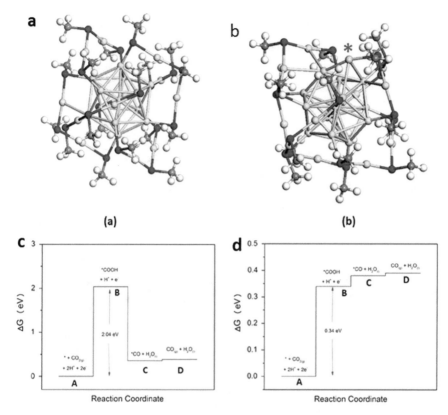

Fig. 2.14 Optimized structure of the model: **a** fully ligand-protected $Au_{25}(SCH_3)_{18}^-$ NC and **b** singly dethiolated $Au_{25}(SCH_3)_{17}^-$ NC. Free energy diagram for electrochemical reduction of CO_2 to CO: **c** over the fully ligand-protected $Au_{25}(SCH_3)_{18}^-$ NC, and **d** over the singly dethiolated $Au_{25}(SCH_3)_{17}^-$ NC. White, gray, blue and golden balls represent H, C, S, and Au atom, respectively. Adapted from Ref. [69]. Copyright 2016 the American Institute of Physics

nanorod comprises two Au_{13} icosahedra fused together by sharing one vertex gold atom, and the rod is protected by five bridging thiolates (–SR–) at the rod's waist, 5 phosphine ligands and one chloride on each end of the nanorod.

The electrochemical results in Fig. 2.16 show that Au_{25} nanosphere has higher CO Faradaic efficiency around 70% than the Au_{25} nanorod (30–60%). The Au_{25} nanosphere also exhibits a much higher CO formation rate. Especially, at high potential of -1.17 V, the CO formation rate of Au_{25} sphere (33.3 μL min^{-1}) is 2.8 times that of Au_{25} nanorod (11.7 μL min^{-1}). The larger CO FE and higher CO formation rate of the Au_{25} nanosphere indicate its high catalytic performance compared with the Au_{25} nanorod.

DFT calculations are used to evaluate the free energy of reaction steps to understand the mechanism of the better performance of the Au_{25} nanosphere (Fig. 2.17). First, the $_\Delta$G values for ligand removal from NCs are calculated. For the Au_{25}

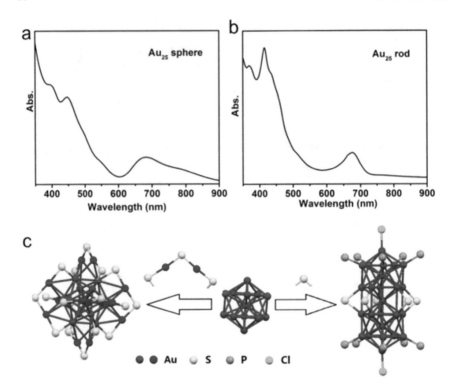

Fig. 2.15 UV-vis spectra of: **a** Au$_{25}$ nanosphere and **b** Au$_{25}$ nanorod; **c** Atom packing structures of Au$_{25}$ nanosphere and nanorod. Adapted from Ref. [71]. Copyright 2018 American Chemical Society

nanosphere, removal of a single –SCH$_3$ is considered, while for the nanorod, removal of –SCH$_3$, –Cl and PH$_3$ is calculated. The results show that the desorption of –PH$_3$ and the removal of –Cl from the Au$_{25}$ nanorod have the same $_\Delta$G: 0.54 eV. For the –SCH$_3$ removal energy of the nanosphere and the nanorod, $_\Delta$G is calculated to be 0.49 eV and 0.95 eV, respectively. These results indicate that the removal of –PH$_3$ and –Cl is more favored for the nanorod. However, the ligand removal from the nanosphere is less endergonic than that from the nanorod. The free energy diagram after ligand removal shows that *COOH, an important intermediate in CO$_2$ reduction to CO on Au, is more stabilized on the Au$_{25}$(SCH$_3$)$_{17}$ nanosphere with one –SCH$_3$ ligand removed compared with any of the other ligand-removed systems of the nanorod. Therefore, it is concluded that the energetically favorable removal of –SCH$_3$ from the Au$_{25}$ nanosphere to expose active sites and the stabilization of *COOH intermediates on the obtained Au$_{25}$(SCH$_3$)$_{17}$ nanosphere contribute to the superior catalytic performance of the Au$_{25}$ nanosphere. This work has successfully correlated the atomic-level morphology with catalytic performance, explaining the factors that determine the CO$_2$RR activities with the aid of DFT calculations. It has shed light on the mechanism for the CO$_2$RR in the future.

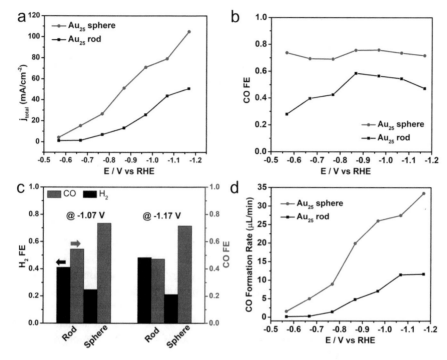

Fig. 2.16 **a** Total current density of CO_2 reduction and **b** Faradaic efficiency (FE) for CO production over the Au_{25} nanosphere and nanorod; **c** FE for CO and H_2 at the potential of -1.07 and -1.17 V over Au_{25} nanosphere and nanorod, **d** CO formation rates over Au_{25} nanosphere and nanorod. Adapted from Ref. [71]. Copyright 2018 American Chemical Society

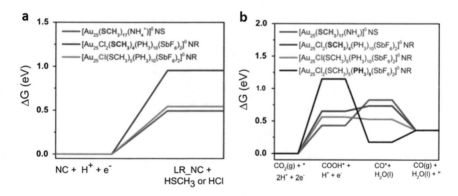

Fig. 2.17 **a** ΔG for ligand removal (eV) from the NCs at 0 V versus RHE, **b** free energy diagrams for CO_2 reduction to CO on the ligand-removed NCs at 0 V versus RHE. Adapted from Ref. [71]. Copyright 2018 American Chemical Society

2.6.3 Cu NC-Catalyzed CO$_2$RR

Cu catalysts are attractive for their ability to reduce CO$_2$ to hydrocarbon products. However, the mechanisms of CO$_2$ reduction on nanostructured Cu catalysts are not well understood yet. In 2017, Tang et al. reported a copper-hydride nanocluster for CO$_2$RR to study the mechanism of Cu catalysts [72]. This Cu$_{32}$H$_{20}$L$_{12}$ NC (L = S$_2$P(OiPr)$_2$) comprises a distorted hexacapped rhombohedral core of 14 Cu atoms sandwiched by two Cu$_9$ triangular cupola fragments of Cu atoms, while the hydrides and L ligands are homogeneously distributed on the surface of the nanocluster (Fig. 2.18). The study on the CO$_2$ reduction mechanism shows that the key initial step of CO$_2$ reduction is where the first hydrogen is added: C or O of CO$_2$. The H addition on C would facilitate the formation of HCOOH, otherwise, CO would occur. Based on the structure of the Cu nanocluster, the authors proposed two possible channels to form the critical HCOO* intermediate: (1) the non-electrochemical absorption of CO$_2$ on lattice hydrides (lattice-hydride channel); (2) the electrochemical CO$_2$ reaction with proton and electron (proton-reduction channel). The free energy diagram of both channels for HCOOH and CO production shows that in both cases the lattice-hydride mechanism exhibits more energy downhill compared with the proton-reduction mechanism, suggesting the CO$_2$ reduction favors the lattice-hydride channel over this Cu NC. After the confirmation of reaction channel, the free energy diagram of CO and HCOOH formation is calculated following the lattice-hydride mechanism (Fig. 2.19a). The results indicate the HCOOH pathway is more favorable than the CO pathway over the Cu cluster.

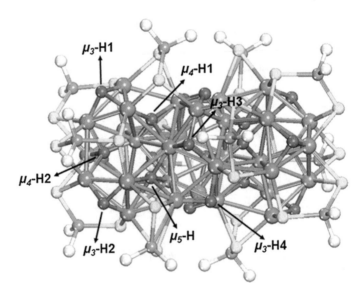

Fig. 2.18 Atomic structure of the Cu$_{32}$H$_{20}$L$_{12}$ NC (L = S$_2$PH$_2$). Orange, Cu; green, hydride; yellow, S; purple, P; white, H. Adapted with permission from Ref. [72]. Copyright 2017 American Chemical Society

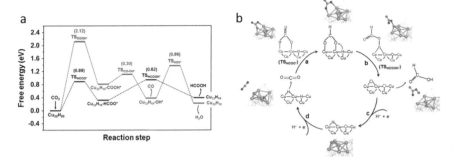

Fig. 2.19 a Free energy diagrams for HCOOH and CO formation on the $Cu_{32}H_{20}L_{12}$ NCs via the lattice-hydride mechanism, **b** overall mechanism of HCOOH formation from CO_2 reduction on $Cu_{32}H_{20}L_{12}$ NCs via the lattice-hydride channel, orange, Cu; green, hydride; red, oxygen; gray, carbon. Adapted with permission from Ref. [72]. Copyright 2017 American Chemical Society

Based on all the DFT calculation results, the authors proposed a complete catalytic mechanism as shown in Fig. 2.19d. It can be seen that the HCOOH is formed through the non-electrochemical lattice-hydride pathway. Additionally, the Cu NC can be recovered by the electrochemical reaction with two protons and electrons. Electrochemical tests are conducted to verify the theoretical prediction that HCOOH is favored over the Cu NC (Fig. 2.20). It can be seen that HCOOH is the dominant product at low potential with selectivity higher than 80%, while H_2 becomes dominant at high potential due to the competing HER. Only a small amount of CO is formed throughout the potential window. Therefore, these electrochemical results have successfully verified the accuracy of theoretical prediction. This work has demonstrated the methods of mechanism study using atomically precise metal

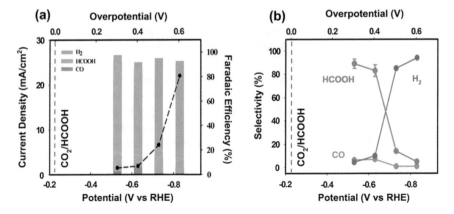

Fig. 2.20 a Average current densities (black) and cumulative FE for H_2, HCOOH and CO, **b** product selectivity for H_2, HCOOH and CO. Adapted with permission from Ref. [72]. Copyright 2017 American Chemical Society

NCs by combining DFT calculation and electrochemical experiment. Compared with traditional nanoparticles, the precise structure of nanoclusters makes the computational modeling much more facile and convincing. It is anticipated that the application of metal NCs in CO_2RR can offer insights in the mechanism study in the future.

2.7   Summary and Future Perspective

In this chapter, we have summarized the literature work about metal NCs as electrochemical catalysts. Compared with the metallic nanoparticles, the metal NCs exhibit discrete electronic energy levels due to the quantum size effect. This unique electronic property, plus atomically precise structures, as well as their various atom-packing structures, render the nanoclusters great potential in catalytic applications.

Metal nanoclusters (homogold and doped ones) have been demonstrated in several important electrochemical reactions including HER, OER, ORR, and CO_2RR. The results show that doping effects, synergetic effects, size effects, thermostability effects, charge effects, and morphology effects all play important roles in the catalytic effects. One of the most important advantages of nanoclusters in catalysis is the feasibility of computational modeling due to the available structure. For example, the catalytic mechanism of CO_2RR over $Cu_{32}H_{20}L_{12}$ NCs are predicted by DFT calculations and successfully verified by the electrochemical experiment. Metal nanoclusters are expected to be a promising class of model catalysts for correlating the structure and properties, providing exciting opportunities for the understanding of catalysis mechanism at the atomic level.

Future work in NCs electrocatalysis should investigate the following aspects:

(i)   The doping effects. Several bimetallic NCs have been reported. However, the catalytic properties of bimetallic NCs are still rarely studied. The investigation of bimetallic NCs may be helpful to understand the impact of doping atoms, further revealing the fundamental catalytic mechanisms;

(ii)   The synergetic effects. Au NCs-loaded composite materials are reported to show enhanced catalytic activity. It is essential to understand the interaction between supporting materials and metal NCs with more precise interfaces;

(iii)   Metal NCs in other catalytic reactions. Owing to the characterized structure and tunable properties, metal NCs should be broadened to other catalytic reactions to study the fundamental catalytic mechanisms, such as nitrogen reduction reactions (NRR). The new electrocatalytic reactions remain to be explored.

References

1. Jin R, Zeng C, Zhou M, Chen Y (2016) Atomically precise colloidal metal nanoclusters and nanoparticles: fundamentals and opportunities. Chem Rev 116(18):10346–10413
2. Lu Y, Chen W (2012) Sub-nanometre sized metal clusters: from synthetic challenges to the unique property discoveries. Chem Soc Rev 41(9):3594–3623
3. Jin R (2010) Quantum sized, thiolate-protected gold nanoclusters. Nanoscale 2(3):343–362
4. Zeng C, Chen Y, Kirschbaum K, Lambright KJ, Jin R (2016) Emergence of hierarchical structural complexities in nanoparticles and their assembly. Science 354(6319):1580–1584
5. Qian H, Zhu M, Wu Z, Jin R (2012) Quantum sized gold nanoclusters with atomic precision. Acc Chem Res 45(9):1470–1479
6. Li G, Jin R (2013) Atomically precise gold nanoclusters as new model catalysts. Acc Chem Res 46(8):1749–1758
7. Turner JA (2004) Sustainable hydrogen production. Science 305(5686):972–974
8. Stambouli AB (2011) Fuel cells: The expectations for an environmental-friendly and sustainable source of energy. Renew Sustain Energy Rev 15(9):4507–4520
9. Das V, Padmanaban S, Venkitusamy K, Selvamuthukumaran R, Blaabjerg F, Siano P (2017) Recent advances and challenges of fuel cell based power system architectures and control–A review. Renew Sustain Energy Rev 73:10–18
10. Gasteiger HA, Kocha SS, Sompalli B, Wagner FT (2005) Activity benchmarks and requirements for Pt, Pt-alloy, and non-Pt oxygen reduction catalysts for PEMFCs. Appl Catal B 56(1–2):9–35
11. Costentin C, Robert M, Savéant J-M (2013) Catalysis of the electrochemical reduction of carbon dioxide. Chem Soc Rev 42(6):2423–2436
12. Qiao J, Liu Y, Hong F, Zhang J (2014) A review of catalysts for the electroreduction of carbon dioxide to produce low-carbon fuels. Chem Soc Rev 43(2):631–675
13. Vasileff A, Xu C, Jiao Y, Zheng Y, Qiao S-Z (2018) Surface and interface engineering in copper-based bimetallic materials for selective CO_2 electroreduction. Chem 4(8):1809–1831
14. Hori Y, Wakebe H, Tsukamoto T, Koga O (1994) Electrocatalytic process of CO selectivity in electrochemical reduction of CO_2 at metal electrodes in aqueous media. Electrochim Acta 39(11–12):1833–1839
15. Zhu W, Michalsky R, Metin ON, Lv H, Guo S, Wright CJ, Sun X, Peterson AA, Sun S (2013) Monodisperse Au nanoparticles for selective electrocatalytic reduction of CO_2 to CO. J Am Chem Soc 135(45):16833–16836
16. Lee H-E, Yang KD, Yoon SM, Ahn H-Y, Lee YY, Chang H, Jeong DH, Lee Y-S, Kim MY, Nam KT (2015) Concave rhombic dodecahedral Au nanocatalyst with multiple high-index facets for CO_2 reduction. ACS Nano 9(8):8384–8393
17. Zhu W, Zhang Y-J, Zhang H, Lv H, Li Q, Michalsky R, Peterson AA, Sun S (2014) Active and selective conversion of CO_2 to CO on ultrathin Au nanowires. J Am Chem Soc 136(46):16132–16135
18. Back S, Yeom MS, Jung Y (2015) Active sites of Au and Ag nanoparticle catalysts for CO_2 electroreduction to CO. Acs Catalysis 5(9):5089–5096
19. Zhao S, Jin R, Jin R (2018) Opportunities and challenges in CO_2 reduction by gold-and silver-based electrocatalysts: from bulk metals to nanoparticles and atomically precise nanoclusters. ACS Energy Lett 3(2):452–462
20. Zeng C, Chen Y, Kirschbaum K, Appavoo K, Sfeir MY, Jin R (2015) Structural patterns at all scales in a nonmetallic chiral $Au_{133}(SR)52$ nanoparticle. Sci Adv 1(2):e1500045
21. Higaki T, Zhou M, Lambright KJ, Kirschbaum K, Sfeir MY, Jin R (2018) Sharp transition from nonmetallic Au_{246} to metallic Au_{279} with nascent surface Plasmon resonance. J Am Chem Soc 140(17):5691–5695
22. Jin R, Qian H, Wu Z, Zhu Y, Zhu M, Mohanty A, Garg N (2010) Size focusing: a methodology for synthesizing atomically precise gold nanoclusters. J Phys Chem Lett 1(19):2903–2910
23. Zhu M, Aikens CM, Hollander FJ, Schatz GC, Jin R (2008) Correlating the crystal structure of a thiol-protected Au_{25} cluster and optical properties. J Am Chem Soc 130(18):5883–5885

24. Wu Z, Suhan J, Jin R (2009) One-pot synthesis of atomically monodisperse, thiol-functionalized Au 25 nanoclusters. J Mater Chem 19(5):622–626
25. Greeley J, Jaramillo TF, Bonde J, Chorkendorff I, Nørskov JK (2011) Computational high-throughput screening of electrocatalytic materials for hydrogen evolution. In: Materials for sustainable energy: a collection of peer-reviewed research and review articles from Nature Publishing Group. World Scientific, Singapore, pp 280–284
26. Kwak K, Choi W, Tang Q, Kim M, Lee Y, Jiang D-E, Lee D (2017) A molecule-like PtAu$_{24}$ (SC$_6$H$_{13}$) 18 nanocluster as an electrocatalyst for hydrogen production. Nat Commun 8:14723
27. Choi W, Hu G, Kwak K, Kim M, Jiang D-E, Choi J-P, Lee D (2018) Effects of metal-doping on hydrogen evolution reaction catalyzed by MAu$_{24}$ and M$_2$Au$_{36}$ Nanoclusters (M=Pt, Pd). ACS Appl Mater Interfaces 10(51):44645–44653
28. Rountree ES, McCarthy BD, Eisenhart TT, Dempsey JL (2014) Evaluation of homogeneous electrocatalysts by cyclic voltammetry. ACS Publications
29. Valdez CN, Dempsey JL, Brunschwig BS, Winkler JR, Gray HB (2012) Catalytic hydrogen evolution from a covalently linked dicobaloxime. Proc Natl Acad Sci 109(39):15589–15593
30. Zhao S, Jin R, Song Y, Zhang H, House SD, Yang JC, Jin R (2017) Atomically precise gold nanoclusters accelerate hydrogen evolution over MoS$_2$ nanosheets: the dual interfacial effect. Small 13(43):1701519
31. Du Y, Xiang J, Ni K, Yun Y, Sun G, Yuan X, Sheng H, Zhu Y, Zhu M (2018) Design of atomically precise Au$_2$Pd$_6$ nanoclusters for boosting electrocatalytic hydrogen evolution on MoS$_2$. Inorganic Chem Front 5(11):2948–2954
32. Tahir M, Pan L, Idrees F, Zhang X, Wang L, Zou J-J, Wang ZL (2017) Electrocatalytic oxygen evolution reaction for energy conversion and storage: a comprehensive review. Nano Energy 37:136–157
33. Xia Z (2016) Hydrogen evolution: guiding principles. Nat Energy 1(10):16155
34. Gong M, Li Y, Wang H, Liang Y, Wu JZ, Zhou J, Wang J, Regier T, Wei F, Dai H (2013) An advanced Ni–Fe layered double hydroxide electrocatalyst for water oxidation. J Am Chem Soc 135(23):8452–8455
35. Zhuang Z, Sheng W, Yan Y (2014) Synthesis of monodispere Au@Co$_3$O$_4$ core-shell nanocrystals and their enhanced catalytic activity for oxygen evolution reaction. Adv Mater 26(23):3950–3955
36. Li Z, Ye K, Zhong Q, Zhang C, Shi S, Xu C (2014) Au–Co$_3$O$_4$/C as an efficient electrocatalyst for the oxygen evolution reaction. ChemPlusChem 79(11):1569–1572
37. Zhao S, Jin R, Abroshan H, Zeng C, Zhang H, House SD, Gottlieb E, Kim HJ, Yang JC, Jin R (2017) Gold nanoclusters promote electrocatalytic water oxidation at the nanocluster/CoSe$_2$ interface. J Am Chem Soc 139(3):1077–1080
38. Ramaswamy N, Mukerjee S (2011) Influence of inner-and outer-sphere electron transfer mechanisms during electrocatalysis of oxygen reduction in alkaline media. J Phys Chem C 115(36):18015–18026
39. Zhao S, Zhang H, House SD, Jin R, Yang JC, Jin R (2016) Ultrasmall palladium nanoclusters as effective catalyst for oxygen reduction reaction. ChemElectroChem 3(8):1225–1229
40. He Q, Cairns EJ (2015) Recent progress in electrocatalysts for oxygen reduction suitable for alkaline anion exchange membrane fuel cells. J Electrochem Soc 162(14):F1504–F1539
41. Guo S, Zhang S, Sun S (2013) Tuning nanoparticle catalysis for the oxygen reduction reaction. Angew Chem Int Ed 52(33):8526–8544
42. Cui C-H, Yu S-H (2013) Engineering interface and surface of noble metal nanoparticle nanotubes toward enhanced catalytic activity for fuel cell applications. Acc Chem Res 46(7):1427–1437
43. Li J, Yin H-M, Li X-B, Okunishi E, Shen Y-L, He J, Tang Z-K, Wang W-X, Yücelen E, Li C (2017) Surface evolution of a Pt–Pd–Au electrocatalyst for stable oxygen reduction. Nat Energy 2(8):17111
44. Sankarasubramanian S, Singh N, Mizuno F, Prakash J (2016) Ab initio investigation of the oxygen reduction reaction activity on noble metal (Pt, Au, Pd), Pt3M (M=Fe Co, Ni, Cu) and Pd3M (M=Fe Co, Ni, Cu) alloy surfaces, for LiO$_2$ cells. J Power Sources 319:202–209

45. Huang X, Zhao Z, Cao L, Chen Y, Zhu E, Lin Z, Li M, Yan A, Zettl A, Wang YM (2015) High-performance transition metal–doped Pt$_3$Ni octahedra for oxygen reduction reaction. Science 348(6240):1230–1234
46. Bu L, Zhang N, Guo S, Zhang X, Li J, Yao J, Wu T, Lu G, Ma J-Y, Su D (2016) Biaxially strained PtPb/Pt core/shell nanoplate boosts oxygen reduction catalysis. Science 354(6318):1410–1414
47. Zhang H, Jin M, Xiong Y, Lim B, Xia Y (2012) Shape-controlled synthesis of Pd nanocrystals and their catalytic applications. Acc Chem Res 46(8):1783–1794
48. Tang Z, Wu W, Wang K (2018) Oxygen reduction reaction catalyzed by noble metal clusters. Catalysts 8(2):65
49. Tang W, Lin H, Kleiman-Shwarsctein A, Stucky GD, McFarland EW (2008) Size-dependent activity of gold nanoparticles for oxygen electroreduction in alkaline electrolyte. J Phys Chem C 112(28):10515–10519
50. Zhang J, Sasaki K, Sutter E, Adzic R (2007) Stabilization of platinum oxygen-reduction electrocatalysts using gold clusters. Science 315(5809):220–222
51. Yin H, Tang H, Wang D, Gao Y, Tang Z (2012) Facile synthesis of surfactant-free Au cluster/graphene hybrids for high-performance oxygen reduction reaction. ACS Nano 6(9):8288–8297
52. Inasaki T, Kobayashi S (2009) Particle size effects of gold on the kinetics of the oxygen reduction at chemically prepared Au/C catalysts. Electrochim Acta 54(21):4893–4897
53. Lee Y, Loew A, Sun S (2009) Surface-and structure-dependent catalytic activity of Au nanoparticles for oxygen reduction reaction. Chem Mater 22(3):755–761
54. Wang L, Tang Z, Yan W, Yang H, Wang Q, Chen S (2016) Porous carbon-supported gold nanoparticles for oxygen reduction reaction: effects of nanoparticle size. ACS Appl Mater Interfaces 8(32):20635–20641
55. Chen W, Chen S (2009) Oxygen electroreduction catalyzed by gold nanoclusters: strong core size effects. Angew Chem Int Ed 48(24):4386–4389
56. Jones TC, Sumner L, Ramakrishna G, Hatshan M, Abuhagr A, Chakraborty S, Dass A (2018) Bulky t-Butyl thiolated gold nanomolecular series: synthesis, characterization, optical properties, and electrocatalysis. J Phys Chem C 122(31):17726–17737
57. Negishi Y, Chaki NK, Shichibu Y, Whetten RL, Tsukuda T (2007) Origin of magic stability of thiolated gold clusters: a case study on Au$_{25}$(SC$_6$H$_{13}$)$_{18}$. J Am Chem Soc 129(37):11322–11323
58. Zhu M, Eckenhoff WT, Pintauer T, Jin R (2008) Conversion of anionic [Au$_{25}$(SCH$_2$CH$_2$Ph)$_{18}$]$^-$ cluster to charge neutral cluster via air oxidation. J Phys Chem C 112(37):14221–14224
59. Zhu M, Aikens CM, Hendrich MP, Gupta R, Qian H, Schatz GC, Jin R (2009) Reversible switching of magnetism in thiolate-protected Au$_{25}$ superatoms. J Am Chem Soc 131(7):2490–2492
60. Qian H, Sfeir MY, Jin R (2010) Ultrafast relaxation dynamics of [Au$_{25}$(SR)$_{18}$]^q nanoclusters: effects of charge state. J Phys Chem C 114(47):19935–19940
61. Lu Y, Jiang Y, Gao X, Chen W (2014) Charge state-dependent catalytic activity of [Au$_{25}$(SC$_{12}$H$_{25}$)$_{18}$] nanoclusters for the two-electron reduction of dioxygen to hydrogen peroxide. Chem Commun 50(62):8464–8467
62. Mills G, Gordon MS, Metiu H (2003) Oxygen adsorption on Au clusters and a rough Au (111) surface: the role of surface flatness, electron confinement, excess electrons, and band gap. 118(9):4198–4205
63. Vickers JW, Alfonso D, Kauffman DR (2017) Electrochemical carbon dioxide reduction at nanostructured gold, copper, and alloy materials. Energy Technol 5(6):775–795
64. Kim D, Resasco J, Yu Y, Asiri AM, Yang P (2014) Synergistic geometric and electronic effects for electrochemical reduction of carbon dioxide using gold–copper bimetallic nanoparticles. Nat Commun 5:4948
65. Geng Z, Kong X, Chen W, Su H, Liu Y, Cai F, Wang G, Zeng J (2018) Oxygen vacancies in ZnO nanosheets enhance CO$_2$ electrochemical reduction to CO. Angew Chem Int Ed 57(21):6054–6059

66. Zhang L, Zhao ZJ, Gong J (2017) Nanostructured materials for heterogeneous electrocatalytic CO_2 reduction and their related reaction mechanisms. Angew Chem Int Ed 56(38):11326–11353
67. Xie H, Wang T, Liang J, Li Q, Sun S (2018) Cu-based nanocatalysts for electrochemical reduction of CO_2. Nano Today 21:41–54
68. Kauffman DR, Alfonso D, Matranga C, Qian H, Jin R (2012) Experimental and computational investigation of Au_{25} clusters and CO_2: a unique interaction and enhanced electrocatalytic activity. J Am Chem Soc 134(24):10237–10243
69. Alfonso DR, Kauffman D, Matranga C (2016) Active sites of ligand-protected Au_{25} nanoparticle catalysts for CO_2 electroreduction to CO. J Chem Phys 144(18):184705
70. Kauffman DR, Thakkar J, Siva R, Matranga C, Ohodnicki PR, Zeng C, Jin R (2015) Efficient electrochemical CO_2 conversion powered by renewable energy. ACS Appl Mater Interfaces 7(28):15626–15632
71. Zhao S, Austin N, Li M, Song Y, House SD, Bernhard S, Yang JC, Mpourmpakis G, Jin R (2018) Influence of atomic-level morphology on catalysis: The case of sphere and rod-like gold nanoclusters for CO_2 electroreduction. ACS Catalysis 8(6):4996–5001
72. Tang Q, Lee Y, Li D-Y, Choi W, Liu C, Lee D, Jiang D-E (2017) Lattice-hydride mechanism in electrocatalytic CO_2 reduction by structurally precise copper-hydride nanoclusters. J Am Chem Soc 139(28):9728–9736

Chapter 3
Catalysts via Atomic Layer Deposition

Rong Chen, Bin Shan, Xiao Liu, and Kun Cao

Abstract Heterogeneous catalysis is crucial to chemical industries, environmental protection, energy storage, and conversion. The demand for catalysts with high activity, selectivity, and stability drives the development of controlled and precise synthesis of catalysts. An atomic-level control of catalyst structure will not only provide better catalytic performance, but also help understanding the fundamental catalytic mechanism and the associated structure–property relationship. Recently, atomic layer deposition (ALD) has attracted great interest as an effective method of catalyst design and synthesis due to its high controllability and uniformity for fabricating complex structures at the atomic level. Herein, the ALD technique for tailoring active sites and composite structures of catalysts will be introduced and discussed, which cover both supported metal catalysts and metal/oxide composite catalysts. In particular, various strategies by modifying ALD processes will be presented for the size, composition, and structure control of supported metal, alloy, and core–shell nanoparticles. Several metal oxide composite structures are developed by adjusting the metal oxide ALD processes, including porous overcoating structures, confined coating, and selective coating structures. Finally, we wrap up the chapter with the latest developments in ALD reactor design for catalysts synthesis and a summary and perspectives of ALD method for catalysts synthesis and applications.

Keywords Atomic layer deposition · Catalyst design · Bimetallic catalysts · Oxide coating structures · Selective deposition · Controllable synthesis · Composite catalyst

R. Chen (✉) · B. Shan · X. Liu · K. Cao
Huazhong University of Science and Technology, Wuhan, People's Republic of China
e-mail: rongchen@mail.hust.edu.cn

© Springer Nature Switzerland AG 2020   69
P. W. N. M. van Leeuwen and C. Claver (eds.), *Recent Advances in Nanoparticle Catalysis*,
Molecular Catalysis 1, https://doi.org/10.1007/978-3-030-45823-2_3

3.1 Introduction of ALD for Catalyst Synthesis

Heterogeneous catalysts are essential and widely used in chemical industries, environmental protection, energy storage, and conversion, for example, Fischer–Tropsch synthesis of olefin, catalytic converter for automobiles, and methane reforming reaction for hydrogen production [1–3]. Heterogeneous catalysts with designed structures perform a key role for many applications, whereas the precise fabrication of such structures remains the Holy Grail in catalysis research. Generally, the performance of catalysts strongly depends on their composition and structure, especially for the surface and hetero-interfaces. In the past decades, a variety of wet chemistry synthesis methods, such as impregnation, precipitation, and hydrothermal method, have been developed to prepare heterogeneous catalysts. Significant progress has been made in this field, such as the size control of metal nanoparticles, morphology control, and surface doping. These methods lead to remarkable improvement in the activity, stability, and selectivity of catalysts, yet there is still a gap between current synthesis technology and the demand of catalyst fabrication with atomic accuracy. To drive the construction of precise configurations with direct reactive sites modulation and understand the catalytic mechanism with a clear structure–property relationship [4], methods for catalyst synthesis with control over catalytic structures with the atomic-level precision are urgently needed [5, 6].

Atomic layer deposition (ALD) has been recently explored as an effective method of heterogeneous catalysts synthesis. ALD technology, considered as a unique chemical vapor deposition method, was firstly reported as atomic layer epitaxy in the 1970s by Dr. Tuomo Suntola in Finland to prepare ZnS thin film for electroluminescent displays [7]. The self-limiting chemical reaction of alternative gaseous precursors on the surface of substrate has been utilized to deposit thin films. A typical ALD process consists of two self-limiting half reactions by precursor A and B as shown in Fig. 3.1. The first half reaction occurs after precursor A is introduced into the reactor, which will react with active sites on the substrate until reaching a saturated chemical adsorption. Then, the inert gas is introduced to purge the excess unreacted precursor A and reacted by-products. Subsequently, precursor B reacts with the residual ligand of A on the substrate to complete the second half reaction, with the same purging process of inert gas followed. Here, we give a typical ALD reaction of Al_2O_3 as listed in Eq. (3.1) and (3.2) for better understanding of ALD process [8]:

$$AlOH^* + Al(CH_3)_3 \rightarrow AlOAl(CH_3)_2^* + CH_4 \qquad (3.1)$$

$$AlCH_3^* + H_2O \rightarrow AlOH^* + CH_4 \qquad (3.2)$$

With the alternating saturated adsorption and reaction of precursors, the film thickness can be accurately controlled by adjusting the ALD cycles. The self-limiting nature of ALD also facilitates the conformal deposition on complex, high aspect ratio structures. Until now, more than a thousand ALD processes have been developed and the applications have been expanded for catalysis field [9–11]. ALD, with

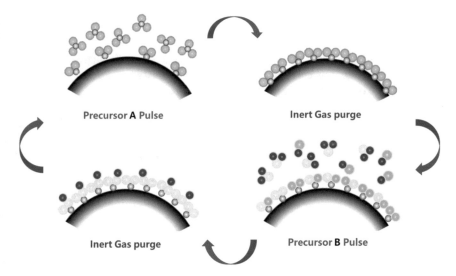

Fig. 3.1 Schematic diagram of ALD technology

a prominent advantage in composition control, has been shown to be effective at modifying metal and metal oxide sites to improve catalytic activity, selectivity, and stability [12]. In the past few years, various ALD strategies have been developed to synthesize catalysts with designed nanostructures, such as supported metal catalysts with different sizes, bimetallic alloys, and core–shell catalysts, as well as metal oxide coating structures. These nanostructures have been applied in different catalytic reactions, which could promote the overall performance such as activity, selectivity, and stability. For instance, high activity and selectivity of bimetallic catalysts can be achieved for specific catalytic reactions by precisely controlling the location of each metal atom during the preparation process via the ALD method [13]. For the further development of catalysts by the ALD method, the relationship of atomic structures and catalytic performance will be an important factor. The chapter will describe not only the ALD processes, but also the composition, morphology, and nanostructures of catalysts prepared by ALD. The rest of this chapter is arranged as follows: Sect. 3.2 describes the supported metal catalysts prepared by the ALD method, including the deposition processes of monometallic nanoparticles, size control, and bimetallic catalysts. Section 3.3 reviews the ALD processes of metal oxides, as well as the porous and selective oxide coatings for various catalytic applications. Section 3.4 gives a brief overview of four typical ALD reactors for catalyst synthesis. Finally, we conclude with perspective and outlook in Sect. 3.5.

3.2 Supported Metal Catalysts

Metal nanoparticles have been widely considered as catalytic active centers in supported catalysts. For catalyst synthesis, ALD processes of Pt, Pd, Ir, Rh, Ru, Ni, Cu, and Co have been applied to deposit metals onto various supports. The metal nanoparticle size can be concisely controlled by adjusting the deposition processes, which relies on the self-limiting nature of surface reactions. To further obtain improved activity and selectivity of a catalyst, bimetallic catalysts in the form of core–shell and alloy structures can also be prepared to utilize the synergistic effect between metals.

3.2.1 ALD Recipes of Metals

Various metals have been grown by ALD as summarized in Fig. 3.2 periodic table. Metal ALD processes on various oxides have been widely investigated. In the nucleation stage, metals tend to follow an island growth model and form nanoparticles on oxide supports. Only a portion of metals are reported to be catalysts, and they are mainly noble metals. The corresponding deposition recipes, supports, and applied catalytic reactions for the catalysts prepared by ALD are listed in Table 3.1.

For most noble metals, the ALD mechanisms are either combustion reactions of metal precursor by active oxidizing agents, such as platinum and rhodium, or ligand exchange reactions, such as palladium [14]. For catalytic applications, a long

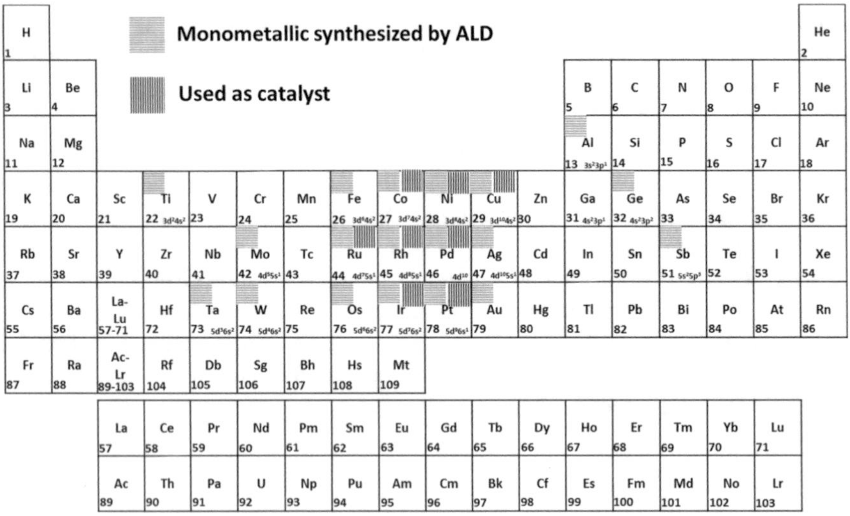

H 1		Monometallic synthesized by ALD															He 2
Li 3	Be 4	Used as catalyst										B 5	C 6	N 7	O 8	F 9	Ne 10
Na 11	Mg 12											Al 13 $3s^23p^1$	Si 14	P 15	S 16	Cl 17	Ar 18
K 19	Ca 20	Sc 21	Ti 22 $3d^24s^2$	V 23	Cr 24	Mn 25	Fe 26 $3d^64s^2$	Co 27 $3d^74s^2$	Ni 28 $3d^84s^2$	Cu 29 $3d^{10}4s^1$	Zn 30	Ga 31 $4s^23p^1$	Ge 32 $4s^23p^2$	As 33	Se 34	Br 35	Kr 36
Rb 37	Sr 38	Y 39	Zr 40	Nb 41	Mo 42 $4d^55s^1$	Tc 43	Ru 44 $4d^75s^1$	Rh 45 $4d^85s^1$	Pd 46 $4d^{10}$	Ag 47 $4d^{10}5s^1$	Cd 48	In 49	Sn 50	Sb 51 $5s^25p^3$	Te 52	I 53	Xe 54
Cs 55	Ba 56	La-Lu 57-71	Hf 72	Ta 73 $5d^36s^2$	W 74 $5d^46s^2$	Re 75	Os 76 $5d^66s^2$	Ir 77 $5d^76s^2$	Pt 78 $5d^96s^1$	Au 79	Hg 80	Tl 81	Pb 82	Bi 83	Po 84	At 85	Rn 86
Fr 87	Ra 88	Ac-Lr 89-103	Rf 104	Db 105	Sg 106	Bh 107	Hs 108	Mt 109									

La 57	Ce 58	Pr 59	Nd 60	Pm 61	Sm 62	Eu 63	Gd 64	Tb 65	Dy 66	Ho 67	Er 68	Tm 69	Yb 70	Lu 71
Ac 89	Th 90	Pa 91	U 92	Np 93	Pu 94	Am 95	Cm 96	Bk 97	Cf 98	Es 99	Fm 100	Md 101	No 102	Lr 103

Fig. 3.2 Overview of ALD for metals

Table 3.1 ALD processes for metal catalysts and the corresponding catalytic reactions

Metal catalyst	Precursor	Deposition temperature	Support	Catalytic reaction
Pt [15, 16, 20, 21, 161–164]	MeCpPtMe$_3$–O$_2$ [15, 21, 161, 163] MeCpPtMe$_3$–O$_3$ [16, 162, 164]	270 °C [162] 280 °C [164] 300 °C [15, 16, 21, 163] 320 °C [161]	Al$_2$O$_3$ [15] TiO$_2$ [162] MoS$_2$ [21] SrTiO$_3$ [163] carbon nanotube (CNT) [16, 164] Carbon aerogel [161]	CO oxidation [161] CO, NO, C$_x$H$_y$ oxidation [162] Water–gas shift [15] Styrene hydrogenation [16] Acrolein hydrogenation [163] Ammonia borane hydrolysis [164] Hydrogen evolution reaction [21] Oxygen reduction reaction [20]
Pd [25–30]	Pd(hfac)$_2$ – CH$_2$O [25–30]	100 °C [26] 150 °C [29] 200 °C [25, 27, 28, 30]	C [30] Al$_2$O$_3$ [25, 26, 29] ZnO [25] Ni(OH)$_2$ [27] CNT [28]	Methanol decomposition [25, 26] Ethanol oxidation [27] Formate oxidation [28] 1,3-butadiene hydrogenation [29] Oxygen reduction reaction [30]
Ir [33, 34]	Ir(acac)$_3$–O$_2$ [33, 34]	200 °C [33, 34] 250 °C [33]	γ-Al$_2$O$_3$ [33] Zeolite [34]	Toluene hydrogenation [33] Decalin conversion [34]
Rh [36]	Rh(acac)$_3$–O$_3$–H$_2$ [36]	250 °C [36]	γ-Al$_2$O$_3$ [36]	DRM [36]

(continued)

Table 3.1 (continued)

Metal catalyst	Precursor	Deposition temperature	Support	Catalytic reaction
Ni [37, 38, 41, 42]	$NiCp_2$–H_2O [41] $NiCp_2$–H_2 [37, 42] $NiCp_2$–O_3-reduction [38]	200 °C [38] 250 °C [41] 300 °C [37, 42]	Al_2O_3 [37] SiO_2 [41, 42] Al_2O_3 nanotube (ANT) [38]	DRM [41] Propylene hydrogenation and hydrogenolysis [37] Hydrogenation of aryl nitro compounds [42] Selective hydrogenation of cinnamaldehyde [38]
Cu [45, 46, 165]	$Cu(thd)_2$–O_2–H_2 [45, 165] [$Cu(^sBu$-amd)]–H_2 [46]	160 °C [45, 165] 185 °C [46]	SiO_2 [45, 46, 165]	Water–gas shift [165] Reverse water–gas shift [45] Phenol degradation [46]
Co [47]	$Co(acac)_2$–O_3 [47]	400 °C [47]	γ-Al_2O_3 [47]	Fischer–Tropsch synthesis [47]

pulse time is usually essential to achieve surface saturation and good conformity or overcome the slow nucleation rates on the supports with highly specific surface area such as Al_2O_3 or carbon materials [15, 16].

Platinum

Platinum is the most widely investigated noble metal catalysts prepared by ALD. Most Pt nanoparticles or films prepared by ALD utilize the alternating pulse of trimethylplatinum(methylcyclopentadienyl) ($MeCpPtMe_3$, Cp=Cyclopentadiene) and oxidizing agent of ozone or oxygen at 200–300 °C. The $MeCpPtMe_3$ is the dominant platinum ALD precursor, the ligands of which can be completely removed by oxidizing agents (O_2 or O_3) during the other half reaction of the ALD cycle [7]. The corresponding growth rate usually varies from 0.03 to 0.05 nm/cycle at 200 and 300 °C [17]. It has been reported that the Pt precursor could hardly be removed by O_2, when the deposition temperature is lower than 200 °C. At higher temperature around 350 °C, this particular Pt precursor suffers from thermal decomposition, which will lead to a slow growth rate of Pt [18, 19]. Besides the oxidizing agents, the reducing H_2 is also utilized to remove the ligands in Pt ALD, although such chemical process has not been reported for the synthesis of Pt catalysts. Recently, CO molecules are introduced into Pt ALD process as a passivation agent to modify the surface energy of already deposited Pt. The CO treatment is beneficial for Pt's direct deposition on the carbon support, leading to a 40% promotion in Pt surface-to-volume ratio [20]. Sufficient adsorption of gaseous Pt precursor could generate a high distribution of Pt cluster on the carbon substrate even when the Pt size is continuously increasing, which can be hardly achieved by traditional wet chemistry methods [16]. Pt ALD has been conducted on a number of different supports, such as alumina, carbon, molybdic sulfide, and $SrTiO_3$ which can be used in various catalytic reactions like CO oxidation, water–gas shift reaction, styrene hydrogenation, and oxygen reduction reaction [15, 16, 20, 21].

Palladium

Similar to the Pt precursor, palladium hexafluoroacetylacetonate ($Pd(hfac)_2$) is the major metal precursor in Pd ALD process to react with the reducing agent of formalin or H_2. Different from most noble metal ALD processes, the Pd processes rely mostly on the reducing agents rather than the common molecular oxygen. Due to the instability of $Pd(hfac)2$ above 230 °C, formalin was used to react with $Pd(hfac)_2$ to complete one Pd ALD cycle at 200 °C, with a growth rate of 0.02–0.03 nm per cycle [22]. At lower temperature of 80–100 °C, limited by its activity, formalin can be replaced by H_2 in the ALD process with a similar growth rate generated on SiO_2 or TaO_x [23, 24]. The distribution of Pd nanoparticles could be easily controlled by changing the exposure time of Pd precursor onto the support in the ALD procedure [25], which is more difficult to achieve by traditional impregnation or other deposition methods. The Pd catalysts synthesized by ALD have been widely used in different reactions like methanol decomposition, ethanol oxidation, and formate oxidation [25–30].

Iridium

Unlike Pt and Pd, many kinds of Ir precursors are used in the thermal Ir ALD process including Ir(acac)$_3$ (acac=acetylacetone), (ethylcyclopentadienyl)-(1,5-cyclooctadiene)iridium (EtCpIr(COD)), and (ethylcyclopentadienyl)(1,3-cyclohexadiene)iridium (Ir(MeCp)(CHD), where the ligands are removed by O$_2$ at 225–400 °C with a rate of 0.02–0.08 nm/cycle [31, 32]. The highly active iridium catalysts prepared by ALD using Ir(acac)$_3$ and O$_2$ show smaller metal particle size and higher activity for decalin conversion and toluene hydrogenation than that prepared by traditional wet chemistry methods [33, 34].

Rhodium

For Rh ALD, Rh(acac)$_3$ is the exclusive precursor, followed by O$_2$ combustion, and the temperature window shows a wide range of 225–325 °C with a growth rate of 0.05–0.19 nm per cycle [35]. To our knowledge, Rh catalyst prepared by ALD had not been reported till Li et al. deposited Rh NPs on Al$_2$O$_3$ used for methane dry reforming reaction (DRM) by using Rh(acac)$_3$ and O$_3$ [36]. Compared with the sample prepared by impregnation method, the Rh nanoparticles prepared by ALD method exhibit a narrower size distribution. Other noble metals, such as Ru, Ag, and Au, which have been successfully synthesized by ALD, might also be prepared as catalysts in the future.

Non-noble metals

Besides the direct reaction between metal precursor and the reducing agent like H$_2$ or NH$_3$, some non-noble metal nanoparticles could be formed from the corresponding oxides by a reduction process in H$_2$ flow considering their strong affinity with oxygen. Taking Ni ALD as an example, two deposition routes are viable, including Ni(Cp)$_2$–H$_2$ and Ni(Cp)$_2$–O$_2$–H$_2$ [37, 38]. The Ni nanoparticles have been deposited on alumina by using Ni(acac)$_2$ and air as precursors. The temperature window for Ni ALD is in the range of 200–300 °C with a growth rate of about 0.5–1.2 nm/cycle [39, 40]. By using Ni(Cp)$_2$ and water as precursor in the Ni ALD process, the formation temperature of Ni nanoparticles is lowered (under 250 °C), which is usually above 400 °C in impregnation methods [41]. Ni catalysts prepared by ALD are usually used as hydrogenation and reforming catalysis [37, 38, 41, 42].

Cu ALD has also attracted great attention due to the variety of industrial applications, and various Cu ALD processes have been developed: CuCl–H$_2$, Cu(thd)$_2$–H$_2$ (thd=2,2,6,6-tetramethyl-3,5-heptanedionate), Cu(acac)$_2$–H$_2$, Cu(dialkylacetamidinate)$_2$–H$_2$, etc. [43]. In an ALD process using Cu(^{s}Bu-amd))$_2$ (^{s}Bu-amd=N,N'-disec-butylacetamidinate) and hydrogen as precursors, the growth rate varies from 1.5 to 2 Å/cycle on SiO$_2$ or Si$_3$N$_4$ at a relative low temperature range of 150–190 °C [44]. Highly dispersed Cu catalysts prepared by ALD can be applied in (reverse) water–gas shift reaction and photodegradation of methylene blue and phenol [45, 46].

Co catalysts are prepared by ALD where the pulse of Co(acac)$_2$ and O$_3$ precursors was performed at 200 and 400 °C [47]. The ALD catalysts show up to 2.3 times higher dispersion than cubic cobalt nanoparticles. The Co catalysts prepared by

ALD method have a narrower size distribution than spherical ones of the catalysts prepared by impregnation methods. Flat cubic cobalt nanoparticles in ALD catalysts exhibit more uniform cobalt species and a stronger interaction with Al_2O_3.

3.2.2 Monometallic Nanoparticle, Cluster, and Single Atom Synthesis

ALD was utilized to fabricate highly dispersed monometallic (Ru, Pd, Pt, Ir, Ni, etc.) nanoparticles utilizing the nucleation growth stage of the ALD process [48–52]. The nanoparticle size (typically 1–3 nm with small deviations) and density distribution can be precisely tuned through the deposition temperature and the number of deposition cycles. It has also become an important method to synthesize single atom catalysts. The size optimization of metal catalysts is a key parameter to evaluate the utilization of catalysts since catalytic reactions occur almost exclusively on the surface. Moreover, the utilizations for the noble metals are extraordinarily important due to their high costs and limited reserves. It has been reported that downsizing the metal catalysts to nanoscale can greatly increase the number of surface sites for reactants' adsorption and reaction [53, 54]. The metal nanoparticles can also possess unique catalytic performances due to the interfacial interactions between metal and supports [55]. Due to the stronger cohesive energy of metal atoms as compared to the binding energy between metal and supports, the initial stage of metal ALD is usually in the island growth mode instead of layer-by-layer growth. Aroused by such nucleation phenomena, many researchers have utilized the first few ALD cycles to synthesize metal nanoparticles [50, 56]. Compared to conventional wet chemistry methods, the gas-phase-based ALD method with the self-limiting nature of surface reactions can satisfy the high dispersion requirement of supported metal nanoparticles. Moreover, the size of metal nanoparticles can be readily controlled by ALD recipes.

Up to now, several parameters in ALD process such as ALD cycles, deposition temperature, ALD precursor, and surface structure of substrate (shown in Fig. 3.3) have been utilized to control the size of nanoparticles. Varying ALD cycles is the most straightforward way of controlling the size of nanoparticles at the nucleation stage [56]. The average sizes of Pt and Pd nanoparticles have been reported to show linear correlation from 1 to 5 nm by selecting the appropriate number of ALD cycles [57, 58]. Besides the number of ALD cycles, the deposition temperature can also affect the average size of deposited metal nanoparticles by controlling the aggregation process of metal nanoparticles on the substrate. It has been reported that a low deposition temperature usually results in narrow particle size distribution [59]. Since the oxidizing atmosphere could lead to atom or cluster mobility on the surface, the ripening of metal nanoparticles can be promoted during ALD growth. It has been reported that the surface diffusion phenomena can be suppressed by replacing the oxidative precursor with reductive or inert precursors. Consequently, the size

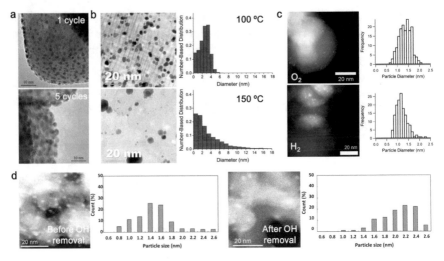

Fig. 3.3 Control of metal nanoparticle size by ALD process: **a** varying ALD cycles, **b** deposition temperature, **c** deposition atmosphere, and **d** surface functional groups of supports. **a** is reprinted with permission from Ref. [56]. Copyright 2011. American Chemical Society. **b** is reprinted with permission from Ref. [57]. Copyright 2017. Royal Society of Chemistry. **c** is reprinted with permission from Ref. [59]. Copyright 2015. American Chemical Society. **d** is reprinted with permission from Ref. [60]. Copyright 2011. American Chemical Society

of metal particles can be controlled by ALD cycle with constant particle density [60, 61]. The surface structures and functional groups can also affect the size of deposited metal nanoparticles by affecting the adsorption of precursors and diffusion of deposited metal atoms. The high-temperature annealing treatment, alcohol, and TMA pretreatment on the Al_2O_3 substrate have been reported to reduce the surface hydroxyls and suppress the growth of Pd nanoparticles.

Recently, the sub-nanometer cluster or single atom catalysts have aroused much attention due to their high utilization of metal atoms [62, 63]. The highly dispersed Pt sub-nanoclusters are synthesized via ALD on $SmMn_2O_5$ mullite-type oxides. The Pt clusters of the as-prepared composite catalysts are on a sub-nanometer scale of 0.5–0.9 nm. Superior CO oxidation activity has been observed with a significantly lowered light-off temperature due to the strong interfacial interactions between Pt clusters and $SmMn_2O_5$ supports (Fig. 3.4a, b) [64]. CeO_2-supported Pt single atoms have also been successfully synthesized using ALD method [65]. The atomically dispersed Pd on graphene is reported to be fabricated using the ALD method, which has shown about 100% butenes selectivity at 95% conversion in selective hydrogenation of 1,3-butadiene, at a mild reaction condition of about 50 °C (Fig. 3.4c, d) [66, 67]. Moreover, some atomically dispersed transition metal catalysts have also been reported to enhance their activity and selectivity [68–70]. For instance, the single-site Co_1-N_4 composite has been prepared by Co ALD process. The composite shows excellent photocatalytic performance with a robust H_2 production activity up

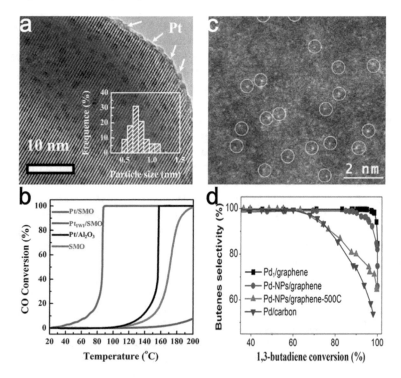

Fig. 3.4 **a, b** Pt_n/$SmMn_2O_5$ for CO oxidation and **c, d** Pd_1/graphene for selective hydrogenation of 1,3-butadiene. **a, b** are reprinted with permission from Ref. [64]. Copyright 2018. Royal Society of Chemistry. **c, d** are reprinted with permission from Ref. [66]. Copyright 2015. American Chemical Society

to 10.8 mmol h^{-1}, which is higher than other referred samples due to the coordinated donor nitrogen in the active site.

3.2.3 Bimetallic Core–Shell and Alloy Nanoparticles Synthesis

By adjusting ALD processes on the monometallic catalysts, it provides ways to synthesize bimetallic core–shell and alloy NPs with controlled sizes and compositions. Researchers have synthesized multilayers and stacking structures to improve the performance of semiconductor materials, leading to the concept of core–shell NPs. In recent years, bimetallic nanoparticle catalysts fabricated with ALD methods have aroused researchers' interests. Generally, core–shell NPs have enhanced performance because of the interaction, electronic modification, or lattice mismatch between two materials. Compared with their corresponding single metal particles, enhanced catalytic selectivity and activity have been shown for the core–shell structured NPs. In addition, the core–shell structured NPs cost less if the core material is replaced with a cheaper material, as the consumption of noble metal can be reduced. The main challenge during the fabrication is to achieve selective deposition, to ensure the growth of shell materials exclusively on the preformed cores. In typical metal ALD processes, the second metal will deposit on the substrate. As a result, monometallic and core–shell NPs are both formed during this process. Up to now, three types of ALD strategies have been reported to fabricate core–shell structured catalysts, which are focused on the selective deposition of the second metal on the first metal core but not on the substrate to prevent the formation of mixture compositions.

It was observed that the Pt growth via ALD using MeCpPtMe$_3$ and O$_2$ at 300 °C on Al$_2$O$_3$ substrate was suppressed when the O$_2$ partial pressure was decreased to 7.5 mTorr [71]. It was found that even after 600 cycles, Pt growth on Al$_2$O$_3$ was still inhibited. However, the growth of Pt on Pd was immediately occurred without nucleation delay and the growth rate reached to 0.45 Å/cycle with the reduced O$_2$ partial pressure ALD process. After Pt deposition, the particles' density per unit area was not changed indicating no new nuclei were formed during Pt ALD process. Thus, selective deposition of Pt on Pd was achieved. The presence of the Pt shell was confirmed by the TEM images (Fig. 3.5a) [72].

Another process that could realize selective growth to fabricate core–shell NPs was tuning the deposition temperature which strongly affected the nucleation and growth rate. On the other hand, bimetallic alloy NPs could be fabricated by dosing two metal precursors with an ABC-type ALD process. For example, Pd and Pt precursors can be dosed on the substrate alternately to form Pd/Pt alloys [73]. The composition and the ratio of the two metals in the alloy can be tuned by adjusting the ALD cycles of two precursors.

Yet another approach to fabricate core–shell NPs was utilizing the area selective ALD (AS-ALD) method on the substrate modified with octadecyltrichlorosilane (ODTS) self-assembled monolayer (SAMs). SAMs was formed by immersing the substrate in ODTS solution. The growth time of SAMs was controlled before a continuous and defect-free SAMs layer was formed. In this way, SAMs with nanoscale pinholes could be formed on the substrate, these pinholes were not covered by ODTS group, and the reactive hydroxyl functional groups of the substrate were exposed. Thus, these pinholes on the substrate could act as the nucleation sites in the ALD reaction. On the contrary, the other part of the substrate was covered by continuous ODTS SAMs that was inert toward ALD reaction, because the methyl end groups of ODTS SAMs could block precursor chemisorption. The core and shell materials growth could only take place at pinhole sites to form core–shell NPs as shown in Fig. 3.5b [74]. Compared with the substrate without ODTS modification, Pt also deposited homogeneously on the substrate rather than exclusively on Pd cores without special ALD process control.

The ability of controlling the composition and structures of core–shell NPs with atomic-scale accuracy opens new opportunities to gain fundamental understanding of

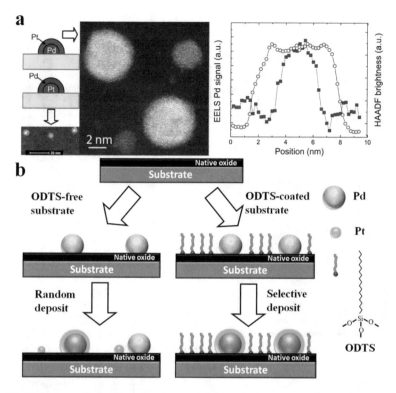

Fig. 3.5 Pd@Pt core–shell NPs synthesized **a** through reducing precursor's partial pressure and **b** on the pinholes sites of self-assembled monolayers. **a** is reprinted with permission from Ref. [72]. Copyright 2012. American Chemical Society. **b** is reprinted with permission from Ref. [74]. Copyright 2015. Springer Nature

the catalytic structure–property relationship. Lu et al. fabricated Pd-coated Au core–shell nanoparticles via selective ALD (Fig. 3.6a) and have found that the shell thickness of the Au@Pd core–shell NPs could influence the catalytic performance toward benzyl alcohol oxidation reaction [75]. The catalytic activity showed a volcano-like trend with shell thickness as shown in Fig. 3.6b. The maximum activity with turnover frequency (TOF) and specific activity at 27,600 and 9800 h^{-1} were realized with 8 cycles Pd grown on Au, and the corresponding shell thickness was ~0.8 nm. At further increasing the Pd ALD cycles, the catalytic performance decreased. These results indicated that benzyl alcohol oxidation reaction was sensitive to the catalytic structure. For the core–shell NPs, the Pd shell would draw electrons from Au and this synergistic effect through electronic modification was helpful for the improvement of catalytic activities. Similarly, Pd@Pt core–shell NPs showed enhanced catalytic activity and selectivity toward CO preferential oxidation in H_2 compared with Pd, Pt NPs, and the Pd/Pt alloys. It was found that core–shell NPs with atomic monolayer Pt on Pd performed the highest activity in this reaction. Based on density functional theory (DFT) and activation energy calculations, it was found that the CO oxidation

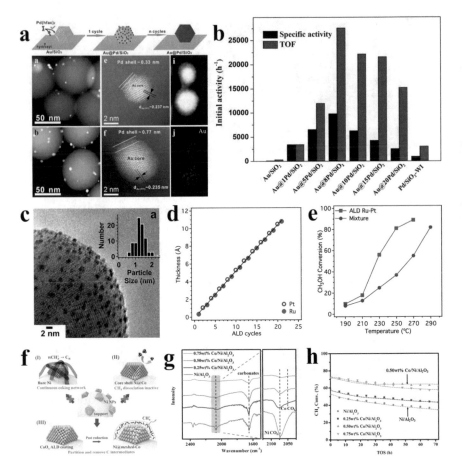

Fig. 3.6 a, b Au@Pd core–shell nanoparticles fabricated with reduced temperature ALD process toward solvent-free aerobic oxidation of benzyl alcohol. **c–e** Ru–Pt bimetallic alloy nanoparticles prepared by alternately dosing Ru and Pt precursors during ALD. **f–h** Meshed Co-coated Ni nanoparticles prepared by ALD for dry reforming of methane. **a, b** are reprinted with permission from Ref. [75]. Copyright 2015. Elsevier. **c–e** are reprinted with permission from Ref. [76]. Copyright 2010. American Chemical Society. **e–g** are reprinted with permission from Ref. [77]. Copyright 2019. Elsevier

barrier significantly decreased after Pt coating. The Pt shell with atomic monolayer thickness showed the lowest CO oxidation energy barrier.

Elam et al. have fabricated Ru/Pt alloyed catalysts by alternately dosing Ru and Pt precursors during ALD as shown in Fig. 3.6c [76]. The Ru/Pt ratio in the NPs was tuned by controlling the ALD cycles of Ru and Pt precursors during ALD process (Fig. 3.6d). Compared with Pt, Ru, as well as the physical mixture of Ru and Pt nanoparticles (Fig. 3.6e), the Ru/Pt alloy was reported to have higher methanol decomposition activity due to the electronic interaction between two elements. The interaction between Ru and Pt in the alloy nanoparticles was confirmed by the Ru K-edge X-ray absorption spectroscopy.

Co/Ni catalysts were fabricated via ALD for the dry reforming of methane reaction (DRM) with enhanced coking resistance and activity as schematically shown in Fig. 3.6f [77]. Co coating layer on Ni NPs exhibited a meshed-like structure confirmed with CO chemisorption FTIR measurements (Fig. 3.6g). This structure was realized by a post-reduction treatment after CoO_x ALD on Ni. The meshed coating structure enabled the formation of Co–Ni interfaces and simultaneously exposed the Ni surface, which was beneficial to break C–H bonds and enhance CO_2 activation. Furthermore, the coating layer with meshed-like configuration partitioned the Ni surface that suppressed continuous carbon tube formation and enhanced coking resistance. The coking amount at 650 °C could be reduced to 2.9%. The Co/Ni catalyst exhibited sintering resistance and good durability (Fig. 3.6h).

3.3 Metal Oxide-Modified Catalysts

Besides metal ALD processes, ALD of metal oxides has demonstrated a great potential in fabricating catalysts with enhanced activity, selectivity, and stability. The oxides can also modulate the electronic properties and morphology of active metal sites via interfacial electron donation and metal oxide interaction. Some oxides also act as active component in catalytic reactions. Thus, oxide coatings can provide additional opportunities to further promote the catalytic performance. The coating structures of oxide on metal nanoparticle also improve the catalytic stability with confinement effect against sintering and leaching under harsh reaction conditions.

3.3.1 ALD Recipes of Metal Oxides

The atomic-level precision control over oxide film thickness with high uniformity on high-surface-area materials makes ALD an ideal tool in metal oxide composite catalysts engineering. We have summarized the oxides that can be grown by ALD reported in literatures as listed in Fig. 3.7, most of which have been used as catalytic components to modulate the catalytic performances in terms of activity, selectivity, and stability (Table 3.2). In the following paragraphs, we introduce several metal oxides grown by ALD in detail according to their specific applications.

Al_2O_3

Al_2O_3 ALD using precursors of trimethylaluminum (TMA) and H_2O is one of the most commonly used ALD procedures and has been thoroughly investigated. The mechanism for Al_2O_3 ALD on oxide surfaces is well understood: In the first half reaction, TMA reacts with hydroxyl groups on the substrate to form $Al(CH_3)_x{}^*$ ($x = 1–2$) surface species and CH_4 gaseous product; in the second half reaction,

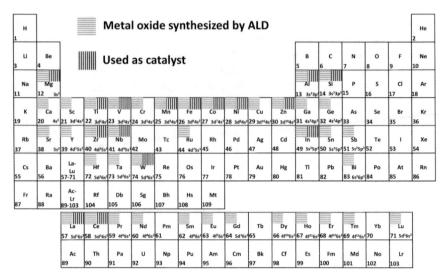

Fig. 3.7 Overview of ALD for metal oxides

exposed H_2O transforms the $Al(CH_3)_x^*$ terminated surface to $Al(OH)_x^*$ ($x = 1–2$) terminated surface with releasing of CH_4 product. Al_2O_3 prepared by ALD has shown its potential applications to enhance the catalytic activity, selectivity, and stability of catalysts in several reactions, such as hydrogenation reaction, methanol decomposition, dehydrogenation of ethane, photocatalysis, CO oxidation, and DRM reaction [78–91].

ZnO

ZnO ALD procedure using the precursors of diethylzinc (DEZ) and H_2O is similar to that of Al_2O_3 ALD. The reaction mechanism is as follows: First, DEZ reacts with hydroxyl groups on the starting surface forming $Zn(C_2H_5)^*$ surface species and C_2H_6 gaseous product; next, the exposed H_2O transforms the $Zn(C_2H_5)^*$ terminated surface to an $Zn(OH)^*$ terminated surface and again releases C_2H_6. ZnO prepared by ALD has been demonstrated in the reactions such as aqueous-phase reforming, photochemical catalysis, and the Chichibabin reaction [92–96].

TiO₂

Different from Al_2O_3 and ZnO ALD, TiO_2 ALD has been reported using many types of precursors as Ti source. Zhang et al. reported TiO_2 ALD on Cu using alternating pulse–purge cycles of titanium tetraisopropoxide (TTIP) and deionized water at 200 °C [86]. Biener et al. performed the well-established titanium tetrachloride ($TiCl_4/H_2O$) ALD processes in a warm wall reactor (wall and stage temperature of 110 °C) to deposit TiO_2 on Au. The normalized mass gain increases approximately linear with the number of ALD cycles as was observed by in situ quartz crystal monitor (QCM), and the growth rates were ~0.7 Å per cycle. Titanium tetrakis(dimethylamide) (TDMAT) and H_2O are also used as precursors to deposit

Table 3.2 ALD processes for metal oxide catalyst and the corresponding catalytic reaction

Metal oxide catalyst	Precursor	Deposition temperature	Support	Catalytic reaction
Al_2O_3 [30, 78–91]	TMA-H_2O [30, 78–91]	30 °C [81] 100 °C [89] 150 °C [30, 80, 90] 160 °C [79] 200 °C [78, 82–88, 91]	Zeolites [78] ANTs [30, 80] CNT [79] Al_2O_3 [81, 83–85, 88, 91] SiO_2 [82, 87] $CuCr_2O_4$–CuO [86] TiO_2 [89, 90]	n-alkane hydroconversion [78] Cinmamaldehyde and nitrobenzene hydrogenation reactions [30, 79] Hydrogenation reactions of 4-nitrophenol [80] Furfural hydrogenation [84, 81, 86–88] Methanol decomposition [82] Oxidative dehydrogenation of ethane to ethylene [83] Hydrogenation of 1,3-butadiene [85] Photooxidation and photoreduction [89] CO oxidation [90] DRM [91]
ZnO [92–96]	DEZ-H_2O [92–96]	150 °C [92, 95] 180°C [94] 200°C [93]	Al_2O_3 [92] TiO_2 [93, 96] ZSM-5 [94] CuO [95]	Aqueous phase reforming of 1-propanol [92] Photo-induced degradation [93] Chichibabin reaction [94] CO_2 photoreduction [95] Photochemical catalysis [96]

(continued)

Table 3.2 (continued)

Metal oxide catalyst	Precursor	Deposition temperature	Support	Catalytic reaction
TiO_2 [79, 86, 93, 97–107]	TTIP-H_2O [79, 86, 99–101, 103, 104, 107] TiCl$_4$–H_2O [98, 102, 105] TDMAT-H_2O [93, 97]	110 °C [102] 120 °C [97] 150 °C [98–101, 104, 105] 160°C [79] 180 °C [103] 200°C [86, 93, 107]	Cu-chromite [86] TiO_2 [98, 99, 106] Al_2O_3 [100, 101] Au [102] Ni [103, 104] C [105] ZnO [93] CNCs [79] SrTiO$_3$ [107] In$_2O_5$Sn [97]	The selective hydrogenation of furfural [86, 98] Semihydrogenation of alkyne to olefin [99] CO oxidation [100–102] Carbon dioxide reforming of methane [103, 104] Electrochemical water oxidation [105] Photo-induced degradation efficiency of the CSHJ [93] Nitrobenzene hydrogenation [79] 1-propanol reforming [106] Acrolein hydrogenation selectivity [107] Solar-driven hydrogen evolution [97]

(continued)

Table 3.2 (continued)

Metal oxide catalyst	Precursor	Deposition temperature	Support	Catalytic reaction
CoO$_x$ [108–115]	Co(Cp)$_2$–O$_3$ [108–115]	150 °C [108, 111, 113–115] 280 °C [109, 110] 465 °C [112]	SBA-15 [109] TiO$_2$ [110, 113] Al$_2$O$_3$ [111] Anodic Al$_2$O$_3$ [112] Au [114]	Epoxidation reaction of styrene [109] Photocatalytic hydrogen production [110] CO oxidation [108, 111] Oxidative MMP decomposition reactions [112] Photoelectrochemical [113] electrochemical [114] Photoelectrochemical water splitting [115]
FeO$_x$ [116–120]	FeCp$_2$–O$_3$ [116, 118] FeCp$_2$–O$_2$ [117, 120] Fe$_2$(O^{t}Bu)$_6$–H$_2$O [119]	150 °C [117, 120] 180 °C [119] 230 °C [118]	Al$_2$O$_3$ [116, 117] CNT [118] TiSi$_2$ [119] C [120]	The selective hydrogenation of cinnamaldehyde [116] 1,3-butadiene hydrogenation [117] Selective reductive coupling reaction to synthesize imine [118] Solar water splitting [119] Dehydrogenation of formic acid [120]
NiO$_x$ [121–123]	Ni(Cp)$_2$–O$_2$ [121, 122] Ni(CP)$_2$–O$_3$ [123]	150 °C [123] 260 °C [121, 122]	SiO$_2$ [121] Al$_2$O$_3$ [122, 123]	CO oxidation [121–123]
MgO [124]	Cp$_2$Mg–H$_2$O [124]	200 °C [124]	Al$_2$O$_3$ [124]	Furfural hydrogenation [124]
In$_2$O$_3$ [125]	InCp–H$_2$O$_2$ [125]	200 °C [125]	Si [125]	Electrochemical [125]

(continued)

Table 3.2 (continued)

Metal oxide catalyst	Precursor	Deposition temperature	Support	Catalytic reaction
MnO$_x$ [126]	Mn(EtCp)$_2$–H$_2$O [126]	175 °C [126]	SiO$_2$ [126]	Syngas conversion to higher oxygenates [126]
CeO$_x$ [106, 127]	Ce(thd)$_4$–O$_3$ [127]	200 °C [127]	Al$_2$O$_3$ [106]	1-propanol reforming [106] CO oxidation [127]
La$_2$O$_3$ [128]	La(thd)$_3$–O$_3$ [128]	300 °C [128]	Fe$_2$O$_3$ [128]	Photoelectrochemical water oxidation [128]
ZrO$_2$ [129, 130]	Zr(NMe$_2$)$_4$–H$_2$O [129, 130]	180 °C [129] 150–250 °C [130]	Al$_2$O$_3$ [129] GNS [130]	Methane oxidation [129] Electrochemical capacitance [130]
VO$_x$ [131, 132]	VTIP-H$_2$O$_2$–H$_2$0 [131] VO(acac)$_2$)–O$_2$ [132]	100 °C [131] 250 °C [132]	Al$_2$O$_3$ [131] SiO$_2$/Al$_2$O$_3$ [132]	Oxidative dehydrogenation of cyclohexane [131] Propane dehydrogenation [132]
WO$_3$ [133, 134, 166]	WCl$_6$-air [133] (tBuN)$_2$(Me$_2$N)$_2$W -H$_2$O [166] W(CO)$_6$-oxidants [134]	350 °C [166] 195–205 °C [134]	SBA-15 [133] TiSi$_2$ [166] ZrO$_2$ [134]	2-butanol dehydration [133] Water splitting [166] 2-propanol dehydration [134]
NbO$_x$ [87, 135]	Nb(OCH$_2$CH$_3$)$_5$–H$_2$O [129, 135]	135 °C [135] 200 °C [87]	SBA-15 [135] spSiO$_2$ [87]	2-propanol dehydration [135] Furfural hydrogenation [87]
SiO$_2$ [136, 137]	Si(OCH$_3$)$_4$–H$_2$O [136] Si(OC$_2$H$_5$)$_4$–H$_2$O [137]	50 °C [136] 150 °C [137]	TiO$_2$ [136] Al$_2$O$_3$ [137]	CO oxidation [136] Cyclohexanol dehydration [137]

TiO_2 at 120 °C [97]. TiO_2 is utilized as catalytic promoters in hydrogenation reaction, CO oxidation, carbon dioxide reforming of methane, photoelectrochemical catalysis, and 1-propanol reforming reaction [79, 86, 93, 97–107].

CoO_x, FeO_x, NiO_x

The most commonly used precursors for CoO_x, FeO_x, and NiO_x ALD are $Co(Cp)_2$, $Fe(Cp)_2$, and $Ni(Cp)_2$ with O_3 or O_2. The CoO_x film's thickness and growth rate as functions of deposition temperature are monitored in the range of 100–300 °C. The growth rate is relatively low when the deposition temperature is below 150 °C, due to the lack of thermal activation energy. Then, it reaches a steady growth rate of 0.37 Å per cycle in the temperature range from 150 °C to 250 °C. The growth rate increases abruptly when the temperature goes beyond 250 °C, partially due to the thermal decomposition of precursors [108]. CoO_x prepared by ALD has shown potential in the epoxidation reaction of styrene, photoelectrochemical catalysis, CO oxidation, and MMP (1-methoxy-2-methyl-2-propanol) decomposition reaction [108–115]. FeO_x prepared by ALD has been investigated in hydrogenation and dehydrogenation reaction, reductive coupling reaction, and photoelectrochemical water splitting [116–120]. NiO_x prepared by ALD is investigated in CO oxidation reaction [121–123].

Other metal oxides

There are still a large number of oxides prepared by ALD as shown in Table 3.2 that we have not covered in the above discussion. MgO [124], In_2O_3 [125], MnO_x [126], CeO_x [106, 127], La_2O_3 [128], ZrO_2 [129, 130], VO_x [131, 132], WO_3 [133, 134], NbO_x [135, 87], SiO_2 [136, 137], CrO_x [138, 139], etc., were used as catalysts in several reactions, such as furfural hydrogenation, electrochemistry, syngas conversion to higher oxygenates, 1-propanol reforming, CO oxidation, photoelectrochemical water oxidation, methane oxidation, electrochemical capacitance, oxidative dehydrogenation of cyclohexane, propane dehydrogenation, 2-butanol dehydration, water splitting, 2-propanol dehydration, cyclohexanol dehydration. For example, Liu et al. fabricated photoelectrodes with WO_3 coated on a Mn catalyst via ALD. The soaking of WO_3 occurs in water with pH of 7, where the oxygen is generated through the water splitting under illumination. This is the first time demonstrating the stable WO_3 photoelectrodes in neutral solution [126].

3.3.2 Porous Oxide Coating Catalysts Structures

Besides oxides acting as support in catalysts, the ultrathin oxide overcoating layer fabricated with ALD can also act as decoration layer to modify NPs to improve catalytic activity and selectivity. Meanwhile, the coating layer could also anchor the NPs and enhance the thermal stability. One method to create porous coating layer fabricated via ALD involves calcination post-treatment after the oxide coating. The residual organic groups in the oxide layer will be oxidized, and nanotunnels will be created. In the calcination process, the stress relaxation of the coating layers can also be helpful for the formation of pores.

For example, Stair et al. fabricated 8 nm porous oxide-coated Pd metal NPs with enhanced catalytic activity and selectivity toward dehydrogenation of ethane reaction [83]. Firstly, the 8 nm conformal Al_2O_3 layer was deposited on the catalyst. After ALD coating, the as-prepared catalyst was calcined under high temperature at oxidation environment to create ~2 nm pores. After calcination, CO chemisorption FTIR measurements indicated the porous structure was formed, the metallic Pd sites were re-exposed, and the porous layer could improve the sintering and coking resistance at high temperature. The porous Al_2O_3 coated Pd catalyst showed improved catalytic selectivity of ethylene (23%), and the by-product yields of CH_4, CO_2, and CO were reduced to 0.9%, 3.9%, and 5.1%, respectively. At the same time, the carbon formation was also reduced significantly by 94%. For uncoated Pd catalyst, after the catalytic reaction tests, heavy coking was observed and the activity was degraded. The porous coating structures formed via ALD have also utilized for other metal-based catalysts like Cu as shown in Fig. 3.8a (especially The Cu/γ-Al_2O_3 system) [86], Co for solid oxide fuel cell cathode, aqueous-phase hydrogenation reactions [89], Au for CO oxidation [130], Ni in the dry reforming of methane [94, 135], Pd for methane oxidation [121, 137], and Ag for plasmonic photocatalysis [14].

An alternative way of creating porous structures was to utilize the initial nucleation growth of metal oxide via ALD. It was reported that in the initial growth stage, porous Al_2O_3 was formed and part of the active metal surface was exposed to the reactants (Fig. 3.8b). FTIR measurements confirmed that Al_2O_3 started its growth on low coordinated sites of Pd NPs in the first few cycles, instead of forming a conformal coating layer (Fig. 3.8c) [82, 140, 141]. The Al_2O_3 protective layers with an optimized thickness showed enhanced catalytic activity, while as the coating layer thickness increased and became continuous that blocked the reactants reaching to the Pd surface, the catalytic activity also decreased (Fig. 3.8d). The ALD Al_2O_3 also effectively improved the stability of the Pd NPs at 500 °C. On the other hand, Detavernier et al. elaborated the formation of porous coating structures with tailored pore sizes by applying TiO_2 ALD [142]. Weimer et al. investigated the molecular layer deposition (MLD) using TMA and ethylene glycol to fabricate alumina alkoxide hybrid films on Pt NPs (~2 nm). The alumina alkoxide hybrid overcoatings could be transformed to porous Al_2O_3 film on Pt NPs since the contained hydrocarbon groups were combusted at high temperature. Although some metal surface sites were re-exposed, it was found that the porous layer still decreased the catalytic performance which might be caused by the small pore size [143].

3.3.3 Site-Selective ALD Coating of Metal Oxide

A more delicate way to perform ALD is exploiting inherently selective ALD process to directly modify the active sites of nanoparticles. Lu et al. showed that during the TiO_2 ALD on Au particles, the TiO_2 layer firstly covered Au's low coordinated sites

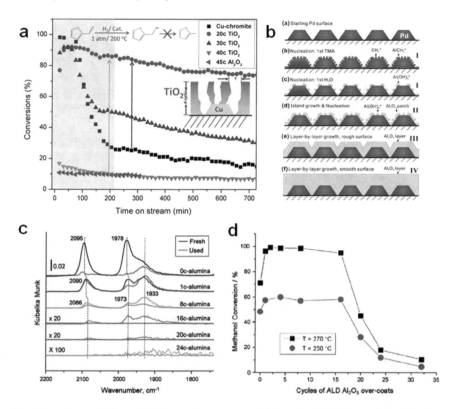

Fig. 3.8 **a** Porous TiO$_2$-coated Cu catalysts structure fabricated with ALD coupled calcination post-treatment, **b** the schematic diagram of the Al$_2$O$_3$ growth stages on Pd nanoparticles with increasing ALD cycles, **c** FTIR measurements of the porous coating structure with increasing ALD cycles, and **d** the catalytic performance toward methanol conversion. **a** is reprinted with permission from Ref. [86]. Copyright 2015. Elsevier. **b** is reprinted with permission from Ref. [140]. Copyright 2012. American Chemical Society. **c**, **d** are reprinted with permission from Ref. [82]. Copyright 2011. American Chemical Society

and the activity of Au catalysts toward CO oxidation was improved Fig. 3.9a [100]. ALD was used to control the Au–TiO$_2$ interface without changing the Au size. Qin et al. also demonstrated that Fe$_2$O$_3$ firstly deposited on Pt's low coordinated sites, and verified by FTIR measurements and DFT simulations as shown in Fig. 3.9b. This structure showed improved selectivity of COL (cinnamyl alcohol) yields from 45% for unmodified Pt to 84% for Fe$_2$O$_3$-coated Pt catalysts [116].

A facet-selective ALD method was developed by exploiting different binding energies of precursors chemisorbed on the nanoparticles' crystal planes. Chen et al. reported the CeO$_x$ deposited selectivity on Pt (111) facets and exposed Pt (100) facets (shown in Fig. 3.9c) [127]. Ce(thd)$_4$ and O$_3$ were utilized as precursors for ALD process. From DFT simulations, the binding energies calculated of Ce precursor fragments (Ce(thd)$_3^-$) followed the sequence of Pt (111) > CeO$_2$ (111) > Pt (100),

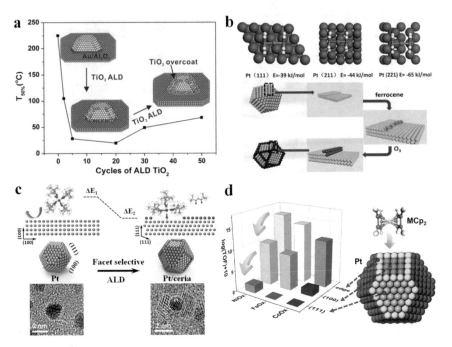

Fig. 3.9 Oxides-coated metal nanoparticles catalysts fabricated with the inherently selective ALD methods. **a** TiO$_2$-coated Au nanoparticles and catalytic activity toward CO oxidation, **b** DFT calculations of adsorption energy of Fe(Cp)$_2$ precursor on Pt (111), Pt(211) and Pt (221), **c** nanofence-like CeO$_x$-coated Pt nanoparticle catalysts fabricated with facet-selective ALD, **d** chemisorption rate calculations of NiO$_x$, FeO$_x$, and CoO$_x$ on Pt surfaces. **a** is reprinted with permission from Ref. [100]. Copyright 2016. American Chemical Society. **b** is reprinted with permission from Ref. [116]. Copyright 2017. Elsevier. **c** is reprinted with permission from Ref. [127]. Copyright 2017. WILEY-VCH Verlag GmbH & Co. KGaA, Weinheim. **d** is reprinted with permission from Ref. [144]. Copyright 2019. American Chemical Society

indicating the preferential growth on Pt (111) facet. This facet-selective coating structure exposed Pt active facets and formed Pt-CeO$_2$ interfaces at the same time which were helpful for activity enhancement toward CO oxidation. The CeO$_2$-coated catalyst was also stable under calcination at 700 °C due to the physical confinement effect of the coating layer that suppressed NPs migration.

In the following works, it was found that NiO_x ALD performed with $Ni(Cp)_2$ and O_3 as the precursors preferred to initiate growth on the edge sites of Pt, indicating the complexity involved in the driving force of the selective ALD. During the nucleation stage, Pt's low coordinated sites were selectively passivated by NiO_x and these sites usually were the most unstable sites and tended to gasify at high temperature in oxidizing environment [123]. The $NiO_x/Pt/Al_2O_3$ catalysts showed improved activity toward CO oxidation. On the other hand, the thermal stability of the catalysts could be also improved because of the selective passivation of the unstable edge sites. As a result, after calcination at 700 °C, the Pt NPs size and catalytic activity could be maintained.

Theoretical simulations are helpful to gain more insights of selective growth origins. For example, Al_2O_3 deposited with TMA showed different binding energies on Pd's facets and low coordinated sites. It was found that TMA dissociative adsorption preferably took take place thermodynamically at the low coordinated sites of Pd rather than at the Pd (111) facets. On the other hand, the binding energy differences on Pt were much smaller than those on Pd surfaces. Thus, the selective growth of Al_2O_3 could be achieved on the Pd surface, and continuous coating layers would form on Pt surface instead. The selective ALD process of MO_x (M=Co, Ni, Fe) on Pt based on DFT simulations is summarized in Fig. 3.9d [144]. The reaction barriers and rates calculations showed that the $M(Cp)_2$ precursors prefer to start growth on the edge sites of Pt. Based on the micro-kinetics analysis, the growth rate of MO_x on Pt sites followed the order of edge > (100) > (111). On the other hand, the activities of $M(Cp)_2$ precursors on edge sites of Pt followed the order of $Ni(Cp)_2$ > $Fe(Cp)_2$ > $Co(Cp)_2$. The DFT calculations also indicated that this selectivity was temperature dependent. Upon increasing the deposition temperature, edge sites selectivity would decrease since the growth rate differences between edge and (100) decreased.

3.3.4 Catalysts Synthesis by ALD with Confined Structure

The confined catalytic structure was developed to achieve a trade-off between the catalytic stability and activity. The structure could not only give rise to active metal oxide interfaces, but also expose a large amount of metal NPs surfaces to reactants compared with the overcoated structure. There were several methods to fabricate the confined catalysts through ALD. For example, organic groups were utilized to selectively chemisorb on the metal NPs before ALD; thus, the organic layer prevented the deposition of oxide materials directly on the metal NPs. After ALD process, the organic groups were removed to re-expose the metal NPs again and nanobowl-like oxides around the metal NPs were formed (Fig. 3.10a). Stair et al. reported the blocking ability of different organic groups during Al_2O_3 ALD. The blocking ability of the organic agents was influenced by chain lengths, functional groups, as well as steric hindrance effects [145]. Chen et al. exploited ODT (1-octadecanethiol) to selectively block Pt and successfully fabricated the Co_3O_4 nanotrap structure around Pt NPs that simultaneously improved the activity and sintering resistance

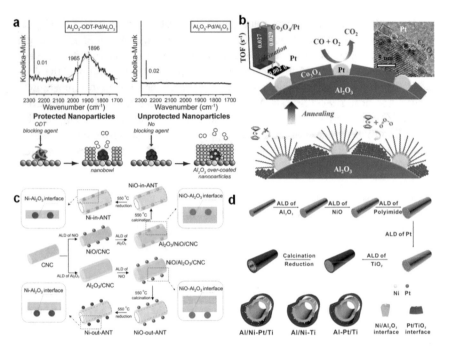

Fig. 3.10 Confined catalysts: **a** Pd nanoparticles embedded in Al_2O_3 nanobowl, the CO chemisorption FTIR characterization were performed on the confined catalyst and fully coated catalyst, **b** nanotrap CoO_x anchored Pt nanoparticles, **c** fabrication process of Ni-out-nanotubes and Ni-in-nanotubes catalysts with carbon nanocoils templates, **d** double layers nanotube confined catalysts fabricated by ALD. **a** is reprinted with permission from Ref. [145]. Copyright 2012. American Chemical Society. **b** is reprinted with permission from Ref. [111]. Copyright 2017. WILEY-VCH Verlag GmbH & Co. KGaA, Weinheim. **c** is reprinted with permission from Ref. [38]. Copyright 2015. WILEY-VCH Verlag GmbH & Co. KGaA, Weinheim. **d** is reprinted with permission from Ref. [79]. Copyright 2016. WILEY-VCH Verlag GmbH & Co. KGaA, Weinheim

of Pt NPs (Fig. 3.10b) [111]. Notestein et al. reported the fabrication of Al_2O_3 sieving layers with 'nanocavities' on a TiO_2 photocatalyst [89]. Sun et al. developed a similar method to encapsulate the Pt NPs in a ZrO_2 nanocage by selective ALD. The encapsulated catalytic structure showed enhanced activity (6.4 times) and stability (10 times) than the commercial Pt/C catalysts for oxygen reduction reaction [146].

Another method to confine catalysts utilized nanotemplates including zeolites, porous materials, and metal-organic frameworks. For porous materials that possessed large aspect ratios and small pore sizes, it was usually difficult to realize conformal deposition into the porous materials since the diffusion of precursor molecules was limited. Al_2O_3 and TiO_2 were widely used to modify the porous templates [78]. Qin et al. also reported the fabrication of Pt nanoclusters with controlled size and high dispersion on zeolites substrate by ALD [147]. On the other hand, the catalysts with confined structure can be fabricated by the template-assisted method via ALD. For example, the confined Ni catalyst and an unconfined Ni catalyst were fabricated based

on the carbon nanocoils templates as shown in Fig. 3.10c [38]. The confined Ni catalyst showed enhanced activity and stability for the hydrogenations of cinnamaldehyde
and nitrobenzene compared to unconfined catalysts. Their group also developed many
confined structures with nanotubes as templates. The multiple interfaces (Ni/Al_2O_3
and Pt/TiO_2) were fabricated by template-assisted ALD as shown in Fig. 3.10d [79].
The synergistic effect of the two interfaces accelerated the catalytic reaction, and the
confined space of the catalysts could adjust the transport of reactants that tuned the
reaction paths. The ALD method has been proved to be a powerful tool for the design
and fabrication of metal NPs. The size, shape, and composition of the catalysts could
be precisely tuned by the ALD process [148].

3.4 ALD Reactor for Catalysts Preparation

As the catalysts are usually in the form of nanoparticles or microparticles with larger
specific surface area than planar substrates, their synthesis by the ALD method
generally requires powder reactors with unique structures. Until now, several new
types of powder reactors have been developed for ALD on the surface of catalytic
particles. Three typical types of ALD reactors are the fluidized bed reactor (FBR),
the rotary reactor, and the fluidized bed coupled rotary reactor as shown in Fig. 3.11.

The powder reactor based on the fluidized bed principle has been reported in
2004 by Weimer et al. [149], the structure of which has been improved in the past
several years. As shown in Fig. 3.11a, the reactor consists of the dosing zone, the
particles bed, the flash zone, the mechanical bed agitation, and a residual gas analyzer.
The dosing zone allows for the controlled delivery of a variety of precursors. The

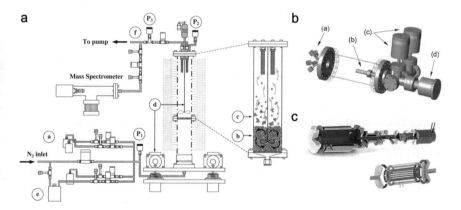

Fig. 3.11 Schematic of **a** fluidized bed reactor, **b** rotary reactor, and **c** fluidized bed coupled rotary
reactor. **a** is reprinted with permission from Ref. [149]. Copyright 2007. Elsevier. **b** is reprinted
with permission from Ref. [153]. Copyright 2007. AIP Publishing. **c** is reprinted with permission
from Ref. [157]. Copyright 2015. AIP Publishing

particles bed is fluidized using either the inter purge gas or the reactive precursor itself, and then, particles entering the splash zone fluidize as soft agglomerates. The mechanical bed agitation, using either vibration motors, a magnetically coupled stirring mechanism, or the pulsation of entering fluidization gases, promotes good fluidization behavior for nanoparticles. A residual gas analyzer is attached to the outlet of the FBR, allowing for the real-time detection of both reactants and products of each ALD half reaction. Van Ommen et al. [150], Rauwel et al. [151], and Soria-Hoyo et al. [152] also used the FBR with similar design to prepare ALD films on the surface of particles. Pt ALD catalysts synthesized in FBR reactor demonstrate nearly 100% conversion of CO to CO_2. The photocatalytic activity of TiO_2 particles is effectively quenched using SiO_2 or alucone polymer films deposited in the FBR reactor.

Weimer and George et al. [153] have reported the rotary type reactor for ALD on particles in 2007, which aims to prevent the particles from being agglomerated. As shown in Fig. 3.11b, a porous cylindrical drum with porous metal walls was positioned inside a vacuum chamber, which was rotated by a magnetically coupled rotary feedthrough. By rotating the cylindrical drum to obtain a centrifugal force of less than one gravitational force, the particles could be in a balance situation among the gravity force, the inward viscous drag force for gases entering the porous vessel, and the outward centrifugal force. In addition, an inert N_2 gas pulse helped to dislodge the particles from the porous walls and provided an efficient method to purge reactants and products from the particle bed. With the static exposures during ALD, the rotary reactor provides an effective approach for making large quantities of nanoparticles. Kedar Manandhar et al. [154] and Jinho Ahn et al. [155] also used a similar rotary reactor to deposit ALD films on the surface of particles. In order to inductively couple plasma-enhanced ALD on particles, Longrie et al. [156] have designed a new rotary reactor by using an open glass tube in a vacuum quartz tube that is connected with the pump. The deposited films on different particles via thermal and plasma-enhanced ALD are conformal, uniform, and pinhole free, which is important in the passivation of catalysts.

To improve the uniformity and efficiency of ALD on nanoparticles, Chen et al. have designed a rotary reactor coupled with fluidization based on the above design [157]. As shown in Fig. 3.11c, it consists of five major parts: reaction chamber, dosing and fluidizing section, pumping section, rotary manipulator components, as well as a double-layer cartridge for storage of particles. In the deposition procedure, continuous fluidization of particles enlarges and homogenizes the void fraction in the particle bed, while rotation enhances the gas–solid interactions to stabilize the fluidization. The cylindrical particle cartridge uses fluidization to disperse particles, and it adopts a high-speed rotation to enhance gas–solid interactions for the uniform and stable fluidization of particle bed. Moreover, enlarged interstitials and intense gas–solid contact under sufficient fluidizing velocity and proper rotation speed facilitate the precursor delivery throughout the particle bed and consequently provide a fast coating process. The cartridge can ensure precursors flowing through the particle bed exclusively to achieve high utilization without static exposure operation. The reactor provides an effective approach for catalytic particle ALD with both high efficiency and precursor utilization.

Besides conventional ALD reactors for particles which all need to be in vacuum work conditions, a spatial ALD process on particles has been reported by Van Ommen et al. [158], Spencer et al. [159], and Elam et al. [160]. In a spatial ALD process, the particles are continuously moving through the reaction areas with alternate ALD precursors, which are separated by inert purge gas. The spatial ALD process is suitable for the scale-up of particle surface modification. For instance, platinum (Pt) nanoclusters with the size of about 1 nm are deposited onto titania (TiO_2) P25 nanoparticles resulting in a continuous production of an active photocatalyst (0.12–0.31 wt% of Pt) [158].

3.5 Conclusion and Outlook

This chapter summarizes recent advances in catalyst preparation via ALD processes. The ALD method allows precise catalyst synthesis of size and composition at atomic level and creates well-defined structures, which is the superiority compared with conventional wet chemistry methods. Strategies to control the structures of metal active center and metal/metal oxide composites by ALD have been reviewed, with a focus on enhancing the performance of catalysts. These strategies of selective ALD have demonstrated unique advantages to design and fabricate the catalysts in atomic scale and provided insights to understand the structure–activity relationship in a more direct way. There are still plenty of challenges lying ahead, however. It is essential to focus on the fundamental understanding of surface reaction mechanisms such as interactions and reactions of precursors with different substrates, the reaction energetic routes of the selective ALD processes. The current choice of precursors for ALD approaches is quite limited at this moment. For instance, gold and silver catalysts prepared by ALD are barely reported, which are important in the catalytic family. On the other hand, it is necessary to develop precursors for catalytic nanoparticles. Various oxides and materials rely on plasma-enhanced ALD, which are not quite compatible with large quantity of nanoparticles as well as three-dimensional growth. For selective ALD to construct different structures, it is important to develop a fundamental theoretical understanding to guide the process and obtain various structures. The combination of studies with both in situ and ex situ experiments and characterizations are essential to understand the structural evolution and reaction processes.

Furthermore, the readers need to be aware that most of the reported results are laboratory-scale demonstrations of proofs of concept showing superior catalytic properties of ALD catalysts. The practical industrial applications are needed to continue motivating research on scale-up of this technology. The main problems originate from the huge specific surface area of practical catalysts. The deposition and decoration of catalysts on powder substrates consume long period of time for precursor diffusion and reaction, and agglomeration of nano-/micron powder substrate also deteriorates the deposition uniformity. The ultimate solutions hinge critically on the development of new ALD processes and equipment to accelerate deposition rate and utilization efficiency of precursors.

Acknowledgements The authors gratefully thank Yun Lang, Miao Gong, Jiaming Cai, Kai Qu, and Yuanting Tang for assisting the collection and organization of corresponding figures and references.

References

1. Zhong L, Yu F, An Y et al (2016) Cobalt carbide nanoprisms for direct production of lower olefins from syngas. Nature 538:84–87
2. Li L, Zhang N, Huang X, Liu et al (2018) Hydrothermal stability of core–shell Pd@$Ce_{0.5}Zr_{0.5}O_2$/Al_2O_3 catalyst for automobile three-way reaction. ACS Catal 8:3222–3231
3. García-Diéguez M, Pieta IS, Herrera MC, Larrubia MA, Alemany LJ (2010) Nanostructured Pt-and Ni-based catalysts for CO_2-reforming of methane. J Catal 270:136–145
4. Meng Y, Song W, Huang H, Ren Z, Chen SY, Suib SL (2014) Structure–property relationship of bifunctional MnO_2 nanostructures: highly efficient, ultra-stable electrochemical water oxidation and oxygen reduction reaction catalysts identified in alkaline media. J Am Chem Soc 136:11452–11464
5. Yin P, Yao T, Wu Y, Zheng L et al (2016) Single cobalt atoms with precise N-coordination as superior oxygen reduction reaction catalysts. Angew Chem Int Ed 55:10800–10805
6. Cao K, Cai J, Liu X (2018) Catalysts design and synthesis via selective atomic layer deposition. J Vac Sci Technol, A 36:010801
7. Hämäläinen J (2013) Atomic layer deposition of noble metal oxide and noble metal thin films. University of Helsinki, Finland
8. George SM (2010) Atomic layer deposition: an overview. Chem Rev 110:111–131
9. Puurunen RL (2005) Surface chemistry of atomic layer deposition: A case study for the trimethylaluminum/water process. J Appl Phys 97:121301
10. Miikkulainen V, Leskelä M, Ritala et al (2013) Crystallinity of inorganic films grown by atomic layer deposition: Overview and general trends. J Appl Phys 113:021301
11. Lim BS, Rahtu A, Gordon RG (2003) Atomic layer deposition of transition metals. Nat Mater 2:749–754
12. Johnson RW, Hultqvist A, Bent SF (2014) A brief review of atomic layer deposition: from fundamentals to applications. Mater Today 17:236–246
13. Johansson AC, Larsen JV, Verheijen MA et al (2014) Electrocatalytic activity of atomic layer deposited Pt-Ru catalysts onto N-doped carbon nanotubes. J Catal 311:481–486
14. Hämäläinen J, Rital M, Leskelä M (2013) Atomic layer deposition of noble metals and their oxides. Chem Mater 26:786–801
15. Setthapun W, Williams WD, Kim S et al (2010) Genesis and evolution of surface species during Pt atomic layer deposition on oxide supports characterized by in situ XAFS analysis and water- gas shift reaction. J Phys Chem C 114:9758–9771
16. Li J, Zhang B, Chen Y et al (2015) Styrene hydrogenation performance of Pt nanoparticles with controlled size prepared by atomic layer deposition. Catal Sci Technol 5:4218–4223
17. Aaltonen T, Ritala M, Tung YL et al (2004) Atomic layer deposition of noble metals: Exploration of the low limit of the deposition temperature. J Mater Res 19:3353–3358
18. Zhou Y, King DM, Liang X et al (2010) Optimal preparation of Pt/TiO_2 photocatalysts using atomic layer deposition. Appl Catal B 101:54–60
19. Hämäläinen J, Puukilainen E, Sajavaara et al (2013) Low temperature atomic layer deposition of noble metals using ozone and molecular hydrogen as reactants. Thin Solid Films 531:243–250

20. Xu S, Kim Y, Park J et al (2018) Extending the limits of Pt/C catalysts with passivation-gas-incorporated atomic layer deposition. Nat Catal 1:624
21. Ren W, Zhang H, Cheng C (2017) Ultrafine Pt nanoparticles decorated MoS$_2$ nanosheets with significantly improved hydrogen evolution activity. Electrochim Acta 241:316
22. Elam JW, Zinovev A, Han CY et al (2006) Atomic layer deposition of palladium films on Al$_2$O$_3$ surfaces. Thin Solid Films 515:1664–1673
23. TenEyck GA, Pimanpang S, Bakhru H et al (2006) Atomic layer deposition of Pd on an oxidized metal substrate. Chem Vap Deposition 12:290–294
24. Senkevich JJ, Tang F, Rogers D et al (2003) Substrate-independent palladium atomic layer deposition. Chem Vap Deposition 9:258–264
25. Feng H, Elam JW, Libera JA et al (2010) Palladium catalysts synthesized by atomic layer deposition for methanol decomposition. Chem Mater 22:3133–3142
26. Feng H, Libera JA, Stair PC et al (2011) Subnanometer palladium particles synthesized by atomic layer deposition. ACS Catal 1:665–673
27. Jiang Y, Chen J, Zhang J et al (2017) Controlled decoration of Pd on Ni (OH)$_2$ nanoparticles by atomic layer deposition for high ethanol oxidation activity. Appl Surf Sci 420:214–221
28. Wang J, Liu C, Lushington A et al (2016) Pd on carbon nanotubes-supported Ag for formate oxidation: the effect of Ag on anti-poisoning performance. Electrochim Acta 210:285–292
29. Hong Y, Du H, Hu Y et al (2015) Precisely controlled porous alumina overcoating on pd catalyst by atomic layer deposition: enhanced selectivity and durability in hydrogenation of 1,3-butadiene. ACS Catal 5:2735–2739
30. Lei Y, Lu J, Luo X et al (2013) Synthesis of porous carbon supported palladium nanoparticle catalysts by atomic layer deposition: application for rechargeable lithium-O$_2$ battery. Nano Lett 13:4182–4189
31. Aaltonen T, Ritala M, Sammelselg V et al (2004) Atomic layer deposition of iridium thin films. J. Electrochem Soc 151:489–492
32. Choi BH, Lee JH, Lee HK et al (2011) Effect of interface layer on growth behavior of atomic-layer-deposited Ir thin film as novel Cu diffusion barrier. Appl Surf Sci 257:9654–9660
33. Silvennoinen RJ, Jylhä OJT, Lindblad M et al (2007) Supported iridium catalysts prepared by atomic layer deposition: effect of reduction and calcination on activity in toluene hydrogenation. Catal Lett 114:135–144
34. Vuori H, Silvennoinen RJ, Lindblad M et al (2009) Beta zeolite-supported iridium catalysts by gas phase deposition. Catal Lett 131:7–15
35. Park KJ, Parsons GN (2006) Selective area atomic layer deposition of rhodium and effective work function characterization in capacitor structures. Appl Phys Lett 89:043111
36. Li Y, Jiang J, Zhu C et al (2018) The enhanced catalytic performance and stability of Rh/γ-Al$_2$O$_3$ Catalyst Synthesized By Atomic Layer Deposition (ALD) for methane dry reforming. Materials 11:172
37. Gould TD, Lubers AM, Neltner BT et al (2013) Synthesis of supported Ni catalysts by atomic layer deposition. J Catal 303:9–15
38. Gao Z, Dong M, Wang G et al (2015) Multiply confined nickel nanocatalysts produced by atomic layer deposition for hydrogenation reactions. Angew Chem 54:9006–9010
39. Lindblad M, Lindfors LP, Suntola T (1994) Preparation of Ni/Al$_2$O$_3$ catalysts from vapor phase by atomic layer epitaxy. Catal Lett 27:323–336
40. Jacobs JP, Lindfors LP, Reintjes JGH et al (1994) The growth mechanism of nickel in the preparation of Ni/Al$_2$O$_3$ catalysts studied by LEIS, XPS and catalytic activity. Catal Lett 25:315–324
41. Kim DH, Sim JK, Seo HO et al (2013) Carbon dioxide reforming of methane over mesoporous Ni/SiO$_2$ catalyst. Fuel 112:111–116
42. Jiang C, Shang Z, Liang X (2015) Chemoselective transfer hydrogenation of nitroarenes catalyzed by highly dispersed, supported nickel nanoparticles. ACS Cata 5:4814–4818
43. Mårtensson P, Larsson K, Carlsson JO (2000) Atomic layer epitaxy of copper: an ab initio investigation of the CuCl/H$_2$ process: III. Reaction barriers. Appl Surface Sci 157:92–100

44. Li Z, Rahtu A, Gordon RG (2006) Atomic layer deposition of ultrathin copper metal films from a liquid copper (I) amidinate precursor. J Electrochem Soc 153:C787–C794
45. Chen CS, Lin JH, Lai TW et al (2009) Active sites on Cu/SiO$_2$ prepared using the atomic layer epitaxy technique for a low-temperature water-gas shift reaction. J Catal 263:155–166
46. Gao F, Jiang J, Du L et al (2018) Stable and highly efficient Cu/TiO$_2$ nanocomposite photocatalyst prepared through atomic layer deposition. Appl Catal A 568:168–175
47. Najafabadi AT, Khodadadi AA, Parnian MJ et al (2016) Atomic layer deposited Co/γ-Al$_2$O$_3$ catalyst with enhanced cobalt dispersion and Fischer-Tropsch synthesis activity and selectivity. Appl Catal A 511:31–46
48. Lee D, Yim S, Kim K, Kim J, Kim K (2008) Formation of Ru nanotubes by atomic Llyer deposition onto an anodized aluminum oxide template. Electrochem Solid State Lett 11:K61–K63
49. Lu J, Stair PC (2010) Low temperature ABC-type atomic layer deposition: synthesis of highly uniform ultrafine supported metal nanoparticles. Angew Chem Int Ed 49:2547–2551
50. Li J, Liang X, King DM, Jiang YB, Weimer AW (2010) Highly dispersed Pt nanoparticle catalyst prepared by atomic layer deposition. Appl Catal B 97:220–226
51. Hämäläinen J, Hatanpää T, Puukilainen E, Costelle L, Pilvi T, Ritala M, Leskelä M (2010) (MeCp) Ir (CHD) and molecular oxygen as precursors in atomic layer deposition of iridium. J Mater Chem 20:7669–7675
52. Wang G, Gao Z, Wan G, Lin S, Yang P, Qin Y (2014) High densities of magnetic nanoparticles supported on graphene fabricated by atomic layer deposition and their use as efficient synergistic microwave absorbers. Nano Research 7:704–716
53. Yang X, Wang A, Qiao B et al (2013) Single-atom catalysts: a new frontier in heterogeneous catalysis. Acc Chem Res 44:1740–1748
54. Shekhar M, Wang J, Lee W et al (2012) Size and support effects for the water-gas shift catalysis over gold nanoparticles supported on model Al$_2$O$_3$ and TiO$_2$. J Am Chem Soc 134:4700–4708
55. Cargnello M, Doan-Nguyen V, Gordon T et al (2013) Control of metal nanocrystal size reveals metal-support interface role for ceria catalysts. Science 341:771–773
56. Enterkin J, Setthapun W, Elam J et al (2011) Propane oxidation over Pt/SrTiO$_3$ nanocuboids. ACS Catal 1:629–635
57. Mackus A, Weber M, Thissen N et al (2016) Atomic layer deposition of Pd and Pt nanoparticles for catalysis: on the mechanisms of nanoparticle formation. Nanotechnology 27:034001
58. Goulas A, Van Ommen JV (2013) Atomic layer deposition of platinum clusters on titania nanoparticles at atmospheric pressure. J Mater Chem A 1:4647–4650
59. Bui H, Grillo F, Kulkarni S et al (2017) Low-temperature atomic layer deposition delivers more active and stable Pt-based catalysts. Nanoscale 9:10802–10810
60. Gould T, Lubers A, Corpuz A et al (2015) Controlling nanoscale properties of supported platinum catalysts through atomic layer deposition. ACS Catal 5:1344–1352
61. Dendooven J, Ramachandran R, Solano E et al (2017) Independent tuning of size and coverage of supported Pt nanoparticles using atomic layer deposition. Nat Commun 8:1074
62. Zhang L, Banis M, Sun X (2018) Single-atom catalysts by the atomic layer deposition technique. Natl Sci Rev 5:628–630
63. Cheng N, Stambula S, Wang D et al (2016) Platinum single-atom and cluster catalysis of the hydrogen evolution reaction. Nat Commun 7:13638
64. Liu X, Tang Y, Shen M et al (2018) Bifunctional CO oxidation over Mn-mullite anchored Pt sub-nanoclusters via atomic layer deposition. Chem Sci 9:2469–2473
65. Wang C, Gu X, Yan H et al (2017) Water-mediated Mars-van Krevelen mechanism for CO oxidation on ceria-supported single-atom Pt$_1$ catalyst. ACS Catal 7:887–891
66. Yan H, Cheng H, Yi H et al (2015) Single-atom Pd$_1$/graphene catalyst achieved by atomic layer deposition: remarkable performance in selective hydrogenation of 1,3-butadiene. J Am Chem Soc 137:10484–10487
67. Yan H, Lv H, Yi H et al (2018) Understanding the underlying mechanism of improved selectivity in Pd$_1$ single-atom catalyzed hydrogenation reaction. J Catal 366:70–79

68. Yan H, Zhao X, Guo N et al (2018) Atomic engineering of high-density isolated Co atoms on graphene with proximal-atom controlled reaction selectivity. Nat Commun 9:3197
69. Li Z, Schweitzer N, League A et al (2016) Sintering-resistant single-site nickel catalyst supported by metal-organic framework. J Am Chem Soc 138:1977–1982
70. Cao Y, Chen S, Luo Q et al (2017) Atomic-level insight into optimizing the hydrogen evolution pathway over a Co_1-N_4 single-site photocatalyst. Angew Chem Int Ed 56:12191–12196
71. Weber MJ, Mackus AJM, Verheijen MA et al (2012) Supported core/shell bimetallic nanoparticles synthesis by atomic layer deposition. Chem Mater 24:2973–2977
72. Weber MJ, Verheijen MA, Bol AA et al (2015) Sub-nanometer dimensions control of core/shell nanoparticles prepared by atomic layer deposition. Nanotechnology 26:094002
73. Lu J, Low K, Lei Y et al (2014) Toward atomically-precise synthesis of supported bimetallic nanoparticles using atomic layer deposition. Nat Commun 5:3264
74. Cao K, Zhu Q, Shan B et al (2015) Controlled synthesis of Pd/Pt core shell nanoparticles using area-selective atomic layer deposition. Sci Rep 5:8470
75. Wang H, Wang C, Yan H et al (2015) Precisely-controlled synthesis of Au@Pd core-shell bimetallic catalyst via atomic layer deposition for selective oxidation of benzyl alcohol. J Catal 324:59–68
76. Christensen ST, Feng H, Libera JL et al (2010) Supported Ru-Pt bimetallic nanoparticle catalysts prepared by atomic layer deposition. Nano Lett 10:3047–3051
77. Cao K, Gong M, Yang JF, Cai JM, Chu SQ, Chen ZP, Shan B, Chen R (2019) Nickel catalyst with atomically-thin meshed cobalt coating for improved durability in dry reforming of methane. J Catal 373:351–360
78. Vandegehuchte BD, Thybaut JW, Detavernier C et al (2014) A single-event microkinetic assessment of n-alkane hydroconversion on ultrastable Y zeolites after atomic layer deposition of alumina. J Catal 311:433–446
79. Ge H, Zhang B, Gu X et al (2016) A tandem catalyst with multiple metal oxide interfaces produced by atomic layer deposition. Angew Chem Int Ed 55:7081–7085
80. Wang M, Gao Z, Zhang B et al (2016) Ultrathin coating of confined Pt nanocatalysts by atomic layer deposition for enhanced catalytic performance in hydrogenation reactions. Chem Eur J 22:8438–8443
81. O'Neill BJ, Jackson DHK, Crisci AJ et al (2013) Stabilization of copper catalysts for liquid-phase reactions by atomic layer deposition. Angew Chem Int Ed 52:13808–13812
82. Feng H, Lu JL, Stair PC et al (2011) Alumina over-coating on Pd nanoparticle catalysts by atomic layer deposition: enhanced stability and reactivity. Catal Lett 141:512–517
83. Lu J, Fu B, Kung MC et al (2012) Coking- and sintering-resistant palladium catalysts achieved through atomic layer deposition. Science 335:1205–1208
84. Zhang H, Gu X-K, Canlas C et al (2014) Atomic layer deposition overcoating: tuning catalyst selectivity for biomass conversion. Angew Chem Int Ed 53:12132–12136
85. Yi H, Du H, Hu Y et al (2015) Precisely controlled porous alumina overcoating on Pd catalyst by atomic layer deposition: enhanced selectivity and durability in hydrogenation of 1,3-Butadiene. ACS Catalysis 5:2735–2739
86. Zhang H, Canlas C, Kropf AJ et al (2015) Enhancing the stability of copper chromite catalysts for the selective hydrogenation of furfural with ALD overcoating (II) - Comparison between TiO_2 and Al_2O_3 overcoatings. J Catal 326:172–181
87. Alba-Rubio AC, O'Neill BJ, Shi F et al (2014) Pore structure and bifunctional catalyst activity of overlayers applied by atomic layer deposition on copper nanoparticles. ACS Catal 4:1554–1557
88. O'Neill BJ, Sener C, Jackson DHK et al (2014) Control of thickness and chemical properties of atomic layer deposition overcoats for stabilizing Cu/γ-Al_2O_3 catalysts. Chem Sustain Chem 7:3247–3251
89. Canlas CP, Lu JL, Ray NA et al (2012) Shape-selective sieving layers on an oxide catalyst surface. J Nat Gas Chem 4:1030
90. Wang CL, Lu JL (2016) Sub-nanometer-thick Al_2O_3 overcoat remarkably enhancing thermal stability of supported gold catalysts. Chin J Chem Phys 29:571

91. Zhao Y, Kang YQ, Li H et al (2018) CO_2 conversion to synthesis gas via DRM on the durable Al_2O_3/Ni/Al_2O_3 sandwich catalyst with high activity and stability. Green Chem 20:2781

92. Lei Y, Lee S, Low K-B et al (2016) Combining electronic and geometric effects of ZnO-promoted Pt nanocatalysts for aqueous phase reforming of 1-Propanol. ACS Catal 6:3457–3460

93. Kayaci F, Vempati S, Ozgit-Akgun C et al (2014) Selective isolation of the electron or hole in photocatalysis: ZnO-TiO_2 and TiO_2-ZnO core-shell structured heterojunction nanofibers via electrospinning and atomic layer deposition. Nanoscale 6:5735–5745

94. Jiang F, Huang J, Niu L et al (2015) Atomic layer deposition of ZnO thin films on ZSM-5 zeolite and its catalytic performance in chichibabin reaction. Catal Lett 145:947–954

95. Wang WN, Wu F, Myung Y et al (2015) Surface engineered CuO nanowires with ZnO islands for CO_2 photoreduction. ACS Appl Mater Interfaces 7:5685–5692

96. Liu J, Sun C, Fu M et al (2018) Enhanced photochemical catalysis of TiO_2 inverse opals by modification with ZnO or Fe_2O_3 using ALD and the hydrothermal method. Mater Res Express 5:025509

97. Kim IS, Michael J, Pellin et al (2019) Acid-compatible halide perovskite photocathodes utilizing atomic layer deposited TiO_2 for solar-driven hydrogen evolution. ACS Energy Lett 4:293–298

98. Lee J, Jackson DHK, Li T et al (2014) Enhanced stability of cobalt catalysts by atomic layer deposition for aqueous-phase reactions. Energy Environ Sci 7:1657–1660

99. Liang H, Zhang B, Ge H et al (2017) Porous TiO_2/Pt/TiO_2 sandwich catalyst for highly selective semihydrogenation of alkyne to olefin. ACS Catal 7:6567–6572

100. Wang C, Wang H, Yao Q et al (2016) Precisely applying TiO_2 overcoat on supported Au catalysts using atomic layer deposition for understanding the reaction mechanism and improved activity in CO oxidation. J Phys Chem C 120:478–486

101. Yao Q, Wang C, Wang H et al (2016) Revisiting the Au particle size effect on TiO_2-coated Au/TiO_2 catalysts in CO oxidation reaction. J Phys Chem C 120:9174–9183

102. Biener MM, Biener J, Wichmann A et al (2011) ALD functionalized nanoporous gold: thermal stability, mechanical properties, and catalytic activity. Nano Lett 11:3085–3090

103. Kim DW, Kim KD, Seo HO et al (2011) TiO_2/Ni inverse-catalysts prepared by atomic layer deposition (ALD). Catal Lett 141:854–859

104. Seo HO, Sim JK, Kim KD et al (2013) Carbon dioxide reforming of methane to synthesis gas over a TiO_2-Ni inverse catalyst. Appl Catal A 451:43–49

105. Kim HJ, Jackson DHK, Lee J et al (2015) Enhanced activity and stability of TiO_2-coated cobalt/carbon catalysts for electrochemical water oxidation. ACS Catal 5:3463–3469

106. Lobo R, Marshall CL, Dietrich PJ et al (2012) Understanding the chemistry of H_2 production for 1-propanol reforming: pathway and support modification effects. ACS Catal 2:2316–2326

107. Kennedy RM, Crosby LA, Kunlun D et al (2018) Replication of SMSI via ALD: TiO_2 overcoats increase Pt-catalyzed acrolein hydrogenation selectivity. Catal Lett 148:2223–2232

108. Huang B, Cao K, Liu X et al (2015) Tuning the morphology and composition of ultrathin cobalt oxide films via atomic layer deposition. RSC Adv 7:71816–71823

109. Li Y, Zhao S, Hu Q et al (2017) Highly efficient CoO_x/SBA-15 catalysts prepared by atomic layer deposition for the epoxidation reaction of styrene. Catal Sci Technol 7:2032–2038

110. Zhang J, Yu Z, Gao Z et al (2017) Porous TiO_2 nanotubes with spatially separated platinum and CoO_x cocatalysts produced by atomic layer deposition for photocatalytic hydrogen production. Angew Chem Int Ed 56:816–820

111. Liu X, Zhu Q, Lang Y et al (2017) Oxide-nanotrap-anchored platinum nanoparticles with high activity and sintering resistance by area-Selective atomic layer deposition. Angew Chem Int Ed 56:1648–1652

112. Deng W, Lee S, Libera JA et al (2011) Cleavage of the C–O–C bond on size-selected subnanometer cobalt catalysts and on ALD-cobalt coated nanoporous membranes. Appl Catal A- General 393:29–35

113. Huang B, Yang W, Wen Y et al (2015) Co_3O_4-modified TiO_2 nanotube arrays via atomic layer deposition for improved visible-light photoelectrochemical performance. ACS Appl Mater Interfaces 7:422–431

114. Zhang C, Huang B, Qian LH et al (2016) Electrochemical biosensor based on nanoporous Au/CoO core-shell with synergistic catalysis. ChemPhysChem 17:98–104

115. Du C, Wang J, Liu X et al (2017) Ultrathin CoO_x-modified hematite with low onset potential for solar water oxidation. Phys Chem Chem Phys 19:14178–14184

116. Hu Q, Wang S, Gao Z, Li Y et al (2017) The precise decoration of Pt nanoparticles with Fe oxide by atomic layer deposition for the selective hydrogenation of cinnamaldehyde. Appl Catal B 218:591–599

117. Yi H, Xia Y, Yan H et al (2017) Coating Pd/Al_2O_3 catalysts with FeO_x enhances both activity and selectivity in 1,3-butadiene hydrogenation. Chin J Catal 38:1581–1587

118. Zhang B, Guo X-W, Liang H et al (2016) Tailoring $Pt-Fe_2O_3$ interfaces for selective reductive coupling reaction to synthesize imine. ACS Catal 6:6560–6566

119. Lin YJ, Zhou S, Sheehan SW et al (2011) Nanonet-based hematite heteronanostructures for efficient solar water splitting. J Am Chem Soc 133:2398–2401

120. Li JJ, Lu JL (2017) FeO_x coating on Pd/C catalyst by atomic layer deposition enhances the catalytic activity in dehydrogenation of formic acid. J Chem Phys 30:319–324

121. Jeong MG, Kim IH, Han SW et al (2016) Room temperature CO oxidation catalyzed by NiO particles on mesoporous SiO_2 prepared via atomic layer deposition: influence of pre-annealing temperature on catalytic activity. J Molecular Catal A-Chem 414:87–93

122. Han SW, Kim DH, Jeong M-G et al (2016) CO oxidation catalyzed by NiO supported on mesoporous Al_2O_3 at room temperature. Chem Eng J 283:992–998

123. Cai JM, Zhang J, Cao K et al (2018) Selective passivation of Pt nanoparticles with enhanced sintering resistance and activity toward CO oxidation via atomic layer deposition. ACS Appl Nano Mater 1:522–530

124. O'Neill BJ, Sener C, Jackson DHK et al (2014) Control of thickness and chemical properties of atomic layer deposition overcoats for stabilizing $Cu/\gamma-Al_2O_3$ catalysts. Chemsuschem 7:3247–3251

125. Zhu B, Wu XH, Liu WJ et al (2019) High-performance on-chip supercapacitors based on mesoporous silicon coated with ultrathin atomic layer-deposited In_2O_3 films. ACS Appl Mater Interfaces 11:747–752

126. Yang N, Yoo JS, Schumann J et al (2017) Rh-MnO interface sites formed by atomic layer deposition promote syngas conversion to higher oxygenates. ACS Catal 7:5746–5757

127. Cao K, Shi L, Gong M et al (2017) Nanofence stabilized platinum nanoparticles catalyst via facet-selective atomic layer deposition. Small 13:1700648

128. Peng Q, Wang J, Feng ZJ et al (2017) Enhanced photoelectrochemical water oxidation by fabrication of $p-LaFeO_3/n-Fe_2O_3$ heterojunction on hematite nanorods. J Phys Chem C 121:12991–12998

129. Onn TM, Zhang SY, Arroyo-Ramirez L et al (2015) Improved thermal stability and methane-oxidation activity of Pd/Al_2O_3 catalysts by atomic layer deposition of ZrO_2. ACS Catal 5:5696–5701

130. Liu J, Meng XB, Hu YH et al (2013) Controlled synthesis of zirconium oxide on graphene nanosheets by atomic layer deposition and its growth mechanism. Carbon 52:74–82

131. Feng H, Elam JW, Libera JA et al (2010) Oxidative dehydrogenation of cyclohexane over alumina-supported vanadium oxide nanoliths. J Catal 269:421–431

132. Keranen J, Auroux A, Ek S et al (2002) Preparation, characterization and activity testing of vanadia catalysts deposited onto silica and alumina supports by atomic layer deposition. Appl Catal A 228:213–225

133. Herrera JE, Kwak JH, Hu JZ et al (2006) Synthesis, characterization, and catalytic function of novel highly dispersed tungsten oxide catalysts on mesoporous silica. J Catal 239:200–211

134. Cong W, Xinyu M, Jennifer L et al (2018) A characterization study of reactive sites in ALD-synthesized WO_x/ZrO_2 catalysts. Catalysts 8:292

135. Pagan-Torres YJ, Gallo JMR, Wang D et al (2011) Synthesis of highly ordered hydrothermally stable mesoporous niobia catalysts by atomic layer deposition. ACS Catal 1:1234–1245

136. Ma Z, Brown S, Howe JY et al (2008) Surface modification of Au/TiO_2 catalysts by SiO_2 via atomic layer deposition. J Phys Chem C 112:9448–9457

137. Mouat AR, George C, Kobayashi T et al (2015) Highly dispersed SiO_x/Al_2O_3 catalysts illuminate the reactivity of isolated silanol sites. Angew Chem Int Ed 54:13346–13351
138. Kytökivi A, Jacobs JP, Hakuli A et al (1996) Surface characteristics and activity of chromia/alumina catalysts prepared by atomic layer epitaxy. J Catal 162:190–197
139. Damyanov D, Mehandjiev D, Obretenoy Ts (1975) Preparation of chromium oxides on the surface of silica gel by the method of molecular deposition. Heterogeneous Catalysis, Varna, pp 191–195
140. Lu J, Liu B, Greeley JP et al (2012) Porous alumina protective coatings on palladium nanoparticles by self-poisoned atomic layer deposition. Chem Mater 24:2047–2055
141. Lu J, Liu B, Guisinger NP et al (2014) First-principles predictions and in situ experimental validation of alumina atomic layer deposition on metal surfaces. Chem Mater 26:6752–6761
142. Deng S, Kurttepeli M, Cott DJ, Bals S, Detavernier C (2015) Porous nanostructured metal oxides synthesized through atomic layer deposition on a carbonaceous template followed by calcination. J Mater Chem A 3:2642–2649
143. Liang X, Li J, Yu M, McMurray CN, Falconer JL, Weimer AW (2011) Stabilization of supported metal nanoparticles using an ultrathin porous shell. ACS Catal 1:1162–1165
144. Wen Y, Cai J, Zhang J et al (2018) Edge-selective growth of MCp2 (M=Fe Co, and Ni) precursors on pt nanoparticles in atomic layer deposition: a combined theoretical and experimental study. Chem Mater 31:101–111
145. Ray NA, Van DRP, Stair PC (2012) Synthesis strategy for protected metal nanoparticles. J Phys Chem C 116:7748–7756
146. Cheng N, Banis MN, Liu J et al (2015) Extremely stable platinum nanoparticles encapsulated in a zirconia nanocage by area-selective atomic layer deposition for the oxygen reduction reaction. Adv Mater 27:277–281
147. Xu D, Wu BS, Ren PJ et al (2017) Controllable deposition of Pt nanoparticles into a KL zeolite by atomic layer deposition for highly efficient reforming of n-heptane to aromatics. Catal Sci Technol 7:1342–1350
148. Gao Z, Qin Y (20157) Design and properties of confined nanocatalysts by atomic layer deposition. Accounts Chem Res 50:2309–2316
149. King DM, Spencer JA, Liang X et al (2007) Atomic layer deposition on particles using a fluidized bed reactor with in situ mass spectrometry. Surf Coat Technol 201:9163–9171
150. Azizpour H, Talebi M, Tichelaar FD et al (2017) Effective coating of titania nanoparticles with alumina via atomic layer deposition. Appl Surf Sci 426:480–496
151. Rauwel E, Nilsen O, Rauwel P et al (2012) Oxide coating of alumina nanoporous structure using ALD to produce highly porous spinel. Chem Vap Deposition 18:315–325
152. Soria-Hoyo C, Valverde JM, Van Ommen JR et al (2015) Synthesis of a nanosilica supported CO_2 sorbent in a fluidized bed reactor. Appl Surf Sci 328:548–553
153. Mccormick JA, Cloutier BL, Weimer AW et al (2007) Rotary reactor for atomic layer deposition on large quantities of nanoparticles. J Vac Sci Technol, A 25:67–74
154. Manandhar K, Wollmershauser JA, Boercker JE et al (2016) Growth per cycle of alumina atomic layer deposition on nano- and micro-powders. J Vac Sci Technol, A 34:021519
155. Seong S, Jung YC, Lee T et al (2016) Fabrication of Fe_3O_4-ZnO core-shell nanoparticles by rotational atomic layer deposition and their multi-functional properties. Curr Appl Phys 16:1564–1570
156. Longrie D, Deduytsche D, Haemers J et al (2012) A rotary reactor for thermal and plasma-enhanced atomic layer deposition on powders and small objects. Surf Coat Technol 213:183–191
157. Duan C-L, Liu X, Shan B et al (2015) Fluidized bed coupled rotary reactor for nanoparticles coating via atomic layer deposition. Rev Sci Instrum 86:075101
158. Van Ommen JR, Kooijman D, Niet MD et al (2015) Continuous production of nanostructured particles using spatial atomic layer deposition. J Vac Sci Technol, A 33(2):021513
159. Spencer IJA, Hall RA (2018) U.S. Patent Application No. 15/737023
160. Elam JW, Yanguas-gil A, Libera JA (2017) U.S. Patent Application No. 15/426789

161. King JS, Wittstock A, Biener J et al (2008) Ultralow loading Pt nanocatalysts prepared by atomic layer deposition on carbon aerogels. Nano Lett 8:2405–2409
162. Hoang S, Lu X, Tang W et al (2019) High performance diesel oxidation catalysts using ultra-low Pt loading on titania nanowire array integrated cordierite honeycombs. Catal Today 320:2–10
163. Enterkin JA, Kennedy RM, Lu J et al (2013) Epitaxial stabilization of face selective catalysts. Top Catal 56:1829–1834
164. Zhang J, Chen C, Chen S et al (2017) Highly dispersed Pt nanoparticles supported on carbon nanotubes produced by atomic layer deposition for hydrogen generation from hydrolysis of ammonia borane. Catal Sci Technol 7:322–329
165. Chen CS, Lin JH, You JH et al (2006) Properties of Cu(thd)$_2$ as a precursor to prepare Cu/SiO$_2$ catalyst using the atomic layer epitaxy technique. J Am Chem Soc 128:15950–15951
166. Liu R, Lin YJ, Chou LY et al (2011) Water splitting by tungsten oxide prepared by atomic layer deposition and decorated with an oxygen-evolving catalyst. Angew Chem Int Ed 50:499–502

Chapter 4
MNP Catalysis in Ionic Liquids

Muhammad I. Qadir, Nathália M. Simon, David Rivillo, and Jairton Dupont

Abstract Ionic liquids (ILs) are the green, versatile and alternative fluids in the fabrication of soluble 'surface-clean' nanoparticles (NPs) that act not only as stabilising agents but also as templates to generate NPs of desired size. Ligand-free NPs in ILs can be prepared either by hydrogen-driven chemical methods, i.e., by the simple hydrogenation/decomposition of zero-valent metal complexes or by a physical method (magnetron sputtering). ILs can be used as liquid supports for NPs, providing an encapsulation-like environment for the catalytically active species that may allow the modulation of the stability of the catalyst and the miscibility of the substrates and products at the active catalytic centre. In this chapter, the hydrogen-involved nano-catalytic reactions (hydrogenation, hydroformylation, Fischer-Tropsch synthesis) in ILs will be summarised.

Keywords Ionic liquids · Nanoparticles · Catalysis · Hydrogenation · Hydroformylation · CO_2 hydrogenation

4.1 Introduction

Room temperature ionic liquids (RTILs) are molten salts that consist entirely of ions. They have noteworthy properties, such as negligible vapour pressure, high chemical and thermal stability, a wide electrochemical window, non-flammability, melting points below 100 °C, high ionic conductivity, acceptable biocompatibility, good capability of dissolving various organic/inorganic materials, high boiling points and stability [1–5]. The very first molten salt, ethylammonium nitrate, was prepared in 1914 with a low melting point of 12 °C. Alkylimidazolium aluminate IL (EMIm.AlCl$_4$), which was prepared by mixing 1-ethyl-2-methylimidazolium chloride with aluminium chloride, has a melting point below −88 °C [6]. In 1990, organochloroaluminate IL (BMIm.AlCl$_4$) was used as a solvent for the dimerisation of propene to hexene isomers at −15 °C catalysed by nickel complexes. But,

M. I. Qadir · N. M. Simon · D. Rivillo · J. Dupont (✉)
Institute of Chemistry, Universidade Federal do Rio Grande do Sul, UFRGS, Av. Bento Gonçalves, 9500, Porto Alegre, RS 91501-970, Brazil
e-mail: jairton.dupont@ufrgs.br

© Springer Nature Switzerland AG 2020
P. W. N. M. van Leeuwen and C. Claver (eds.), *Recent Advances in Nanoparticle Catalysis*, Molecular Catalysis 1, https://doi.org/10.1007/978-3-030-45823-2_4

this IL was not inert to many organic solvents, very sensitive to moisture and could easily be degraded to generate other side products including HCl, Al(OH)$_3$ and Al$_2$O$_3$. To circumvent this problem, our group reported the very first air-stable imidazolium-based ILs containing tetrafluoroborate (BF$_4^-$) and hexafluorophosphate (PF$_6^-$) anions [7, 8]. With the preparation of these air-stable and easily handled green solvents, interest in them showed an exponential growth.

Besides imidazolium ILs, phosphonium, pyrrolidinium, pyridinium and amine-based ILs have been prepared and studied. Unlike water and other traditional solvents, imidazolium-based ILs display high self-organisation on the nanomolecular scale. 1,3-Dialkylimidazolium ILs, which form one of the most investigated classes, are better described as hydrogen-bonded polymeric supramolecules of the type $\{[(DAI)_x(X)_{x-n})]^{n+}[(DAI)_{x-n}(X)_x)]^{n-}\}_n$ where DAI is the 1,3-dialkylimidazolium cation and X the anion [12]. This structural pattern is a general trend for the solid phase and is maintained to a great extent in the liquid phase, and even in the gas phase, while upon mixing with other molecules they should be better regarded as nano-structured materials with polar and nonpolar regions rather than homogeneous solvents.

ILs are composed of ions (cations and anions) and are generally asymmetric and flexible with delocalised electrostatic charges. These phenomena result in more complex Coulombic interactions (mutual electrostatic attraction or repulsion of charged particles), van der Waals interactions, polarisation and π–π, dipole–dipole, hydrogen-bonding, and solvophobic interactions. ILs in the bulk state often exhibit richer well-defined nanostructures with segregated regions of polar and nonpolar domains in the bulk phase [9]; for example, long alkyl side chain 1,3-dialkylimidazolium ILs, in which the alkyl chains can segregate to form nonpolar domains, while other parts of the IL form polar domains (Fig. 4.1). As the length of the alkyl chain increases, the nonpolar domains increase in size and become more connected, leading to a microphase separation, such as in liquid crystal formation. Anisotropic-like liquids are observed in this case.

4.2   Stabilisation and Characterisation of NPs in ILs

Highly dispersed soluble NPs are desirable for catalysis and solution-processible optoelectronics. But, NPs are kinetically unstable with respect to agglomeration to the bulk metal. Therefore, they require stabilisation, which can be achieved by either surface-ligating anions or other ligands. In this regard, different types of stabilising agents, such as water-soluble polymers, quaternary ammonium salts, surfactants and polyoxoanions, have been applied to provide electronic and steric protection [13]. These candidates effectively decrease the efficiency of the soluble NPs through their strong binding at the NP surface. In this way, ILs, especially imidazolium-based ILs, have emerged as versatile ionic media that stabilise the formed NPs through Coulombic forces, van der Waals interactions, polarisation and π–π, dipole–dipole, hydrogen-bonding and solvophobic interactions. It is believed that the ILs stabilise

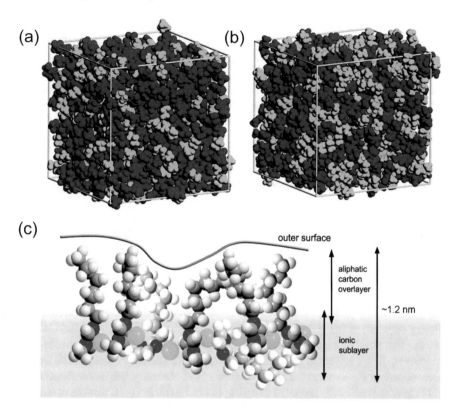

Fig. 4.1 Spatial heterogeneity observed in the MD simulation presenting snapshots of **a** [BMIm][PF$_6$] CPK colouring; **b** [C$_6$MIm][PF$_6$] in the same configuration as in part (**a**) with red/green (charged/nonpolar) colouring [9, 10]; **c** Proposed model of the highly ordered IL-vacuum interface for C$_8$MIm.Cl. The first molecular layer consists of octyl chains protruding mainly to the vacuum (called the 'aliphatic carbon overlayer'), and underneath are the ionic parts of the cation and the chloride anions (called the 'ionic sublayer'). For the larger anions, the first molecular layer is considerably less ordered. The indicated extension of 1.2 nm of the first molecular layer of C$_8$MIm.Cl corresponds roughly to the information depth in ARXPS (angle resolved X-ray photo-electron spectroscopy) setup for an electron emission angle of 80° relative to the normal surface [11]. Reproduced by permission of the American Chemical Society

the NPs by formation of an electric double layer (the Deryagin–Landau–Verwey–Overbeek model, DLVO) in which a first solvation shell of anions surrounds the metal cluster, followed by a less ordered layer of cations, and so on [14]. However, other studies have demonstrated evidence of close interactions of the nanoparticles with the cations by deuterium exchange on positively charged imidazolium rings, and by surface-enhanced Raman spectroscopy on gold nanoparticles in imidazolium liquids [15, 16]. Moreover, ILs also act as 'templates' to prepare the desired NPs. The size and size distribution of NPs synthesised in ILs are affected by the physicochemical properties of the ILs, which affect NP stabilisation.

The metallic nanoparticles in ILs are solvated preferentially by the charged moieties of the ions, with an interface layer that is one ion thick. Therefore, both cations and anions are present in contact with the metal. Pensando et al. used density functional theory (DFT) methods to study the mechanism of solvation and stabilisation of Ru NPs through the interactions between 1,3-dimethylimidazolium bis(trifluoromethanesulfonyl)amide (MMIm.NTf$_2$), and the Ru (001) surface of a cluster of metal atoms [14]. It was observed that the alkyl side chain of the imidazolium cation is found further away from the NPs, while the orientation of the cations with respect to the NP surface shows that the imidazolium ring most probably lies perpendicular to the NP surface, with the CH$_3$ group attached to N3 closer to the metal. This is a result of the stronger attractive interaction of the charged moieties with the metal surface.

The presence of interaction between the ILs and NPs has been extensively studied by TEM, SAX, XRD, XPS and EXAFS analyses. Dupont et al. reported that XAS analysis demonstrated the formation of an IL protective layer surrounding Pt and Ir NP surfaces with an extended molecular length of around 2.8–4.0 nm depending on the type of anion present [17, 18]. This suggests the presence of semi-organised anionic species composed of supramolecular aggregates of the type $[(BMI)_{x-n}(X)_x)]^{n-}$. This multilayer is probably composed of anions located immediately adjacent to the NP surface providing the Coulombic repulsion and counter-cations that provide the charge balance, i.e., quite close to DLVO-type stabilisation. Moreover, XPS analysis of the isolated Pt, Ir, Co, Ru and Rh NPs also indicates the presence of an IL layer on the surface of NPs [19]. Recently, the association of F1s of NTf$_2$ with Ru/Fe NPs prepared in hydrophobic BMIm.NTf$_2$ IL was observed. The F1s signal showed a peak associated with uncoordinated NTf$_2$ (688.6 eV) and a new component appeared at 684.8 eV [20], which was attributed to the IL interaction with Ru/Fe NPs, as is observed in Au NPs [21].

4.3 Preparation of Soluble NPs by Chemical Methods

To prepare 'soluble' transition-metal NPs in ILs, simple reduction/hydrogenation of organometallic complexes and metal salts in ILs by a chemical method is the most investigated and used method (Table 4.1). The most frequent reducing agents are H$_2$ gas, NaBH$_4$, ascorbic acid, sodium citrate and SnCl$_2$ [22]. However, hydride sources are not likely to be used in ILs, due their basic character, deprotonating the imidazolium cation and generating carbenes that may bind to the metal surface. On the other hand, these reducing agents also produce various by-products such as sodium (Na) and boron (B) compounds, which are difficult to remove from the IL and stick to the surface of the formed NPs [22]. Hydrogen gas can easily reduce the metal complexes and salts into their respective 'surface-clean' NPs in ILs under elevated reaction conditions. In this regard, Dupont et al. were the first to prepare Ir NPs (2.0 ± 0.4 nm) in BMIm.PF$_6$ ILs by the simple reduction of [Ir(cod)Cl]$_2$ using

Table 4.1 Hydrogen-driven chemical preparation of the ligand-free 'soluble' NPs prepared in different types of IL

Entry	IL	Metallic precursor	Reducing agent	NPs	Size (nm)	References
1	BMIm.PF$_6$	[Ir(cod)Cl]$_2$	H$_2$	Ir	2.0 ± 0.4	[13]
2	BMIm.BF$_4$	[Ir(cod)Cl]$_2$	H$_2$	Ir	2.9	[18]
3	BMIm.CF$_3$SO$_4$	[Ir(cod)Cl]$_2$	H$_2$	Ir	2.6	[18]
4	BMIm.BF$_4$	[Ir(cod)$_2$]BF$_4$	H$_2$	Ir	1.9 ± 0.4	[24]
5	C$_1$C$_{10}$Im.BF$_4$	[Ir(cod)$_2$]BF$_4$	H$_2$	Ir	1.9 ± 0.4	[24]
6	BMIm.BF$_4$	[Ir(cod)Cl]$_2$	H$_2$	Ir	2.5 ± 0.5	[24]
7	C$_2$C$_{10}$Im.BF$_4$	[Ir(cod)Cl]$_2$	H$_2$	Ir	3.6 ± 0.9	[24]
8	BMIm.BF$_4$	AgBF$_4$	H$_2$	Ag	2.8 ± 0.8	[25]
9	BMIm.PF$_6$	AgBF$_4$	H$_2$	Ag	4.4 ± 1.3	[25]
10	BMIm.OTf	AgBF$_4$	H$_2$	Ag	8.7 ± 3.4	[25]
11	BMIm.NTf$_2$	AgBF$_4$	H$_2$	Ag	26.1 ± 6.4	[25]
12	BMIm.PF$_6$	Ru(cod)(cot)	H$_2$	Ru	2.6 ± 0.4	[29]
13	BMIm.PF$_6$	RuO$_2$	H$_2$	Ru	2.5 ± 0.4	[30]
14	BMIm.SO$_3$CF$_3$	RuO$_2$	H$_2$	Ru	2.5 ± 0.4	[30]
15	BMIm.BF$_4$	RuO$_2$	H$_2$	Ru	2.0 ± 0.2	[30]
16	BMIm.NTf$_2$	Ru(cod)(cot)	H$_2$	Ru	2.4 ± 0.3	[27]
17	BMIm.NTf$_2$	Ru(cod)(cot)	H$_2$	Ru	2.1 ± 0.5	[31]
18	BMIm.BF$_4$	Ru(cod)(cot)	H$_2$	Ru	2.9 ± 0.5	[31]
19	DMIm.NTf$_2$	Ru(cod)(cot)	H$_2$	Ru	2.1 ± 0.5	[31]
20	(BCN)MIm.NTf$_2$	Ru(cod)(methylallyl)$_2$	H$_2$	Ru	2.2 ± 0.5	[32]
21	BMIm.PF$_6$	RhCl$_3$.H$_2$O	H$_2$	Rh	2.3 ± 0.6	[28]

(continued)

Table 4.1 (continued)

Entry	IL	Metallic precursor	Reducing agent	NPs	Size (nm)	References
21	IL_{PEG750}	$RhCl_3.H_2O$	H_2	Rh	1.3 ± 0.2	[33]
22	$BMIm.PF_6$	$Pt_2(dba)_3$	H_2	Pt	2.0–2.5	[17]
23	$BMIm.PF_6$	PtO_2	H_2	Pt	2.3 ± 0.3	[34]
24	$BMIm.BF_4$	$Pd(acac)_2$	H_2	Pd	4.9 ± 0.8	[35]
25	$BMIm.PF_6$	$Pd(cod)Cl_2$	H_2	Pd	6–8	[36]
26	$BMIm.NTf_2$	$Ni(cod)_2$	H_2	Ni	5.9 ± 1.4	[37]
27	$BMIm.BF_4$	$Ni(cod)_2$	H_2	Ni	5.1 ± 1.2	[37]
28	$C_1C_8Im.NTf_2$	$Ni(cod)_2$	H_2	Ni	5.6 ± 1.3	[38]
29	$C_1C_{10}Im.NTf_2$	$Ni(cod)_2$	H_2	Ni	4.9 ± 0.9	[38]
30	$C_1C_{14}Im.NTf_2$	$Ni(cod)_2$	H_2	Ni	5.1 ± 0.9	[38]
31	$C_1C_{16}Im.NTf_2$	$Ni(cod)_2$	H_2	Ni	5.5 ± 1.1	[38]
32	$BMIm.NTf_2$	$Fe(cod)_2$	H_2	FeO	5.3 ± 1.6	[39]
33	$BMIm.PF_6$	$Co(Cp)_2/Pt_2(dba)_3$	H_2	Co/Pt	4.4 ± 1.9	[40]
34	$BMIm.PF_6$	$Ru(cod)(methylallyl)_2/Pt_2(dba)_3$	H_2	Ru/Pt	2.9 ± 0.2	[41]
35	$BMIm.NTf_2$	$Fe(CO)_5/Pt_2(dba)_3$	H_2	Fe/Pt	1.7 ± 0.2	[42]
36	$BMIm.PF_6$	$Fe(CO)_5/Pt_2(dba)_3$	H_2	Fe/Pt	1.8 ± 0.23	[42]
37	$BMIm.BF_4$	$Fe(CO)_5/Pt_2(dba)_3$	H_2	Fe/Pt	2.5 ± 0.4	[42]
38	$BMIm.NTf_2$	$Ru(cod)(methylallyl)_2/Fe(CO)_5$	H_2	Ru/Fe	1.7 ± 0.3	[20]
39	$BMIm.NTf_2$	$Ru(cod)(methylallyl)_2/Ni(cod)_2$	H_2	Ru/Ni	2.4 ± 0.7	[43]

DMIm = 1-n-decyl-3-methylimidazolium; IL_{PEG750} = N,N-dimethyl-N-(2-(2-methoxyethoxy)ethyl)-N-(2-(2-octyloxyethoxy)ethyl)ammonium methanesulfonate; dba = dibenzylidene acetone; acac = acetylacetonate; Cp = bis(cylopentadienyl)

$R_1-N\overset{+}{\underset{R_2}{\diagup\diagdown}}N-CH_3$ X^-

$R_4\overset{R_1}{\underset{\underset{R_3}{N+}}{\diagup\diagdown}}R_2$ X^-

R_1= Et; R_2= H; X= Br; C_2MIm.Br

R_1= "Prop; R_2= CH_3; X= NTf_2; C_3CNMMIm.NTf_2

R_1= "Prop; R_2= H; X= NTf_2; C_3CNMIm.NTf_2

R_1= "Bu; R_2= CH_3; X= TPPM[a]; BMMIm.TPPM

R_1= "Bu; R_2= CH_3; X= TPPT[a]; (BMMIm)$_3$.TPPT

R_1= "Bu; R_2= H; X= BF_4; BMIm.BF_4

R_1= "Bu; R_2= CH_3; X= BF_4; BMMIm.BF_4

R_1= "Bu; R_2= H; X= PF_6; BMIm.PF_6

R_1= "Bu; R_2= CH_3; X= PF_6; BMMIm.PF_6

R_1= "Bu; R_2= H; X= Br; BMIm.Br

R_1= $CH_2OCH_2CH_2CH_3$; R_2= CH_3; X= Cl; MPMIm.Cl

R_1= "Oc; R_2= H; X= $CF_3CF_2CF_2CF_2SO_3$; C_8MIm.$CF_3(CF_2)_3SO_3$

R_1= $CH_3(CH_2)_7(OCH_2CH_2)_2$; R_2= $CH_3(OCH_2CH_2)_2$; R_3= CH_3;

R_4= CH_3; X= CH_3SO_3; NMPEG.CH_3SO_3

R_1= $CH_3(OCH_2CH_2)_{16}$; R_2= CH_3CH_2; R_3= CH_3CH_2;

R_4= CH_3CH_2; X= CH_3SO_3; NEtPEG.CH_3SO_3

R_1= R_2= R_3= R_4= "Bu; X= Br; Bu_4N.Br

R_1= R_2= R_3= R_4= "Bu; X= OAc; Bu_4N.OAc

Fig. 4.2 Structures and abbreviations of ILs discussed in this chapter

molecular H_2 (Table 4.1, entry 1) [13]. After this first report, numerous other metal NPs of different sizes and shapes were similarly prepared in different types of ILs (Fig. 4.2).

The same group also studied the effect of ILs on the size and interaction of Ir NPs in BMIm.BF_4, BMIm.PF_6 and BMIm.CF_3SO_4 ILs. Smaller-sized NPs (2.0 nm) were obtained in a hydrophobic IL (BMIm.PF_6), while relatively larger NPs (2.9 and 2.6 nm) were observed in hydrophilic BMIm.BF_4 and BMIm.CF_3SO_4 ILs, respectively [18]. In situ XAS analysis of Ir NPs in ILs revealed the formation of IL protective layer surrounding the NP surface with an extended molecular length of around 2.8–4.0 nm depending on the type of anion, suggesting the presence of semi-organised anionic species composed of supramolecular aggregates of the type $[(BMI)_{x-n}(X)_x)]^{n-}$, as usually observed in the structural organisation of imidazolium ions in the solid, liquid and gas phase and in solution (Fig. 4.3). This multilayer is probably composed of anions located immediately adjacent to the nanoparticle surface, providing the Coulombic repulsion and counter-cations that provide the charge balance, i.e., quite close to DLVO-type stabilisation. Similar interactions were also observed between the IL and Pt (0) NPs prepared by the decomposition of the [Pt$_2$(dba)$_3$] organometallic complex in the presence of H_2 in the BMIm.PF_6 IL (Table 4.1, entry 22) [17].

It was observed that the stability of the Ir (0) NPs in imidazolium-based ILs is due to the surface-attached carbenes that can be formed due to the reaction of Ir nano-clusters with the imidazolium moiety. This was confirmed by the incorporation kinetics of deuterium into the 2-H position of the imidazolium cation [15, 23].

The morphology and size of the iridium NPs prepared in imidazolium ILs depend on the nature (ionic or neutral) of the metallic precursor used and on the environment of the ILs in which they are dissolved. Ir (cod)BF_4 tends to dissolve in polar ILs, in contrast to Ir (cod)Cl. Spherical-shaped Ir NPs are obtained in BMIm.BF4 IL, while worm-like NPs of 1.9 ± 0.4 nm size are found in C_1C_{10}Im.BF_4 IL (Table 4.1, entries

(a)

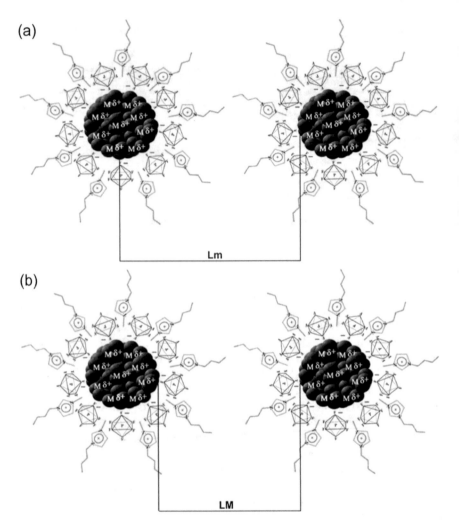

(b)

Fig. 4.3 Schematic illustration of the proposed two-phase model, when **a** is defined as first minimum of the correlation function and **b** as the first maximum of the interface distribution function: **a** Lm is the most probable distance between the centre of gravity of an Ir (0) nanoparticle and its adjacent region in the correlation function, and **b** can be estimated as the disordered ionic liquid-phase thickness in the interface distribution function [18]

4–5) [24]. The worm-like Ir NPs are polycrystalline, resulting from the aggregation of individual 'spherical' nanoparticles of around 1.9 nm. The volume of the polar domains of the imidazolium ILs may be controlled by changing the size of the anions and therefore the size of metal NPs prepared from ionic metal precursors. For example, Ag NPs obtained by H_2 reduction of silver salts (AgX, X = BF_4, PF_6, OTf) dissolved in 1-*n*-butyl-3-methylimidazolium-based ILs (BMIm.X; X = BF_4, PF_6, OTf and NTf_2) and in the presence of *n*-butylimidazole have a size distribution

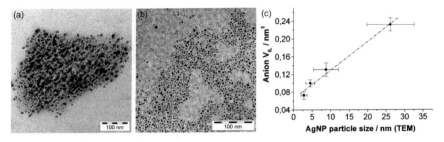

Fig. 4.4 Silver NPs produced from **a** AgOTf in BMIm.OTf without scavenger; **b** AgBF₄ in BMIm.BF₄ using BMIm scavenger; and **c** correlation between the observed Ag NP size (from TEM) and the molecular volume of IL anion (VIL anion) [25]. Reproduced by permission of the American Chemical Society

in the diameter range 2.8–26.1 nm (Table 4.1, entries 8–11) [25]. It was observed that the diameter of the Ag NPs increases linearly with the molecular volume of the IL anion or, more likely, of the anionic aggregate (Fig. 4.4).

In contrast, the neutral character of a Ru organometallic precursor ([Ru(cod)(cot)], cod = 1,5-cyclooctadiene and cot = 1,3,5-cyclooctatriene) was found as a key factor in the size and shape of prepared NPs, since these neutral compounds may concentrate in nonpolar regions of the ILs [26]. In such cases, the nanoparticle growth process is probably controlled by the local concentration of the precursor and is consequently limited to the size and shape of IL nonpolar domains that are imposed by the length of the N-alkyl side chain. Indeed, a linear relationship between the NP size and the length of the N-alkyl chain in ILs was observed. Moreover, the temperature also has a profound effect on the size of Ru NPs prepared in BMIm.NTf₂ using a Ru complex. The size of NPs increases with increasing temperature from 0 to 75 °C. Smaller NPs 0.9 ± 0.4 nm were obtained at 0 °C, whereas sizes of 2.4 ± 0.3 and 2.6 ± 0.4 nm were obtained at 25 and 75 °C, respectively [27].

Rh NPs can also be prepared by the reduction of RhCl₃ in ILs [28]. Monodispersed Rh NPs (2.3 ± 0.6 nm) were prepared in BMIm.PF₆ using RhCl₃.H₂O, but care should be taken to avoid the presence of small amount of water in IL. The presence of water causes the partial decomposition of the IL (BMIm.PF₆) with the formation of phosphates (identified by ³¹P NMR and IR), and the evolution of HF and rhodium fluorides isolated together with the metal NPs. It should be noted that the ionic liquid decomposition occurs only in the presence of both water and the transition-metal precursor, i.e., RhCl₃, which indicates that the transition metal is involved in the hydrolysis of the PF₆ anion.

The development of a reliable method for the generation of soluble and stable Pd NPs is related to the solubility of the Pd precursor in ILs. PdCl₂ has low solubility, usually leading to a very broad size and shape distribution, since the precursors are distributed heterogeneously in the media. On the other hand, efficient synthesis of Pd NPs (4.9 ± 0.8 nm) by the reduction of Pd(acac)₂ by molecular H₂ using non-functionalised ILs as unique stabilisers was demonstrated earlier (Table 4.1, entry 24) [35]. However, these nanoparticles are not stable, and the use of different ligands (phenanthroline) and functionalised ILs, such as nitrile-functionalised or 2,2'-dipyridyl-amine-functionalised imidazolium salts, have been used for the stabilisation and dispersion of Pd NPs in ILs [44–47].

Dupont et al. reported that the colloidal suspensions of cubic fcc Ni(cod)$_2$ NPs of 5.0–6.0 nm were achieved in BMIm.NTf$_2$ and BMIm.BF$_4$ ILs using H$_2$ at 75 °C (Table 4.1, entries 26–27) [37]. These ILs acted as stabilising agents for the NPs. X-ray diffraction showed clearly Ni(0) NPs embedded in these ILs, avoiding their agglomeration, whereas X-ray absorption spectroscopy evidenced that Ni(0) NPs were surrounded by a cap layer due to steric and electrostatic interactions with the ILs. They also observed that there were slight decreases in both the NP diameter and the size distribution with an increase in the carbon numbers of the alkyl side chain in the imidazolium cation up to 14 carbons [38].

Alloys and core–shell Co/Pt [40], Ru/Pt [41], Ru/Fe [20], Fe/Pt [42] and Ru/Ni [43] bimetallic NPs are also easily accessible via simple decomposition of their organometallic complexes in ILs using H$_2$. The advantage of the use of these organometallic complexes is that they possess only hydrocarbon-containing ligands which, by reduction/decomposition, generate organic by-products with poor coordinating properties to the surface of the NPs and that can easily be removed from the reaction mixture under vacuum. Consequently, ILs may provide adequate templates for the generation of these bimetallic NPs. For example, Fe/Pt core–shell NPs prepared in BMIm.NTf$_2$ have a diameter of 1.7 ± 0.2 nm, whereas the diameters of those prepared in BMIm.PF$_6$ and BMIm.BF$_4$ IL are 1.8 ± 0.3 nm and 2.5 ± 0.4 nm, respectively (Table 4.1, entries 35–37). Moreover, the thickness of the Pt shell layer has a direct correlation with the water stability of the anion and increases in the order PF$_6$ > BF$_4$ > NTf$_2$, yielding the metal compositions Pt$_4$Fe$_1$, Pt$_3$Fe$_2$ and Pt$_1$Fe$_1$, respectively [42].

4.4 Preparation of Naked NPs by Physical Methods

The fabrication of well-defined 'surface-clean' metal NPs still remains a challenge for the chemical and engineering community [48]. In this respect, magnetron sputtering deposition has emerged as a compromise physical technique that is a simple, clean and provides an easy approach to prepare such nano-devices that may demonstrate quite unique properties for several applications in fields such as electronics, sensors, biomedicine and catalysis [49]. Indeed, this versatile method allows the fabrication of 'naked' NPs in a single step, generating highly pure metal-supported NPs using both solid and liquid supports, as opposed to the chemical and electrochemical methods that usually need further purification steps [50–54]. In particular, the fabrication of metal nanoparticles in ILs or hybrid IL materials is a simple and controllable process for several applications, with huge advantages compared to the classical wet methods [21, 55–57]. In this technique, bombardment under vacuum of the metal target (ultrapure, >99.99%) with energetic gaseous argon ions causes the physical ejection of surface atoms and/or metal clusters. The generated sputtered metal species

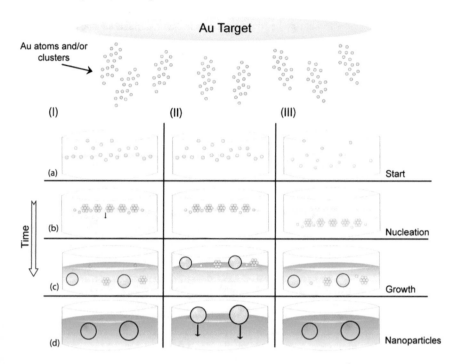

Fig. 4.5 Possible mechanisms for the nucleation and growth of sputtered gold nanoparticles into IL substrates [58]. Reproduced by permission of the American Chemical Society

inject into the ILs without remarkable gas-phase collisions in the space between the target and the IL solution. This results in their coalescence to form a dispersion of larger metal NPs with diameters of several nanometres (Fig. 4.5) [58].

This approach flourished in the pioneering works of Kuwabata and Torimoto [59], who used ILs to decorate the Au and Ag NPs [60, 61]. Small-sized Au NPs with diameters of 5.5 ± 0.86 nm (Table 4.2, entry 1) were obtained by applying the non-functionalised IL 1-ethyl-3-methylimidazolium tetrafluoroborate (EMIm.BF$_4$). Dupont et al. studied extensively the effect of the presence of functionalised moieties (–CN, –SH, –CO$_2$H) in imidazolium-based ILs on the size, shape and location of Au and Ag NPs (sputtering time 150 s and 40 mA current). They observed that small-sized NPs were obtained when hydrophobic 1-butyronitrile-methylimidazolium bisimide [(BCN)MIm.NTf$_2$] IL was used (Table 4.2, entries 3–4,7), while relatively larger NPs were observed in 1-methoxyethyl-methylimidazolium bis(trifluoromethanesulfonyl)imide [(MeOE)MIm.NTf$_2$] and 1-thioethyl-methylimidazolium bis(trifluoromethanesulfonyl)imide [(HSE)MIm.NTf$_2$] ILs. They proposed that the chemical configuration of the IL surface controls the size and shape of the NPs. The small size of NPs in (BCN)MIm.NTf$_2$ IL may be due to the presence of a –CN group, which possesses two possible effects: (a) the dipole moment of the nitrile group and (b) the orientation of the nitrile-functionalised

Table 4.2 Naked mono- and bimetallic NPs prepared by magnetron sputtering into different types of IL

Entry	IL	NPs	Size (nm)	References
1	EMIm.BF$_4$	Au	5.5 ± 0.86	[59]
2	BMIm.PF$_6$	Au	2.6 ± 0.3	[66]
3	(BCN)MIm.NTf$_2$	Au	4–8	[67]
4	(BCN)MIm.NTf$_2$	Au	5–7.2	[67]
5	(HSE)MIm.NTf$_2$	Au	8–22.9	[67]
6	(MeOE)MIm.NTf$_2$	Au	4–6.8	[67]
7	(BCN)MIm.NTf$_2$	Ag	5–14	[68]
8	(HSE)MIm.NTf$_2$	Ag	8.7–12	[68]
9	(MeOE)MIm.NTf$_2$	Ag	8.2–15.5	[68]
10	BMIm.PF$_6$	Ag	2–3	[68]
11	(BCN)MIm.NTf$_2$	Pd	1.4–2.9	[69]
12	BMIm.PF$_6$	Pd	2–4.5	[70]
13	Me$_3$PrN.NTf$_2$	Pt	2–3	[71]
14	BMIm.PF$_6$	Au/Ag	2.6 ± 0.33	[60]
15	TMPA.TFSA	Au/Pt	1.5 ± 0.4	[72]
16	EMIm.BF$_4$	Au/Cu	2.6–3.4	[73]
17	HyEMIm.BF$_4$	Pd/Au	4.4–8.6	[74]

(TMPA.TFSA) = 1-(2-hydroxyethyl)-3-methylimidazolium tetrafluoroborate (HyEMI-BF$_4$)

side chain towards the vacuum/IL interface [62]. It is very probable that in (BCN)MIm.NTf$_2$ IL, the surface is enriched with –CN groups (pointed towards the vacuum/IL interface), which drive the anisotropic nucleation and growth of the NPs [62], as opposed to the less anisotropic [(MeOE)MIm.NTf$_2$] IL, in which the functional groups are located preferentially away from the liquid surface. Moreover, they also observed by High-Sensitivity Low-Energy Ion Scattering (HS-LEIS) analysis that the sputtered NPs tend to migrate deeper into the regions of the ILs containing lower coordinating groups (–OMe), whereas the small nanoparticles are present close to the fluid surface, having a single monolayer of IL surrounded by ILs associated with strong sigma-donating groups (–CN and –SH). A similar study was also carried out for Pd NPs [63].

It is well known that the discharge time does not affect the shape and size distribution of the NPs in ILs. Only the concentration of NPs increases with the increase in sputtering time. On the other hand, the discharge current can affect the size distribution. The size of Au, Ag and Pd NPs increases with an increase in discharge current [60, 64]. This tendency may be correlated with an increase in the deposition rate of sputtered atoms. When higher deposition rates are used, more metal atoms/clusters hit the surface of the IL per unit of time, changing the kinetics of particle growth on the surface of the IL. As the molecular arrangement of the surface of the IL gradually changes towards the bulk conformation, the environment in which the NPs grow will change dramatically depending upon the depth that the sputtered atoms penetrate into the IL surface. Therefore, a threshold energy must exist for sputtering atoms to start growing into NPs, as the deposited atoms are not isoenergetic, but their energies follow a Boltzmann distribution [65]. Hence, increasing the discharge current of the sputtering process will increase the average translational energy of the sputtered atoms as well the fraction of metal atoms.

The relation between the size of Au NPs with the anion and cation of imidazolium-based ILs at different temperatures (20–80 °C) was studied by Nishikawa et al. [75]. The 1-*n*-butyl-3-methylimidazolium cation with different anions (TFSA, FAS, OTf, PF_6, BF_4) was studied. It was observed that the sizes of Au NPs may be determined by the competition between the collision frequency of the ejected Au atoms and the stabilising capability of the anions that form the first coordination shell around the NPs. The Au NP sizes are closely related to the anion volume. As the anion volume increased in the order TFSA > FAS > OTf > PF_6 > BF_4, the diameter of the NPs also increased. Furthermore, bimetallic alloy NPs have also been prepared by sequential and simultaneous sputter deposition onto ILs (Table 4.2, entries 14–17) [60, 72–74].

4.5 Metal Nanoparticle Catalysis in ILs

It is believed that the IL creates a layer around the NPs that acts as a catalytic membrane-like device. Hence, the geometric and electronic properties of metal NPs supported in ILs can be tuned by the proper choice of the IL cations and anions, along with NPs that have a strong influence on the residence time/diffusion of the reactants, intermediates and products in the nano-environment (Scheme 4.1) [20, 21, 57]. In the remainder of this chapter, a short preview of the catalytic application of the non-supported NPs in ILs in hydrogenation, hydroformylation, Fischer-Tropsch and carbon–carbon coupling reactions will be given.

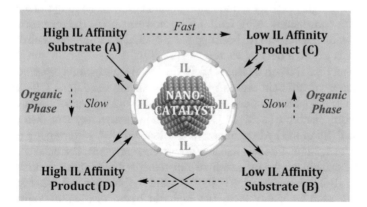

Scheme 4.1 Metal-supported NPs in IL membrane-like device [57]. Reproduced by permission of the American Chemical Society

The 'soluble' NPs in ILs usually behave as homogeneous-like catalysts for the hydrogenation of alkenes, alkynes and arenes. Among these hydrocarbons, the semi-hydrogenation of benzene in imidazolium-based ILs has been studied extensively. Thermodynamic and kinetic barriers strongly control the semi-hydrogenation of benzene. Thermodynamically, the semi-hydrogenation of benzene to cyclohexene is not favourable, since cyclohexane is at least 75 kJ mol^{-1} more stable than cyclohexene [76]. The kinetics of the liquid-phase hydrogenation of benzene is still under debate, as the mechanism concerns the chemical state of the catalytically active hydrogen and the role of the reaction intermediates. One of the serious problems with benzene is its solubility in classical solvents. In this regard, 1-n-butyl-3-methylimidazolium hexafluorophosphate (BMIm.PF$_6$) IL has been used as a biphasic media in which benzene is highly soluble while its hydrogenated products (cyclohexene and cyclohexane) are insoluble. Dupont et al. were able to hydrogenate benzene by using Ru NPs (2.6 ± 0.4 nm) in BMIm.PF$_6$ and observed a high selectivity of <39% to cyclohexene at very low benzene (1%) conversion. The conversion of benzene reached 73% with a 100% selectivity to cyclohexane (Table 4.3, entry 1) [29]. It is assumed that BMIm.PF$_6$ generates an IL-cage around the Ru NPs that not only stabilise the NPs, but could also repel the cyclohexene formed and so decrease the re-adsorption and further hydrogenation to cyclohexane [29]. Pt NPs were also applied in the same reaction conditions (75 °C, 4 bar H$_2$) using BMIm.PF$_6$ IL, which yielded a 46% conversion of benzene with 100% conversion to cyclohexane [17]. By the incorporation of Ru into Pt NPs, Ru/Pt bimetallic NPs in BMIm.PF$_6$ produced the unexpected 1,3-cyclohexadiene with 21% selectivity at 5% benzene conversion (Table 4.3, entry 1) [41].

Mono-dispersed RuO$_2$ NPs (2–3 nm) in BMIm.PF$_6$ presented a higher activity in the hydrogenation of benzene (97%) to cyclohexane with 43 turnover frequency (Table 4.3, entry 3) [30], whereas only 3% benzene conversion was obtained when BMIm.BF$_4$ IL was used. This catalytic system (RuO$_2$/BMIm.PF$_6$) effectively hydrogenated cyclohexene and 1-methyl-1-cyclohexene with 99 and 97% conversion to cyclohexane and methyl-cyclohexane, respectively. Low-cost Ni NPs (11 nm) demonstrated higher activity in the hydrogenation of benzene to cyclohexane, with 31 and 34% conversions using BMIm.NTf$_2$ and BPy.NTf$_2$ ILs, respectively [77]. Different types of additive have been applied to hydrogenate benzene with the desired selectivity. For further details, see reviews [78, 79].

Ru-immobilised NPs in BMIm.NTf$_2$ IL drove the liquid–liquid biphasic catalytic hydrogenation of different arenes under mild reaction conditions (50–90 °C and 4 bar H$_2$). The apparent activation energy of $E_a = 42.0$ kJ mol^{-1} was estimated for the hydrogenation of toluene [31]. The same catalytic system also presented efficient activity in the hydrogenation of o-xylene with 32% conversion and 100% selectivity to 1,2-dimethylcyclohexane (Table 4.3, entry 13). Dyson et al. revealed that Rh NPs stabilised the imidazolium-functionalised bipyridine compounds 4,4'-bis-[7-heptyl]-2,2'-bipyridine^{2+} ([BIHB]$^{2+}$) and 4,4'-bis(methyl)-2,2'-bipyridine^{2+} ([BIMB]$^{2+}$) used as catalysts in the biphasic hydrogenation of various arene substrates [80]. The catalytic activity was strongly influenced by the stabiliser employed and followed the trend [BIHB]$^{2+}$ > bipy > [BIMB]$^{2+}$ (Table 4.3, entries

Table 4.3 Miscellaneous reactions involving H$_2$ catalysed by NPs in ILs without additional additives

Entry	Reaction	NPs	IL	Substrate	Major product	Conv./sel (%)	References
1	Hydrogenation	Ru	BMIm.PF$_6$	Benzene	Cyclohexane	73/100	[29]
2	Hydrogenation	Ru	BMIm.BF$_4$	Benzene	Cyclohexane	30/100	[29]
3	Hydrogenation	RuO$_2$	BMIm.PF$_6$	Benzene	Cyclohexane	97/100	[30]
4	Hydrogenation	RuO$_2$	BMIm.BF$_4$	Benzene	Cyclohexane	3/100	[30]
5	Hydrogenation	Ru	BMIm.NTf$_2$	Benzene	Cyclohexane	100/100	[31]
6	Hydrogenation	Ru/Pt	BMIm.PF$_6$	Benzene	1,3-Cyclohexadiene	5/21	[41]
7	Hydrogenation	Rh	BMIm.PF$_6$	Benzene	Cyclohexane	100/100	[29]
8	Hydrogenation	Ir	BMIm.PF$_6$	Benzene	Cyclohexane	100/100	[29]
9	Hydrogenation	Ni	BMIm.NTf$_2$	Benzene	Cyclohexane	15/100	[77]
10	Hydrogenation	Ni	BPy.NTf$_2$	Benzene	Cyclohexane	18/100	[77]
11	Hydrogenation	Pt	BMIm.PF$_6$	Benzene	Cyclohexane	46/100	[17]
12	Hydrogenation	Ru	BMIm.NTf$_2$	Toluene	Methylcyclohexane	85/100	[31]
13	Hydrogenation	Ru	BMIm.NTf$_2$	o-Xylene	1,2-Dimethylcyclohexane	32/100	[31]
14	Hydrogenation	Rh	BMIm.NTf$_2$/BIHB.NTf$_2$	Toluene	Methylcyclohexane	100/100	[80]
15	Hydrogenation	Rh	BMIm.NTf$_2$/bipy	Toluene	Methylcyclohexane	68/100	[80]
16	Hydrogenation	Rh	BMIm.NTf$_2$/BIMB.$_2$	Toluene	Methylcyclohexane	18/100	[80]
17	Hydrogenation	Rh	BMI.BF$_4$/poly[NVP-co-VBIM.Cl]	Ethylbenzene	Ethylcyclohexane	100/>99	[81]
18	Hydrogenation	Pd	BMMIm.PF$_6$	Styrene	Ethylcyclohexane	100/100	[46]
19	Hydrogenation	Ru	BMMIm.TPPM/BMMIm.BF$_4^b$	Quinoline	1,2,3,4-Tetrahydroquinoline	95/98	[82]
20	Hydrogenation	Ni	C$_3$CNMMIm.NTf$_2$	Diphenylacetylene	(Z)-Stilbene	100/90	[83]
21	Hydrogenation	Ni	C$_3$CNMMIm.NTf$_2$	4-Octyne	(Z)-Oct-4-ene	100/82	[83]
22	Hydrogenation	Ni	C$_3$CNMMIm.NTf$_2$	Ethynyl-benzene	Styrene	100/79	[83]
23	Hydrogenation	Pd	C$_3$CNMMIm.NTf$_2$	Diphenylacetylene	(Z)-Stilbene	87/98	[84]
24	Hydrogenation	Pd	C$_3$CNMMIm.NTf$_2$	Ethynyl-benzene	Styrene	97/95	[84]

(continued)

Table 4.3 (continued)

Entry	Reaction	NPs	IL	Substrate	Major product	Conv./sel (%)	References
25	Hydrogenation	Pd	(BMMIm)$_3$.TPPT/BMMIm.PF$_6^b$	Nitrobenzene	Aniline	100/100	[45]
26	Hydrogenation	Ru	(BMMIm)$_3$.TPPT/BMMIm.PF$_6^b$	o-Nitroaceto-phenone	o-Aminoaceto-phenone	100/100	[45]
27	Hydroformylation	Rh	NEtPEG.CH$_3$SO$_3$	1-Octene	Aldehyde(nonanal 2-methyloctanal)	99/91	[85]
28	Hydroformylation	Rh	NEtPEG.CH$_3$SO$_3$	1-Octene	Aldehyde(nonanal 2-methyloctanal)	100/91	[33]
29	Fischer-Tropsch	Co	BMIm.NTf$_2$	CO/H$_2$	HCs (C$_7$–C$_{30}$)	34/–	[86]
30	Fischer-Tropsch	Ru/Fe	BMIm.NTf$_2$	CO$_2$/H$_2$	HCs (C$_1$–C$_{21}$)	12/–	[20]
31	Fischer-Tropsch	Ru/Ni	BMIm.NTf$_2$	CO$_2$/H$_2$	HCs (C$_1$–C$_6$)	30/–	[43]

14–16). NPs stabilised by BIHB.NTf$_2$, with a C$_7$ alkyl chain separating the imidazolium functionality from the pyridine backbone, were considerably more active than the bipy-stabilised system (with increases in conversion of as much as 68% being observed), whereas NPs protected by the BIHB.NTf$_2$ stabiliser with one CH$_2$ group between the imidazolium and the pyridine resulted in the lowest activity.

A reduction in the aromatic character of arenes is an important field in industrial hydrogenation processes. The chemoselective hydrogenation of aromatic compounds (aromatic ketones, aromatic aldehydes, quinolines, etc.) by Ru NPs in BMIm.BF$_4$ IL improved the catalytic performance: acetophenone could be hydrogenated to 1-phenylethanol with 77% conversion and 99% chemoselectivity. It was essential to use 1-butyl-2,3-dimethylimidazolium hydroxide (BMMIm.OH) IL as a base to improve the catalytic performance. Konnerth and Prechtl developed a low-cost system using Ru NPs stabilised by EGMMIm.NTf$_2$ as catalyst. The best performance was achieved in the hydrogenation of quinoline to 1,2,3,4-tetrahydroquinoline (THQ) with up to 99% selectivity [82].

Metal NPs prepared in nitrile-functionalised ILs have been tested as catalysts in the hydrogenation of alkynes, and they showed excellent potential for the production of alkenes. A catalyst system formed by Ni NPs in C$_3$CNMMIm.NTf$_2$ achieved high selectivity in the hydrogenation of diphenylacetylene to the alkene using the very mild reaction conditions of 30 °C and 1 bar H$_2$. Also, Pd NPs in C$_3$CNMIm.NTf$_2$ presented similar results. The nitrile group is crucial for the alkene selectivity. Moreover, coordination on the nanoparticle surface involving the nitrile group is suggested. In both Ni and Pd systems, the methods employed are applicable to internal aliphatic alkynes as well as to terminal phenylalkynes (Table 4.3, entries 21–24) [83]. Furthermore, the catalysts are recyclable with stable conversion rates and selectivity for at least four runs. Interestingly, there is a noticeable difference between the two systems: the application of higher hydrogen pressure (4 bar) does not affect the selectivity reached by Ni NPs in C$_3$CNMMIm.NTf$_2$, while Pd NPs in C$_3$CNMIm.NTf$_2$ tended towards the formation of alkanes.

ILs, especially imidazolium-based ones, have also been used as solvents for liquid–liquid biphasic catalysis of hydroformylation. Organometallic Rh complexes have frequently been used, and their activity depends strongly on the structure of the ILs used. Rh NPs were investigated for the hydroformylation of 1-octene in thermoregulated IL/organic biphasic systems composed of [CH$_3$(OCH$_2$CH$_2$)$_{16}$ N$^+$Et$_3$][CH$_3$SO$_3^-$] (IL$_{PEG750}$) IL, producing aldehyde yields above 85% at 99% conversion [85]. The use of PEG-functionalised ILs and different organic solvents enabled the separation of the catalysts from the products for reuse. This IL revealed unique solubility in organic solvent depending on the temperature. It is immiscible with the toluene/n-heptane mixture at room temperature but becomes homogeneous when the temperature is increased to a certain point. Consequently, the reaction proceeds in a virtually homogeneous system under heating, and upon cooling to room temperature separates into a biphasic system composed of an IL phase containing the Rh catalyst and an organic phase containing the products. This NEtPEG.CH$_3$SO$_3$/toluene catalytic system also showed efficiency with other olefins (cyclohexene and styrene). Wang reported on small-sized Rh NPs (2.4 ± 0.3 nm) in a quaternary ammonium-based IL (N,N-dimethyl-N-(2-(2-methoxyethoxy)ethyl)-N-ethyl)ammonium methanesulfonate) in the hydroformylation of 1-octene, 1-decene, 1-dodecene and cyclohexene, yielding aldehydes at above 98% with 99% conversion [33].

The hydrogenation of CO, known as Fischer-Tropsch synthesis (FTS), is a promising route to valuable chemicals and fuels. Co-, Ru- and Fe-based nano-catalysts have been successfully applied in FTS [87–90]. Cobalt-based catalysts are considered the best candidates owing to their wide availability and high activity. Not much work has been performed involving ILs. Dupont et al. demonstrated that Co NPs (4.7 ± 1.2 nm) immobilised in BMIm.NTf$_2$ IL under 20 atm of syngas (CO/H$_2$, 1:2) at 210 °C produced mainly higher hydrocarbons (HCs) in liquid phase [86]. The reaction followed the Anderson-Schulz-Flory (AFS) type of distribution, where a 0.90 ASF factor was estimated for the products (C$_8$–C$_{30}$).

A possible way to facilitate more benign FTS reactions is to use nontoxic, abundant and inexpensive CO$_2$ as reactant in place of CO. Catalytic hydrogenation of CO$_2$ to higher hydrocarbons follows two consecutive processes, in which CO generated via the reverse water gas shift (RWGS) reaction [91, 92] undergoes the Fischer-Tropsch process to yield methane [93], or heavier HC/oxygenates [94]. Iron-based catalysts are considered to be ideal candidates for CO$_2$ hydrogenation due to their intrinsic WGS and RWGS activity [95]. Methane and HCs can be produced selectively by CO$_2$ hydrogenation, either via CO or the higher HCs pathways depending on the delicate thermodynamic/kinetic balance and the fine-tuning of the properties of the electronic and geometric catalysts. In this regard, ILs have emerged as ideal candidates through which not only the conversion of CO$_2$ can be increased, but also the selectivity of the hydrocarbons controlled (Fig. 4.6). There have been many reports on the use of CO$_2$-FTS, based mainly on Fe catalysts [96–99]. Melo et al. were the first to investigate Ru NPs immobilised in 1-octyl-3-methylimidazolium bis(trifluoromethanesulfonyl)imide, (OMIm.NTf$_2$) IL for CO$_2$ hydrogenation into methane at high temperatures (150 °C) [100]. By the combination of Fe and Ru, Dupont et al. synthesised Ru/Fe bimetallic NPs that effectively hydrogenated CO$_2$

Fig. 4.6 CO$_2$ hydrogenation driven by RuFe NPs in BIMm.NTF$_2$ and BMIm.OAc ILs [20]. Reproduced by permission of the American Chemical Society

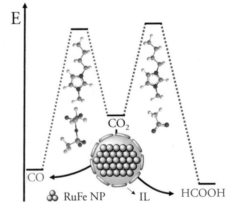

into higher hydrocarbons ($>C_1$) and selectively could be controlled through the proper choice of IL [20]. Among different hydrophobic and hydrophilic ILs (BMIm.BF$_4$, BMIm.OAc, BMIm.FAP and BMIm.NTf$_2$), Fe/Ru NPs dispersed in the hydrophobic IL (BMIm.NTf$_2$) displayed outstanding abilities in the formation of long-chain HCs (57% C_2–C_4, 31% C_5–C_6 and 10% C_7–C_{21}) with efficient catalytic activity (12% conversion), while in the hydrophilic ILs (BMIm.BF$_4$ and BMIm.OAc), RuFe NPs revealed high production of CO ($>39\%$). This effect may arise due to the hydrophilic nature of these ILs, which not only reduces the rate of CO_2 hydrogenation but may also reduce the FT catalytically active surface species with the dominant CO path formation. Moreover, when the IL with basic anion (acetate) in DMSO/H$_2$O was used, formic acid was generated with a molar ratio of 2.03 M, with 400 TONs. Recently, a high CO_2 conversion (30%) was observed when RuNi bimetallic core–shell NPs were explored using the hydrophobic BMIm.NTf$_2$ IL, which efficiently generated low-molecular-weight C_1–C_6 hydrocarbons (79% alkanes and 16% olefins) with 5% CH$_4$ [43].

References

1. Earle MJ, Esperança JMSS, Gilea MA, Canongia Lopes JN, Rebelo LPN, Magee JW, Seddon KR, Widegren JA (2006) Nature 439:831–834
2. Chambreau SD, Schneider S, Rosander M, Hawkins T, Gallegos CJ, Pastewait MF, Vaghjiani GL (2008) J Phys Chem A 112:7816–7824
3. Dupont J, Spencer J (2004) Angew Chem Int Ed 43:5296–5297
4. Shiddiky MJA, Torriero AAJ (2011) Biosens Bioelectron 26:1775–1787
5. Gomes JM, Silva SS, Reis RL (2019) Chem Soc Rev 48:4317–4335
6. Hurley FH (1951) WIer TP. J Electrochem Soc 98:203–206
7. Chauvin Y, Mussmann L, Olivier H (1996) Angew Chem Int Ed 34:2698–2700
8. Suarez PAZ, Dullius JEL, Einloft S, De Souza RF, Dupont J (1996) Polyhedron 15:1217–1219
9. Canongia Lopes JNA, Pádua AAH (2006) J Phys Chem B 110:3330–3335
10. Dupont J (2011) Acc Chem Res 44:1223–1231
11. Kolbeck C, Cremer T, Lovelock KRJ, Paape N, Schulz PS, Wasserscheid P, Maier F, Steinrück HP (2009) J Phys Chem B 113:8682–8688
12. Dupont J (2004) J Braz Chem Soc 15:341–350
13. Dupont J, Fonseca GS, Umpierre AP, Fichtner PFP, Teixeira SR (2002) J Am Chem Soc 124:4228–4229
14. Pensado AS, Pádua AAH (2011) Angew Chem Int Ed 50:8683–8687
15. Ott LS, Cline ML, Deetlefs M, Seddon KR, Finke RG (2005) J Am Chem Soc 127:5758–5759
16. Schrekker HS, Gelesky MA, Stracke MP, Schrekker CML, Machado G, Teixeira SR, Rubim JC, Dupont J (2007) J Colloid Interface Sci 316:189–195
17. Scheeren CW, Machado G, Dupont J, Fichtner PFP, Texeira SR (2003) Inorg Chem 42:4738–4742
18. Fonseca GS, Machado G, Teixeira SR, Fecher GH, Morais J, Alves MCM, Dupont J (2006) J Colloid Interface Sci 301:193–204
19. Scholten JD, Leal BC, Dupont J (2012) ACS Catal 2:184–200
20. Qadir MI, Weilhard A, Fernandes JA, de Pedro I, Vieira BJC, Waerenborgh JC, Dupont J (2018) ACS Catal 8:1621–1627
21. Luza L, Rambor CP, Gual A, Alves Fernandes J, Eberhardt D, Dupont J (2017) ACS Catal 7:2791–2799

22. Dupont J, Scholten JD (2010) Chem Soc Rev 39:1780–1804
23. Scholten JD, Ebeling G, Dupont J (2007) Dalton Trans 5554–5560
24. Migowski P, Zanchet D, Machado G, Gelesky MA, Teixeira SR, Dupont J (2010) Phys Chem Chem Phys 12:6826–6833
25. Redel E, Thomann R, Janiak C (2008) Inorg Chem 47:14–16
26. Gutel T, Santini CC, Philippot K, Padua A, Pelzer K, Chaudret B, Chauvin Y, Basset J-M (2009) J Mater Chem 19:3624–3631
27. Gutel T, Garcia-Antõn J, Pelzer K, Philippot K, Santini CC, Chauvin Y, Chaudret B, Basset J-M (2007) J Mater Chem 17:3290–3292
28. Fonseca GS, Umpierre AP, Fichtner PFP, Teixeira SR, Dupont J (2003) Chem Eur J 9:3263–3269
29. Silveira ET, Umpierre AP, Rossi LM, Machado G, Morais J, Soares GV, Baumvol IJ, Teixeira SR, Fichtner PF, Dupont J (2004) Chem Eur J 10:3734–3740
30. Rossi LM, Dupont J, Machado G, Fichtner PFP, Radtke C, Baumvol IJR, Teixeira SR (2004) J Braz Chem Soc 15:904–910
31. Prechtl MHG, Scariot M, Scholten JD, Machado G, Teixeira SR, Dupont J (2008) Inorg Chem 47:8995–9001
32. Prechtl MHG, Scholten JD, Dupont J (2009) J Mol Catal A Chem 313:74–78
33. Xu Y, Wang Y, Zeng Y, Jiang J, Jin Z (2012) Catal Lett 142:914–919
34. Scheeren CW, Domingos JB, Machado G, Dupont J (2008) J Phys Chem C 112:16463–16469
35. Umpierre AP, Machado G, Fecher GH, Morais J, Dupont J (2005) Adv Synth Catal 347:1404–1412
36. Durand J, Teuma E, Malbosc F, Kihn Y, Gómez M (2008) Catal Commun 9:273–275
37. Migowski P, Teixeira SR, Machado G, Alves MCM, Geshev J, Dupont J (2007) J Electron Spectrosc Relat Phenom 156–158:195–199
38. Migowski P, Machado G, Texeira SR, Alves MCM, Morais J, Traverse A, Dupont J (2007) Phys Chem Chem Phys 9:4814–4821
39. Leal BC, Scholten JD, Alves MCM, Morais J, de Pedro I, Fernandez Barquin L, Dupont J (2016) Inorg Chem 55:865–870
40. Silva DO, Luza L, Gual A, Baptista DL, Bernardi F, Zapata MJM, Morais J, Dupont J (2014) Nanoscale 6:9085–9092
41. Weilhard A, Abarca G, Viscardi J, Prechtl MHG, Scholten JD, Bernardi F, Baptista DL, Dupont J (2017) ChemCatChem 9:204–211
42. Adamski J, Qadir MI, Serna JP, Bernardi F, Baptista DL, Salles BR, Novak MA, Machado G, Dupont J (2018) J Phys Chem C 122:4641–4650
43. Qadir MI, Bernardi F, Scholten JD, Baptista DL, Dupont J (2019) Appl Catal B Environ 252:10–17
44. Leal BC, Consorti CS, Machado G, Dupont J (2015) Catal Sci Tech 5:903–909
45. Wu Z, Jiang H (2015) RSC Adv 5:34622–34629
46. Hu Y, Yu Y, Hou Z, Li H, Zhao X, Feng B (2008) Adv Synth Catal 350:2077–2085
47. Huang J, Jiang T, Han B, Gao H, Chang Y, Zhao G, Wu W (2003) Chem Commun 1654–1655
48. Bussamara R, Eberhardt D, Feil AF, Migowski P, Wender H, de Moraes DP, Machado G, Papaléo RM, Teixeira SR, Dupont J (2013) Chem Commun 49:1273
49. Torimoto T, Tsuda T, Okazaki K-I, Kuwabata S (2010) Adv Mater 22:1196–1221
50. Arimotoa S, Kageyama H, Torimoto T, Kuwabata S (2008) Electrochem Commun 10:1901–1904
51. Redel E, Walter M, Thomann R, Hussein L, Kruger M, Janiak C (2010) Chem Commun 46:1159–1161
52. Lahiri A, Endres F (2017) J Electrochem Soc 164:D597–D612
53. Wegner S, Janiak C (2017) Top Curr Chem 375:65
54. Wender H, Andreazza ML, Correia RRB, Teixeira SR, Dupont J (2011) Nanoscale 3:1240–1245
55. Luza L, Gual A, Rambor CP, Eberhardt D, Teixeira SR, Bernardi F, Baptista DL, Dupont J (2014) Phys Chem Chem Phys 16:18088–18091

56. Foppa L, Luza L, Gual A, Weibel DE, Eberhardt D, Teixeira SR, Dupont J (2015) Dalton Trans 44:2827–2834
57. Luza L, Rambor CP, Gual A, Bernardi F, Domingos JB, Grehl T, Brüner P, Dupont J (2016) ACS Catal 6:6478–6486
58. Wender H, de Oliveira LF, Migowski P, Feil AF, Lissner E, Prechtl MHG, Teixeira SR, Dupont J (2010) J Phys Chem C 114:11764–11768
59. Torimoto T, Okazaki K-I, Kiyama T, Hirahara K, Tanaka N, Kuwabata S (2006) Appl Phys Lett 89:243117
60. Okazaki K-I, Kiyama T, Hirahara K, Tanaka N, Kuwabata S, Torimoto T (2008) Chem Commun 691–693
61. Engemann DC, Roese S, Hövel H (2016) J Phys Chem C 120:6239–6245
62. Wender H, Migowski P, Feil AF, de Oliveira LF, Prechtl MHG, Leal R, Machado G, Teixeira SR, Dupont J (2011) Phys Chem Chem Phys 13:13552–13557
63. Huang J, Song Y, Ma D, Zheng Y, Chen M, Wan H (2017) Chin J Catal 38:1229–1236
64. Bernechea M, de Jesus E, Lopez-Mardomingo C, Terreros P (2009) Inorg Chem 48:4491–4496
65. Stuart RV, Wehner GK (1964) J Appl Phys 35:1819–1824
66. Khatri OP, Adachi K, Murase K, Okazaki K-I, Torimoto T, Tanaka N, Kuwabata S, Sugimura H (2008) Langmuir 24:7785–7792
67. Kauling A, Ebeling G, Morais J, Padua A, Grehl T, Brongersma HH, Dupont J (2013) Langmuir 29:14301–14306
68. Qadir MI, Kauling A, Calabria L, Grehl T, Dupont J (2018) Nano-Struct Nano-Objects 14:92–97
69. Qadir MI, Kauling A, Ebeling G, Fartmann M, Grehl T, Dupont J (2019) Aust J Chem 72:49–54
70. Oda Y, Hirano K, Yoshii K, Kuwabata S, Torimoto T, Miura M (2010) Chem Lett 39:1069–1071
71. Tsuda T, Kurihara T, Hoshino Y, Kiyama T, Okazaki K-I, Torimoto T, Kuwabata S (2009) Electrochemistry 77:693–695
72. Suzuki S, Suzuki T, Tomita Y, Hirano M, Okazaki K-I, Kuwabata S, Torimoto T (2012) Cryst Eng Commun 14:4922–4926
73. Suzuki S, Tomita Y, Kuwabata S, Torimoto T (2015) Dalton Trans 44:4186–4194
74. Hamada T, Sugioka D, Kameyama T, Kuwabata S, Torimoto T (2017) Chem Lett 46:956–959
75. Hatakeyama Y, Judai K, Onishi K, Takahashi S, Kimura S, Nishikawa K (2016) Phys Chem Chem Phys 18:2339–2349
76. Ullmann F, Gerhartz W, Yamamoto YS, Campbell FT, Pfefferkorn R, Rounsaville JF (1995) Ullmann's encyclopedia of industrial chemistry. VCH
77. Wegner S, Rutz C, Schutte K, Barthel J, Bushmelev A, Schmidt A, Dilchert K, Fischer RA, Janiak C (2017) Chem Eur J 23:6330–6340
78. Foppa L, Dupont J (2015) Chem Soc Rev 44:1886–1897
79. Chacón G, Dupont J (2019) ChemCatChem 11:333–341
80. Dykeman RR, Yan N, Scopelliti R, Dyson PJ (2011) Inorg Chem 50:717–719
81. Léger B, Denicourt-Nowicki A, Olivier-Bourbigou H, Roucoux A (2008) Inorg Chem 47:9090–9096
82. Jiang H-Y, Zheng X-X (2015) Catal Sci Technol 5:3728–3734
83. Konnerth H, Prechtl MH (2016) Chem Commun (Camb) 52:9129–9132
84. Venkatesan R, Prechtl MHG, Scholten JD, Pezzi RP, Machado G, Dupont J (2011) J Mater Chem 21:3030–3036
85. Zeng Y, Wang Y, Xu Y, Song Y, Zhao J, Jiang J, Jin Z (2012) Chin J Catal 33:402–406
86. Silva DO, Scholten JD, Gelesky MA, Teixeira SR, Dos Santos ACB, Souza-Aguiar EF, Dupont J (2008) ChemSusChem 1:291–294
87. Bezemer GL, Bitter JH, Kuipers HPCE, Oosterbeek H, Holewijn JE, Xu X, Kapteijn F, van Dillen AJ, de Jong KP (2006) J Am Chem Soc 128:3956–3964
88. de Smit E, Weckhuysen BM (2008) Chem Soc Rev 37:2758–2781
89. Jacobs G, Das TK, Zhang Y, Li J, Racoillet G, Davis BH (2002) Appl Catal A Gen 233:263–281

90. Ojeda M, Nabar R, Nilekar AU, Ishikawa A, Mavrikakis M, Iglesia E (2010) J Catal 272:287–297
91. Posada-Pérez S, Ramírez PJ, Evans J, Viñes F, Liu P, Illas F, Rodriguez JA (2016) J Am Chem Soc 138:8269–8278
92. Roiaz M, Monachino E, Dri C, Greiner M, Knop-Gericke A, Schlögl R, Comelli G, Vesselli E (2016) J Am Chem Soc 138:4146–4154
93. Kattel S, Yan B, Yang Y, Chen JG, Liu P (2016) J Am Chem Soc 138:12440–12450
94. Prieto G (2017) ChemSusChem 10:1056–1070
95. Newsome DS (1980) Catal Rev Sci Eng 21:275–318
96. Gnanamani MK, Jacobs G, Hamdeh HH, Shafer WD, Davis BH (2013) Catal Today 207:50–56
97. Landau MV, Vidruk R, Herskowitz M (2014) ChemSusChem 7:785–794
98. Owen RE, O'Byrne JP, Mattia D, Plucinski P, Pascu SI, Jones MD (2013) ChemPlusChem 78:1536–1544
99. Dorner RW, Hardy DR, Williams FW, Willauer HD (2010) Energy Environ Sci 3:884
100. Melo CI, Szczepańska A, Bogel-Łukasik E, Nunes da Ponte M, Branco LC (2016) ChemSusChem 9:1081–1084

Chapter 5
Covalent Assemblies of Metal Nanoparticles—Strategies for Synthesis and Catalytic Applications

Yuanyuan Min, M. Rosa Axet, and Philippe Serp

Abstract Metal nanoparticles' (NP) covalent assemblies exhibit interesting structural, electronic, and photonic features of interest for applications in catalysis. In these structures, the ligands play a fundamental role on constructing the NP network and defining their chemical environment. Two types of strategies to produce NP assemblies are discussed in this chapter: (i) the direct cross-linking method, which is simple and well-controlled, and involves a chemical reaction between the metal nanoparticle surface and the ligand; and (ii) the indirect cross-linking method in which the chemical reaction necessary for network building does not directly involve the surface of the metallic nanoparticle. Additionally, the formation of reversible covalent networks is also discussed, which allows switching between a covalent network of NP and isolated NP by applying diverse stimuli. Generally, there is a lack of attention paid on the catalytic application of metal NP covalent assemblies, and this despite of the interesting properties of such assemblies for catalysis, for instance: (i) a confined environment, (ii) the possibility to finely tune the metal/ligand interaction, and (iii) the potential robustness of the structure. Various reactions catalyzed by NP networks have been investigated such as reduction, oxidation, or water-splitting, most of them focusing on Au NP, and also few other metals (Ag, Pd, Pt, Ru). As illustrated in some cases, the organized networks show better catalytic performances than dispersed or aggregated NP due to stability or confinement effect.

Keywords Metallic nanoparticles · Covalent assemblies · Ditopic ligand · Gold · Catalysis

5.1 Introduction

Metal nanoparticles (NP) have attracted the interest of the catalysis community, as it has been shown that their inherent properties, such as size, shape, crystallographic structure, in addition to surface modifiers (ligands and/or supports) have a

Y. Min · M. R. Axet · P. Serp (✉)
LCC-CNRS, Université de Toulouse, INPT, UPS, 205 Route de Narbonne, 31077 Toulouse CEDEX 4, France
e-mail: philippe.serp@ensiacet.fr

© Springer Nature Switzerland AG 2020
P. W. N. M. van Leeuwen and C. Claver (eds.), *Recent Advances in Nanoparticle Catalysis*, Molecular Catalysis 1, https://doi.org/10.1007/978-3-030-45823-2_5

remarkable impact on their catalytic properties [1–4]. In this sense, a lot of effort has been devoted to understanding and tuning these characteristics. On the other hand, supramolecular catalysis allows creating unique catalyst–substrate interactions that can be tailored to direct substrates along particular reaction paths and selectivities [5–7]. The assembly of metal NP could also permit to direct substrates, or to create confined spaces in order to produce better catalysts. Nevertheless, the assembly of metal NP has received relatively little attention for applications in catalysis. Several approaches have been described for the directed self-assembly of metallic NP [8] that involves manipulation of: (i) molecular interaction forces such as van der Waals (vdW) [9], electrostatic interactions [10], zwitterion-type electrostatic interactions, [11–14] hydrogen bonding, [15–19] host–guest interactions [20], or DNA-assisted assemblies [21], and (ii) covalent interactions (i.e., coordination bonds, bifunctional linkage), or external fields [22] (i.e., capillary forces, magnetic [23] and electric field). Herein, we will focus on the creation of covalent interactions between NP, since they may provide robust NP networks, which are desirable for applications in catalysis. Covalent assemblies of metal NP have been obtained following several methodologies: direct cross-linking methods, indirect cross-linking methods, and stimuli-responsive reversible covalent networks of NP, which are described below. Finally, their applications in catalysis are discussed.

5.2 Synthesis of Covalent Assemblies of Metal Nanoparticles

We can distinguish two approaches for creating covalent bonds between metallic NP, the direct and indirect cross-linking methods, described herein.

5.2.1 Direct Cross-Linking Methods

This section provides an overview of metal NP covalent assemblies through directed linkage, i.e., the ligand molecules bearing two or more anchoring groups (dithiol, dicarboxylic, etc.) bound on the surface of adjacent metal NP to establish various dimensional networks (1-D, 2-D, 3-D). The covalent metal NP assemblies have been investigated for more than thirty years. For a large portion of the research reported, the synthesis process involved a ligand exchange procedure. Thus, a ditopic ligand achieving cross-linking would replace the monofunctional ligand stabilizing the isolated metal NP (monolayer-protected NP). Different methods can be used to build covalent NP networks such as the layer-by-layer method (LBL), the two-phase method, the one-phase method, Langmuir–Blodgett (LB) method, among others. Among them, the first three methods (Fig. 5.1) are the most frequently employed ones.

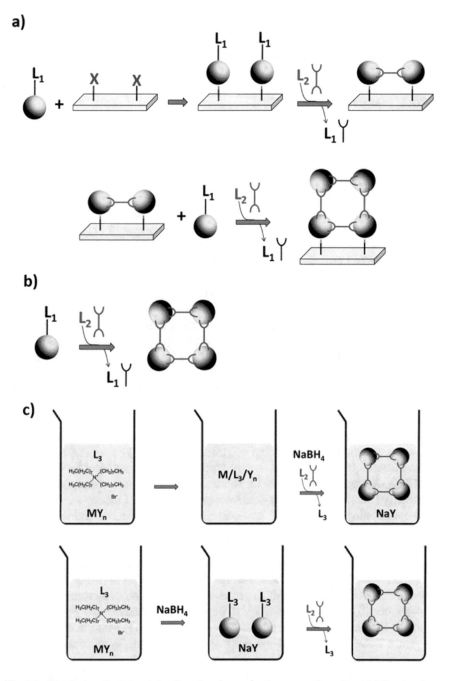

Fig. 5.1 Classical methods involving ligand exchange for the preparation of metal NP networks: **a** layer-by-layer; **b** one-phase; and **c** two-phase methods

The review of the metal nanostructures will be classified and presented according to the synthesis method. Table 5.1 shows representative examples of metal NP networks produced by direct cross-linking methods.

5.2.1.1 Layer-by-Layer Method

Along the development of metal NP assemblies, LBL self-assembly of building blocks has played a significant role, as it allows controllable growth over a substrate turning into a nanocomposite demonstrating specific optical and electrical properties [42, 43]. The LBL technique is a versatile approach to create ultrathin surface coatings on a wide range of surfaces. It involves a ligand/linker exchange process during the assembly cycle, by immersion of substrates into solution of NP and linker molecule.

In 1996, Andres et al. [24] reported an approach to prepare 2-D self-assembly of Au NP connected by ditopic ligands, in which a solid substrate acted as a template to obtain the 2-D growth. In this process, isolated metal NP were first produced in the gas phase and protected by alkylthiols in solution. After that, being coated on a substrate placed into aryl dithiol or di-isonitrile ligand solution, the NP are linked into an assembly thanks to the exchange of ligands (Fig. 5.2). The authors have demonstrated that by modifying the NP size or composition, the length and chemical structure of the linker, and the nature of the substrate, a wide range of electronic behavior can be achieved. The electrical conductance of NP assemblies (78 nS) was lower than the one of unlinked NP (133 nS).

Zhong et al. have described a novel strategy toward the assembly of bimetallic Au-Ag NP via carboxylate-Ag^+ binding at selective sites on the NP surface, giving an assembly of composition $Au_{23}Ag_{77}$ (Fig. 5.3) [44]. The possibility to modify the composition of the Au-Ag NP as well as the chain length of the dicarboxylic acids allows to modulate the optical and electronic properties. The Au-Ag but also Au NP assemblies [45] with different inter-particle distances, adjusted by varying the X-$(CH_2)_n$-X length, were applied for vapor sensing, revealing a correlation between sensitivity (electric conductivity) and inter-particle spacing [46].

Another alloyed Au-Ag NP assembly produced from the 4-aminothiophenol (PATP) linker on substrate was investigated by Raman scattering [47]. The LBL assembly was produced first by using the protonated pyridine groups of a polyvinylpyridine-functionalized glass as anchoring sites for negatively charged Au NP. PATP ligands were then adsorbed on Au NP through the formation of Au–S bonds. The PATP molecule, which possesses two resonance structures (benzenoid and quinonoid), can interact with Ag NP through the quinoid form. It was concluded from the Raman study that the b2 vibrational mode of PATP, which is characteristic of charge transfer between the metal NP and PATP ligands, is enhanced by charge transfer from the Ag to Au NP by tunneling through PATP.

Alkanedithiols with different carbon chain length (C_6, C_9, C_{12}, C_{16}), which are typical linkers for Au NP assemblies, were investigated for sensors applications [48]. X-ray photoelectron spectroscopy (XPS) analyses of Au NP assembly 2-D films confirmed that dodecylamine ligands on Au NP were quantitatively exchanged

Table 5.1 Representative examples of metal NP networks produced by the direct cross-linking methods

Metal	Ligand	Method	NP size (nm)	Inter-NP distance (nm)[a]	References
Au	1,4-di(4-thiophenylethynyl)-2-ethylbenzene	LBL	3.7	1.7	[24]
Au	DMAAB/DMAAcH/PBDT/cHBDT	LBL	~4	~1	[25]
Au	Octanedithiols/biphenyldithiols	Two-phase	1–5	~1	[26]
Au	MeSi(CH$_2$SMe)$_3$/Si(CH$_2$SMe)$_4$)	Two-phase	6.4 ± 0.8	~2	[27]
Au	RAFT oligomers	Two-phase	4.6 ± 1.5	0.6	[28]
Au	C$_{60}$ thiol polymer	Two-phase	10.8 ± 1.5	3.3 ± 0.8	[29]
Au	V-, Y-, and X-shaped methylthio aryl ethynyls	Two-phase	5.1 ± 0.9	1.1–1.7	[30]
Au	X-MTA X'-MTA	Two-phase	5–6	1.5 1.3	[31]
Ag	Dithiol	Two-phase	4.31 ± 0.28	2.1 ± 0.2	[32]
Au	Fluorenyl dithiol	Two-phase	3–7	1.2–3.0	[33]
Au	α,ω-dithiol	One-phase	2/5	1.5	[34]
Pd	1,4-phenylene diisocyanide, 4,4'-bis(diphenylphosphino)biphenyl, 1,3,5-tris([2,2':6',2''-terpyridin]-4'-ylethynyl)benzene	One-phase	~2.4	~2	[35]

(continued)

Table 5.1 (continued)

Metal	Ligand	Method	NP size (nm)	Inter-NP distance (nm)[a]	References
Au	1,8-octanedithiol	One-phase	5	0.59–1.4	[36]
Au	$HSC_6H_4SC_6H_4SH$	LB	5.1	1.6	[37]
Au-Ag	9,9-didodecyl-2,7-bis-thiofluorene	Two-phase	3–5	1.5	[38]
Pd	Tetrakis(terpyridine)	One-phase	2.7	1.2	[39]
Au	Two pyridine-N-oxide	One-phase	5.2 ± 0.3	5.7/6.0 (center–center)	[40]
	One pyridine-N-oxide and one ArF-I group			7.26 (center–center)	
Ru	$C_{66}(COOH)_{12}$	Single-step wet chemical synthesis	1.5	2.85 (center–center)	[41]

[a]Except for Refs. [40] and [41], the distance between NPs is the edge-to-edge distance

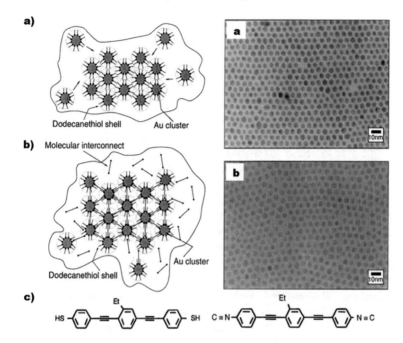

Fig. 5.2 **a** An unlinked array of Au NP and the respective TEM micrograph; **b** a linked network and the respective TEM micrograph; and **c** ditopic ligands used for NP assembly. Reproduced with permission from Ref. [24]

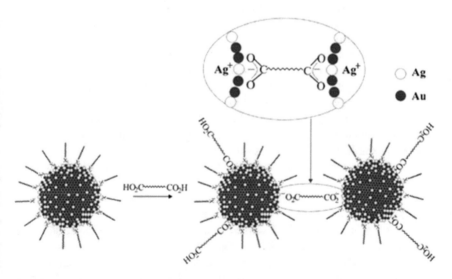

Fig. 5.3 Schematic representation of the surface binding sites and the selective linkage for the assembly of Au-Ag NP mediated by dicarboxylic acid ligands. Reproduced with permission from Ref. [44]

by alkanedithiols, of which 60% of the alkanedithiol are bound to NP by both groups, while 40% are bound with only one thiol group. All films showed ohmic $I-V$ characteristics and Arrhenius-type activation of charge transport. Interlinked NP assemblies help to enhance the conductivity of the films compared to free NP. Indeed, the electron-tunneling decay constant β_N was much lower for non-covalently linked NP film. A fast and fully reversible increase of film resistance was measured upon exposure to vapors of toluene or tetrachloroethylene. The resistance increases exponentially with increasing alkanedithiol chain length at a given concentration. This type of film was forecasted as a promising material for sensing application, which was studied later extending the ligand to alkanedithiol-C_{12}, 4,4'-terphenyldithiol and [4]-staffane-3,3'''-dithiol to make chemiresistors. This study has revealed that the flexibility and resistivity of the interlinkage has a profound impact on the response characteristics of the sensors [49]. Thus, the use of the flexible 1,12-dodecanedithiol as linker induces the interlinked film to respond with an increase in resistance. Oppositely, with the rigid staffane linker the interlinked film responded with a decrease in resistance.

In a similar attempt of investigating the influence of dithiol ligands, Wessels et al. synthesized Au NP assembly films from six different dithiols, which were classified into three groups according to the nature of the group in the middle and at the end (Table 5.2) [25]. The optical and electrical properties of the films were highly influenced by the nature of the ligand. Especially, the conductivity increased by one order of magnitude for linkers that contain a cyclohexane ring instead of a benzene ring. As the molecule consists in non-conjugated and conjugated parts, according to the electron-tunneling decay constant (β_{N-CON}), the conductivity can be tuned from the insulating to the metallic limit, regardless of the inter-particle spacing (4 ± 0.8 nm, $d_{NP} \sim 1$ nm).

Daskal et al. followed the Au NP assembly deposition using a quartz crystal microbalance. This is a suitable method as it can sense material deposition in the nanogram range [50]. The assembly efficiency is higher for shorter alkyldithiols and ligands with more strongly interacting functional groups such as alkylbisdithiocarbamates, which outperformed alkyldiamines. Comparing plasmon resonance, a typical and well-studied feature of Au NP, the spectrum of the bisdithiocarbamate composite exhibits a distinct blue shift, which was attributed to more delocalized electron charge at the NP, because the linker possesses more bulky groups allowing a longer structure.

The number of the deposition cycles performed to produce multilayer films on substrate counts importantly for properties and applications. Thus, enhanced localized surface plasmon resonance sensing was obtained with multilayer structures fabricated with 1,10-decanedithiol as linker from four NP deposition cycles [51]. This was reflected by a ~fourfold improvement of the sensitivity to the changes of the environmental refractive index compared to the submonolayer structure. Dhar et al. produced Au and Ag NP composite thin films with a carboxyl-functionalized chitosan polymer bearing COO^- and $-NH_2$ groups (Scheme 5.1), which were deposited on flat quartz substrates [52]. The process of assemblies allows producing up to more than 40 layers, the growth of which was monitored by UV-visible spectroscopy, atomic

Table 5.2 An overview of the DT linker molecules and the distances between S atoms and S⁻ atoms [25]. DMAAB/DMAAcH/PBDT/cHBDT add abbreviations used in Table 5.1

Ligand	Length (Å)	$\Delta(E_{\text{HOMO-LUMO}})$	
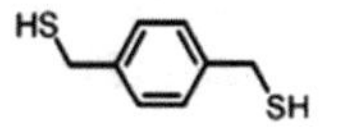 HS⬡SH	7.7	4.0	Bis-mercaptomethylenes
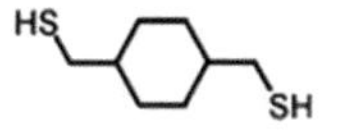 HS⬡SH	8.2	5.0	
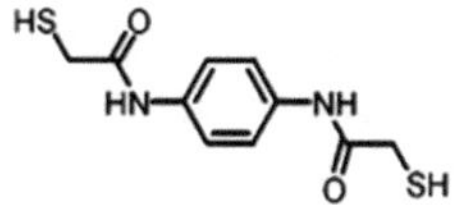 HS–CO–HN⬡NH–CO–SH	14.8	3.5	Bis-acetamidothiols
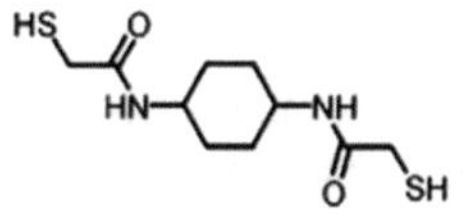 HS–CO–HN⬡NH–CO–SH	14.9	4.7	
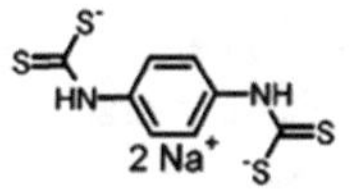 S=C(S⁻)–HN⬡NH–C(=S)S⁻ 2 Na⁺	10.7	2.7	Bis-dithiocarbamates
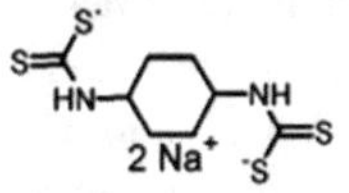 S=C(S⁻)–HN⬡NH–C(=S)S⁻ 2 Na⁺	9.6	3.2	

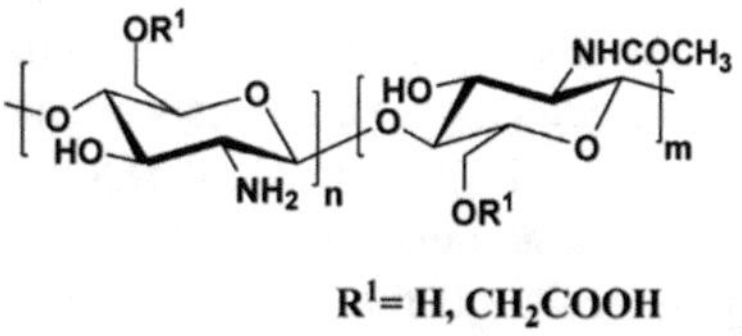

Scheme 5.1 A carboxyl-functionalized chitosan polymer. Reproduced with permission from Ref. [52]

$$R^1 = H, CH_2COOH$$

force microscopy (AFM), and scanning electron microscopy (SEM). The catalytic activity of these composite films was investigated for an organic electron transfer reaction. The kinetic data obtained suggest that the reaction rates are directly related to the NP size and porosity of the membrane. The reaction rates were five and six times higher, respectively, with films of 10 and 40 layers of Au NP (5 nm). Mainly, the NP located on the film surface are the ones that take part in the reaction, since the accessibility of the embedded NP decreases as a function of number of layers.

5.2.1.2 Two-Phase Method

In the method of Brust's two-phase synthesis [53, 54], the linker is added to a solution of the metal precursor with phase transfer molecules, and reduction afforded NP self-assemblies [26]. This two-phase method is called mediator–template strategy [27], as the transfer molecule capped with hydrophobic chain works like a template between NP, and as linker to mediate the assembly. Thus, spherical Au NP assemblies (~20–300 nm diameter) were produced taking advantage of linker molecules with tri- and quart-ending groups, $Si(CH_2SMe)_3$ or $Si(CH_2SMe)_4$ (Fig. 5.4) [27, 55, 56]. In these works, multidentate thioether ligands were used as molecular mediators and tetraalkylammonium-capped Au NP (5 nm) as templates. The assemblies can be disassembled by adding decanethiolate, as reflected by the optical response, because the binding of thiolates is stronger than that of thioethers, which may pave the way to drug delivery applications.

When deposited on a support, the spheres can spread to the hydrophobic surface, indicating the soft nature of the outmost layer of the assembly. In a subsequent work [57], the self-assembly process was assessed from the kinetic and thermodynamic point of view. It was concluded that the process is enthalpy-driven. The enthalpy change (-1.3 kcal mol^{-1}) was close to the magnitude of the van der Waals interaction energy for alkyl chains and the condensation energy of hydrocarbons.

Gold NP network formation mediated by multifunctional synthetic polymers is also a versatile strategy [58]. Stemmler et al. describe a protocol to produce a Au NP assembly after introducing polyphenylene dendrimers bearing 16 lipoic acid end-groups (Fig. 5.5) [59]. The loosely coordinated network rearranged rapidly within hours, showing strong optical coupling between Au NPs and a corresponding spectral shift detected by UV-vis-NIR spectroscopy.

Another example of using polymers is described by Rossner et al., who produced spherical shape Au NP network with multifunctional polymers of styrene bearing multiple trithiocarbonate groups as linkers to interconnect Au NP (Fig. 5.6) [28, 60, 61]. The authors demonstrated that the establishment of assemblies relies more on polymer length than anchoring group number. The distance between NP can be tuned in a quite wide range when regulating the length of the polymer chain. The internal

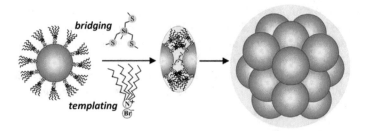

Fig. 5.4 Schematic illustration of Au NP assembly formation via mediator–template strategy. Reproduced with permission from Ref. [27]

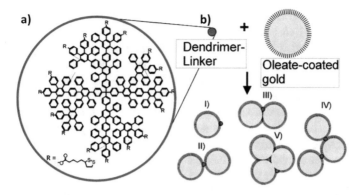

Fig. 5.5 **a** Dendrimer used as linker; and **b** schematic representation of the formation of aggregates from dendrimers (red spheres) and Au NP. Reproduced with permission from Ref. [59]

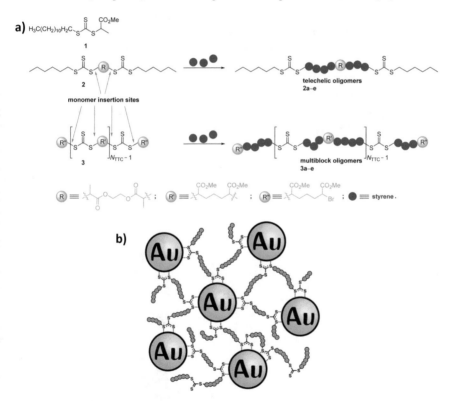

Fig. 5.6 **a** Synthetic scheme of the RAFT polymer. Reproduced with permission from Ref. [28]. **b** The structures of the Au NP assembly cross-linked by di- and multifunctional RAFT oligomers. Reproduced with permission from Ref. [60]

structures of the 3-D networks were characterized by small-angle X-ray scattering (SAXS), dynamic light scattering (DLS), and UV-vis extinction spectroscopy. These analyses have revealed that a large fraction of Au NP network can only be obtained when the number of styrene units between two anchoring sites in RAFT oligomers is below a crucial threshold, which is probably due to the formation of loops that may sterically impede cross-linking with different NP. Their work provided useful information to build self-assembled NP networks with respect to the degrees of polymerization of polymers. It was also shown in the case of hyperbranched polymer-Au NP assemblies that the degree of branching of the linker polymer, in addition to the concentration and number of anchoring groups, has a strong influence on the self-assembly process [62].

Disulfide-functionalized C_{60} polymers produced by reacting bis-2-aminoethyl disulfide with C_{60} constitute another example of the use of a polymer to produce Au NP 2-D self-assemblies through Au–S bond on a substrate (Fig. 5.7) [29]. In that case, part of the S–S bonds in the polymeric chain are broken, forming Au–S bonds. The 2-D films obtained on electrodes were used to fabricate single electron devices, which exhibited Coulomb blockade-type current–voltage characteristics.

The influence of the number of ending groups in linker ligands on tuning interparticle interactions and structures was investigated by Zhong et al. [30]. Four different rigid aryl ethynyl molecules with V, Y, and X shapes attached to 2,3,4 methylthio-end-groups were used as linkers for Au NP (Fig. 5.8), showing an influence on the optical and spectroscopic properties. The measured average edge-to-edge

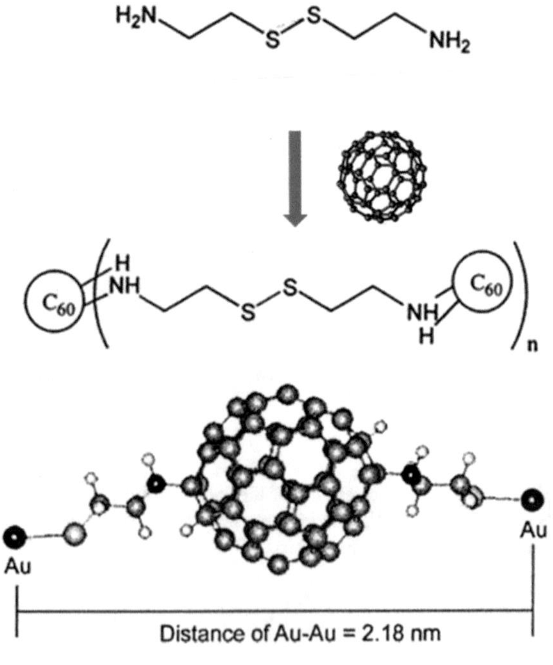

Fig. 5.7 C_{60} polymer structure and the bonding between two Au NP. Reproduced with permission from Ref. [29]

Distance of Au-Au = 2.18 nm

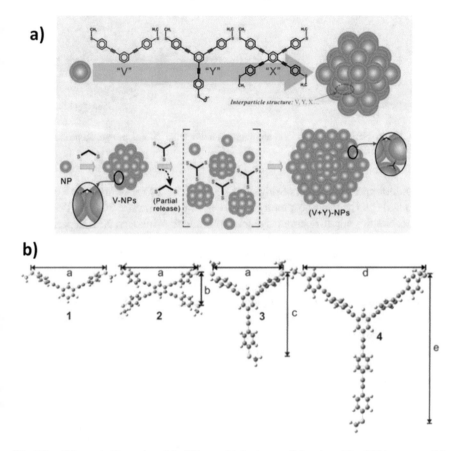

Fig. 5.8 a Schematic illustration of the NP assembly by two mediators (e.g., V and Y) in a sequential addition process; and **b** structures of V-, X-, and Y-shaped methylthio arylethynes. Reproduced with permission from Ref. [30]

inter-particle distance ranged between 1.1 and 1.7 nm according to the nature of the ligand, which are in good agreement with molecular modeling results for the inter-particle orientations. It was shown that sequential addition of ligands can induce an initial partial disassembly. Thus, if a network is built from a weakly coordinating ligand, the addition of a ligand that binds strongly causes an initial disassembly due to exchange of ligands, before the second ligand mediated the formation of mixed ligand assemblies. For X-shaped arylethynes, it was also shown that their molecular rigidity, π-conjugation, and importantly the tunability in terms of size, shape, and binding strength affect the size, kinetics, and optical and spectroscopic properties of the assembly [31].

The group of Fratoddi has shown that Ag NP networks can be produced by means of the organometallic compound *trans, trans*-[CH$_3$CO–S–Pt(PBu$_3$)$_2$(C≡C–C$_6$H$_4$–C$_6$H$_4$–C≡C)–Pt(PBu$_3$)$_2$–S–COCH$_3$] [32]. The organometallic bifunctional

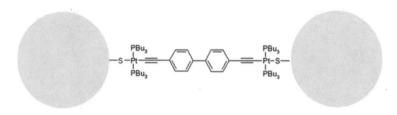

Fig. 5.9 Silver NP linked by a Pt-containing organometallic dithiol bridge. Reproduced with permission from Ref. [63]

thiol-containing Pt(II) centers bridge the Ag NP. The 2-D networks were obtained through –SH derivatives covalently bound to Ag NP surface without Pt–S bond cleavage. Employing an accurate synchrotron radiation-induced XPS experiment, a semi-quantitative analysis of the chemical interactions between Ag and the complex was performed. The percentage of completely and partially covalent thiols was 54.6% (direct covalent bond) and 45.4% (one terminal grafting involving non-covalent interactions between NP), respectively. A subsequent work was performed to characterize the material by X-ray absorption fine structure spectroscopy (XAFS) and XPS techniques, providing deeper insight at the atomic level into the physicochemical properties of the material [63]. XAS (S and Ag K-edges) analysis supply further support for the interaction of the dithiol with Ag atoms on the surface of the NP (Fig. 5.9).

Dithiol organic molecules with different chain lengths and various middle groups have been used to tune the structure and optical and electric properties of Au NP networks. For instance, Au NP networks were produced from 9,9-didodecyl-2,7-bis(acetylthio)fluorene (Au NP-1) and 9,9-didodecyl-2,7-bis(acetylthiophenylethynyl)fluorene (Au NP-2) playing the role of linkers [33]. A red shift of the emission band for Au NP-1 in emission spectroscopy was observed compared to the free thioester, suggesting an electron density flow to the gold center through the S-bridge, whereas the effect was quenched in Au NP-2, probably due to the increased distance of the fluorenyl chromophore from the gold core. The same type of molecule (π–π conjugated 9,9-didodecyl-2,7-bis-thiofluorene, FL) was reported to give a structural reorganization of the Au NP assembly after thermal treatments (Fig. 5.10) [38].

With the help of grazing incidence X-ray diffraction technique, a proposition was put forward that a partial transition from a hexagonal-like to a cubic-like packing of Au NP occurred during the thermal annealing (up to 100 °C with a rate of 10 °C/min). The electric properties of the Au NP assemblies and deposited films were also probed [64]. Current–voltage (I/V) response curve follows a non-ohmic relation for Au NP few layer thin films, with a conduction mechanism that strongly depends on polarons and bipolarons along π bridges of the carbon chain of the fluorene bridge. For multilayer thick films, the properties followed approximately an ohmic law. In contrast, Au NP few layer thin film possesses a resistance two orders of magnitude smaller. The overall resistance decreases upon illumination thanks to the organic linkers in the network, and optoelectronic properties appeared.

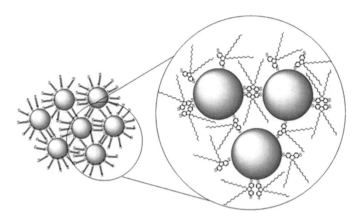

Fig. 5.10 Schematic representation of Ag NP assembly mediated by FL molecule. Reproduced with permission from Ref. [64]

Another example of studying the influence of the dithiol ligands involves the use of 4,4′-dithiol-biphenyl (BI), 4,4′-dithiolterphenyl (TR) and 4,4′-dithiol-trans-stilbene (ST) [65]. XPS analyses showed that both physisorbed and chemisorbed thiols are present on the NP surface. Ag NP showed a lower quantity of physisorbed thiol and a higher tendency to form interconnected networks. Furthermore, the Au NP-BI and Au NP-TR possess high stability both in solution and under thermal stressing conditions (500 °C) [66].

Besides Au and Ag NP networks, there are reports on Pt NP organization produced with the two-phase method. Morsbach et al. reported Pt NP porous three-dimensional network built with bifunctional amines from "unprotected" NP through a two-phase method [67–69]. The accessibility of the metal sites in the assembly was characterized by cyclic voltammetry for determining the electrochemical surface area (ECSA), showing that 50% of the metal surface atoms are ligand-free. This means that more possible catalytically active sites are present than in NP produced using mono-amine ligands with a much higher coverage. IR characterizations demonstrated that Pt NP are coordinated by the amine. The presence of adsorbed CO originated from the solvent (ethanol decomposition) indicates that ligand-free adsorption sites are present for all ligand-linked NP network. On the other hand, the absence of CO in capped NP indicates the formation of a full monolayer of hexadecylamine on Pt NP [68].

5.2.1.3   One-Phase Method

This method is similar to the two-phase solution method but carried out with the ligand exchange or linkage in one pot in order to simplify the procedure. In the work published by Leibowitz et al., the one-phase method was also called one-step exchange-cross-linking-precipitation route [34]. The materials synthesized had comparable electrical properties to those prepared by the LBL method. The ligand exchange process took place in 1,9-nonanedithiol solution with 2 and 5 nm decanethiol capped Au NP, followed by loading on a support, which resulted in covalently linked thin films. The surface plasmon resonance displayed by the assemblies having inter-particle distances shorter than the NP size shifted at longer wavelength. The electrochemical properties are different (gold–sulfur binding and partial blockage electron transfer barrier), probably influenced by NP size and crystallinity, which was important for further catalytic application [70]. Following the same preparation route, Au–Pt alloyed NP assembly cross-linked by 1,9-nonanedithiol was produced and applied for methanol electrooxidation [71]. If controlling assembly in 2-D or 3-D has been achieved with success, then the 1-D assembly is still a challenge. Nonetheless, the use of nonanedithiol produced size-controlled 1-D linear assemblies [72]. The length and branches of these gold nanochains were tuned by varying the ratio of dithiols to Au NP, resulting in dimers, oligomers, and branched linear shapes.

The use of aromatic compounds bearing bidentate or tridentate coordination sites as cross-linkers (1,4-phenylene diisocyanide, 4,4′-bis(diphenylphosphino)biphenyl, 1,3,5-tris([2,2′:6′,2″-terpyridin]-4′-ylethynyl)benzene) allows producing Pd NP networks [35]. The resulting assemblies have shown better hydrogen storage capacities than isolated Pd NP. Also, the tetrakis(terpyridine) linker (Fig. 5.11a) was used to interconnect Pd NP to well-organized self-assemblies as a result of the rigid tetrahedral core [39].

Supra-spheres (80–320 nm) containing large amounts of metal NP and involving more than two different kinds of metals (nearly monodisperse (5.5 nm) Au, Ag, Pd, or Pt NP stabilized by dodecylamine and didodecyldimethylammonium bromide) were produced from 1,8-octanedithiol [73]. These supra-spheres can be converted into a new material (nanoporous metals) by removing the organic ligands, extending the potential of metal NP assembly for material design.

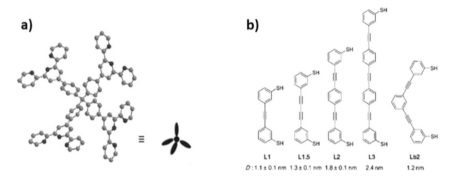

Fig. 5.11 **a** Structure of the tetra(terpyridine) linker used for Pd NP network synthesis. Reproduced with permission from Ref. [39]. **b** Chemical structures of L_n ($n = 1, 1.5, 2, 3$) and L_{b2}. Reproduced with permission from Ref. [74]

Neouze et al. presented a new approach to build networks of Pt NP using both a carboxylic acid-functional thiol ligand (mercaptopropionic acid) and an alcohol-functional thiol ligand (mercaptoethanol) before introducing them into a TiO_2 matrix [75]. Pt NP stabilized by the two different ligands (Au–S coordination) were cross-linked. FTIR analyses have shown that the –COOH groups did not react with the –OH groups, and that the pending –COOH groups interact, as –COO$^-$ groups with neighboring Pt NP. Furthermore, there were residual pending carboxylic acids of the Pt–Pt network that can further coordinate to the titanium centers of the TiO_2 matrix, ensuring stability to the coating film.

A work of Boterashvili et al. reports the use of different building blocks, single-crystalline and multiple-twinned Au NP and cross-linkers possessing two binding sites (pyridine-N-oxide and/or tetrafluoro-iodoaromatics, ArF-I) (Fig. 5.12) to control the aggregation of Au NP via halogen bonding [40]. The authors demonstrated that N-oxide moieties contribute more to the formation of organized Au NP networks, whereas the ArF-I moieties differ in bonding reactivity with different facets. The observed reactivity of the single-crystalline and multiple-twinned Au NP to assembly illustrates the importance of NP crystallinity to control the self-assembly degree, providing aspects to consider for designing material in future work.

Even though the assembly of Au NP with dithiol molecules has been investigated in detail, the production of ordered super-lattices is still a significant challenge. Nayak

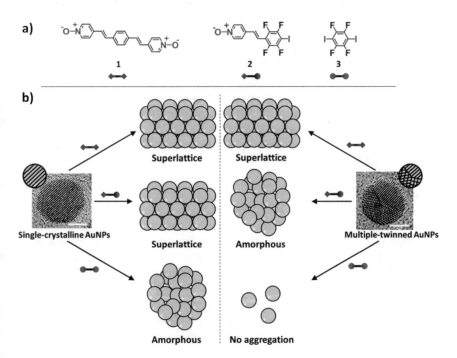

Fig. 5.12 **a** Molecular structures of cross-linkers **1–3**; and **b** schematic representation of the reactions between Au NP and the cross-linkers (**1–3**). Reproduced with permission from Ref. [40]

et al. found that using the one-phase route to produce self-assemblies, short-ranged FCC crystals can be formed from the cross-linked network for specific dithiol linker-length and NP size [75]. The stability of the lattices was evaluated based on geometrical considerations and numerical simulations as a function of ligand length and number of connected nearest neighbors, and a phase diagram of super-lattice formation was provided. The methods employed provide perspective in further exploration for well-crystallized lattices. Further management of Au NP assemblies was investigated, and an interesting electric device was constructed, consisting of an electric platform of nano-electrode–molecule–NP bridge, showing a significant improvement of reproducibility of electrical measurements (Fig. 5.13) [36]. The Au NP (5 nm) situated between the nano-electrode (19 nm) were connected with 1,8-octanedithiol, where the outer thiol groups are protected by triphenylmethyl (ω-trityl), which lead to reproducible and stable metal–molecule bonds replacing the physisorbed metal–molecule junction. Correspondingly, the conductivity increased and the spread of the resistance histogram reduced by one order of magnitude, which demonstrated that the use of this platform can be a potential method for characterization of Au NP networks and furthermore molecule levels current transport application.

Langmuir–Blodgett (LB) Method

The LB method makes use of the surface pressure to initiate NP cross-linking deposited onto substrate. One example reported by Chen uses 4,4'-thiobisbenzenethiol as bifunctional linker that facilitated cross-linking of monolayer-protected Au NP [37]. Two-dimensional NP networks were prepared by using the LB method, where neighboring NP were chemically bridged by the bifunctional linker at the air/water interface. The fluctuations of surface pressure were studied during the formation of the assembly, showing that high surface pressures help to activate cross-linking, which resulted in long-range ordered and robust NP networks. The typical surface plasmon band of Au NP redshifted by 30 nm, which was attributed to the electronic coupling interactions between neighboring NP. Ordered arrays of quantum dots can also be produced by photo-oxidation to remove the organic part in the network [76].

Light-Triggered Self-assemblies

A light adjustable Au NP network was produced by using azobenzene-thiol derivatives as inter-particle linkers. It was shown that the spacing between NP can be reversibly controlled by *trans-cis* isomerization of the azobenzene moiety induced by UV and visible light (Fig. 5.14) [77]. Analogously, another photoactive *trans*-ligand 4,4'-bis(11-mercaptoundecanoxy)azobenzene (ADT) containing a photoswitchable azobenzene unit was used to assemble metal NP into assembly triggered by UV irradiation modified by light of different wavelengths [78, 79]. In that case, the self-assembly was irreversible for high ligand surface concentrations, resulting in NP organized into permanently cross-linked structures by dithiols.

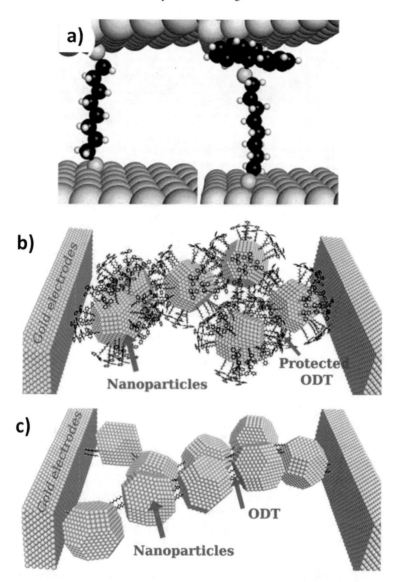

Fig. 5.13 Schematic representation of **a** 1,8-octanedithiol chemisorbed at two nearby gold surfaces and triphenylmethyl protected 1,8-octanedithiol chemisorbed at one end and physisorbed at other end; **b** trapped ω-trityl protected 1,8-octanedithiol-coated (ODT) Au NP; and **c** the junction after removal of the trityl protective resulting in formation of chemisorbed junctions at both ends of 1,8-octanedithiol. Reproduced with permission from Ref. [36]

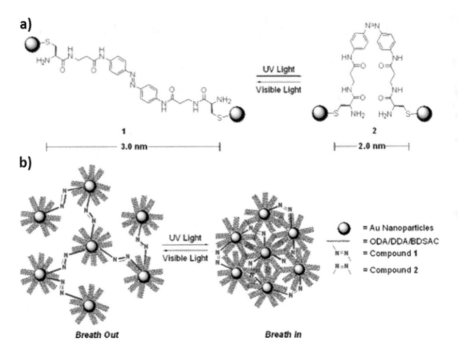

Fig. 5.14 **a** Scheme of photoactive linker molecule; and **b** reversible tunable Au NP assembly. Reproduced with permission from Ref. [77]

Single-Step Wet Chemical Synthesis

Besides the ligand exchange strategy, we can also mention a directed covalent assembly produced in one-step, in which the self-assembly process by linker ligand is constructed when the metal NP are growing without any other monocapped ligand/stabilizers. As described by Serp et al., $C_{66}(COOH)_{12}$ hexa-adduct can be successfully used as a building block to construct 3-D networks via carboxylate bridges with very homogeneous and well-crystallized sub-1.8 nm Ru NP, displaying a Ru NP-NP distance of 2.85 nm (Fig. 5.15a, b) [41]. Furthermore, IR, SSNMR, and XPS point out that the substituted fullerene coordinates to the Ru NP via carboxylate groups, which was corroborated by DFT calculations.

Spherical Ru@C_{60} objects consisting of Ru fullerides decorated with small-size Ru NP (1.5 nm (Fig. 5.15c)) were also synthesized using unmodified C_{60} fullerene as linker. A net charge transfer from Ru to C_{60} in the Ru@C_{60} nanostructures was measured [80–82]. A similar work was published on Pd–Ni NP assemblies with C_{60} derivatives bearing carboxylic acid and amine groups, which were used as catalysts for nitrobenzene hydrogenation [83].

Solovyeva et al. prepared Ag NP assemblies in a controllable way with diaminostilbene as linker for surface-enhanced Raman scattering (SERS) detection, with the intention that the well-interlinked NP can form hot spots providing intensive SERS spectra [84]. Indeed, only interconnected Ag NP at sub- or monolayer surface

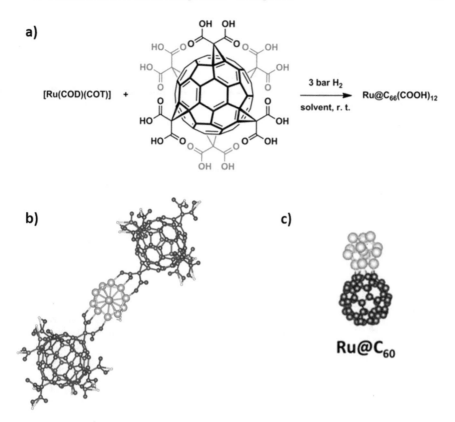

Fig. 5.15 **a** Synthesis of $C_{66}(COOH)_{12}$-mediated covalent assembly of Ru NP; and **b** optimized structure of the $C_{66}(COOH)_{12}$–Ru_{13}–$C_{66}(COOH)_{12}$ species by DFT. Reproduced with permission from Ref. [41]. **c** Correlation model of $Ru@C_{60}$. Reproduced with permission from Ref. [80]

coverage formed hot spots. The Ag aggregates obtained demonstrate a high SERS activity, suggesting that such substrates could be used for SERS detection of analytes at a single-molecule level.

In this single-step method, the effect of the multidentate rigid ligands on the size of Au NP was also investigated with different length of aryl ethynyls dithiol molecules (L_n and L_{b2}, Fig. 5.11b) [74]. It was revealed that the size of the Au NP and the final structure as NP or assemblies are synergistically regulated by ligand length, ratio of ligand to metal, and ligand coordination mode on the NP surface.

Novel Strategies Producing NP Assemblies

Covalent organic frameworks (COF) are porous materials with well-defined 2- or 3-dimensional structures with organic molecule building blocks, which have been used as supports for metal NP with rather evenly distribution [85–88]. Lu et al. have produced Pt NP@COF hybrid material with well-dispersed average NP using thioether-containing COF (Fig. 5.16) [89]. The strong anchoring groups inside the pores facilitate the binding of metal NP and COF, as supported to some extent by XPS characterization, which provide a potential pathway to get well-organized covalent NP assemblies by designing anchoring groups on COF.

The use of DNA allowed programmed assembly through different length of DNA sequence. Au NP stabilized by alkanethiol-capped oligonucleotides were linked into short-range and long-range ordered networks by complementary linker oligonu-cleotide (DNA) strands (Fig. 5.17), in which the optical properties are affected by aggregate size as well as inter-particle distance [90, 91].

An interesting strategy to interconnect Pd NP was proposed by Simon et al. [92]. The 3-D network was built by insertion of the bifunctional linker molecules 4,4'-diamino-1,2-diphenylethane to $Pd_{561}phen_{36}$ clusters, which was prepared by deoxygenation of $Pd_{561}phen_{36}O_{200}$ clusters by H_2. The resulting material showed an increase of the charging energy from 0.02 to 0.05 eV and a decrease of the electrical capacitance between the clusters compared to dense aggregates before insertion of the linkers.

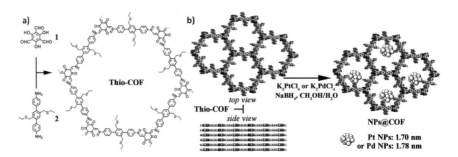

Fig. 5.16 **a** Synthesis of Thio-COF; and **b** schematic representation of the synthesis of Thio-COF supported Pt NP@COF and Pd NP@COF. Reproduced with permission from Ref. [89]

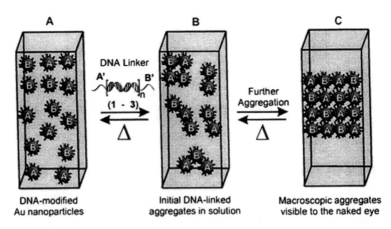

Fig. 5.17 Scheme showing the DNA-based nanoparticle assembly strategy. Reproduced with permission from Ref. [91]

From this analysis of the literature dealing with direct cross-linking methods, it is clear that the direct covalent bonding of metal NP into assemblies is highly impacted by the choice of the multidentate ligand. The chain length, the binding group, and middle backbone tune the properties of the final material at different levels. Deep insight into the self-assembly process was obtained at both the nano- and atomic levels by a vast panel of characterization techniques, as well as modeling. A large proportion of work involved Au and Ag, particularly because of potential applications in optics, sensors, and electronics. Less work has been devoted to other noble metals of interest for catalysis, such as Pt, Pd, or Ru, meaning that additional efforts are needed to develop reliable methods to produce covalent networks of NP with these metals.

5.2.2 Indirect Cross-Linking Methods

Besides the direct cross-linking methods that often involved the use of ditopic ligands to directly link metal NP, several indirect methods have also been reported. These methods involved the reaction between NP stabilized by specific ligands, which allow a coupling reaction between them to create the metal NP network. Various reactions can occur (click chemistry, Diels–Alder reaction, nucleophilic substitution…) either under thermal- or photo-activation (see Table 5.3 for representative examples).

5.2.2.1 Click Chemistry

Click chemistry, a group of chemical reactions with favorable reaction rate and orthogonality, is an efficient way for NP modification [100], including their assembly. Trials with NP have been reported based on click chemistry, including copper-catalyzed cycloaddition, strain-promoted azide–alkyne cycloaddition [101], and inverse-demand Diels–Alder reaction [102].

Metallic NP can be modified by azide- and/or alkyne-functional groups, and then assembled into organized arrays through the basic Cu(I)-catalyzed azide/alkyne-"click" chemistry approach (Scheme 5.2).

Two procedures have been followed, the use of a mixture of NP (one population being azide-functionalized and the other one alkyne-functionalized) and the reaction of azide- or alkyne-functionalized NP with a dialkyne or a diazide (tri- and tetra-azides have also been used), respectively. The Huisgen 1,3-cycloaddition reaction between azide and ethynyl groups has been employed to obtain assemblies (nanochains) of Au nanorods [103]. Using a 1:1 mixture of the nanorods stabilized by azidoalkane- and alkyne-thiols allows the formation of chain-like assemblies. The preferred end-to-end assembly of the Au nanorods could be attributed to the preferential ligand displacement at the (111) faces at the end of the nanorods. In the case of Au NP (2 nm), Au NP networks with an inter-particle distance of 2 nm (which fits with the molecular length of 1.8 nm for the rigid azobenzene unit) were obtained by

Table 5.3 Representative examples of metal NP networks produced by the indirect cross-linking methods

Metal	Ligand/Reactants	Method	Size (nm)	Inter-NP distance (nm)[a]	References
Au	1-Azide-1-undecane thiol Dialkyne derivative/CuSO₄	Click chemistry	2	2	[93]
Au	Octadentate thioethers	Acetylene oxidative coupling	1.1	0.8–2.5	[94]
Au	Norbornene thiols Ru catalyst	Ring-opening metathesis polymerization	4.3	7	[95]
Au	Terminal amines on polyethylene glycol (NP) Thiolated amine (nanorods) Traut's reagent	Dithiol coupling	2.5 and 1.8 (NP) 50 × 5 (nanorods)	0.9 and 1.6	[96]
CdSe	2-Aminoethanethiol Glutaraldehyde	Condensation	2.7	3.2–7.6	[97]
Ag	Polyvinyl alcohol Dicarboxylic acids	Esterification	10.4	0.93 (oxalic acid) 1.74 (sebacic acid)	[98]
Pt	Aluminum-organic-stabilizer Bifunctional alcohols	Protonolytic cross-linking reaction	1.2	0.98 (hydroquinone) 1.23 (4,4′-dihydroxybiphenyl)	[99]

[a]The distance between NP is the edge–edge distance if not noted specially

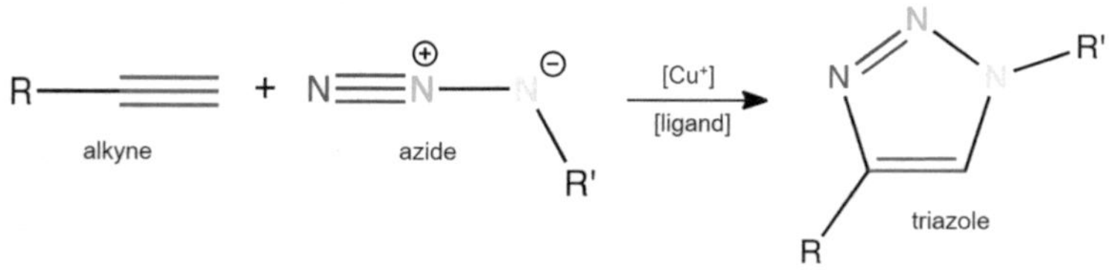

Scheme 5.2 Azide–alkyne 1,3-dipolar cycloaddition reaction

click reactions of azide-functionalized Au NP and dialkyne-terminated functional molecules (Fig. 5.18a–c) [93]. A similar procedure (use of azide-tagged Au NP and a dialkyne cross-linker) was also followed with Au NP of ca. 18 nm diameter [104]. The aggregation rate was found to depend on Cu concentration, but in all cases, aggregation was clearly visible within 20 min from the beginning of the reaction. The reaction of gold NP (1.3 nm) stabilized by a single dendritic thioether ligand comprising an alkyne function with di-, tri- and tetra-azide linker molecules has also been reported [105]. In that case, dimers, trimers, and tetramers could be selectively produced after the click reaction with the corresponding linker. All measured inter-particle distances were significantly shorter than the calculated maximum possible spacing. This behavior was due to the fact that the NP rearrange in a more folded geometry. Mixtures of gold NP that have azide- and alkyne-terminated groups have also been used to produce covalent networks [106–108]. This click reaction was also used for the covalent bonding of functionalized Au NP onto surfaces [109], or to produce multilayers of covalently bonded Au NP onto surfaces (Fig. 5.18d) [110, 111]. Interestingly, it was possible to use the click cycloaddition to prepare assemblies of Au nanorods and Ag NP [112]. For this reaction, new disulfides with azide (for Ag NP) or alkyne (for Au nanorods) terminations were used.

Covalent assemblies of other metals than gold have also been prepared by click chemistry, such as Fe NP [113], or core-shell CdSe/ZnS quantum dots [114]. The case of iron NP is particularly interesting since it highlights the limit of the method in the case of NP that can suffer oxidation. The authors stated that the solutions for avoiding oxidation, while assembling such NP, are: (i) the functionalization and assembly should be extremely efficient and fast to avoid surface oxidation; (ii) the solvents used should not be a substantial source of oxygen for NP oxidation; and (iii) ideally, purely thermal or photochemical methods that do not require catalysts (and in the case of "click" reaction water-based solvents to dissolve such catalysts) should be used.

Finally, it is worth mentioning that 1,3-dipolar cycloaddition reactions, such as those forming 1,2,3-triazoles can be reversed, but the necessary conditions make the potential use of the reverse reactions rather unrealistic at this point [115].

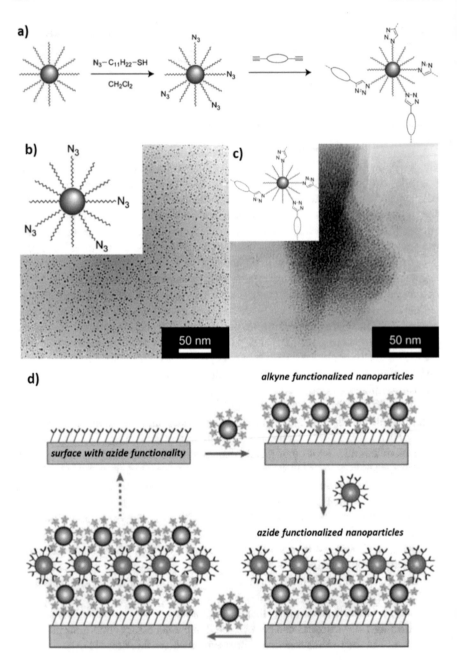

Fig. 5.18 **a** Typical procedure to obtain Au NP network using click chemistry; **b** TEM micrograph of azide-functionalized Au NP; and **c** TEM micrograph of the Au NP network prepared by the click reaction. Reproduced with permission from Ref. [93]. **d** Basic scheme of NP multilayers deposition. Reproduced with permission from Ref. [110]

5.2.2.2 Covalent Coupling Reactions

In addition to click reactions, diverse organic transformations have been used to connect metal NP into 1-D, 2-D, or 3-D assemblies. An amide-coupling reaction was used to organize Au NP into linear chains [116]. The originality of this work lies in the functionalization of the polar singularities that must form when a curved NP surface is coated with ordered monolayers, such as a phase-separated mixture of ligands leading to point defects (Fig. 5.19b). This interesting approach overcomes the disability of NP to bind along specific directions as atoms and molecules do. Gold NP coated with a binary mixture of 1-nonanethiol and 4-methylbenzenethiol were first produced, and their reaction with 1,6-diaminohexane produces linear chains of Au NP (Fig. 5.19a–c). These results suggest that polar singularities react faster than other defects in the ligand shell.

Fiałkowski et al. used a naphthalene dianhydride derivative as cross-linking agent that forms amide bonds with the aminothiolate ligand that stabilized Au NPs (5.6 nm) under biphasic conditions [117]. In this system, the thiol group is coordinated to Au NP surface, and the unbound amino group has the capability of binding protons in a reversible way so that an equilibrium between the protonated and non-protonated

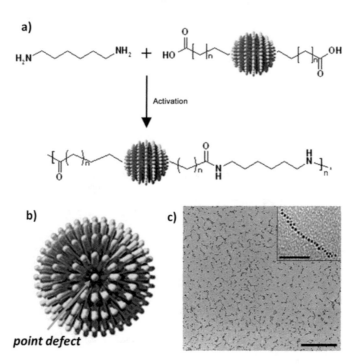

Fig. 5.19 **a** Schematic representation of the chain formation reaction; **b** idealized drawing of a top view of a rippled particle showing the two polar defects that must exist to allow the alternation of concentric rings; and **c** TEM images of Au NP chains. Scale bars 200 nm, inset 50 nm. Reproduced with permission from Ref. [116]

amino groups is established depending on the pH. To carry out the cross-linking reaction, the Au NP have to be brought close enough to one another. To do so, the Au NP are transferred from the bulk aqueous phase to the oil–water interface that takes place only at specific (basic) pH levels. This technique allows the formation of freestanding monolayer of covalently bonded Au NP at the water–oil interface.

Azo (–N=N–) or amido (CONH) linkages were also used to covalently attach gold and CdS NPs (Fig. 5.20a) [118]. These reactions were carried out under very dilute conditions to control the assembly and avoid the polymerization. Indeed, at high ligand concentrations, the NP formed a network of chains for both amido- and azo-functionalized particles. Gold NP (1.2 nm)-CdSe–ZnS quantum dots (5.6, 6.8, and 7.8 nm) assemblies were formed via covalent coupling of carboxyl-functionalized Au NP, which were activated with N-hydroxysuccinimide, with amine-functionalized quantum dots [119].

Esterification reactions were used for the covalent binding of OH-terminated (polyvinylalcool, PVA) Ag NP (10.4 nm) by means of dicarboxylic acids (oxalic, l-(+)-tartaric, d-glucaric acid and sebacic acid) with a defined molecular length [98]. PVA plays a key role in this reaction because it is the reducing agent of Ag^I, the surface stabilizer of the resulting NP, and acts as a mild Lewis acid that can exhibit oxophilic character toward the carbonyl group, which forms the basis for the catalysis

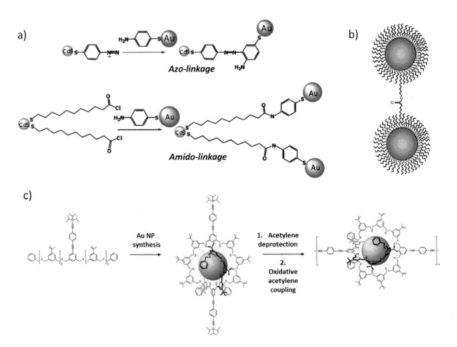

Fig. 5.20 **a** Azo and amido linkage. Reproduced with permission from Ref. [118]. **b** Ester coupling. Reproduced with permission from Ref. [120]. **c** Oxidative acetylene coupling for the assembly of Au NP. Reproduced with permission from Ref. [94]

of the esterification reaction. As expected, the increase in the molecular length of the diacids results in larger inter-particle spacing (Fig. 5.21).

Ester coupling (Fig. 5.20b) of monolayer-protected Au NP (11-mercaptoundecanoic acid exchanged and 11-mercaptoundecanol exchanged) was performed with ester coupling reagents such as 1,3-dicyclohexylcarbodiimide and 4-(dimethylamino) pyridine [120]. Bifunctionalized Au NP (1.1 nm) with two octadentate thioether ligands were covalently interlinked [94]. Since two ligands are required to cover the NP surface, exactly two triisopropylsilyl-protected acetylenes are present as functional groups on the periphery of the NP, allowing their interlinking by wet chemistry: (i) deprotection of the acetylene by fluoride ions and (ii) rapid (15 min.) oxidative acetylene coupling (Fig. 5.20c). In these

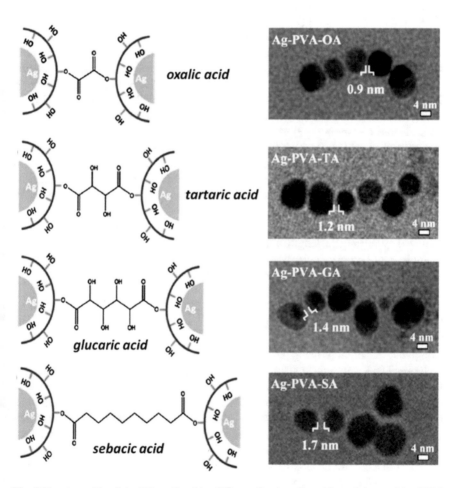

Fig. 5.21 Assembly of Ag NP mediated by different dicarboxylic acids, and the resulting TEM images, where the average inter-particle distances are indicated. Reproduced with permission from Ref. [98]

assemblies, the inter-particle distance (0.8–2.5) that depends on the length of the oligo(phenyleneethynyl) rod is generally shorter than the linker length (1.4–2.7), which may indicate a tangential arrangement of the rigid rod structure at the Au NP's surface. Dithiol coupling was reported as an efficient method to covalently link Au NP and Au nanorods (length = 50 nm, width 5 nm) [96]. In this work, the control of the position, spacing, and quantity of Au NP per nanorod is achieved through anisotropic surface functionalization of the nanorod with aminoalkylthiols of increasing size. The dithiol coupling was performed by converting the terminal amines on polyethylene glycol-stabilized Au NP into thiols through an excess of a cyclic thioimidate (Traut's reagent, 2-iminothiolane HCl). Au NP (4.3 nm) functionalized with norbornenethiol were cross-linked by ring-opening metathesis polymerization (ROMP) with the use of a water-soluble pyridine-substituted ruthenium benzylidene catalyst [95]. The ROMP of the Au NP can be achieved across a large area, stitching the NP crystalline domains together.

The amination reaction of fullerene C_{60} with amine-functionalized Au NP [121] was used to produce C_{60}-linked Au NP [122]. For the assembly, the C_{60} were reacted with 4-aminothiophenoxide/hexanethiolate-protected Au NP. Although resulting from a non-covalent bond, the affinity of streptavidin for biotin results in the strongest biological interaction known, and it was used for the end-to-end linkages of gold nanorods [123]. A protonolytic cross-linking reaction of aluminum-organic-stabilized Pt NP with bifunctional alcohols (ethylene glycol, hydroquinone, 4,4′-dihydroxybiphenyl, and 1,10-decanediol) was reported by Bönnemann et al. to build 3-D Pt NP networks [99]. The key feature of this synthesis is the formation of an organometallic colloidal protecting shell around the Pt NP. When reacting with bifunctional ligands, a cross-linking of the NP occurs that leads to the formation of a 3-D NP network (Fig. 5.22). The control of the inter-particle distance can be achieved by varying the length of the spacer molecules (Table 5.3). Finally, cadmium selenide NP (2.7 nm) modified with 2-aminoethanethiol were cross-linked with the homobifunctional amine-reactive cross-linker glutaraldehyde [97].

5.2.2.3 Biomolecular Coupling

Compared to linking using small organic ligands, linking with biomolecules including DNA, supramolecular protein [124], and viruses offers several advantages. First, these biomolecules can be tailored to specific lengths by varying the number of base pairs. Additionally, the double-helix structure of DNA is rigid, enabling the precise control of the spacing. Thus, NP network engineering with DNA gives independent control of three important design parameters (NP size, lattice parameters, and crystallographic symmetry) by separating the identity of the particle from the variables that control its assembly [125]. One-, two-, and three-dimensional assemblies of metal NP have been obtained with DNA. There is a very rich literature on that subject, and we will discuss in this section some basic principles of NP assembly with DNA. The readers particularly interested in this subject, which has already found some applications in (electro)catalysis (sometimes the DNA can be removed through

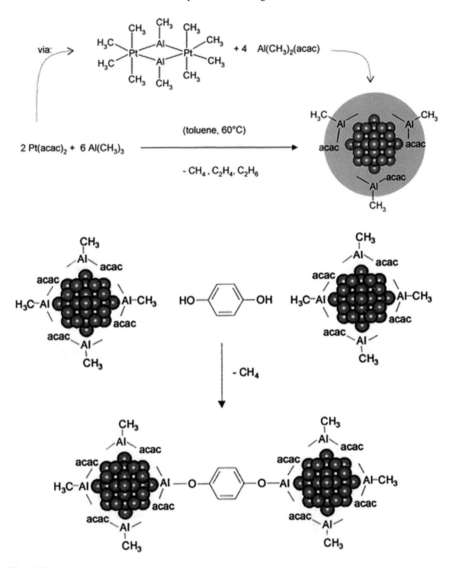

Fig. 5.22 Assembly of Pt NP by a protonolytic cross-linking reaction. Reproduced with permission from Ref. [99]

calcination to generate robust porous NP super-lattices [126]), [126–132] are invited to read some recent reviews [21, 133–135].

Although many modifications are available for functionalizing DNA to bind to metals [136], it can be used as a ligand by itself [137]. The coordination chemistry of DNA nucleosides (2′-deoxyadenosine (dA), 2′-deoxycytidine (dC), 2′-deoxyguanosine (dG), and 2′-deoxythymidine (dT)) on Au NP was probed by surface-enhanced Raman spectroscopy and is shown in Fig. 5.23 [138]. The coordi-

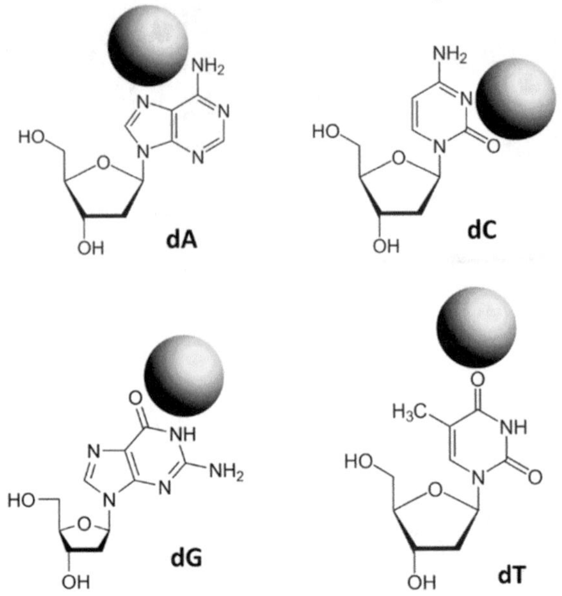

Fig. 5.23 Proposed structures of DNA nucleosides bound to the gold nanoparticles. Reproduced with permission from Ref. [138]

nation of the dA, dC, and dG is stronger to Au NP than that of dT. The dA mainly binds to Au NP via an N atom of the imidazole ring, and the NH₂ group participates in the coordination process. The dC binds to the Au surface via an N atom of the pyrimidine ring with a partial contribution from the oxygen of C=O group. The coordination of dG implicates both the N atom and the oxygen of the C=O group of the pyrimidine ring. Only dT binds to the Au surfaces via the oxygen of C=O group of the pyrimidine ring.

One-dimensional DNA@Au NP wires were generated by a method involving the incorporation of a functionalized Au NP (psoralen-modified NP, 3 nm) into double-stranded DNA, followed by the photochemical cross-linking ($\lambda = 360$ nm) of the ligand to the DNA matrix [139]. Under these conditions, psoralen undergoes a photo-induced $2\pi + 2\pi$ cycloaddition with the thymine residues, a process that leads to the covalent attachment of the ligand to the DNA. Labean et al. used a combination of self-assembly, molecular recognition, and templating, which rely on an oligonucleotide covalently bounded to a high-affinity gold-binding peptide. After integration of the peptide-coupled DNA into a self-assembling super-structure, the templated peptides recognize and bind the Au NP [140]. Gold nanoparticles, 1.4 nm in diameter, were assembled in 2-D arrays with inter-particle spacing of 4 and 64 nm. The NP formed precisely integrated components, which are covalently bonded to the DNA scaffolding. For the self-assembly of NP into 3-D lattices, the use of DNA origami frames (rigid and with well-defined geometries) has emerged as a promising solution [141]. Thus, the 3-D organization of Au NP (7, 10, and 15 nm) spatially arranged in pre-determined positions was reported using DNA origami octahedron as frame [142]. The octahedra can serve as programmable inter-particle linkers. 2-D organization of Au NP has also been reported by wrapping them in flower-shaped DNA origami structures [143].

5.2.2.4 Metal-Ion-Induced Nanoparticle Assembly

Modification of NP with metal coordination compounds [144, 145] offers the possibility of their assembly on surfaces [146], but also to create metal NP networks with coordination bonds connecting the NP. This strategy has been mainly employed for the development of plasmon-based colorimetric sensors for ultrasensitive molecular diagnostics [147]. The technique is based on controlled aggregation of NP in the presence of metal ions. Aggregation of metal NP in the presence of analyte ions changes the color of the NP solution. Covalent assemblies have been built with various families of ligands such as carboxylates, amino acids (Fig. 5.24a), amine ligands, or crown ethers (Fig. 5.24b). While most of the reports rely on irreversible metal NP assemblies, attempts to use metal ions as chemical stimuli for reversible assembly have also been reported (see Sect. 5.2.3.2). Table 5.4 shows representative examples obtained with various ligands and cations for gold, Ni_2P, and CdTe NP.

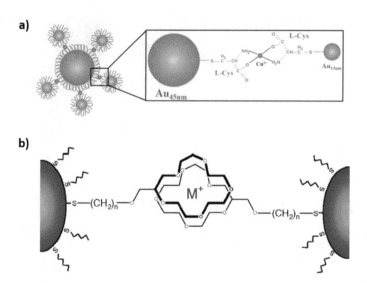

Fig. 5.24 **a** Illustration of Cu^{2+} mediated assembly of Au NP after self-assembly. Reproduced with permission from Ref. [148]. **b** Crown ether-metal ion-crown ether "sandwich" inter-particle bridge. Reproduced with permission from Ref. [149]

Table 5.4 Representative examples of NP networks obtained with various ligands and cations for gold, Ni_2P, and CdTe NP

Metal	Ligand	Metal ion	NP diameter	TEM	References
Au^a	Amino acid (L-cysteine)	Cu^{2+}	13 and 15 nm	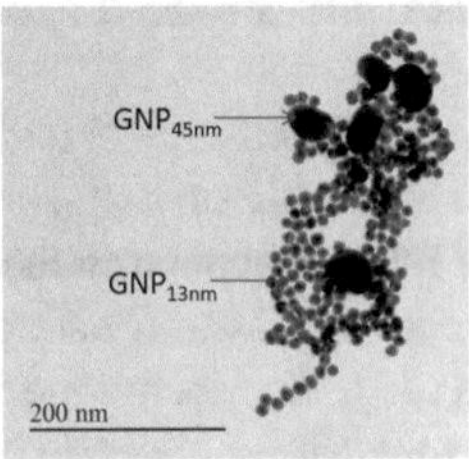	[148]
Au	Amino acid (L-cysteine)	Hg^{2+}, Cd^{2+}, Fe^{3+}, Pb^{2+}, Al^{3+}, Cu^{2+}, and Cr^{3+}	30 nm	–	[150]
Au^b	Amino acid (histidine)	Fe^{3+}	18 nm	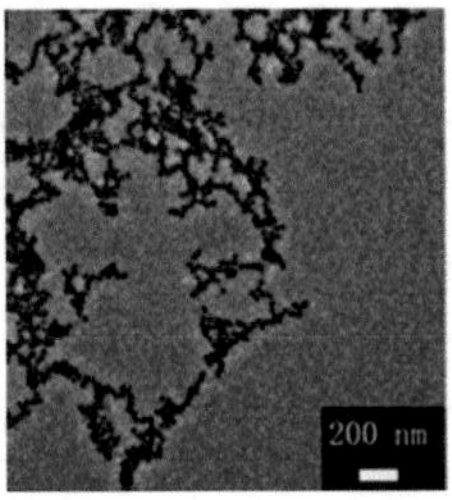	[151]
Au	Crown ether	K^+, Na^+, and Ag^+	Au_{140}, Au_{220}, and Au_{309}	–	[149]
Au	Crown ether	K^+ and Na^+	18 nm	–	[152]
Au^c	Crown ether	K^+	18 nm	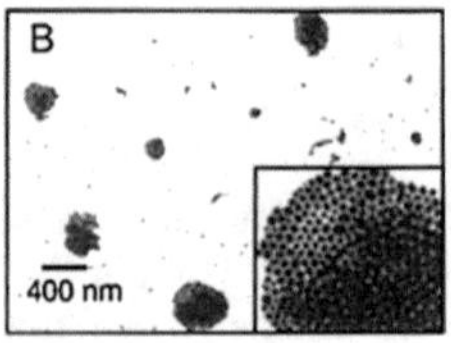	[153]
Au	Pyridine	Cu^{2+}	2 nm	–	[154]

(continued)

Table 5.4 (continued)

Metal	Ligand	Metal ion	NP diameter	TEM	References
Au[d]	Terpyridine	Fe^{2+}, Zn^{2+}, Cu^+, Ag^+	2 nm	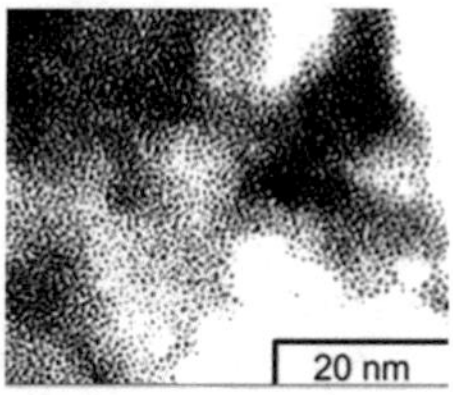 20 nm	[155]
Au	Phenanthroline	Os^{3+}	90 nm	–	[156]
Au[e]	1,10-Phenanthroline	Li^+	4 and 32 nm	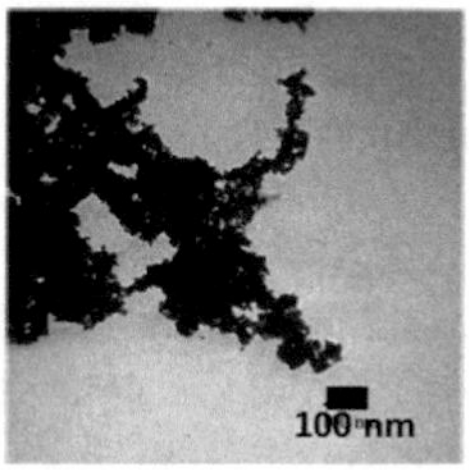 100 nm	[157]
Au	Carboxylate (citrate)	Na^+, Ca^{2+}, Mg^{2+}, Cu^{2+}, Zn^{2+}, and Al^{3+}	18 nm	–	[158]
Au[f]	Carboxylate (valine)	Pb^{2+}	22–75 nm	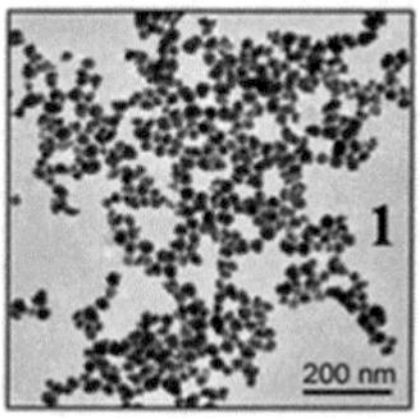 1 200 nm	[159]
Au	Carboxylate	Pb^{2+}, Hg^{2+}, and Cd^{2+}	14 nm	–	[160]
Au	Bishydroxamate	Zr^{4+}	6 nm	–	[161]
Ni_2P[g]	Carboxylate	Ni^{2+}	8–9 nm	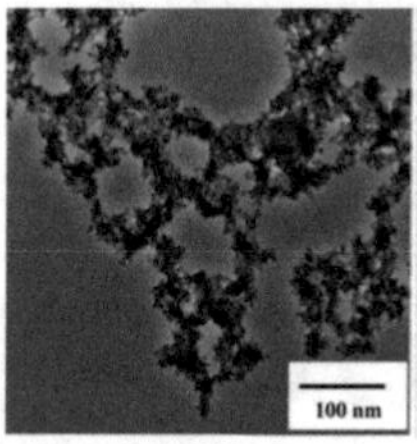 100 nm	[162]

(continued)

Table 5.4 (continued)

Metal	Ligand	Metal ion	NP diameter	TEM	References
CdTe	Mercaptomethyltetrazole	Cd^{2+}	3 nm	–	[163]

[a]Figure reproduced with permission from Ref. [148]
[b]Figure reproduced with permission from Ref. [151]
[c]Figure reproduced with permission from Ref. [153]
[d]Figure reproduced with permission from Ref. [155]
[e]Figure reproduced with permission from Ref. [157]
[f]Figure reproduced with permission from Ref. [159]
[g]Figure reproduced with permission from Ref. [152]

5.2.2.5 Light-Induced Self-assembly of Irreversible Nanoparticle Networks

The use of externally manipulated light offers excellent spatial and temporal control for the signal molecule's activation, degradation, or creation/disruption of a self-assembled system [164]. Light-induced self-assembly (LISA) of metal NP have thus found applications into the biomedical field. Although most of the work deals with non-covalent interactions (mainly NP aggregation due to changes in electrostatic stabilization), some works have reported the irreversible and photochemically activated formation of covalent NP networks (Fig. 5.25).

Zhao and coworkers reported that UV light irradiation triggers Au NP that are, respectively, functionalized with o-nitrobenzyl alcohol and benzylamine to proceed with a covalent ligation reaction [165]. Indeed, o-nitrobenzyl alcohol and benzylamine can undergo an aldehyde-amine ligation-like coupling reaction under UV irradiation. This reaction leads to the assembly of Au NP into anisotropic 1-D arrays in aqueous solution via indazolone linkages (Fig. 5.25a). The formation of anisotropic 1-D Au NP arrays was attributed to the anisotropic electrostatic repulsions during the self-assembling of the charged NP. The self-assembly of Au NP or nanorods into chains and networks was also achieved by a UV light in the presence of silver ions; taking advantage of the easy photoreduction property of silver nitrate [168]. The authors used CTAB (or citrate)-capped gold NP or nanorods that were mixed with the required concentrations of silver salt (the assembly was strongly dependent on the AgNO$_3$ content), and then exposed to a 365 nm UV lamp. From HRTEM and XPS analyses, the authors proposed the formation of a core–shell Ag@Au structure (near identical lattice constants of Au (0.408 Å) and Ag (0.409 Å)) on both NP and nanorods. The group of Shi studied the light-triggered covalent coupling of gold NP [166, 167]. They have shown that Au NP (20 nm) stabilized with photolabile diazirine terminal group of PEG5000 (polyethyleneglycol, Mn = 5000) can be covalently cross-linked upon 405 nm laser irradiation (Fig. 5.25b) [166]. The surface diazirine group is first transformed into a carbene species upon laser excitation at 405 nm. Then, the resulting reactive carbene moieties formed covalent bonds with ligands of adjacent Au NP through C–C, C–H, O–H, and X–H (X = heteroatom) insertions, leading

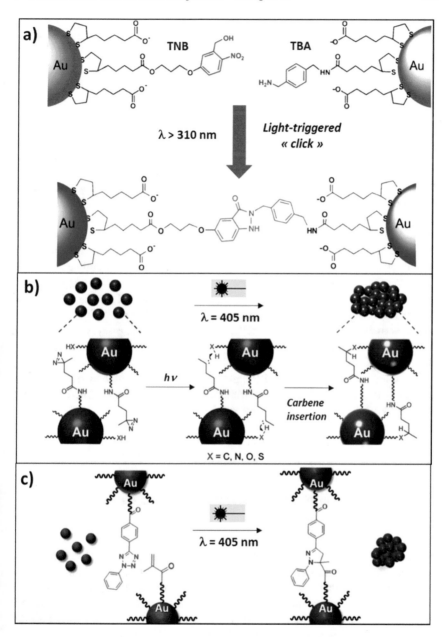

Fig. 5.25 Some examples of light-induced self-assembly of irreversible NP networks: **a** UV light irradiation promotes the irreversible assembly of Au NPs functionalized with o-nitrobenzyl alcohol and benzylamine. Reproduced with permission from Ref. [165]. **b** Photolabile Au NPs covalently cross-linkable with a 405 nm laser irradiation via a diazirine terminal group of PEG5000 ligands. Reproduced with permission from Ref. [166]. **c** Photolabile Au NPs covalently cross-linkable with a 405 nm laser irradiation between 2,5-diphenyltetrazole and methacrylic groups attached PEG5000 ligands. Reproduced with permission from Ref. [167]

Scheme 5.3
Photodimerization of
thymine units

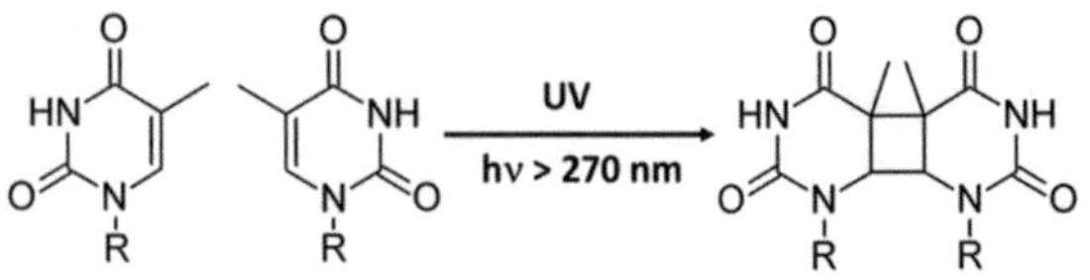

to the formation of covalently cross-linked Au NP aggregates. It is worth mentioning that the water susceptibility of the diazirine group will suppress the photo-cross-linking efficiency. Another strategy reported by the same group involves the use of photolabile Au NP that can effectively form cross-linked aggregates upon 405 nm laser irradiation, between 2,5-diphenyltetrazole and methacrylic groups attached to PEG5000 on the surfaces of 23 nm Au NP (Fig. 5.25c) [167]. The tetrazole groups on the NP surface firstly undergo a facile cyclo-reversion reaction upon laser excitation at 405 nm to release N_2 and to generate nitrile imine dipoles that cyclize spontaneously with the alkene moieties of methacrylic acid on the adjacent Au NP to afford pyrazoline cycloadducts, leading to the formation of covalently cross-linked NP networks. The degree of connection of the Au NP was strongly dependent on the exposure time. Finally, the photochemical assembly of 3.2 nm Au NP was also realized by using the photodimerization of thymine (Scheme 5.3) [169]. In that work, mixtures of 11,11′-dithiobis(undecanoic acid 2-(thymine-1-yl)ethyl ester) and 1-dodecanethiol were used as stabilizers to suppress intramolecular photoreaction of the thymine units on the Au NP surface.

The diameter of the obtained NP aggregates became larger with increasing photo-irradiation time: 0.15, 0.25, and 1 μm were observed after 6, 22, and 72 h, respectively. In a study dealing with 2 and 7 nm Au NP, Ralston et al. have shown that the photodimerization of thymine, which is a [2 + 2] cycloaddition reaction, was mainly influenced by particle size, surface charge, and solvent type [170, 171].

5.2.3 Stimuli-Responsive Reversible Covalent Networks of Nanoparticles

Stimuli-responsive nanomaterials have been particularly studied for biomedical applications, such as drug delivery [172]. In the field of catalysis with metal nanoparticles, artificial switchable catalysts usually rely on non-covalent interactions between NP that induce aggregation [173–175]. Higher activity is achieved when the NP are homogeneously distributed in the reaction medium, while it is lowered after aggregation. Association of metallic NP with stimuli-responsive gels/polymers is another strategy, for which catalytic performances can be modified by organizing and confining metal NP, which goes beyond the scope of this chapter [176, 177].

Diverse stimuli have successfully been used to direct reversible NP covalent networks [178]. They can be classified as chemical stimuli (dynamic covalent chemistry [179], metal ions) and physical stimuli (mainly light).

5.2.3.1 Dynamic Covalent Chemistry

Dynamic covalent chemistry usually relates to chemical reactions carried out reversibly under conditions of equilibrium control, which are employed by chemists to make complex supramolecular assemblies from discrete molecular building blocks [180].

Borsley and Kay have demonstrated that gold NP can be covalently and reversibly assembled using boronate ester chemistry (Fig. 5.26) to form covalent bonds with 1,2-dihydroxybenzenes (catechols) [181]. After the addition of different ditopic catechols in the presence of N-methylmorpholine, the aggregation of NP occurs and leads to an insoluble material. Indeed, boronic acids react with various dihydroxy compounds to yield boronate esters in the presence of Lewis bases. Treating the resulting precipitate

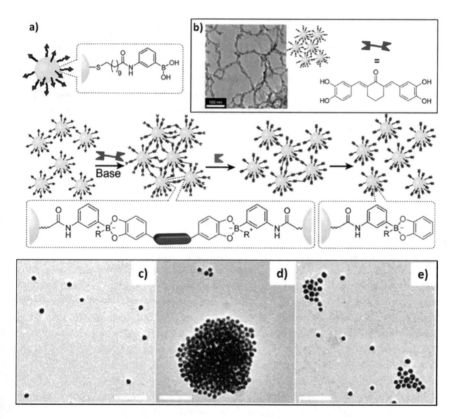

Fig. 5.26 **a** Schematic representation of boronate ester-driven dynamic covalent Au NP assembly and disassembly on sequential addition of a bifunctional linker and a monofunctional capping unit (R = N-methylmorpholinium); and **b** chemical structures of a bifunctional catechol linker and representative TEM image of the assemblies formed on treating Au NP with the linker. Reproduced with permission from Ref. [181]. **c** TEM image of D-Lac-Au NP; **d** TEM images of the assembly of D-Lac-Au NP; and **e** TEM images of the disassembly of D-Lac-Au NP. Scale bar: 100 nm. Reproduced with permission from Ref. [182]

with a monofunctional catechol makes it possible to break the covalent bonds between NP and form free NP in solution. This phenomenon is nevertheless slow (5 days for aggregation and 42 days for disaggregation). Qu and coworkers also took advantage of the reversibility of boron chemistry to produce Au NP networks that are composed of covalent and readily reversible spiroborate diester linkages [182]. Au NP stabilized with thioctic amides terminated by D-lactose (D-Lac) were cross-linked with borate ions by forming dynamic covalent spiroborates between the *cis*-vicinal diol sites of D-Lac under alkaline conditions. In the presence of external chemical stimulus, such as *cis*-vicinal diols that compete for bonding with the borate, the spiroborate linkages between the Au NP can be broken. TEM micrographs of the initial D-Lac stabilized Au NPs and of their assembly and disassembly are shown on Fig. 5.26c–e.

The thermally reversible Diels–Alder reaction between furan and maleimides has been also investigated to connect and disconnect Au NP (Fig. 5.27a) [102]. Maleimide-modified monolayer-protected Au NP (2-Au) were produced from protected furan-maleimide NP (1-Au) via a thermally reversible Diels–Alder reaction. These maleimide-NP served as a general platform for a Diels–Alder reaction with furan-modified Au NP (3-Au) to prepare 3D networks reversibly. A similar strategy was followed by Xia and coworkers, but in that case the reaction occurred through mild Diels–Alder cross-linking between maleimide bearing oligo(p-phenylenevinylene)- OPV- and 2 nm furan-functionalized Au NP (Au-f) as depicted in Fig. 5.27b [183].

5.2.3.2 Metal Ions as Chemical Stimuli

The interest of assembling (plasmonic) NP using metal ions comes initially from the development of sensors to detect these often-toxic ions.

The self-assembly, disassembly, and reassembly of Au nanorods mediated by [(disulfide-terminated tpy)$_2$-M^{II}] complexes (M = Fe, Cd) was investigated by Newcome et al. [184]. The side faces of the Au nanorods are protected more strongly by the stabilizer (cetyl trimethylammonium bromide, CTAB) than their tips, where the ligand exchange reaction occurs preferentially with the M^{II}-based cross-linker. Facile disassembly occurred upon NaOH addition for the Fe^{II} linker and Cd(NO$_3$)$_2$.4H$_2$O addition for the Cd^{II} linker. The process was not reversible in the case of iron since subsequent Fe^{II} addition results in the chelation of Fe^{II} by two terpyridine units on the same nanorod. Reversibility was achieved using Cd^{II}, since cadmium complexes with terpyridine ligands are weaker and more labile.

Metal-ion-induced reversible self-assembly of carboxylated peptide-functionalized gold NP was reported by Mandal et al. [185]. The extent of assembly (2-D and 3-D structures) is dependent on the amount of metal ions (Pb^{II}, Cd^{II}, Cu^{II}, and Zn^{II}) present in aqueous solution. The process is completely reversible by addition of alkaline ethylenediaminetetraacetic acid (EDTA) solution (Fig. 5.28).

In order to achieve the formation of stable NP aggregation, which is difficult to control by binding divalent metal ions with carboxylate ligands because of charge

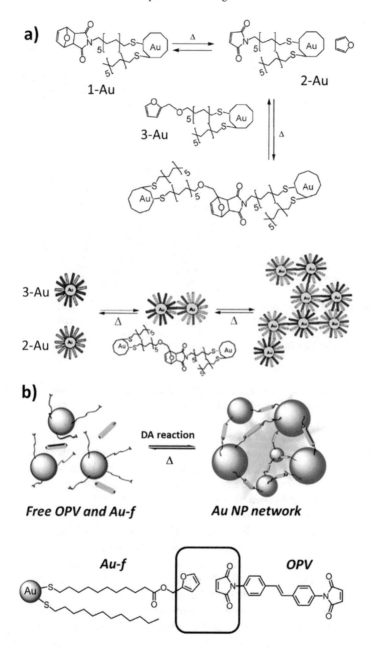

Fig. 5.27 **a** Forward and retro Diels–Alder reactions between maleimide-modified Au NP (2-Au) and furan-modified Au NP (3-Au). Reproduced with permission from Ref. [102]. **b** Representative illustration for the thermally reversible Diels–Alder reaction leading to ordered Au NP self-assemblies. Reproduced with permission from Ref. [183]

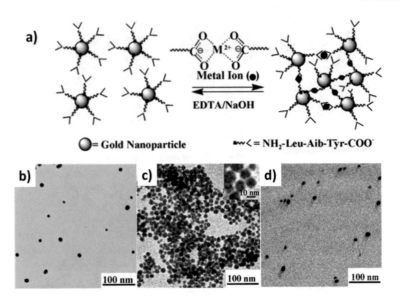

Fig. 5.28 **a** Metal-ion-induced reversible self-assembly of carboxylated peptide-functionalized Au NP; and TEM images of a NP suspension **b** before addition of metal ions, **c** 10 min after addition of Pb^{II} ions, and **d** 5 min after addition of alkaline EDTA solution to the solution of Au NP containing Pb^{II} ions. Reproduced with permission from Ref. [185]

neutralization on the NP, Pillai et al. studied how the colloidal stability is affected by the incorporation of different amounts of thiols terminated with NMe_3^+ groups within monolayers of COO^--terminated ligands (Fig. 5.29) [186]. After addition of Pb^{II}, NP containing 80% of a COO^-- and 20% of NMe_3^+-terminated thiol formed aggregates. Including the NMe_3^+ species within the protective coating was essential for reversing the self-assembly. The subsequently added NaOH sequestered Pb^{II} from NP aggregates, resulting in the disassembly. Without NMe_3^+ species, Pb^{II} ions were bound too strongly and could not be removed using NaOH.

The use of Au NP capped with a zwitterionic peptide $(AuNP-(EK)_3)$ has allowed Surareungchai et al. to trap Ni^{II} ions [187]. Zwitterionic polypeptide-capped Au NP contain alternate carboxylic/amine groups. The zwitterionic peptide can function dually by being able to trap metal ions and maintain colloidal stability. The authors have demonstrated that the aggregation mechanism is due to the interactions between the $-NH_2$ group of the peptide and Ni^{II}, and the aggregation process is reversible because the pH controls the protonation/deprotonation of the $-COOH/-NH_2$ groups, and therefore the ligand affinity. Spiropyran-terminated alkanethiols were used to stabilized gold NP [188]. Under visible light, the spiropyran (photoswitchable organic molecules) exists in their non-planar and closed form (A), which isomerized to the planar and open form (merocyanine, B) with UV light irradiation heterocyclic ring cleavage (Scheme 5.4). The open form has a phenolate group that can bind with Cu^{II} metal ions, and the chelated metal ions can be released by visible light irradiation.

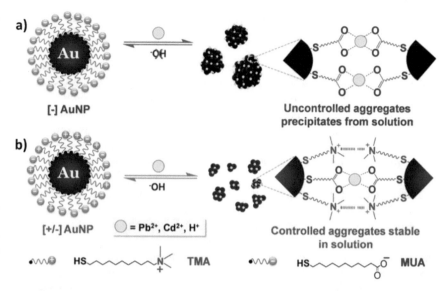

Fig. 5.29 Concept of regulating inter-particle forces to achieve controlled aggregation in charged NP. The interactions of **a** [−] and **b** [±] Au NP with triggering ions resulting in the formation of unstable and stable Au NP-ion aggregates, respectively. The colloidal stability of [±] Au NP is retained in the aggregates due to the electrostatic repulsions experienced from the like charged head groups on adjacent NP. Reproduced with permission from Ref. [186]

Scheme 5.4 Spiropyran (A) to merocyanine (B)

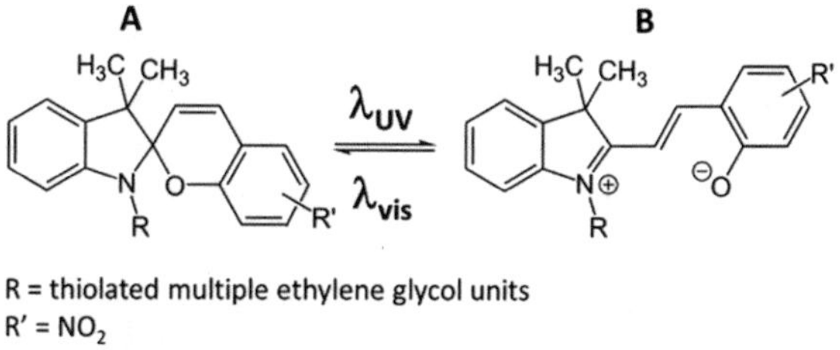

R = thiolated multiple ethylene glycol units
R' = NO$_2$

Finally, other chemical stimuli can be investigated. Thus, in the case of covalent cross-linking of gold NP with dithiol molecules, which is normally considered to be an irreversible self-assembly process, it was recently shown that the process may be reversible by oxidizing the dithiolate linkers with ozone [189]. Ozone quantitatively oxidizes the dithiolate groups to weakly bound sulfonates, thereby destroying the cross-linking and freeing the NPs to redisperse as charge-stabilized colloids. The process is repeatable by adding additional dithiol to reinitiate the assembly.

5.2.3.3 Dynamic Covalent Chemistry Assisted by Light as Physical Stimulus

The use of light-induced reactions as external stimuli offers rapid and precise spatial control in closed systems [190]. A simple way to render NP photoresponsive is to functionalize their surface with ligands terminated by light-switchable moieties, which can lead to light-induced reversible self-assembly (LIRSA). Photoreversible [2 + 2] photocycloaddition [190, 191] or reversible Diels–Alder reaction employing a diarylethene as the diene and a maleimide as the dienophile [192] have been used for exerting photocontrol over the connection and disconnection of dynamic covalent bonds.

The successful LIRSA of gold NP stabilized with coumarin ligands was reported by Zhan et al. [193]. Due to the thiolated coumarin derivative, Au NP can be self-assembled by light irradiation at 365 nm via a [2 + 2] photocycloaddition, and the resulting purple NP network can be disassembled back to the initial deep red disperse state by a relatively short exposure to UV light, as shown in Fig. 5.30a. This LIRSA cycle was repeated four times (Fig. 5.30b). A similar procedure was used by Lin et al. with amphiphilic star-like poly(acrylicacid)-block-poly(7-methylacryloyloxy-4-methylcoumarin) diblock copolymers [194].

The use of 3-cyanovinylcarbazole as a photochemical switch to reversibly ligate gold NP assemblies using light was reported by Kanaras et al. [195]. In this work, gold NP stabilized with oligonucleotides are hybridized under the appropriate conditions to form NP networks. The sequences used in this work are reported in Table 5.3. Each DNA sequence has a thiol modification, a spacer part of 15 thymine bases, and a complementarity region of 15 bases with carbazole modification. The oligonucleotide-coated NP contains a 3-cyanovinylcarbazole modification, which can react upon irradiation at $\lambda = 365$ nm with an adjacent thymine in the complementary strand via a [2 + 2] photocycloaddition, causing the formation of a cyclobutane

Fig. 5.30 **a** LIRSA behavior of Au NP; and **b** reversible change in the wavelength maximum during assembly of Au NP with illumination at $\lambda = 365$ nm (72 h) and disassembly at $\lambda = 254$ nm (60 min). Reproduced with permission from Ref. [193]

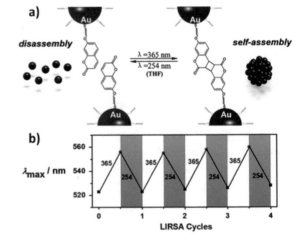

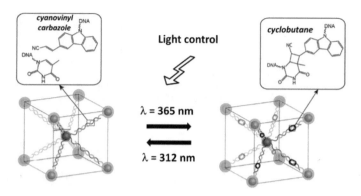

Fig. 5.31 Reversible photochemical ligation of Au nanoparticle networks. Reproduced with permission from Ref. [195]

(Fig. 5.31). This photochemical process can be reversed upon irradiation with light at $\lambda = 312$ nm.

All the indirect cross-linking methods we have seen up to now rely on the use of organic ligands to direct the NP assembly that acts as inter-particle spacers supporting the macroscale assemblies. Thus, the NP interactions are mediated by these ligands, and their presence can be detrimental to electrical transport, inter-particle coupling, and thermal stability, properties of importance for catalysis. Therefore, the synthesis of self-supported NP networks with direct inter-particle linkages (metallic and not covalent bonding) has also been investigated.

The primary method to construct direct interfacial linkages of metal NP involves the controlled oxidation of the ligands via a modified sol-gel synthesis using a destabilizer (ethanol, hydrogen peroxide) leading to non-supported metal aerogels [196]. To gelate the stable sols (aqueous colloidal metal solutions), efficient destabilization is initiated by concentrating the sols, and gel formation is achieved by addition of the destabilizer. Clear differences between mono-(Au, Pt, Ag) and bimetallic (Au–Ag, Pt–Ag) gel formations were observed in their effective destabilization agents and timescales of formation. Strongly increased reproducibility was obtained with bimetallic systems (see Fig. 5.32a for the Pt–Ag system). In this process, it was proposed that as the oxidation occurs, low-coordinated surface sites are created, which then react with similar surface sites of nearby NP to reduce the surface energy [197]. Au/Ag alloy aerogels (Fig. 5.32b) were also produced via oxidative self-assembly of colloidal NP using tetranitromethane as destabilizer [198]. This new technique allows the production of high surface area (50–70 m^2 g^{-1}), self-supported, bimetallic super-structures via the controlled oxidation of the ligands.

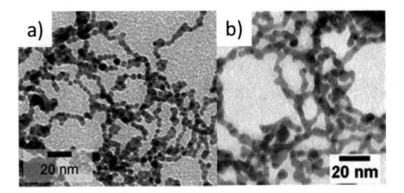

Fig. 5.32 TEM micrographs of: **a** Pt/Ag. Reproduced with permission from Ref. [196]. **b** Au/Ag alloy aerogel. Reproduced with permission from Ref. [198]

5.3 Metal Nanoparticle Assemblies as Catalysts

It has been shown that assemblies of metal NP can be beneficial for catalytic applications. For instance, FePt NP assembled on graphene were more active and robust than the same NP or commercial Pt NP deposited on a commercial carbon support for the oxygen reduction reaction [199]. Also, Ru nanochains produced by self-assembling of Ru nanoseeds (3.5 nm) were more efficient as CO oxidation catalyst than the respective isolated Ru nanoseeds (3.5 nm) or Ru NP (6 nm) [200]. The self-assembly of the Ru NP was accomplished by reacting the Ru NP seeds with cetyltrimethylammonium bromide in water, which allowed the authors to obtain interconnected networks of Ru NP, the length of which varies from a few tens of nanometers to the micron scale. Photoresponsive Au NP (6 nm), Fe_3O_4 NP (11 nm), and SiO_2 NP (17 nm) led to a reversible self-assembly by using the azobenzene anchored to their surface and light [201]. In this particular example, the cavity in between the NP was exploited to trap selectively some molecules, including enantiomeric compounds, and also to perform reactions more efficiently in the confined space. The acid-catalyzed hydrolysis of an acetal or the anthracene dimerization reaction proceeded more efficiently compared to unconfined reactions (without the assemblies in the reaction media). The dimensions of the confined space could be modulated by the NP size as well as by the lengths of the surface ligands. Even though those examples do not concern specifically covalently bound metal NP, they show that assemblies of metal NP can find applications as catalyst and that their assembly affects in a positive way their activity, robustness, or both. Covalent assemblies of metal NP have also been used as catalysts; however, a limited work on the field has been described, and in many cases, the effect of the assemblies compared to non-aggregated NP has not been discussed. Nevertheless, some works have evidenced that the covalent assemblies of metal NP have a positive effect in the catalytic performances with respect to other catalytic systems [202, 203]. Also, the possibility to use dynamic (switchable) assemblies triggered by different stimuli (pH, light, temperature) has been exploited

[173, 174, 177], which, in some cases are reversible, thus allowing one to turn on/off the catalytic reaction at will [202].

As stated in the previous sections, a large volume of scientific contributions in the field of covalent assemblies of metal NP is devoted to Au and at certain extent to Ag, mainly because of their possible applications in several fields such as optics, sensors, and electronics [204, 205]. Au assembled NP have also found applications as catalysts, mainly in reduction [52, 130, 202, 206, 207] and oxidation reactions [45, 70, 71, 111, 126, 208]. Even if predominately metal NP assemblies pivot around Au NP, other metals have been discussed and investigated. Metals such as Pd, Pt, and Ru that possess excellent catalytic properties have also been studied for the production of NP assemblies, which in turn can be used as catalysts [205, 209]. Some examples of the use of metal NP assemblies as catalyst in reduction, oxidation, water splitting as well as in other reactions are summarized in Tables 5.5, 5.6, and 5.7, respectively, and representative systems are discussed below.

Reduction Reactions

Thin films of Au NP prepared by means of the LBL method using O-carboxymethyl chitosan as stabilizer were used as catalysts for the reduction 4-nitrophenol by $NaBH_4$ [52]. The layers of the catalysts deposited on glass plates (10, 20, or 40 layers) and the size of the Au NP (5 nm or 10–20 nm) were crucial for the measured catalytic activity. The rate of reaction was higher with catalysts with less layers (ten layers) and smaller NP size (5 nm), hence proportional to the available catalytic metal surface, related to the porosity and NP size. SEM analyses were performed after catalysis, showing that the most active catalyst was the least stable, but no leaching tests were reported.

Au NP inside DNA hydrogel were synthesized by reduction of $HAuCl_4$ adsorbed in the DNA hydrogel by $NaBH_4$ (Fig. 5.33) [130]. The Au nanocomposite displayed a weak absorbance band at $\lambda = 550$ nm, with an average NP size of 2.8 ± 1.0 nm. Reduction of 4-nitrophenol to 4-aminophenol by $NaBH_4$ was monitored spectroscopically. The calculated rate constant was 1.5×10^{-3} s^{-1}, which according to the authors was faster than the rates of already reported analogous systems. However, no recycling test or characterization of the catalysts after catalysis was given to evaluate the robustness of this Au nanocomposite.

Two-dimensional self-assembled AuCu NP were synthesized from $CuCl_2$ and $HAuCl_4$ in the presence of hexadecylamine by reduction with glucose, to give well-organized assemblies into highly flexible ribbon-like structures [206]. The morphology and the self-assembly were highly dependent on the synthetic procedure. These assemblies were active in the photocatalytic degradation of 4-nitrophenol by $NaBH_4$, which was carried out with a 515 nm continuous laser.

1-D or 3-D assemblies of Au NP were synthesized in a controlled manner by adjusting the pH in the reduction of $HAuCl_4$ by α-cyclodextrin. While alkaline solutions promoted the synthesis of isolated Au NP, higher pH produced the Au NP assemblies. 1-D assemblies could also be obtained by the host–guest interaction between α-cyclodextrin capped Au NP and toluene [210]. The Au NP were tested as catalyst in the reduction of 4-nitrophenol by $NaBH_4$. An effect of the size on

Table 5.5 Reduction reactions catalyzed by NP assemblies

Metal	Ligand	Method	Reaction	References
Au	O-Carboxymethyl chitosan	LBL	Reduction of 4-nitrophenol by $NaBH_4$	[52]
Au	DNA cross-linked hydrogel	DNA-assisted	Reduction of 4-nitrophenol by $NaBH_4$	[130]
AuCu	Hexadecyl amine	One-phase	Photocatalytic reduction of 4-nitrophenol by $NaBH_4$	[206]
Au	α-cyclodextrin	Indirect cross-linking	Reduction of 4-nitrophenol by $NaBH_4$	[210]
Au/Fe_3O_4	Oleic acid and α-cyclodextrin	Indirect cross-linking	Reduction of methylene blue by $NaBH_4$	[207]
Au/TiO_2	Amine and aldehyde compounds	Dynamic self-assembly	Photocatalytic reduction of methylene blue	[202]
Ru	C_{60}	One-phase	Hydrogenation of nitrobenzene and cinnamaldehyde by H_2	[80, 81, 211]
Ru	$C_{66}(COOH)_{12}$	One-phase	Hydrogenation of nitrobenzene by H_2	[41]
Pd	COF	Two-phase	Reduction of 4-nitrophenol by $NaBH_4$	[88]
Pt	Thio-COF	One-phase	reduction of 4-nitrophenol by $NaBH_4$	[89]
Pt	DNA	One-phase	reduction of nitroarenes by $NaBH_4$	[128]
$NiWO_4$	DNA	One-phase	$K_3[Fe(CN)_6]$ to $K_4[Fe(CN)_6]$ in the presence of $Na_2S_2O_3$ under UV light	[131]
AuPt	Peptide-SWCNT	One-phase	Electrocatalytic oxygen reduction	[127]
Au	Thiol	LBL	Electrocatalytic oxygen reduction	[45]

Table 5.6 Oxidation reactions catalyzed by NP assemblies

Metal	Ligand	Method	Reaction	References
Au	DNA	DNA-assisted	Aerobic 4-hydroxybenzyl alcohol oxidation	[126]
ZnWO$_4$	DNA	DNA-assisted	Benzyl alcohol oxidation by H$_2$O$_2$	[132]
Au	Thiol	Two-phase	Electrooxidation of CO	[70]
Au	Thiol	LBL	Electrooxidation of CO and MeOH	[45]
AuPt	Thiol	Two-phase	Electrooxidation of MeOH	[71]
Au	Azide + alkyne	Click chemistry	Electrooxidation of MeOH	[111]

Table 5.7 Miscellaneous reactions catalyzed by NP assemblies

Metal	Ligand	Method	Reaction	References
Au	Azide + alkyne	Click chemistry	Electrochemical water splitting	[111]
CdSe	Thiol	Light-triggered	Photocatalytic water splitting	[203]
Pt	Amine	Two-phase	Catalytic H$_2$ gas sensing	[67, 68]
Pt	Amine	Two-phase	Catalytic H$_2$ gas sensing.	[67, 69]
Pt	Thio-COF	One-phase	Suzuki–Miyaura coupling reaction	[89]
Au	Amines and azobenzene-thiols	Light-triggered	Hydrosilylation of 4-methoxybenzaldehyde	[175]
Au	Thiol and azobenzene-thiols	Light-triggered	Ester hydrolysis	[212]

the catalytic performances was evidenced, which was mainly attributed to the fact that smaller NP have a larger catalytically active metallic surface, but the effect of the assembly was not studied. The α-cyclodextrin was also used to create Au/Fe$_3$O$_4$ assemblies, which were used as reduction catalyst of methylene blue with NaBH$_4$ [207]. The assemblies were synthesized by mixing oleic acid-decorated Fe$_3$O$_4$ NP with α-cyclodextrin-decorated Au NP thanks to the host–guest binding of oleic acid and α-cyclodextrin. The synthesized bimetallic structures spontaneously assembled into spherical architectures in water by controlling several parameters: a high ratio and small size of α-cyclodextrin-decorated Au NP allowed to produce the assembly. The nanocomposite was catalytically active for the reduction of methylene blue, and due to the presence of Fe$_3$O$_4$ NP, it was easily recovered and reused up to seven times.

Stimuli-responsive NP assemblies of Au and surface-modified TiO$_2$ display a reversible dynamic self-assembly, by the reversible covalent bond due to Schiff base formation, responding to an acid/base stimulus [202]. Water-soluble, ammonium-modified TiO$_2$ NP (TiO$_2$–NH$_2$ NP) and aldehyde-modified Au NP (Au–CHO NP)

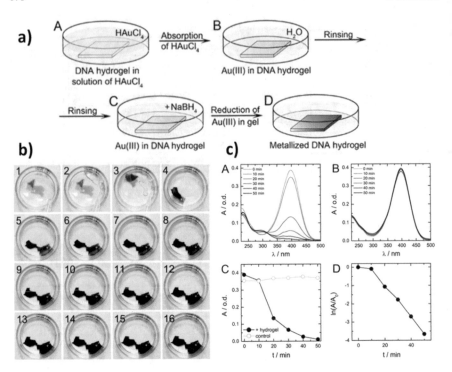

Fig. 5.33 **a** Protocol for preparation of hybrid hydrogel; **b** reduction of HAuCl₄ in the DNA hydrogel. Photographic images of the DNA hydrogel film containing HAuCl₄ after addition of 5 mL of a solution of 10 mM NaBH₄. The time interval between snapshots is 1.5 s; and **c** (A and B) time-dependent changes in UV-vis absorbance spectra of 4-nitrophenol solution (0.2 mM) with (A) and without (B) added hybrid hydrogel (0.1 g) after addition of NaBH₄ (1 mM) in 1 mL of water solution, (C) time dependence of nitrophenol absorbance at $\lambda = 400$ nm in solution with (filled circles) and without (open circles) hybrid hydrogel, and (D) time dependence of the normalized nitrophenol absorbance at $\lambda = 400$ nm built in logarithmic coordinates. Reproduced with permission from Ref. [130]

did not form the corresponding Schiff base as the amine groups are protonated in water. The increase of the pH to 11 led to the deprotonation of the ammonium groups on the TiO₂ and in consequence, to the efficient formation of the Schiff base and the assembly of both NP, TiO₂–NH₂ and Au-CHO NP. Under basic conditions, the catalyst that was effectively connected was more efficient for the photocatalytic degradation of methylene blue than the unassembled system under neutral conditions. Furthermore, the addition of the macrocycle cucurbit [6] uril (CB [6]) under acid conditions led to a non-covalent self-assembly of TiO₂–NH₂ and Au–CHO NP, which in turn behaved differently in the photocatalytic degradation reaction. Due to the high aggregation of the NP in this case, the catalyst was almost inactive (Fig. 5.34). The control of the pH leads to an artificial switchable photocatalyst, where the three described different states of the catalyst had remarkably different photocatalytic performances. The authors pointed out that these differences are the result of the

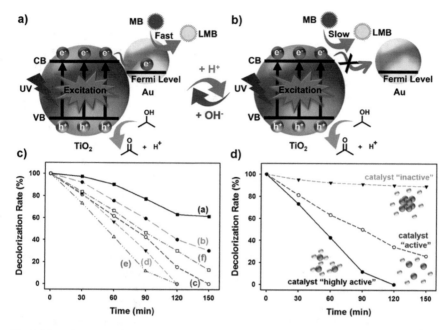

Fig. 5.34 **a, b** pH-switching photocatalytic mechanism of the dynamic covalent heteroassembly of Au-CHO and TiO_2-NH_2 NP; **c** photocatalytic decolorization of methylene blue (MB) curves of the Au–TiO_2 complexed in basic conditions with different amounts of Au–CHO NP: (a) 0; (b) 20; (c) 40; (d) 100; (e) 160; (f) 220 μL; (d) photocatalytic decolorization of the tri-stable system under different pH conditions. Reproduced with permission from Ref. [202]

control of the electron-transfer efficiency between Au and TiO_2 NP (controlled by the covalent dynamic assembly), and the accessibility of the catalytic active surface (non-covalent self-assembly).

C_{60} and C_{60} adducts were used for the synthesis of assembled NP using a one-step procedure [41, 80–82]. Ru NP assembled over Ru-fulleride spheres have demonstrated to be a highly selective catalyst for the reduction of nitrobenzene. Indeed, this catalyst is able to hydrogenate the nitro group first, and subsequently the aromatic ring, which is in contrast with other heterogeneous catalysts based on Ru over carbon [80, 211]. Ru@C_{60} provides electro-deficient Ru species, which could be responsible for such a selective system. DFT calculations together with experimental results point out that the reaction selectivity is mainly governed by surface hydride coverage onto the Ru NP surface. The reactivity can also be modulated by the ligands present on the surface (C_{60}, amine, carbene or a polymer) [211]. Ru NP assemblies displaying a short-range order were obtained with the use of a hexakis fullerene C_{60} adduct [41]. The $C_{66}(COOH)_{12}$ multitopic ligand bearing –COOH anchoring groups, robustly coordinated to the Ru NP surfaces, the high symmetry of which giving the possibility to create a 3-D assembly. Ru@$C_{66}(COOH)_{12}$ assemblies were active in the hydrogenation of nitrobenzene, and TEM analysis performed after catalysis has shown that the 3-D assembly is maintained. However, the lack of porosity of the assembly induces a lower catalytic activity, when compared to other Ru-based systems already reported, which is due to the poor accessibility to the Ru NP surface. PdNi bimetallic NP over C_{60}, C_{60}-malonic acid, and C_{60}-ethanediamine were also active for this reaction. Pd–6Ni–N–C_{60} exhibited good catalytic performances for the reduction of the nitro group and could be reused six times [83].

Pd NP embedded in a COF structure were successfully used as catalyst for the reduction of 4-nitrophenol with NaBH$_4$ [88]. Pd NP encapsulated in a sacrificial MOF were used to create a COF in a second-step reaction, which coated the Pd/MOF species. The subsequent selective etching of the MOF allowed encapsulation of multiple Pd NP inside the cavities of the COF. The reduction of 4-nitrophenol was monitored by UV-vis absorption spectroscopy, the constant rate calculated to be 0.41 min^{-1}. The catalytic performance of this catalyst was compared to other Pd NP systems, which evidenced the superior catalytic activity of the encapsulated Pd NP, and an effect of the shell thickness of the COF. The catalyst was characterized by electron microscope techniques after the catalysis, which revealed no change. Filtration and recycling tests (up to four cycles), together with the characterization after catalysis suggested that Pd NP are effectively protected from aggregation and leaching by the COF shell. The successful synthesis of a thiol-modified COF allowed the synthesis of Pd or Pt NP within the pores [89]. K_2PtCl_4 or K_2PdCl_4 were first embedded in the thiol-modified COF and successively, the metallic salts were reduced with NaBH$_4$. Using this straightforward procedure, ultra-small NP (<2 nm) of both metals were confined in the COF structure. Pt NP were successfully used as catalyst in the reduction of 4-nitrophenol by NaBH$_4$. The catalyst was reused six times, displaying up to 90% of conversion in each cycle. TEM image shows that the recycled catalyst does not undergo aggregation. Similarly, Pd NP within the thiol-modified COF were used as catalyst in the Suzuki–Miyaura coupling reaction. Excellent catalytic activities were also observed. The catalytic performances were compared with Pd NP and with the $[PdCl_2(PPh_3)_2]$ complex, evidencing the positive effect of the encapsulation, as lower activities were obtained with these latter systems. The catalyst exhibited excellent stability and recyclability under the catalytic reaction conditions, as evidenced by TEM analyses and recycling tests (up to five cycles).

In another example, chain-like aggregates of NiWO$_4$ were used as catalyst in the reduction of ferricyanide to ferrocyanide in the presence of $Na_2S_2O_3$ under light [131].

The electrocatalytic oxygen reduction reaction was used to confirm the controlled arrangement of AuPt NP assemblies [127]. Peptide-coated single-walled carbon nanotubes (SWCNT) were prepared, to further accommodate AuPt NP, synthesized by reduction of HAuCl$_4$ and PtCl$_6$ with NaBH$_4$ in the presence of several peptide-coated SWCNT supports. The peptide-coated SWCNT supports connect efficiently the conducting SWCNT with the electrocatalytically active NP. Engineering the peptide allowed the efficient control of the distance between the bimetallic NP, creating controlled assemblies. The electrochemical performances were governed by the size and inter-particle distance, which was further endorsed by theoretical calculations.

Oxidation Reactions

A three-step process was used to synthesize catalytically active super-lattices of Au NP, which consisted of: (i) adsorption of citrate-capped Au NP on thiol-DNA, followed by the formation of body-centered cubic (*bcc*) super-lattices by addition of different DNA linker strands, (ii) silica embedding of the Au NP DNA super-lattices, and (iii) calcination in air for 2 h at 350 °C (Fig. 5.35) [126]. This procedure gave rise to the production of robust porous structures containing arranged Au NP, which maintained the *bcc* super-lattice structure. The Au NP were active in the aerobic oxidation of 4-hydroxybenzyl alcohol, while DNA-functionalized Au NP (unsupported) as well as the uncalcined super-lattice Au NP were inactive, pointing out the importance of the porosity. Nevertheless, the recycling test showed almost not catalytic activity, which was attributed by the authors to the poisoning of the catalyst by the products of the catalytic reaction.

ZnWO$_4$ NP chain-like assemblies [132] were synthesized similarly to the chain-like aggregates of NiWO$_4$ [131] described by the same group. ZnWO$_4$ NP assemblies were prepared by reaction of Zn(NO$_3$)$_2$.6H$_2$O salt with Na$_2$WO$_4$.2H$_2$O and DNA by microwave heating. NP size and chain lengths were controlled by tuning the reaction parameters. ZnWO$_4$ NP assemblies were active in benzyl alcohol oxidation using H$_2$O$_2$ as oxidizing agent. The catalyst was recycled five times with a slight decrease of the activity.

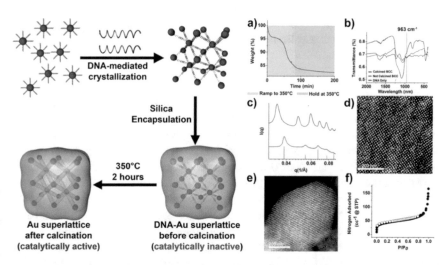

Fig. 5.35 Left: Schematic representation of the synthesis of DNA-Au NP super-lattices; and right) **a** TGA trace of silica-embedded DNA-Au NP super-lattices; **b** FTIR spectrum of pure DNA only (blue), super-lattices before calcination (red), and super-lattices after calcination (black); **c** SAXS data of the as-synthesized solution-phase *bcc* super-lattices (black) compared to the silica-embedded lattices after calcinations (red). Theoretical scattering from a perfect bcc lattice is shown in gray; **d, e** TEM images of silica-encapsulated super-lattices after calcination at 350 °C for 2 h; **f** nitrogen adsorption (filled circles)/desorption (hollow circles) isotherms of calcined nanoparticle super-lattices. Reproduced with permission from Ref. [126]

Thiolate-encapsulated gold NP assembled as thin films on electrodes were efficient catalysts for the electrooxidation of CO [70]. Decanethiolate was used to produce 2 and 5 nm Au NP on glassy carbon electrodes by cross-linking with 1,9-nonanedithiol. The 2 nm Au NP are more active for CO oxidation than the 5 nm Au NP. The same authors further studied the 2 nm Au assembled NP, by investigating the effect of the number of layers deposited (up to 20) on different supports (glassy carbon and carbon black), in the electrooxidation of CO, MeOH and, electroreduction of oxygen [45]. Similarly, AuPt NP were used as electrocatalyst for MeOH oxidation [71].

Azide- and alkyne-terminated groups have been used to produce covalent networks by click chemistry and LBL, allowing the production of assemblies of Au NPs onto several surfaces (Fig. 5.36) [111]. Alkyne-functionalized substrates (titania, silica, tin oxide, glass, stainless steel substrates) were immersed in a 3.5 wt% solution of azide-functionalized NP in THF for 12 h, and then immersed in an aqueous solution of copper sulfate and ascorbic acid for another 6 h to catalyze the click reaction. To produce multilayers, the Au NP assembly was subsequently soaked in an alkyne-functionalized NP suspension as many times as required. The authors point out that the growth rate of the monolayer was inversely proportional to the size of the NP and the density of the particles depended on the solvent, as well as on the substrate, being higher on TiO_2. Cyclic voltammetry profiles (up to 150 cycles) performed during methanol electrooxidation with the Au NP monolayers on silica, titania, ITO, and stainless steel substrates in electrolyte solution (0.1 M NaOH + 2.5 M MeOH) have confirmed the high stability of the catalytic systems. The same catalytic systems have been also tested in water splitting and photocatalytic degradation of Rhodamine B dye.

Miscellaneous Reactions

Au NP deposited onto several surfaces discussed above (Fig. 5.36) were active catalysts for the electrochemical water splitting [111]. The onset potential at approximately 0.7 V (vs. SCE) for all four monolayer Au NP assemblies on silicon, titania, ITO, and stainless steel substrates shows the oxygen-evolution current at approximately 0.6, 11, 2.3, and 45 mA/cm^2, respectively, which is higher than that of the bare substrates. The authors claim that the overall current density is dependent on the NP density, the electrocatalytic activity of the substrate, and the metal-NP support interactions (Fig. 5.37). The comparison of these systems to other Au NP based catalysts pointed out a positive effect of the assemblies in terms of stability.

Photocatalytic water splitting was investigated by using CdSe NP assemblies as catalyst [203]. Capped oleylamine/thiol Au, Pd, Pt, CdSe NP can form NP assemblies by the photo-oxidation of the capping thiol ligands. The assemblies consisted of vesicles, the size of which could be tuned by the size of the NP, and the thickness (number of NP layers) by the solvents used. Indeed, the type and the polarity of the solvent are crucial in the formation of the vesicles by light. Concerning the catalytic properties, the photocatalytic rate using the prepared CdSe nanovesicles was 1.5 times of that using individual CdSe NP, due to the enhanced light absorption of the assembly. Also, the catalytic activity remained unchanged after 10 h of visible light

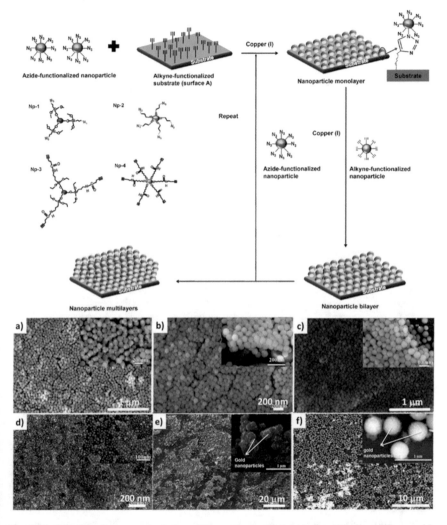

Fig. 5.36 Up: Schematic representation of the synthesis of an azide-functionalized NP assembly on alkyne-functionalized substrates via click chemistry; and bottom) SEM images of **a** two (scale bar 1 μm, inset 200 nm), **b** three (scale bar 200 nm, inset 200 nm), and **c** four layers of silica nanoparticles (scale bar 1 μm, inset 100 nm); and **d** bilayer of gold nanoparticles (scale bar 200 nm, inset 100 nm), **e** Au@TiO$_2$@silicon (scale bar 20 μm, inset 1 μm), and **f** Au@SiO$_2$@silicon (scale bar 10 μm, inset 1 μm). Reproduced with permission from Ref. [111]

irradiation; this is clearly in contrast with the activity of isolated CdSe NP, the activity of which decreased to less than 40% of the initiate value. The authors attributed this robustness to the mechanical strength in aqueous solution of the vesicles.

Unprotected Pt NP were assembled by the addition of several amino compounds (mono and diamines), using a two-phase method [67–69]. The chemical structure of the ligands plays a crucial role in the performance of the Pt-based catalyst in the

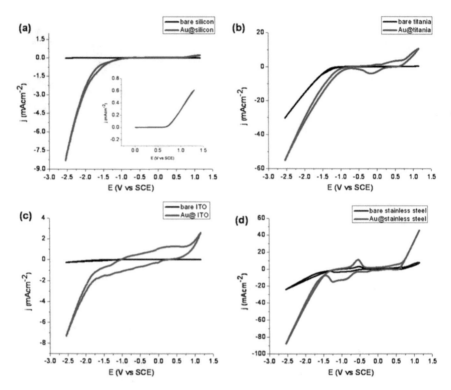

Fig. 5.37 Cyclic voltammetric results for water splitting on **a** Au@silicon, **b** Au@titania, **c** Au@ITO, and **d** Au@stainless steel substrates. Reproduced with permission from Ref. [111]

catalytic H_2 gas sensing. Monoamine assemblies were less stable and sintered, while diamine assemblies show better stability, which was related to the reduction of the ligand desorption rates and degradation. Ligand backbone also plays a role; aromatic structures remained stable during H_2 oxidation, while alkyl fragments oxidized and decomposed.

Photoswitchable catalysis mediated by the assembly of Au NP has been reported (Fig. 5.38). The formation of the assembly triggered by light is used in two different ways. In a first example, the assembly of Au NP led to an inactive catalyst when the assembly is triggered by light; [175] in the second example, the disassembly of the Au NP allows the Zn^{II}-coordinated β-cyclodextrin dimer to be free to coordinate to the substrates and the catalytic reaction proceeds [212].

Au NP capped with a mixture of dodecylamine (DDA) and a photoswitchable azobenzene-terminated alkane thiols remain unaggregated in the absence of light, therefore exposing a large catalytic active surface area (catalysis on). Upon UV light irradiation, Au NP assembled and were no longer able to catalyze the hydrosilylation reaction (catalysis off) (Fig. 5.38). The reaction could be turned on and off several times over time as depicted in Fig. 5.39. Even though three on-off cycles could be efficiently performed, after that, the aggregation of Au NP was less efficient and as

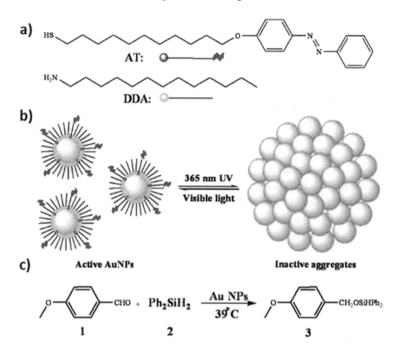

Fig. 5.38 **a** Molecular structures of the "background" DDA surfactant and of the photoresponsive azobenzene-thiol ligand, AT. **b** Schematic representation of a photoswitchable Au NP system. Dispersed NP are catalytically active; aggregated NP are catalytically inactive. **c** Hydrosilylation of 4-methoxybenzaldehyde catalyzed by Au NP in dry toluene at 39 °C and under argon. Reproduced with permission from Ref. [175]

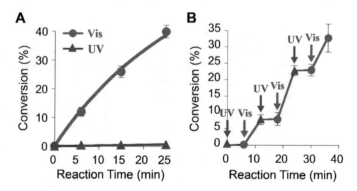

Fig. 5.39 **a** Percent conversion for the Au NP-catalyzed hydrosilylation of 4-methoxybenzaldehyde as a function of time under irradiation with visible light (red markers) and under 365 nm UV irradiation (blue markers). Surface coverage of azobenzene ligands on the Au NP was $\chi = 0.3$. **b** Hydrosilylation of 4-methoxybenzaldehyde can be switched "on" by visible light (red portions of the curve) and "off" by UV (blue portions). For the experiments shown, $\chi = 0.31$. Error bars were calculated based on standard deviations from three NP batches, three samples per batch. Reproduced with permission from Ref. [175]

a consequence, the catalyst displayed some catalytic activity when off. The authors point out that the possible reason is the displacement of some surface ligands by coordination of reagents or products, which leads to the loss of the photoresponsive azobenzene-thiol ligand.

The photoisomerization of the azobenzene unit from the *trans* to the *cis* form under UV-light irradiation was also exploited to generate another photoswitchable catalyst [212]. In this case, the host–guest interaction of Au NP containing azobenzene units with a β-cyclodextrin dimer was controlled by light. Under UV-light irradiation, the *cis* form of the azobenzene weakened the binding with the β-cyclodextrin, allowing the host to encapsulate the substrates and catalyze the reaction. Under visible light, *trans* azobenzene forms the inclusion complex with the β-cyclodextrin, turning off the catalytic reaction. Several catalytic cycles of the catalyzed ester hydrolysis could be achieved with this system, and in addition, the reaction rate of this system was higher than that of other Au-based catalysts.

5.4   Conclusion

A considerable effort of the catalysis community is focused on creating more active, selective, and robust catalysts. In particular, in heterogeneous catalysis, both experimental and theoretical detailed studies have led to a rational catalyst development. The knowledge obtained from these works make that some (but still few) heterogeneous catalysts are almost comparable to well-defined homogeneous systems, in terms of understanding of their catalytic activity, and selectivity, thereby opening the door for improvement. However, for most heterogeneous systems, important parameters such as NP size, inter-particle distance, and surface interactions with the support at the atomic level are not well-controlled yet.

On the other hand, metal NP are nowadays well-accepted catalysts, which to a certain extent are in the borderline of heterogeneous and homogeneous systems. In line with the heterogeneous catalysts, in recent years the fine-tuning of their properties (size, shape, exposed crystallographic phase, surface ligands, among others) permitted us to better understand and develop solutions for the catalysis of tomorrow. These systems, however, suffer from a difficult separation (if not impossible for large-scale processes). In that context, the formation of covalent NP assemblies from a bottom-up approach constitutes an interesting way to produce heterogeneous catalysts with controlled NP size, inter-particle distances, and surface coordination chemistry.

In this chapter, the covalent assembly of the metal NP has been reviewed, focusing on the methodologies of synthesis and their applications in catalysis. As far as the synthesis of covalent assemblies of metal NP is concerned, while the first assemblies were produced with simple methods, efficient and more complex systems have been engineered and applied recently. In particular, the formation of reversible covalent networks of metal NP is now possible by application of diverse stimuli. This concept, which has been rarely employed in catalysis, definitively deserves further studies. If the metal–ligand interaction is strong enough, the principal advantages of covalent metal NP assemblies in catalysis are their stability and robustness, which led in some cases to a better recycling, when compared to unassembled metal NP. Furthermore, in some cases assembled metal NP displayed better catalytic performances than isolated metal NP. However, if the length and structure of the linker are not appropriate, the lack of porosity, especially in 3-D networks, can be detrimental to the use of all metal NP active surface. This effect, which can be regarded as a disadvantage, has been exploited in switchable catalysts to create confined spaces or to turn off the catalysts. These recent advances provide alternatives for better catalysts in the future.

Acknowledgements Funding from the Agence Nationale de la Recherche (ANR project ANR-16-CE07-0007-01, Icare-1) is gratefully acknowledged.

References

1. Serp P, Philippot K (eds) (2013) Nanomaterials in catalysis. Wiley-VCH Verlag GmbH & Co. KGaA
2. Jin R, Zeng C, Zhou M, Chen Y (2016) Atomically precise colloidal metal nanoclusters and nanoparticles: fundamentals and opportunities. Chem Rev 116:10346–10413
3. Jin R, Pei Y, Tsukuda T (2019) Controlling nanoparticles with atomic precision. Acc Chem Res 52:1
4. Astruc D (ed) (2008) Nanoparticles and catalysis. Wiley Interscience, New York
5. Meeuwissen J, Reek JNH (2010) Supramolecular catalysis beyond enzyme mimics. Nat Chem 2:615–621
6. Liu J, Chen L, Cui H, Zhang J, Zhang L, Su C-Y (2014) Applications of metal-organic frameworks in heterogeneous supramolecular catalysis. Chem Soc Rev 43:6011–6061
7. Brown CJ, Toste FD, Bergman RG, Raymond KN (2015) Supramolecular catalysis in metal-ligand cluster hosts. Chem Rev 115:3012–3035
8. Grzelczak M, Vermant J, Furst EM, Liz-Marzán LM (2010) Directed self-assembly of nanoparticles. ACS Nano 4:3591–3605
9. Wei W, Bai F, Fan H (2019) Surfactant-assisted cooperative self-assembly of nanoparticles into active nanostructures. iScience 11:272–293
10. Sastry M, Rao M, Ganesh KN (2002) Electrostatic assembly of nanoparticles and biomacromolecules. Acc Chem Res 35:847–855
11. Lim IIS, Ip W, Crew E, Njoki PN, Mott D, Zhong C-J, Pan Y, Zhou S (2007) Homocysteine-mediated reactivity and assembly of gold nanoparticles. Langmuir 23:826–833
12. Lim IIS, Ouyang J, Luo J, Wang L, Zhou S, Zhong C-J (2005) Multifunctional fullerene-mediated assembly of gold nanoparticles. Chem Mater 17:6528–6531
13. Sutradhar S, Patnaik A (2017) Structure and dynamics of a N-methylfulleropyrrolidine-mediated gold nanocomposite: a spectroscopic ruler. ACS Appl Mater Interfaces 9:21921–21932
14. Lim IIS, Pan Y, Mott D, Ouyang J, Njoki PN, Luo J, Zhou S, Zhong C-J (2007) Assembly of gold nanoparticles mediated by multifunctional fullerenes. Langmuir 23:10715–10724
15. Tricard S, Said-Aizpuru O, Bouzouita D, Usmani S, Gillet A, Tassé M, Poteau R, Viau G, Demont P, Carrey J, Chaudret B (2017) Chemical tuning of Coulomb blockade at room-temperature in ultra-small platinum nanoparticle self-assemblies. Mater Horiz 4:487–492
16. Nonappa, Haataja JS, Timonen JVI, Malola S, Engelhardt P, Houbenov N, Lahtinen M, Häkkinen H, Ikkala O (2017) Reversible supracolloidal self-assembly of cobalt nanoparticles to hollow capsids and their superstructures. Angew Chem Int Ed 56:6473–6477

17. Gomez S, Erades L, Philippot K, Chaudret B, Collière V, Balmes O, Bovin J-O (2001) Platinum colloids stabilized by bifunctional ligands: self-organization and connection to gold. Chem Commun 1474–1475
18. Boal AK, Ilhan F, DeRouchey JE, Thurn-Albrecht T, Russell TP, Rotello VM (2000) Self-assembly of nanoparticles into structured spherical and network aggregates. Nature 404:746–748
19. Yucknovsky A, Mondal S, Burnstine-Townley A, Foqara M, Amdursky N (2019) Use of photoacids and photobases to control dynamic self-assembly of gold nanoparticles in aqueous and nonaqueous solutions. Nano Lett 19:3804–3810
20. Li D, Qi L (2018) Self-assembly of inorganic nanoparticles mediated by host-guest interactions. Curr Opin Colloid Interface Sci 35:59–67
21. Julin S, Nummelin S, Kostiainen MA, Linko V (2018) DNA nanostructure-directed assembly of metal nanoparticle superlattices. J Nanopart Res 20:119
22. Watanabe K, Kuroda K, Nagao D (2018) External-stimuli-assisted control over assemblies of plasmonic metals. Materials 11:794
23. Singamaneni S, Bliznyuk VN, Binek C, Tsymbal EY (2011) Magnetic nanoparticles: recent advances in synthesis, self-assembly and applications. J Mater Chem 21:16819–16845
24. Andres RP, Bielefeld JD, Henderson JI, Janes DB, Kolagunta VR, Kubiak CP, Mahoney WJ, Osifchin RG (1996) Self-assembly of a two-dimensional superlattice of molecularly linked metal clusters. Science 273:1690–1693
25. Wessels JM, Nothofer HG, Ford WE, von Wrochem F, Scholz F, Vossmeyer T, Schroedter A, Weller H, Yasuda A (2004) Optical and electrical properties of three-dimensional interlinked gold nanoparticle assemblies. J Am Chem Soc 126:3349–3356
26. Lin G, Wang Y, Zhang Q, Zhang X, Ji G, Ba L (2011) Controllable formation and TEM spatial visualization of cross-linked gold nanoparticle spherical aggregates. Nanoscale 3:4567–4570
27. Maye MM, Lim IIS, Luo J, Rab Z, Rabinovich D, Liu T, Zhong C-J (2005) Mediator–template assembly of nanoparticles. J Am Chem Soc 127:1519–1529
28. Rossner C, Glatter O, Saldanha O, Köster S, Vana P (2015) The structure of gold-nanoparticle networks cross-linked by di- and multifunctional RAFT oligomers. Langmuir 31:10573–10582
29. Shih S-M, Su W-F, Lin Y-J, Wu C-S, Chen C-D (2002) Two-dimensional arrays of self-assembled gold and sulfur-containing fullerene nanoparticles. Langmuir 18:3332–3335
30. Yan H, Lim SI, Zhang Y-J, Chen Q, Mott D, Wu W-T, An D-L, Zhou S, Zhong C-J (2010) Molecularly-mediated assembly of gold nanoparticles with interparticle rigid, conjugated and shaped aryl ethynyl structures. Chem Commun 46:2218–2220
31. Lim IIS, Vaiana C, Zhang Z-Y, Zhang Y-J, An D-L, Zhong C-J (2007) X-shaped rigid arylethynes to mediate the assembly of nanoparticles. J Am Chem Soc 129:5368–5369
32. Matassa R, Fratoddi I, Rossi M, Battocchio C, Caminiti R, Russo MV (2012) Two-dimensional networks of Ag nanoparticles bridged by organometallic ligand. J Phys Chem C 116:15795–15800
33. Quintiliani M, Bassetti M, Pasquini C, Battocchio C, Rossi M, Mura F, Matassa R, Fontana L, Russo MV, Fratoddi I (2014) Network assembly of gold nanoparticles linked through fluorenyl dithiol bridges. J Mater Chem C 2:2517–2527
34. Leibowitz FL, Zheng W, Maye MM, Zhong C-J (1999) Structures and properties of nanoparticle thin films formed via a one-step exchange–cross-linking–precipitation route. Anal Chem 71:5076–5083
35. Yamamoto Y, Miyachi M, Yamanoi Y, Minoda A, Maekawa S, Oshima S, Kobori Y, Nishihara H (2013) Synthesis and hydrogen storage properties of palladium nanoparticle-organic frameworks. J Inorg Organomet Polym 24:208–213
36. Jafri SH, Lofas H, Blom T, Wallner A, Grigoriev A, Ahuja R, Ottosson H, Leifer K (2015) Nano-fabrication of molecular electronic junctions by targeted modification of metal-molecule bonds. Sci Rep 5:14431
37. Chen SW (2000) Two-dimensional crosslinked nanoparticle networks. Adv Mater 12:186–189

38. Fontana L, Fratoddi I, Venditti I, Ksenzov D, Russo MV, Grigorian S (2016) Structural studies on drop-cast film based on functionalized gold nanoparticles network: the effect of thermal treatment. Appl Surf Sci 369:115–119
39. Yamanoi Y, Yamamoto Y, Miyachi M, Shimada M, Minoda A, Oshima S, Kobori Y, Nishihara H (2013) Nanoparticle assemblies via coordination with a tetrakis(terpyridine) linker bearing a rigid tetrahedral core. Langmuir 29:8768–8772
40. Boterashvili M, Shirman T, Popovitz-Biro R, Wen Q, Lahav M, van der Boom ME (2016) Nanocrystallinity and direct cross-linkage as key-factors for the assembly of gold nanoparticle-superlattices. Chem Commun 52:8079–8082
41. Leng F, Gerber IC, Lecante P, Bentaleb A, Muñoz A, Illescas BM, Martín N, Melinte G, Ersen O, Martinez H, Axet MR, Serp P (2017) Hexakis [60]fullerene adduct-mediated covalent assembly of ruthenium nanoparticles and their catalytic properties. Chem Eur J 23:13379–13386
42. Richardson JJ, Cui J, Björnmalm M, Braunger JA, Ejima H, Caruso F (2016) Innovation in layer-by-layer assembly. Chem Rev 116:14828–14867
43. Xiao F-X, Pagliaro M, Xu Y-J, Liu B (2016) Layer-by-layer assembly of versatile nanoarchitectures with diverse dimensionality: a new perspective for rational construction of multilayer assemblies. Chem Soc Rev 45:3088–3121
44. Kariuki NN, Luo J, Hassan SA, Lim IIS, Wang L, Zhong CJ (2006) Assembly of bimetallic gold–silver nanoparticles via selective interparticle dicarboxylate–silver linkages. Chem Mater 18:123–132
45. Maye MM, Luo J, Han L, Kariuki NN, Zhong C-J (2003) Synthesis, processing, assembly and activation of core-shell structured gold nanoparticle catalysts. Gold Bull 36:75–82
46. Wang W, Shi X, Kariuki NN, Schadt M, Wang GR, Rendeng Q, Choi J, Luo J, Lu S, Zhong C-J (2007) Array of molecularly mediated thin film assemblies of nanoparticles: correlation of vapor sensing with interparticle spatial properties. J Am Chem Soc 129:2161–2170
47. Zhou Q, Li X, Fan Q, Zhang X, Zheng J (2006) Charge transfer between metal nanoparticles interconnected with a functionalized molecule probed by surface-enhanced Raman spectroscopy. Angew Chem Int Ed 45:3970–3973
48. Joseph Y, Besnard I, Rosenberger M, Guse B, Nothofer H-G, Wessels JM, Wild U, Knop-Gericke A, Su D, Schlögl R, Yasuda A, Vossmeyer T (2003) Self-assembled gold nanoparticle/alkanedithiol films: preparation, electron microscopy, XPS-analysis, charge transport, and vapor-sensing properties†. J Phys Chem B 107:7406–7413
49. Joseph Y, Peic A, Chen X, Michl J, Vossmeyer T, Yasuda A (2007) Vapor sensitivity of networked gold nanoparticle chemiresistors: importance of flexibility and resistivity of the interlinkage. J Phys Chem C 111:12855–12859
50. Daskal Y, Tauchnitz T, Guth F, Dittrich R, Joseph Y (2017) Assembly behavior of organically interlinked gold nanoparticle composite films: a quartz crystal microbalance investigation. Langmuir 33:11869–11877
51. Ye J, Bonroy K, Nelis D, Frederix F, D'Haen J, Maes G, Borghs G (2008) Enhanced localized surface Plasmon resonance sensing on three-dimensional gold nanoparticles assemblies. Colloids Surf A 321:313–317
52. Dhar J, Patil S (2012) Self-assembly and catalytic activity of metal nanoparticles immobilized in polymer membrane prepared via layer-by-layer approach. ACS Appl Mater Interfaces 4:1803–1812
53. Vitale F, Fratoddi I, Battocchio C, Piscopiello E, Tapfer L, Russo MV, Polzonetti G, Giannini C (2011) Mono- and bi-functional arenethiols as surfactants for gold nanoparticles: synthesis and characterization. Nanoscale Res Lett 6:103
54. Brust M, Schiffrin DJ, Bethell D, Kiely CJ (1995) Novel gold-dithiol nano-networks with non-metallic electronic properties. Adv Mater 7:795–797
55. Maye MM, Chun SC, Han L, Rabinovich D, Zhong C-J (2002) Novel spherical assembly of gold nanoparticles mediated by a tetradentate thioether. J Am Chem Soc 124:4958–4959
56. Lim I-IS, Zhong C-J (2007) Molecularly-mediated assembly of gold nanoparticles. Gold Bull 40:59–66

57. Lim IIS, Maye MM, Luo J, Zhong C-J (2005) Kinetic and thermodynamic assessments of the mediator–template assembly of nanoparticles. J Phys Chem B 109:2578–2583
58. Shenhar R, Norsten TB, Rotello VM (2005) Polymer-mediated nanoparticle assembly: structural control and applications. Adv Mater 17:657–669
59. Stemmler MP, Fogel Y, Müllen K, Kreiter M (2009) Bridging of gold nanoparticles by functional polyphenylene dendrimers. Langmuir 25:11917–11922
60. Rossner C, Ebeling B, Vana P (2013) Spherical gold-nanoparticle assemblies with tunable interparticle distances mediated by multifunctional RAFT polymers. ACS Macro Lett 2:1073–1076
61. Rossner C, Ebeling B, Vana P (2015) Design strategies for the fabrication of tailored nanocomposites via RAFT polymerization. In: Controlled radical polymerization: materials (vol 1188). American Chemical Society, pp 293–307
62. Dey P, Blakey I, Thurecht KJ, Fredericks PM (2014) Hyperbranched polymer-gold nanoparticle assemblies: role of polymer architecture in hybrid assembly formation and SERS activity. Langmuir 30:2249–2258
63. Battocchio C, Fratoddi I, Fontana L, Bodo E, Porcaro F, Meneghini C, Pis I, Nappini S, Mobilio S, Russo MV, Polzonetti G (2014) Silver nanoparticles linked by a Pt-containing organometallic dithiol bridge: study of local structure and interface by XAFS and SR-XPS. Phys Chem Chem Phys 16:11719–11728
64. Fratoddi I, Matassa R, Fontana L, Venditti I, Familiari G, Battocchio C, Magnano E, Nappini S, Leahu G, Belardini A, Li Voti R, Sibilia C (2017) Electronic properties of a functionalized noble metal nanoparticles covalent network. J Phys Chem C 121:18110–18119
65. Fontana L, Bassetti M, Battocchio C, Venditti I, Fratoddi I (2017) Synthesis of gold and silver nanoparticles functionalized with organic dithiols. Colloids Surf A 532:282–289
66. Bearzotti A, Papa P, Macagnano A, Zampetti E, Venditti I, Fioravanti R, Fontana L, Matassa R, Familiari G, Fratoddi I (2018) Environmental Hg vapours adsorption and detection by using functionalized gold nanoparticles network. J Environ Chem Eng 6:4706–4713
67. Morsbach E, Kunz S, Bäumer M (2016) Novel nanoparticle catalysts for catalytic gas sensing. Catal Sci Technol 6:339–348
68. Morsbach E, Spéder J, Arenz M, Brauns E, Lang W, Kunz S, Bäumer M (2014) Stabilizing catalytically active nanoparticles by ligand linking: toward three-dimensional networks with high catalytic surface area. Langmuir 30:5564–5573
69. Morsbach E, Brauns E, Kowalik T, Lang W, Kunz S, Bäumer M (2014) Ligand-stabilized Pt nanoparticles (NPs) as novel materials for catalytic gas sensing: influence of the ligand on important catalytic properties. Phys Chem Chem Phys 16:21243–21251
70. Maye MM, Lou Y, Zhong C-J (2000) Core – Shell Gold Nanoparticle Assembly as Novel Electrocatalyst of CO Oxidation. Langmuir 16:7520–7523
71. Lou Y, Maye MM, Han L, Luo J, Zhong C-J (2001) Gold–platinum alloy nanoparticle assembly as catalyst for methanol electrooxidation. Chem Commun 473–474
72. Hussain I, Brust M, Barauskas J, Cooper AI (2009) Controlled step growth of molecularly linked gold nanoparticles: from metallic monomers to dimers to polymeric nanoparticle chains. Langmuir 25:1934–1939
73. Klajn R, Gray TP, Wesson PJ, Myers BD, Dravid VP, Smoukov SK, Grzybowski BA (2008) Bulk synthesis and surface patterning of nanoporous metals and alloys from supra spherical nanoparticle aggregates. Adv Funct Mater 18:2763–2769
74. Sakamoto M, Tanaka D, Teranishi T (2013) Rigid bidentate ligands focus the size of gold nanoparticles. Chem Sci 4:824–828
75. Nayak S, Horst N, Zhang H, Wang W, Mallapragada S, Travesset A, Vaknin D (2018) Ordered networks of gold nanoparticles crosslinked by dithiol-oligomers. Part Part Syst Charact 35:1–7
76. Chen S (2001) Langmuir–Blodgett fabrication of two-dimensional robust cross-linked nanoparticle assemblies. Langmuir 17:2878–2884
77. Sidhaye DS, Kashyap S, Sastry M, Hotha S, Prasad BLV (2005) Gold nanoparticle networks with photoresponsive interparticle spacings. Langmuir 21:7979–7984

78. Klajn R, Bishop KJM, Grzybowski BA (2007) Light-controlled self-assembly of reversible and irreversible nanoparticle suprastructures. Proc Natl Acad Sci 104:10305–10309
79. Klajn R, Bishop KJM, Fialkowski M, Paszewski M, Campbell CJ, Gray TP, Grzybowski BA (2007) Plastic and moldable metals by self-assembly of sticky nanoparticle aggregates. Science 316:261–264
80. Leng F, Gerber IC, Lecante P, Moldovan S, Girleanu M, Axet MR, Serp P (2016) Controlled and chemoselective hydrogenation of nitrobenzene over Ru@C_{60} catalysts. ACS Catal 6:6018–6024
81. Leng F, Gerber IC, Axet MR, Serp P (2018) Selectivity shifts in hydrogenation of cinnamaldehyde on electron-deficient ruthenium nanoparticles. C R Chim 21:346–353
82. Leng F, Gerber IC, Lecante P, Bacsa W, Miller J, Gallagher JR, Moldovan S, Girleanu M, Axet MR, Serp P (2016) Synthesis and structure of ruthenium-fullerides. RSC Adv 6:69135–69148
83. Qu Y, Chen T, Wang G (2019) Hydrogenation of nitrobenzene catalyzed by Pd promoted Ni supported on C_{60} derivative. Appl Surf Sci 465:888–894
84. Solovyeva EV, Ubyivovk EV, Denisova AS (2018) Effect of diaminostilbene as a molecular linker on Ag nanoparticles: SERS study of aggregation and interparticle hot spots in various environments. Colloids Surf A 538:542–548
85. Kalidindi SB, Oh H, Hirscher M, Esken D, Wiktor C, Turner S, Van Tendeloo G, Fischer RA (2012) Metal@COFs: covalent organic frameworks as templates for Pd nanoparticles and hydrogen storage properties of Pd@COF-102 hybrid material. Chem Eur J 18:10848–10856
86. Pachfule P, Panda MK, Kandambeth S, Shivaprasad SM, Díaz DD, Banerjee R (2014) Multi-functional and robust covalent organic framework-nanoparticle hybrids. J Mater Chem A 2:7944–7952
87. Pachfule P, Kandambeth S, Diaz Diaz D, Banerjee R (2014) Highly stable covalent organic framework-Au nanoparticles hybrids for enhanced activity for nitrophenol reduction. Chem Commun 50:3169–3172
88. Cui K, Zhong W, Li L, Zhuang Z, Li L, Bi J, Yu Y (2018) Well-defined metal nanoparticles@covalent organic framework yolk-shell nanocages by ZIF-8 template as catalytic nanoreactors. Small e1804419
89. Lu S, Hu Y, Wan S, McCaffrey R, Jin Y, Gu H, Zhang W (2017) Synthesis of ultrafine and highly dispersed metal nanoparticles confined in a thioether-containing covalent organic framework and their catalytic applications. J Am Chem Soc 139:17082–17088
90. Mirkin CA, Letsinger RL, Mucic RC, Storhoff JJ (1996) A DNA-based method for rationally assembling nanoparticles into macroscopic materials. Nature 382:607
91. Storhoff JJ, Lazarides AA, Mucic RC, Mirkin CA, Letsinger RL, Schatz GC (2000) What controls the optical properties of DNA-linked gold nanoparticle assemblies? J Am Chem Soc 122:4640–4650
92. Simon U, Flesch R, Wiggers H, Schön G, Schmid G (1998) Chemical tailoring of the charging energy in metal cluster arrangements by use of bifunctional spacer molecules. J Mater Chem 8:517–518
93. Kimoto A, Iwasaki K, Abe J (2010) Formation of photoresponsive gold nanoparticle networks via click chemistry. Photochem Photobiol Sci 9:152–156
94. Hermes JP, Sander F, Peterle T, Cioffi C, Ringler P, Pfohl T, Mayor M (2011) Direct control of the spatial arrangement of gold nanoparticles in organic-inorganic hybrid superstructures. Small 7:920–929
95. Kosif I, Kratz K, You SS, Bera MK, Kim K, Leahy B, Emrick T, Lee KYC, Lin B (2017) Robust gold nanoparticle sheets by ligand cross-linking at the air-water interface. ACS Nano 11:1292–1300
96. Nepal D, Drummy LF, Biswas S, Park K, Vaia RA (2013) Large scale solution assembly of quantum dot-gold nanorod architectures with Plasmon enhanced fluorescence. ACS Nano 7:9064–9074
97. Torimoto T, Tsumura N, Miyake M, Nishizawa M, Sakata T, Mori H, Yoneyama H (1999) Preparation and photoelectrochemical properties of two-dimensionally organized CdS nanoparticle thin films. Langmuir 15:1853–1858

98. Abargues R, Albert S, Valdés JL, Abderrafi K, Martínez-Pastor JP (2012) Molecular-mediated assembly of silver nanoparticles with controlled interparticle spacing and chain length. J Mater Chem 22:22204–22211

99. Bönnemann H, Waldöfner N, Haubold HG, Vad T (2002) Preparation and characterization of three-dimensional Pt nanoparticle networks. Chem Mater 14:1115–1120

100. Li N, Binder WH (2011) Click-chemistry for nanoparticle-modification. J Mater Chem 21:16717–16734

101. Escorihuela J, Marcelis ATM, Zuilhof H (2015) Metal-free click chemistry reactions on surfaces. Adv Mater Interfaces 2:1500135

102. Zhu J, Kell AJ, Workentin MS (2006) A retro-Diels–Alder reaction to uncover maleimide-modified surfaces on monolayer-protected nanoparticles for reversible covalent assembly. Org Lett 8:4993–4996

103. Voggu R, Suguna P, Chandrasekaran S, Rao CNR (2007) Assembling covalently linked nanocrystals and nanotubes through click chemistry. Chem Phys Lett 443:118–121

104. Hua C, Zhang WH, De Almeida SRM, Ciampi S, Gloria D, Liu G, Harper JB, Gooding JJ (2012) A novel route to copper(II) detection using 'click' chemistry-induced aggregation of gold nanoparticles. Analyst 137:82–86

105. Sander F, Fluch U, Hermes JP, Mayor M (2014) Dumbbells, trikes and quads: organic-inorganic hybrid nanoarchitectures based on "clicked" gold nanoparticles. Small 10:349–359

106. Zhou Y, Wang S, Zhang K, Jiang X (2008) Visual detection of copper(II) by azide- and alkyne-functionalized gold nanoparticles using click chemistry. Angew Chem Int Ed 47:7454–7456

107. Xu X, Daniel WL, Wei W, Mirkin CA (2010) Colorimetric Cu(2+) detection using DNA-modified gold-nanoparticle aggregates as probes and click chemistry. Small 6:623–626

108. Zhang Y, Li B, Xu C (2010) Visual detection of ascorbic acid via alkyne-azide click reaction using gold nanoparticles as a colorimetric probe. Analyst 135:1579–1584

109. Rianasari I, de Jong MP, Huskens J, van der Wiel WG (2013) Covalent coupling of nanoparticles with low-density functional ligands to surfaces via click chemistry. Int J Mol Sci 14:3705–3717

110. Liu Y, Williams MG, Miller TJ, Teplyakov AV (2016) Nanoparticle layer deposition for highly controlled multilayer formation based on high-coverage monolayers of nanoparticles. Thin Solid Film 598:16–24

111. Upadhyay AP, Behara DK, Sharma GP, Bajpai A, Sharac N, Ragan R, Pala RGS, Sivakumar S (2013) Generic process for highly stable metallic nanoparticle-semiconductor heterostructures via click chemistry for electro/photocatalytic applications. ACS Appl Mater Interfaces 5:9554–9562

112. Locatelli E, Ori G, Fournelle M, Lemor R, Montorsi M, Comes Franchini M (2011) Click chemistry for the assembly of gold nanorods and silver nanoparticles. Chem Eur J 17:9052–9056

113. Liu Y, RamaRao N, Miller T, Hadjipanayis G, Teplyakov AV (2013) Controlling physical properties of iron nanoparticles during assembly by "click chemistry". J Phys Chem C 117:19974–19983

114. Jańczewski D, Tomczak N, Liu S, Han M-Y, Vancso GJ (2010) Covalent assembly of functional inorganic nanoparticles by "click" chemistry in water. Chem Commun 46:3253–3255

115. Bielski R, Witczak Z (2013) Strategies for coupling molecular units if subsequent decoupling is required. Chem Rev 113:2205–2243

116. DeVries GA, Brunnbauer M, Hu Y, Jackson AM, Long B, Neltner BT, Uzun O, Wunsch BH, Stellacci F (2007) Divalent metal nanoparticles. Science 315:358–361

117. Andryszewski T, Iwan M, Hołdyński M, Fiałkowski M (2016) Synthesis of a free-standing monolayer of covalently bonded gold nanoparticles. Chem Mater 28:5304–5313

118. Maneeprakorn W, Malik MA, O'Brien P (2010) Developing chemical strategies for the assembly of nanoparticles into mesoscopic objects. J Am Chem Soc 132:1780–1781

119. Aldeek F, Ji X, Mattoussi H (2013) Quenching of quantum dot emission by fluorescent gold clusters: what it does and does not share with the Förster formalism. J Phys Chem C 117:15429–15437

120. Tognarelli DJ, Miller RB, Pompano RR, Loftus AF, Sheibley DJ, Leopold MC (2005) Covalently networked monolayer-protected nanoparticle films. Langmuir 21:11119–11127

121. Shon Y-S, Choo H (2002) [60]Fullerene-linked gold nanoparticles: synthesis and layer-by-layer growth on a solid surface. Chem Commun 2560–2561

122. Dinh T, Shon Y-S (2009) Direct assembly of photoresponsive C_{60}–gold nanoparticle hybrid films. ACS Appl Mater Interfaces 1:2699–2702

123. Caswell KK, Wilson JN, Bunz UHF, Murphy CJ (2003) Preferential end-to-end assembly of gold nanorods by biotin–streptavidin connectors. J Am Chem Soc 125:13914–13915

124. Pieters BJGE, van Eldijk MB, Nolte RJM, Mecinović J (2016) Natural supramolecular protein assemblies. Chem Soc Rev 45:24–39

125. Macfarlane RJ, Lee B, Jones MR, Harris N, Schatz GC, Mirkin CA (2011) Nanoparticle superlattice engineering with DNA. Science 334:204–208

126. Auyeung E, Morris W, Mondloch JE, Hupp JT, Farha OK, Mirkin CA (2015) Controlling structure and porosity in catalytic nanoparticle superlattices with DNA. J Am Chem Soc 137:1658–1662

127. Kang ES, Kim Y-T, Ko Y-S, Kim NH, Cho G, Huh YH, Kim J-H, Nam J, Thach TT, Youn D, Kim YD, Yun WS, DeGrado WF, Kim SY, Hammond PT, Lee J, Kwon Y-U, Ha D-H, Kim YH (2018) Peptide-programmable nanoparticle superstructures with tailored electrocatalytic activity. ACS Nano 12:6554–6562

128. Sankar SS, Sangeetha K, Karthick K, Anantharaj S, Ede SR, Kundu S (2018) Pt nanoparticle tethered DNA assemblies for enhanced catalysis and SERS applications. New J Chem 42:15784–15792

129. Maeda Y, Akita T, Daté M, Takagi A, Matsumoto T, Fujitani T, Kohyama M (2010) Nanoparticle arrangement by DNA-programmed self-assembly for catalyst applications. J Appl Phys 108:094326

130. Zinchenko A, Miwa Y, Lopatina LI, Sergeyev VG, Murata S (2014) DNA hydrogel as a template for synthesis of ultrasmall gold nanoparticles for catalytic applications. ACS Appl Mater Interfaces 6:3226–3232

131. Nithiyanantham U, Ede SR, Anantharaj S, Kundu S (2015) Self-assembled $NiWO_4$ nanoparticles into chain-like aggregates on DNA scaffold with pronounced catalytic and supercapacitor activities. Cryst Growth Des 15:673–686

132. Ede SR, Ramadoss A, Nithiyanantham U, Anantharaj S, Kundu S (2015) Bio-molecule assisted aggregation of $ZnWO_4$ nanoparticles (NPs) into chain-like assemblies: material for high performance supercapacitor and as catalyst for benzyl alcohol oxidation. Inorg Chem 54:3851–3863

133. Tan LH, Xing H, Lu Y (2014) DNA as a powerful tool for morphology control, spatial positioning, and dynamic assembly of nanoparticles. Acc Chem Res 47:1881–1890

134. Li N, Shang Y, Han Z, Wang T, Wang Z-G, Ding B (2019) Fabrication of metal nanostructures on DNA templates. ACS Appl Mater Interfaces 11:13835–13852

135. Niemeyer CM, Simon U (2005) DNA-based assembly of metal nanoparticles. Eur J Inorg Chem 2005:3641–3655

136. Bandy TJ, Brewer A, Burns JR, Marth G, Nguyen T, Stulz E (2011) DNA as supramolecular scaffold for functional molecules: progress in DNA nanotechnology. Chem Soc Rev 40:138–148

137. Clever GH, Kaul C, Carell T (2007) DNA–Metal Base Pairs. Angew Chem Int Ed 46:6226–6236

138. Jang N-H (2002) The coordination chemistry of DNA nucleosides on gold nanoparticles as a probe by SERS. Bull Korean Chem Soc 23(12):1790–1800

139. Patolsky F, Weizmann Y, Lioubashevski O, Willner I (2002) Au-nanoparticle nanowires based on DNA and polylysine templates. Angew Chem Int Ed 41:2323–2327

140. Carter JD, LaBean TH (2011) Oganization of inorganic nanomaterials via programmable DNA self-assembly and peptide molecular recognition. ACS Nano 5:2200–2205

141. Hong F, Zhang F, Liu Y, Yan H (2017) DNA origami: scaffolds for creating higher order structures. Chem Rev 117:12584–12640

142. Tian Y, Wang T, Liu W, Xin HL, Li H, Ke Y, Shih WM, Gang O (2015) Prescribed nanoparticle cluster architectures and low-dimensional arrays built using octahedral DNA origami frames. Nat Nanotechnol 10:637

143. Schreiber R, Santiago I, Ardavan A, Turberfield AJ (2016) Ordering gold nanoparticles with DNA origami nanoflowers. ACS Nano 10:7303–7306

144. Sato MR, da Silva PB, de Souza RA, dos Santos KC, Chorilli M (2015) Recent advances in nanoparticle carriers for coordination complexes. Curr Top Med Chem 15:287–297

145. Beloglazkina EK, Majouga AG, Romashkina RB, Zyk NV, Zefirov NS (2012) Gold nanoparticles modified with coordination compounds of metals: synthesis and application. Russ Chem Rev 81:65–90

146. Rubinstein I, Vaskevich A (2010) Self-assembly of nanostructures on surfaces using metal-organic coordination. Isr J Chem 50:333–346

147. Tang L, Li J (2017) Plasmon-based colorimetric nanosensors for ultrasensitive molecular diagnostics. ACS Sens 2:857–875

148. Weng Z, Wang H, Vongsvivut J, Li R, Glushenkov AM, He J, Chen Y, Barrow CJ, Yang W (2013) Self-assembly of core-satellite gold nanoparticles for colorimetric detection of copper ions. Anal Chim Acta 803:128–134

149. Pompano RR, Wortley PG, Moatz LM, Tognarelli DJ, Kittredge KW, Leopold MC (2006) Crown ether-metal "sandwiches" as linking mechanisms in assembled nanoparticle films. Thin Solid Films 510:311–319

150. Sener G, Uzun L, Denizli A (2014) Colorimetric sensor array based on gold nanoparticles and amino acids for identification of toxic metal ions in water. ACS Appl Mater Interfaces 6:18395–18400

151. Guan J, Jiang L, Li J, Yang W (2008) pH-dependent aggregation of histidine-functionalized au nanoparticles induced by Fe^{3+} ions. J Phys Chem C 112:3267–3271

152. Lin S-Y, Chen C-H, Lin M-C, Hsu H-F (2005) A cooperative effect of bifunctionalized nanoparticles on recognition: sensing alkali ions by crown and carboxylate moieties in aqueous media. Anal Chem 77:4821–4828

153. Lin S-Y, Liu S-W, Lin C-M, Chen C-H (2002) Recognition of potassium ion in water by 15-crown-5 functionalized gold nanoparticles. Anal Chem 74:330–335

154. Chen S, Pei R, Zhao T, Dyer DJ (2002) Gold nanoparticle assemblies by metal ion–pyridine complexation and their rectified quantized charging in aqueous solutions. J Phys Chem B 106:1903–1908

155. Norsten TB, Frankamp BL, Rotello VM (2002) Metal directed assembly of terpyridine-functionalized gold nanoparticles. Nano Lett 2:1345–1348

156. Zeng Q, Marthi R, McNally A, Dickinson C, Keyes TE, Forster RJ (2010) Host–guest directed assembly of gold nanoparticle arrays. Langmuir 26:1325–1333

157. Obare SO, Hollowell RE, Murphy CJ (2002) Sensing strategy for lithium ion based on gold nanoparticles. Langmuir 18:10407–10410

158. Miller A, Adams S, Zhang JZ, Wang L (2016) Study of the interaction of citrate-capped hollow gold nanospheres with metal ions. J Nanomed Nanotechnol 7:1000371

159. Priyadarshini E, Pradhan N (2017) Metal-induced aggregation of valine capped gold nanoparticles: an efficient and rapid approach for colorimetric detection of Pb(2+) ions. Sci Rep 7:9278–9278

160. Kim Y, Johnson RC, Hupp JT (2001) Gold nanoparticle-based sensing of "spectroscopically silent" heavy metal ions. Nano Lett 1:165–167

161. Wanunu M, Popovitz-Biro R, Cohen H, Vaskevich A, Rubinstein I (2005) Coordination-based gold nanoparticle layers. J Am Chem Soc 127:9207–9215

162. Hitihami-Mudiyanselage A, Senevirathne K, Brock SL (2014) Bottom-up assembly of Ni_2P nanoparticles into three-dimensional architectures: an alternative mechanism for phosphide gelation. Chem Mater 26:6251–6256

163. Lesnyak V, Voitekhovich SV, Gaponik PN, Gaponik N, Eychmüller A (2010) CdTe nanocrystals capped with a tetrazolyl analogue of thioglycolic acid: aqueous synthesis, characterization, and metal-assisted assembly. ACS Nano 4:4090–4096

164. Bao C, Zhu L, Lin Q, Tian H (2015) Building biomedical materials using photochemical bond cleavage. Adv Mater 27:1647–1662
165. Lai J, Xu Y, Mu X, Wu X, Li C, Zheng J, Wu C, Chen J, Zhao Y (2011) Light-triggered covalent assembly of gold nanoparticles in aqueous solution. Chem Commun 47:3822–3824
166. Cheng X, Sun R, Yin L, Chai Z, Shi H, Gao M (2017) Light-triggered assembly of gold nanoparticles for photothermal therapy and photoacoustic imaging of tumors in vivo. Adv Mater 29:1604894
167. Xia H, Gao Y, Yin L, Cheng X, Wang A, Zhao M, Ding J, Shi H (2019) Light-triggered covalent coupling of gold nanoparticles for photothermal cancer therapy. ChemBioChem 20:667–671
168. Zhen SJ, Zhang ZY, Li N, Zhang ZD, Wang J, Li CM, Zhan L, Zhuang HL, Huang CZ (2013) UV light-induced self-assembly of gold nanocrystals into chains and networks in a solution of silver nitrate. Nanotechnology 24:055601
169. Itoh H, Tahara A, Naka K, Chujo Y (2004) Photochemical assembly of gold nanoparticles utilizing the photodimerization of thymine. Langmuir 20:1972–1976
170. Zhou J, Beattie DA, Sedev R, Ralston J (2007) Synthesis and surface structure of thymine-functionalized, self-assembled monolayer-protected gold nanoparticles. Langmuir 23:9170–9177
171. Zhou J, Sedev R, Beattie D, Ralston J (2008) Light-induced aggregation of colloidal gold nanoparticles capped by thymine derivatives. Langmuir 24:4506–4511
172. Blum AP, Kammeyer JK, Rush AM, Callmann CE, Hahn ME, Gianneschi NC (2015) Stimuli-responsive nanomaterials for biomedical applications. J Am Chem Soc 137:2140–2154
173. Blanco V, Leigh DA, Marcos V (2015) Artificial switchable catalysts. Chem Soc Rev 44:5341–5370
174. Vassalini I, Alessandri I (2018) Switchable stimuli-responsive heterogeneous catalysis. Catalysts 8:569
175. Wei Y, Han S, Kim J, Soh S, Grzybowski BA (2010) Photoswitchable catalysis mediated by dynamic aggregation of nanoparticles. J Am Chem Soc 132:11018–11020
176. Díaz Díaz D, Kühbeck D, Koopmans RJ (2011) Stimuli-responsive gels as reaction vessels and reusable catalysts. Chem Soc Rev 40:427–448
177. Zhang J, Zhang M, Tang K, Verpoort F, Sun T (2014) Polymer-based stimuli-responsive recyclable catalytic systems for organic synthesis. Small 10:32–46
178. Grzelczak M, Liz-Marzan LM, Klajn R (2019) Stimuli-responsive self-assembly of nanoparticles. Chem Soc Rev 48:1342–1361
179. Kay ER (2016) Dynamic covalent nanoparticle building blocks. Chem Eur J 22:10706–10716
180. Rowan SJ, Cantrill SJ, Cousins GRL, Sanders JKM, Stoddart JF (2002) Dynamic covalent chemistry. Angew Chem Int Ed 41:898–952
181. Borsley S, Kay ER (2016) Dynamic covalent assembly and disassembly of nanoparticle aggregates. Chem Commun 52:9117–9120
182. Wei W, Wu L, Xu C, Ren J, Qu X (2013) A general approach using spiroborate reversible cross-linked Au nanoparticles for visual high-throughput screening of chiral vicinal diols. Chem Sci 4:1156–1162
183. Liu X, Liu H, Zhou W, Zheng H, Yin X, Li Y, Guo Y, Zhu M, Ouyang C, Zhu D, Xia A (2010) Thermoreversible covalent self-assembly of oligo(p-phenylenevinylene) bridged gold nanoparticles. Langmuir 26:3179–3185
184. Chan Y-T, Li S, Moorefield CN, Wang P, Shreiner CD, Newkome GR (2010) Self-assembly, disassembly, and reassembly of gold nanorods mediated by bis(terpyridine)–metal connectivity. Chem Eur J 16:4164–4168
185. Si S, Raula M, Paira TK, Mandal TK (2008) Reversible self-assembly of carboxylated peptide-functionalized gold nanoparticles driven by metal-ion coordination. ChemPhysChem 9:1578–1584
186. Rao A, Roy S, Unnikrishnan M, Bhosale SS, Devatha G, Pillai PP (2016) Regulation of interparticle forces reveals controlled aggregation in charged nanoparticles. Chem Mater 28:2348–2355

187. Parnsubsakul A, Oaew S, Surareungchai W (2018) Zwitterionic peptide-capped gold nanoparticles for colorimetric detection of Ni^{2+}. Nanoscale 10:5466–5473
188. Liu D, Chen W, Sun K, Deng K, Zhang W, Wang Z, Jiang X (2011) Resettable, multi-readout logic gates based on controllably reversible aggregation of gold nanoparticles. Angew Chem Int Ed 50:4103–4107
189. Luan Z, Salk T, Abelson A, Jean S, Law M (2019) Reversible aggregation of covalently cross-linked gold nanocrystals by linker oxidation. J Phys Chem C 123:23643–23654
190. Frisch H, Marschner DE, Goldmann AS, Barner-Kowollik C (2018) Wavelength-gated dynamic covalent chemistry. Angew Chem Int Ed 57:2036–2045
191. Cardenas-Daw C, Kroeger A, Schaertl W, Froimowicz P, Landfester K (2012) Reversible photocycloadditions, a powerful tool for tailoring (nano)materials. Macromol Chem Phys 213:144–156
192. Göstl R, Hecht S (2014) Controlling covalent connection and disconnection with light. Angew Chem Int Ed 53:8784–8787
193. He H, Feng M, Chen Q, Zhang X, Zhan H (2016) Light-induced reversible self-assembly of gold nanoparticles surface-immobilized with coumarin ligands. Angew Chem Int Ed 55:936–940
194. Chen Y, Wang Z, He Y, Yoon YJ, Jung J, Zhang G, Lin Z (2018) Light-enabled reversible self-assembly and tunable optical properties of stable hairy nanoparticles. Proc Natl Acad Sci 115:E1391–E1400
195. De Fazio AF, El-Sagheer AH, Kahn JS, Nandhakumar I, Burton MR, Brown T, Muskens OL, Gang O, Kanaras AG (2019) Light-induced reversible DNA ligation of gold nanoparticle superlattices. ACS Nano 13:5771–5777
196. Bigall NC, Herrmann A-K, Vogel M, Rose M, Simon P, Carrillo-Cabrera W, Dorfs D, Kaskel S, Gaponik N, Eychmüller A (2009) Hydrogels and aerogels from noble metal nanoparticles. Angew Chem Int Ed 48:9731–9734
197. Arachchige IU, Brock SL (2007) Sol-gel methods for the assembly of metal chalcogenide quantum dots. Acc Chem Res 40:801–809
198. Gao X, Esteves RJA, Nahar L, Nowaczyk J, Arachchige IU (2016) Direct cross-linking of Au/Ag alloy nanoparticles into monolithic aerogels for application in surface-enhanced Raman scattering. ACS Appl Mater Interfaces 8:13076–13085
199. Guo S, Sun S (2012) FePt nanoparticles assembled on graphene as enhanced catalyst for oxygen reduction reaction. J Am Chem Soc 134:2492–2495
200. Sreedhala S, Vinod CP (2015) Surfactant assisted formation of ruthenium nanochains under mild conditions and their catalytic CO oxidation activity. Chem Commun 51:10178–10181
201. Zhao H, Sen S, Udayabhaskararao T, Sawczyk M, Kucanda K, Manna D, Kundu PK, Lee J-W, Kral P, Klajn R (2016) Reversible trapping and reaction acceleration within dynamically self-assembling nanoflasks. Nat Nanotechnol 11:82–88
202. Zhang Q, Wang W-Z, Yu J-J, Qu D-H, Tian H (2017) Dynamic self-assembly encodes A tri-stable Au-TiO_2 photocatalyst. Adv Mater 29:1604948
203. Bian T, Shang L, Yu H, Perez MT, Wu L-Z, Tung C-H, Nie Z, Tang Z, Zhang T (2014) Spontaneous organization of inorganic nanoparticles into nanovesicles triggered by UV light. Adv Mater 26:5613–5618
204. Neouze M-A (2013) Nanoparticle assemblies: main synthesis pathways and brief overview on some important applications. J Mater Sci 48:7321–7349
205. Bouju X, Duguet E, Gauffre F, Henry CR, Kahn ML, Mélinon P, Ravaine S (2018) Nonisotropic self-assembly of nanoparticles: from compact packing to functional aggregates. Adv Mater 30:1706558
206. Bazán-Díaz L, Mendoza-Cruz R, Velázquez-Salazar JJ, Plascencia-Villa G, Ascencio-Aguirre FM, Ojeda-Galván HJ, Herrera-Becerra R, Guisbers G, José-Yacamán M (2018) Synthesis and properties of the self-assembly of gold-copper nanoparticles into nanoribbons. Langmuir 34:9394–9401
207. Chen Z, Li J, Zhang X, Wu Z, Zhang H, Sun H, Yang B (2012) Construction of nanoparticle superstructures on the basis of host-guest interaction to achieve performance integration and modulation. Phys Chem Chem Phys 14:6119–6125

208. Zhong CJ, Maye MM (2001) Core-shell assembled nanoparticles as catalysts. Adv Mater 13:1507–1511
209. Henry CR (2015) 2D-arrays of nanoparticles as model catalysts. Catal Lett 145:731–749
210. Huang T, Meng F, Qi L (2009) Facile synthesis and one-dimensional assembly of cyclodextrin-capped gold nanoparticles and their applications in catalysis and surface-enhanced Raman scattering. J Phys Chem C 113:13636–13642
211. Axet MR, Conejero S, Gerber IC (2018) Ligand effects on the selective hydrogenation of nitrobenzene to cyclohexylamine using ruthenium nanoparticles as catalysts. ACS Appl Nano Mater 1:5885–5894
212. Zhu L, Yan H, Ang CY, Nguyen KT, Li M, Zhao Y (2012) Photoswitchable supramolecular catalysis by interparticle host-guest competitive binding. Chem Eur J 18:13979–13983

Chapter 6
Catalysis with MNPs on N-Doped Carbon

Rajenahally V. Jagadeesh

Abstract In recent years, nitrogen doped carbon supported/or encapsulated metal nanoparticles have emerged as prominent catalysts for sustainable chemical processes involving the synthesis of fine and bulk chemicals as well as life science molecules. This special kind of nanoparticles (metallic or metal oxide or core-shell) can be conveniently prepared by immobilization and pyrolysis of metal-nitrogen complexes or nitrogen-based metal organic frame works on heterogamous supports. Interestingly, nitrogen doping of carbon materials influences the nature, size and dispersion of nanoparticles. In addition, metal-N interactions induce a higher catalytic activity and selectivity of nanoparticles, which constitute successful catalysts for challenging reactions involving functional and structurally diverse molecules. In this chapter, the preparation and catalytic applications of nanoparticles on N-doped carbon are discussed in detail. Mainly, the preparation of non-noble metal nanoparticles-based catalysts and their sustainable applications in hydrogenation, oxidation, amination and other synthetic reactions are discussed.

Keywords Metal nanoparticles · N-doped carbon · Metal-N complexes · Metal organic frameworks

6.1 Supported Iron-Based Nanoparticles Surrounded by Nitrogen Doped Graphene/Carbon Layers

In 2013, Beller and coworkers have reported the preparation of carbon supported Fe_2O_3 nanoparticles surrounded by 3–5 layers of nitrogen doped graphene (Fe_2O_3/NGr@C; NGr-graphene) by the immobilization of in situ generated iron acetate-phenanthroline complex (Fe: ligand ratio = 1:3) on carbon support (Vulcan XC-72R) and subsequent pyrolysis at 800 °C under argon for 2 h (Fig. 6.1) [1, 2]. Remarkably, these iron oxide-based nanoparticles represent highly active and

R. V. Jagadeesh (✉)
Synergy Between Homogeneous and Heterogeneous Catalysis, Leibniz-Institut für Katalyse e. V, Albert-Einstein-Straße 29a, 18059 Rostock, Germany
e-mail: jagadeesh.rajenahally@catalysis.de

© Springer Nature Switzerland AG 2020   199
P. W. N. M. van Leeuwen and C. Claver (eds.), *Recent Advances in Nanoparticle Catalysis*,
Molecular Catalysis 1, https://doi.org/10.1007/978-3-030-45823-2_6

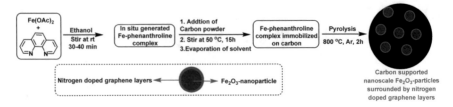

Fig. 6.1 Preparation of carbon supported Fe_2O_3-nanoparticles surrounded by nitrogen doped graphene layers

selective catalysts for chemoselective hydrogenation of nitroarenes [1]. To compare catalytic activities and to know the role of the nitrogen ligand, different material by the pyrolysis of simple iron acetate on carbon without ligand was also prepared (Fe_2O_3@C) and found that this catalyst is completely inactive for the hydrogenation of nitroarenes [1]. Both active and inactive catalysts have been characterized by using TEM, XPS, EPR and Mössbauer spectral analysis. All of these characterization studies revealed that the most active catalyst (Fe_2O_3/NGr@C) contains nanoscale Fe_2O_3 nanoparticles with particle sizes between 20 and 80 nm, which occurred together with smaller particles of 2–5 nm size [1, 2]. XPS analysis of Fe_2O_3/NGr@C determined the presence of nitrogen, which is in pyridinic form and contained Fe-N centers [1, 2]. Interestingly, these iron oxide nanoparticles of the active catalyst are surrounded by a shell of 3–5 nitrogen doped graphene layers. However, the inactive catalyst prepared without phenanthroline ligand, (Fe_2O_3@C), contained well facetted large Fe_2O_3 particles of 100–800 nm size that are not surrounded by nitrogen doped graphene/graphitic layers [1, 2]. In these catalytic materials, the formation of nanoscale iron oxide particles and the generation of nitrogen doped graphene layers, which surrounded nanoparticles are crucial structural features to exhibit high activity and selective. Obviously, the formation of nitrogen doped graphene layers and the nanosized iron oxide nanoparticles were produced by the pyrolysis of Fe-phenanthroline complex. Hence, nitrogen ligated complexes are essential for the generation of more active and selective nanocatalysts [1, 2].

These nitrogen doped carbon activated Fe_2O_3 nanoparticles (Fe_2O_3/NGr@C) constitute excellent catalysts for the hydrogenation of structurally diverse and functionalized nitroarenes to anilines (Scheme 6.1) under industrially variable and scalable conditions [1]. The resulting anilines represent key intermediates and central precursors for the preparation of advanced chemicals, life science molecules and materials [3]. As shown in Scheme 6.1, nitro group was highly selectively hydrogenated in presence of sensitive halides and functional groups such as aldehyde, keto, nitrile, amide, ester, alkene and alkyne groups. Noteworthy, all of these groups were well tolerated without being dehalogenated or reduced (Scheme 6.1) [1]. This catalyst system can be easily recycled and reused up to six times without significant loss of either activity or selectivity. Further, this Fe-based hydrogenation process has been up-scaled for several grams (1–5 g). In these cases yields, similar to those in the small scale (standard 0.5 mmol-scale) experiments were obtained [1].

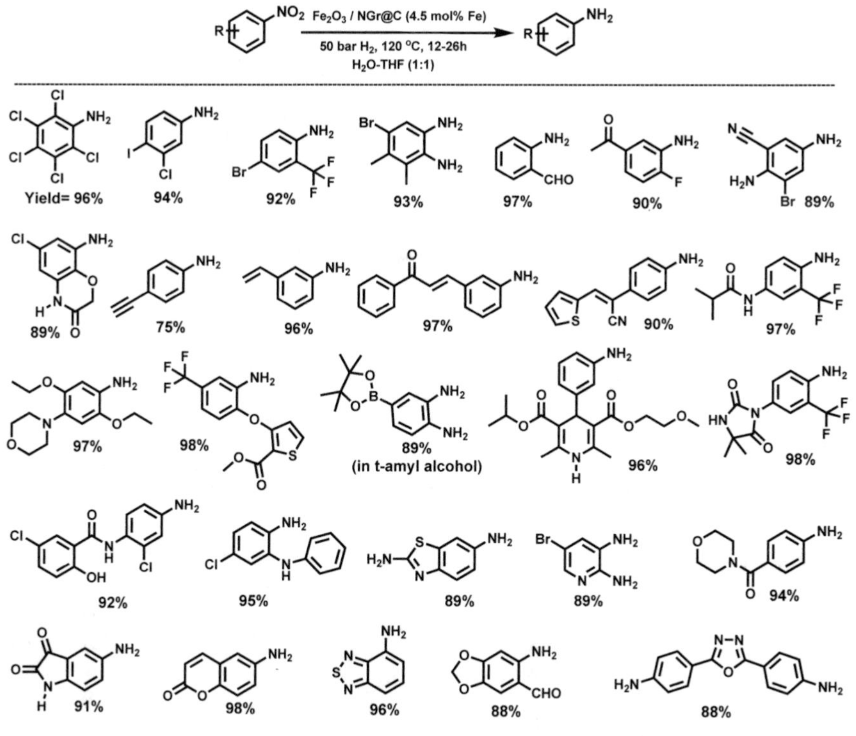

Scheme 6.1 Fe$_2$O$_3$/NGr@C catalyzed hydrogenation of structurally diverse and functionalized nitroarenes to anilines

Noteworthy, this Fe$_2$O$_3$/NGr@C catalyst enabled the reductive *N*-methylation of nitroarenes and amines for the preparation of *N*-methyl amines using paraformaldehyde without external hydrogen (Schemes 6.2 and 6.3) [4]. In this reaction sequence, first paraformaldehyde was decomposed under comparably mild conditions and produced formaldehyde and syngas (CO + H$_2$) (Scheme 6.3). In presence of Fe$_2$O$_3$/NGr@C catalyst, water-gas shift reaction of CO and H$_2$O occurred and produced additional H$_2$ and CO$_2$. Remarkably, small amounts of in situ generated hydrogen enabled the catalytic reduction of nitroarene and the corresponding imine after condensation with formaldehyde to yield mono-*N*-methylated product (Scheme 6.3). Following a second condensation with formaldehyde followed by reduction again gives finally *N,N*-dimethylamine (Scheme 6.3) [4]. Notably, in this reductive *N*-methylation process, paraformaldehyde serves as both methylation and reducing agent for the one-pot reductive amination of nitroarenes [4]. This Fe-based reductive amination protocol was used for the preparation of functionalized and structurally diverse *N*-methyl amines in good to excellent yields. The synthetic applicability of this method was demonstrated for *N*-methylation of life science molecules and for the preparation of existing drugs (Scheme 6.3) [4].

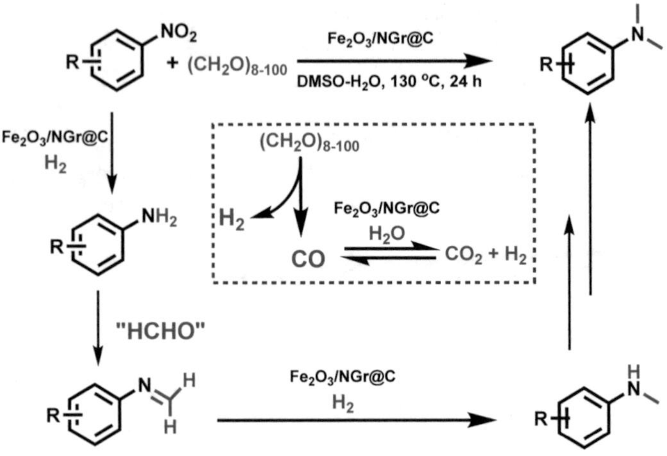
Fe2O3/NGr@C
DMSO-H2O, 130 °C, 24 h
(CH2O)8-100
H2
H2O
CO
CO2 + H2
"HCHO"

Scheme 6.2 Fe_2O_3/NGr@Ccatalyzed reductive *N*-methylation without external hydrogen

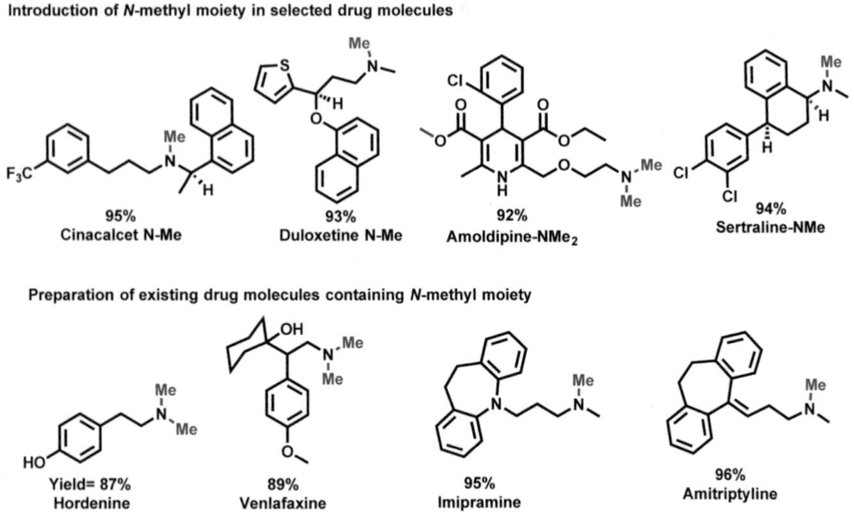
Introduction of N-methyl moiety in selected drug molecules
95%
Cinacalcet N-Me
93%
Duloxetine N-Me
92%
Amoldipine-NMe2
94%
Sertraline-NMe
Preparation of existing drug molecules containing N-methyl moiety
Yield= 87%
Hordenine
89%
Venlafaxine
95%
Imipramine
96%
Amitriptyline

Scheme 6.3 Fe_2O_3/NGr@C catalyzed reductive *N*-amination protocol for methylation of life science molecules and synthesis of existing drugs

In addition to hydrogenation reactions, this Fe_2O_3/NGr@C catalyst also exhibited excellent activity for the green oxidation of alcohols and amines to synthesize various nitriles of industrial and commercial interest using aqueous ammonia and molecular oxygen (Scheme 6.4) [5, 6].

Further, larger nanoparticles from Fe_2O_3/NGr@C were leached out by HCl to create more active catalysts embedded in small iron oxide/nitrogen doped graphene

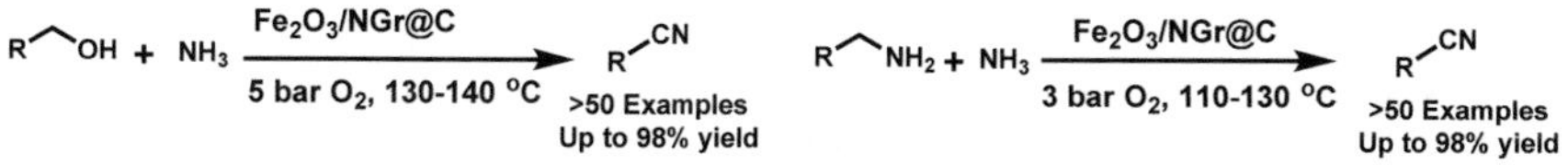

Scheme 6.4 Fe$_2$O$_3$/NGr@C catalyzed synthesis of nitriles from alcohols and amines

core/shell catalysts (FeOx@NGr-C) for oxidative dehydrogenations of several N-heterocycles (Scheme 6.5) [7].

In 2017, Balaraman et al. [8] have also prepared iron oxide nanoparticles supported on N-doped graphene oxides by the immobilization of Fe-phenanthroline complex on exfoliated graphene oxide (EGO) and subsequent pyrolysis at 900 °C under argon. Applying these iron-based catalysts acceptor-free dehydrogenation of N-heterocycles, amines, alcohols has been performed to obtain valuable compounds in good to excellent yields (Scheme 6.6) [8].

Dodelet et al. [9] have prepared carbon supported iron-based catalysts with active sites containing iron cations coordinated by pyridinic nitrogen functionalities in the interstices of graphitic sheets within the micropores as efficient oxygen reduction catalysts. These catalysts are prepared from carbon support, phenanthroline and ferrous acetate by ball-milling and then pyrolyzed twice, first in argon, then in ammonia. This Fe-based electro-catalyst exhibits the current density of a cathode equal to that of a platinum-based cathode with a loading of 0.4 mg of platinum per square centimeter at a cell voltage of $\geq$0.9 V [9].

Apart from phenanthroline, N-aryliminopyridines have also been used as suitable nitrogen ligands for the preparation of iron nanoparticles supported on N-doped carbon [10]. A series of N-aryliminopyridines were prepared using different substituted anilines. In situ generated Fe complexes of these ligands were immobilized on carbon and subsequent pyrolysis produced iron-based nanocatalysts

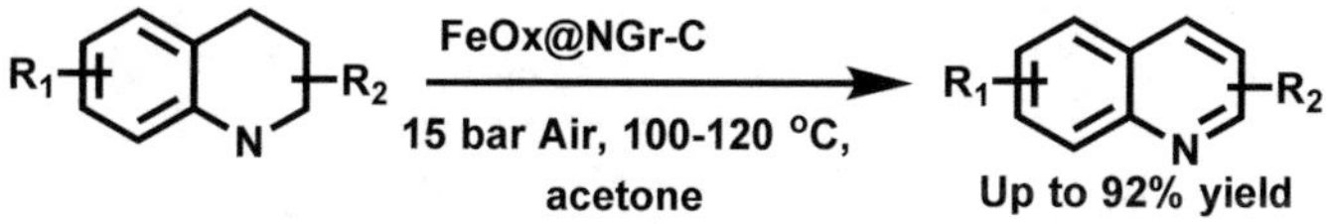

Scheme 6.5 Oxidative dehydrogenation of N-heterocycles using Fe$_2$O$_3$-nitrogen doped core-shell catalysts

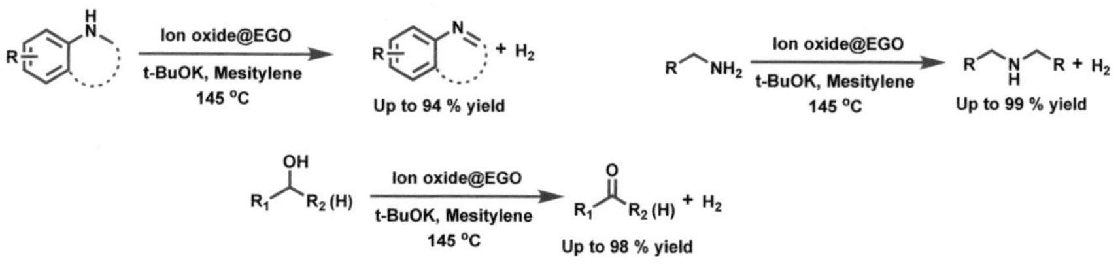

Scheme 6.6 Iron-catalyzed acceptor-free dehydrogenation of N-heterocycles, amines and alcohols

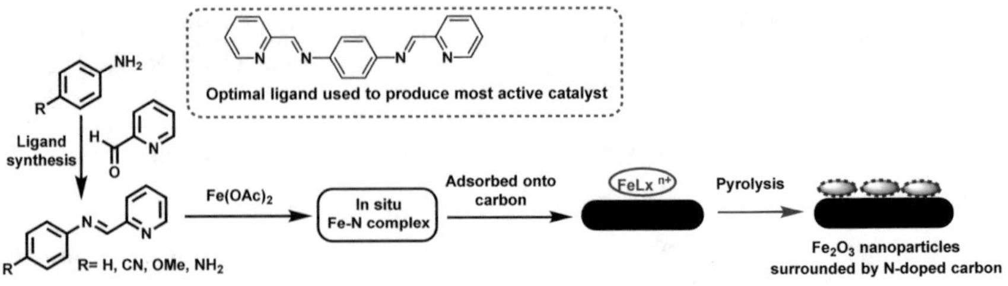

Fig. 6.2 Preparation of iron-based nanoparticle on N-doped carbon by using *N*-aryliminopyridine ligands

(Fig. 6.2) [10]. Among these, the ligand prepared using 1,4-amino benzene and 2-pyridinecarboxaldehyde formed the most active material for the hydrogenation of *N*-heteroarenes [10]. Characterization by TEM, XRD, XPS and Raman spectroscopy revealed that this iron nanomaterial contained Fe(0), Fe_3C and FeN_x in an N-doped carbon matrix (Fig. 6.2) [10].

Applying this iron catalyst, various (iso)quinolines were highly selectively hydrogenated to 1,2,3,4-tetrahydrochinolines, which represent versatile structural motifs present in many pharmaceuticals, natural products and biological molecules [11]. Noteworthy, this catalyst showed excellent chemoselectivity and the tested function groups such as nitriles, halogens, esters and amides are well tolerated. This Fe-based N-heterocyclic hydrogenation protocol has been implemented in the multistep synthesis of natural products and pharmaceutical lead compounds as well as modification of photo-luminescent materials (Scheme 6.7) [10]. As a result, this methodology constitutes the first heterogeneous iron-catalyzed hydrogenation of substituted (iso)quinolines with broad synthetic applicability (Scheme 6.7) [10].

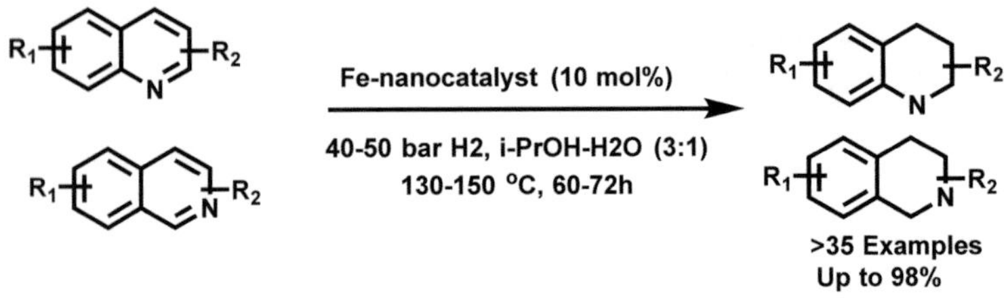

Scheme 6.7 Hydrogenation of quinolines and other *N*-heteroarenes by using N-doped carbon activated iron nanoparticles

6.2 Supported Cobalt-Based Nanoparticles Surrounded by Nitrogen Doped Graphene/Graphitic Shells

In addition to iron nanocatalysts [1] Beller and coworkers have also developed nitrogen doped graphene encapsulated cobalt-oxide nanoparticles supported on carbon (Co_3O_4-NGr@C) [12, 13]. These cobalt nanocatalysts were prepared again by the immobilization and pyrolysis of cobalt-nitrogen complexes on heterogeneous support (Fig. 6.3). In order to prepare active cobalt nanoparticles-based catalysts, different nitrogen ligands such as phenanthroline, terpyridine and 2,6-bis(2-benzimidazolyl)pyridine and pyridine have been used (Fig. 6.3) [12]. Among these, the material prepared using phenanthroline produced the most active catalyst for the hydrogenation of nitroarenes. Materials prepared using terpyridine and 2,6-bis(2-benzimidazolyl)pyridine gave less active catalysts, while the ones prepared using 1,1'-bipyridine and pyridine are completely inactive [12].

Detailed characterization using TEM and EDX of this cobalt-oxide (Co_3O_4-NGr@C) catalyst revealed the formation of Co_3O_4 particles of wide size distribution with a fraction of particles of 2–10 nm, and particles and agglomerates in the range 20–80 nm [12, 13]. XPS analysis showed that nitrogen is present in states such as pyridinic, pyrrolic and quaternary amine (R_4N^+). Like in the case of Fe_2O_3 nanoparticles [1], these cobalt-oxide particles are surrounded by nitrogen doped graphene layers. Carbon supported cobalt-oxide nanoparticles exhibit excellent hydrogenation and oxidation, as well as reductive amination activity [13]. Applying these catalysts, functionalized nitroarenes were highly selectively hydrogenated to produce the corresponding anilines in good to excellent yields (Scheme 6.8) [12].

Further, one-pot reductive amination of nitro compounds and carbonyl compounds for the synthesis of amines has been demonstrated for the preparation of various amines by using a Co_3O_4-NGr@C catalyst (Scheme 6.9) [13]. Here, different nitro compounds as well as aldehydes underwent reaction to form imine first, and then hydrogenation of the corresponding imine to produce the desired secondary amines. This methodology constitutes a green and economic one-step process for the preparation of higher value amines [13].

In addition to hydrogenations, the Co_3O_4/NGr@C catalyst exhibited excellent activity and selectivity for direct oxidative esterification of alcohols to synthesize various esters [14], which represent an abundant class of chemicals used as important building blocks for the fine and bulk chemical industry [15]. This cobalt catalyst enabled both cross and self-rectification of alcohols under mild reaction conditions to obtain different kinds of esters in good to excellent yields by using atmospheric oxygen (Scheme 6.10) [14].

Further by applying Co_3O_4/NGr@C catalyst, green synthesis of nitriles from alcohols and ammonia via aerobic oxidation process has been performed [5].

Organic nitriles constitute major building blocks for organic synthesis and represent a versatile motif found in numerous medicinally and biologically important compounds [16]. In general, functionalized nitriles are synthesized by cyanation

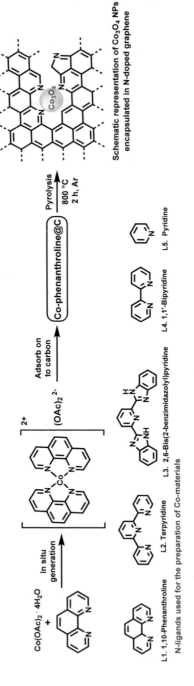

Fig. 6.3 Preparation of carbon supported cobalt-oxide nanoparticles encapsulated in N-doped graphene

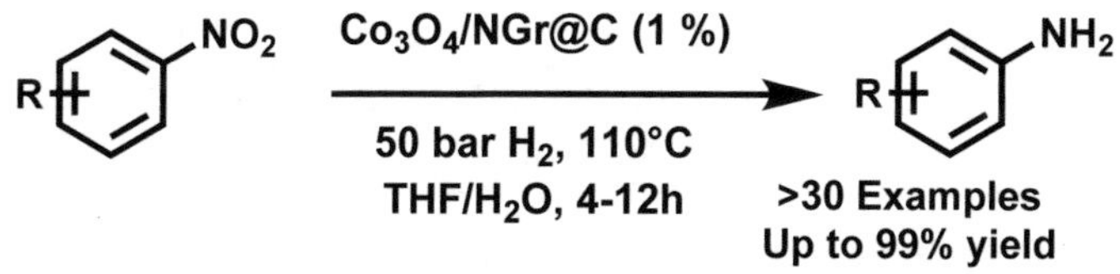

Scheme 6.8 Hydrogenation of nitro compounds using Co_3O_4-NGr@C catalyst

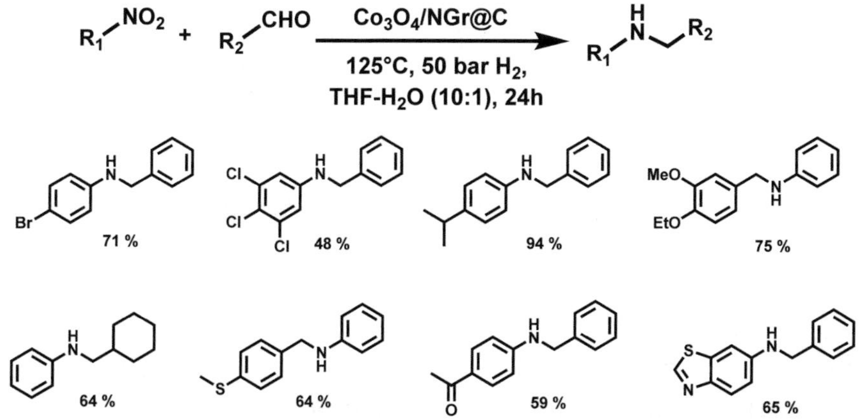

Scheme 6.9 One-pot reductive amination for synthesis of amines using Co_3O_4/NGr@C catalyst

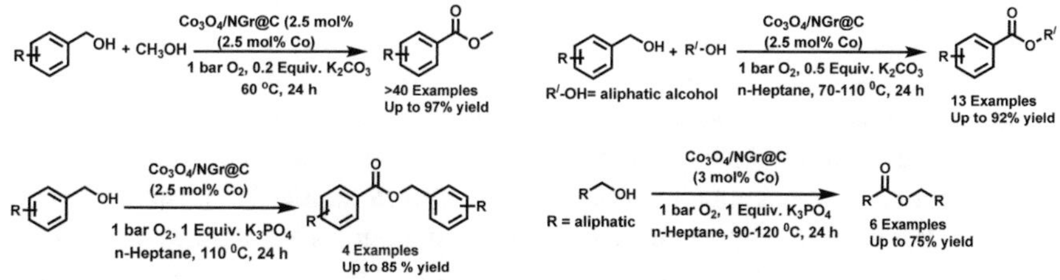

Scheme 6.10 Co-catalyzed direct oxidative esterification of alcohols to esters

procedures [17] using toxic cyanides, while the simple ones are produced by ammoxidation processes in the gas phase [18]. Compared to these traditional processes, the synthesis of nitriles from alcohols and ammonia is considered to be greener and a more convenient process for the preparation of nitriles. Advantageously Co_3O_4/NGr@C empower the synthesis of >80 functionalized and structurally diverse aromatic, heterocyclic, allylic and aliphatic nitriles (Scheme 6.11) [5].

Unfortunately, the use of carbon supported cobalt nanoparticles prepared by the immobilization and pyrolysis of nitrogen ligated Co-complexes is limited to

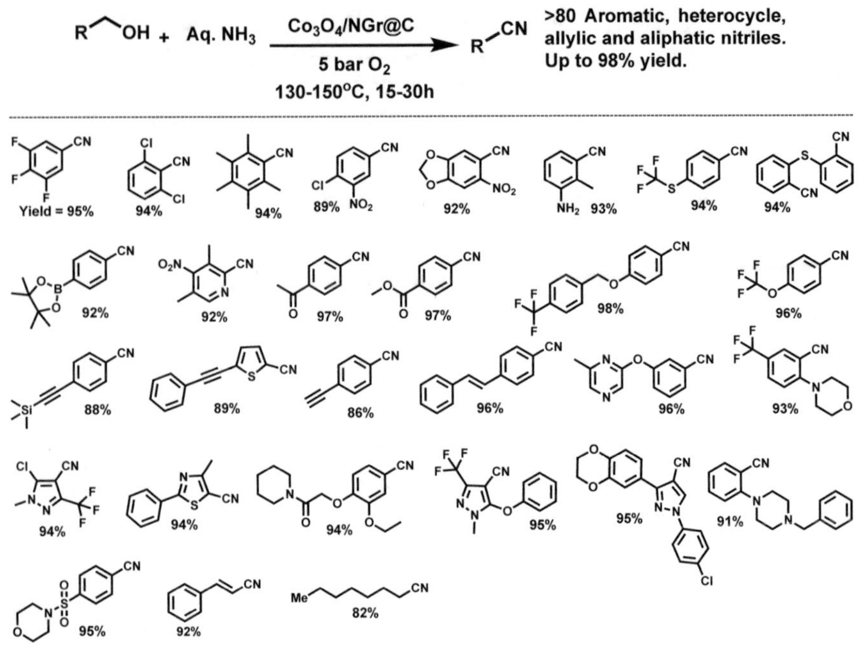

Scheme 6.11 Green synthesis of nitriles using $Co_3O_4/NGr@C$ catalyst

the hydrogenation of nitro compounds and imines [12, 13]. However, the pyrolysis of Co-phenanthroline complexes on oxidic supports such as Al_2O_3 and silica produced more active catalysts, which allow for the hydrogenation of other functional molecules [19–21]. Upon pyrolysis of $Co(OAc)_2$-phenanthroline complex on α-alumina produced cobalt-oxide/metallic cobalt nanoparticles giving rise to a core-shell structure and nitrogen doped graphene layers (Co_3O_4–Co/NGr@α-Al_2O_3) [19]. This supported alumina cobalt nanocatalyst displayed high activity and selectivity for the hydrogenation of quinolines to tetrahydroquinolines and indoles to indolines [19] (Scheme 6.12), nitriles to amines (Scheme 6.13) [20] and carbonyl compounds to alcohols [20] (Scheme 6.13).

Fascinatingly, all of the studied aromatic nitriles have been hydrogenated at 85 °C with 40 bar of hydrogen [20]. Even aliphatic nitriles, which are difficult to reduce, were also smoothly hydrogenated and produced the corresponding aliphatic amines in excellent yields [20]. Outstandingly, hexamethylenediamine, an important feedstock

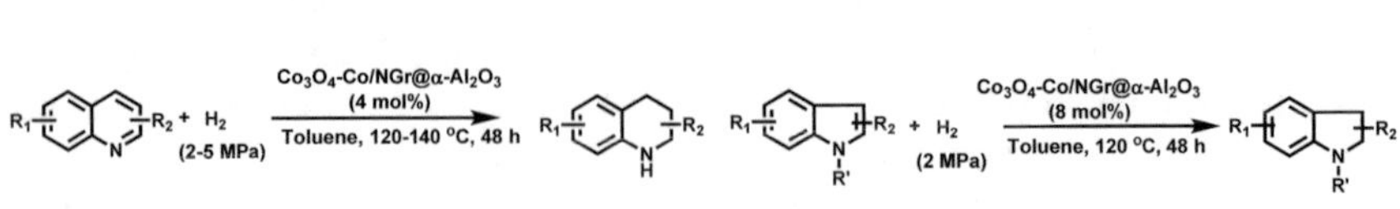

Scheme 6.12 Co_3O_4-Co/NGr@α-Al_2O_3 catalyzed hydrogenation of N-heteroarenes

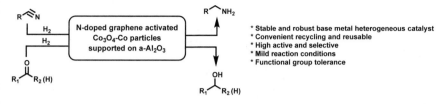

Scheme 6.13 Hydrogenation of nitriles and carbonyl compounds using Co_3O_4-Co/NGr@α-Al_2O_3 catalyst

for nylon polymer, was also prepared in 95% yield. These results are summarized in Table 6.1.

Next ketones and challenging aldehydes were also selectively hydrogenated by using Co_3O_4-Co/NGr@α-Al_2O_3 catalyst [20]. As an example, the keto group in structurally diverse molecules and steroid derivatives has been reduced to an alcoholic group without affecting other structural motifs of the molecules (Table 6.2). In all these hydrogenation reactions, Co_3O_4-Co/NGr@α-Al_2O_3 was recycled and reused up to eight times without significant loss of activity or selectivity [20].

Similarly, immobilization and pyrolysis of a Co-phenanthroline complex on silica produced N-graphitic-modified cobalt nanoparticles, which represent highly active and selective catalysts for semi-hydrogenation of alkynes to alkenes [21]. The selective hydrogenation of alkynes to alkene is considered to be an environmentally benign process applied in both research laboratories and industries [22, 23]. In industry, this transformation is used to "purify" bulk alkenes that serve as central intermediates in the chemical and petrochemical industries [23]. Applying these silica supported nitrogen doped graphic shell encapsulated cobalt particles, semi-hydrogenation of different kinds of internal alkynes have been selectively hydrogenated to produce alkene with Z selectivity up to 93% (Scheme 6.14) [21]. In addition to internal alkynes, terminal alkynes were also semi-hydrogenated by employing these silica supported cobalt nanoparticles [21].

After having demonstrated the synthetic applicability of cobalt nanoparticles on N-doped carbons, Beller et al. also investigated the dehydrogenation of formic acid for the generation of hydrogen [24]. Increasing the ratio of Co-phenanthroline from 1:2 ratio to 1:7 or 1:10 ratio and immobilization of this mixture on carbon followed by pyrolysis at 800 °C under argon atmospheres produced highly dispersed cobalt single atoms with CoNx centers [24]. This cobalt-based material showed excellent activity for the dehydrogenation of formic acid to generate hydrogen (Scheme 6.15) [24]. Hydrogen is considered as the simplest chemical energy carrier and is of particular interest because of its efficient transformation back into electricity through fuel cells with water as a green byproduct [25, 26].

Different catalytic applications of nitrogen doped carbon activated cobalt nanoparticles prepared by the immobilization and pyrolysis of Co-phenanthroline complex on heterogeneous supports are summarized in Table 6.3.

Table 6.1 Hydrogenation of aliphatic nitriles by using cobalt-based catalysts

$$R-CN \; + \; H_2 \; \xrightarrow[\substack{{}^iPrOH \; (2 \; mL), \; NH_3(aq) \\ then \; 1.25 \; M \; HCl_{MeOH}}]{Co_3O_4-Co/NGr@a-Al_2O_3 \; (4 \; mol\%)} \; R-CH_2\overset{+}{N}H_3 \; \overset{-}{Cl}$$

Entry	Substrate	H_2 (bar)	NH_3(aq) (mL)	T (°C)	t (h)	Yield (%)[a]
1	NC (5a)	40	0.1	130	2	94
2		5	0.3	85	24	98
3	CN (5b)	40	0.1	130	2	99
4	CN (5c)	40	0.1	130	2	99
5		5	0.3	85	24	97
6	CN (5d)	40	0.1	130	2	99
7		5	0.3	85	24	96
8	CN (5e)	40	0.1	130	2	91
9	MeO CN (5f)	40	0.1	130	2	85
10	$CH_3(CH_2)_{17}CN$ (5g)	40	0.2	130	2	84
11	$CH_3(CH_2)_{15}CN$ (5h)	40	0.2	130	2	93

$$NC-R-CN \; + \; H_2 \; \xrightarrow[\substack{{}^iPrOH \; (2 \; mL), \; NH_3(aq) \\ then \; 1.25 \; M \; HCl_{MeOH}}]{Co_3O_4-Co/NGr@a-Al_2O_3 \; (4 \; mol\%)} \; H_2NH_2C-R-CH_2NH_2$$

7 ⟶ 8

$\overset{-}{Cl} \; H_3\overset{+}{N} \diagdown\!\!\!\diagup \overset{+}{N}H_3 \; \overset{-}{Cl}$
95%

$H_2N \diagdown\!\!\!\diagup NH_2$
65%[a]

$H_2N \diagdown$ benzene $\diagdown NH_2$
88%

$\overset{-}{Cl} \; H_3\overset{+}{N} \diagdown$ benzene $\diagdown \overset{+}{N}H_3 \; \overset{-}{Cl}$
82%

All of the catalysts presented in Table 6.3 were prepared by the immobilization of in situ generated cobalt-phenanthroline complex by heterogeneous support and subsequent pyrolysis at 800 °C under argon for 2 h.

In 2018, Wang and Li [27] have reported atomically dispersed cobalt single sites on porous nitrogen doped carbon (SAS-Co/OPNC) as efficient catalysts for dehydrogenation and transfer hydrogenation/hydrogenation of N-heterocycles. These cobalt single atom-based catalysts were prepared by the pyrolysis of $CoCl_2$-2,2′-bipyridine

Table 6.2 Lumina supported N-doped graphene activate cobalt nanoparticles-catalyzed hydrogenation of carbonyl compounds [20]

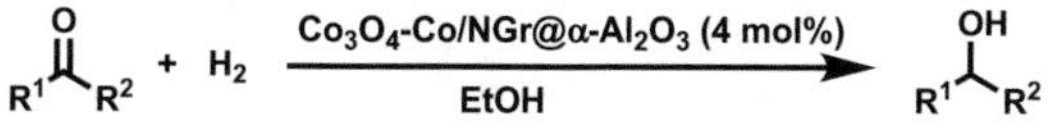

Entry	Substrate	Catalyst (mol%)	H$_2$ (atm)	T (°C)	Product	Yield (%)
1	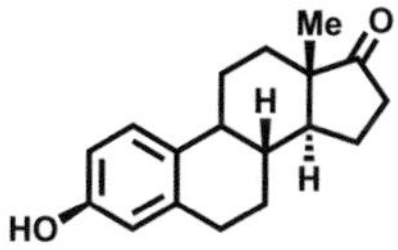	4	20	100	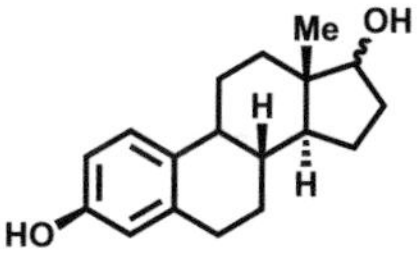	98
2	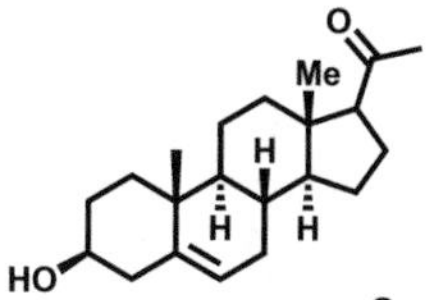	6	30	120	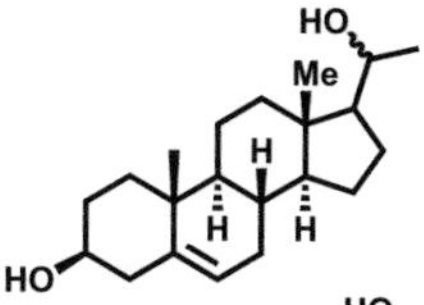	99
3	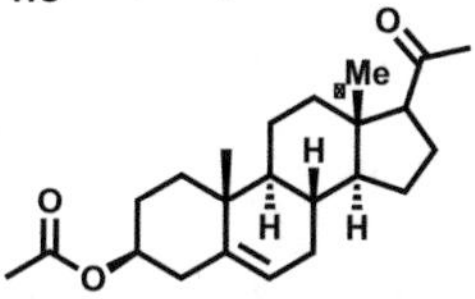	6	30	120	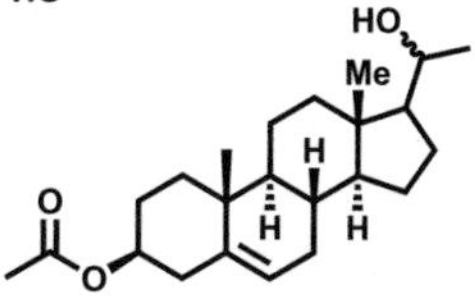	98
4	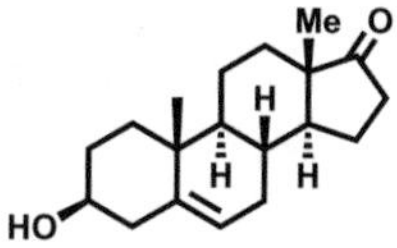	6	30	120	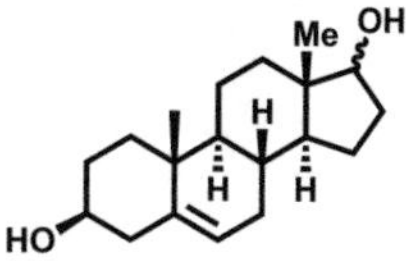	97

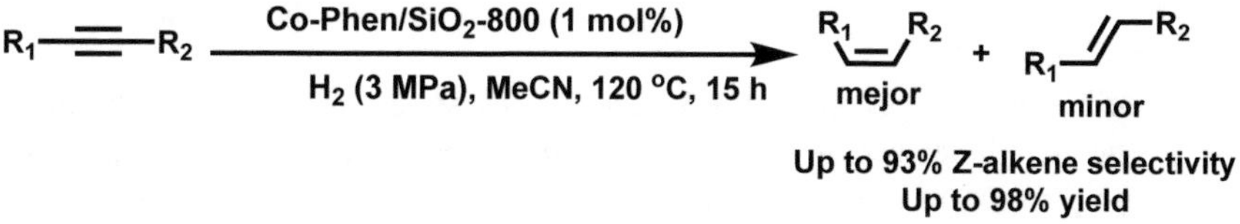

Up to 93% Z-alkene selectivity
Up to 98% yield

Scheme 6.14 Cobalt nanoparticles-catalyzed semi-hydrogenation of alkynes to alkenes

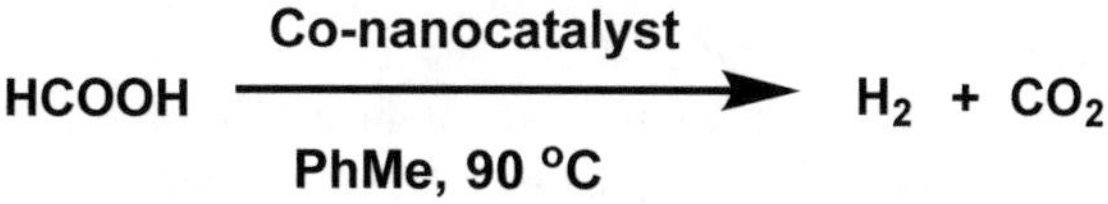

Scheme 6.15 Dehydrogenation of formic acid catalyzed by cobalt nanocatalysts

on SBA-15 silica support at 800 °C under argon for 2 h and subsequent removal of silica template by treating with 6 M sodium hydroxide aqueous solution at 60 °C for 24 h [27]. The resulting catalyst is highly efficient for acceptor-less dehydrogenation of N-heterocycles to release H$_2$ as well as for the reverse transfer hydrogenation

Table 6.3 Cobalt nanoparticles surrounded by N-doped carbon prepared by the pyrolysis of Co-phenanthroline complex on different supports and their catalytic applications

Entry	Ratio of Co-phenanthroline	Support	Pyrolysis conditions (t, Ar, t)	Nature of cobalt particles	Catalytic applications
1	1:2	Carbon	800 °C, Ar, 2 h	Co_3O_4 nanoparticles surrounded by N-doped graphene	Hydrogenation of nitro compounds and reductive amination
2	1:2	Carbon	800 °C, Ar, 2 h	Co_3O_4 nanoparticles surrounded by N-doped graphene	Synthesis of esters and nitriles form alcohols
3	1:2	α-Al_2O_3		Co, Co_3O_4 nanoparticles surrounded by N-doped graphene	Hydrogenation of quinlaines, nitriles and carbonyl compounds
4	1:2	SiO_2		Co, CoO and Co_3O_4 nanoparticles encapsulated in N-doped graphitic shells	Semi-hydrogenation of alkynes to alkenes
5	1:7 or 1:10	Carbon		Cobalt single atoms with CoNx centers	Dehydrogenation of formic acid for the generation of hydrogen

(or hydrogenation) of *N*-heterocycles to store H_2, using formic acid or external hydrogen as a hydrogen source (Scheme 6.16) [27]. Several tetrahydroquinolines and indolines were selectively dehydrogenated producing quinolines and indoles in up to 99% yields [27].

In addition to nitrogen ligated metal complexes, in recent years, nitrogen-based metal organic frameworks (MOFs) were conveniently used as suitable procures for the preparation of N-doped carbon supported nanoparticles-based catalysts by pyrolytic processes [28–31]. Notably, MOFs prepared using different metal ions and organic linkers represent a stable class of porous compounds, which can be assembled in a highly modular manner [32, 33]. Due to their structural tunability, different MOFs can be obtained, which can serve as desired precursors to prepare active and selective nanoparticle-based catalysts.

In general, nitrogen-based MOFs can be directly pyrolyzed [28–30], which act as self-sacrifing templates or can be immobilized/assembled on heterogeneous supports followed by pyrolysis [31] to generate metal nanoparticles supported on nitrogen doped carbons (Scheme 6.17).

Li et al. [30] have shown that direct pyrolysis of Co-DABCO-TPA-TPA (DABCO = 1,4-diazabicyclo[2.2.2]octane; triethylenediamine. TPA = terephthalic acid; 1,4-benzenedioic acid, BDA) at 900 °C under an inert atmosphere produced bifunctional N-doped cobalt nanoparticle (Co@NC-900)-based catalysts. On pyrolysis, this

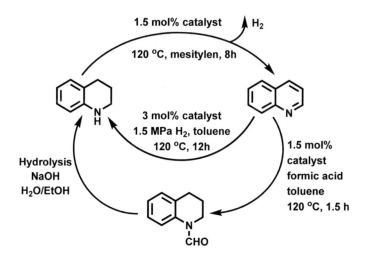

Scheme 6.16 SAS-Co/OPNC catalyzed reversible system for hydrogen storage and release

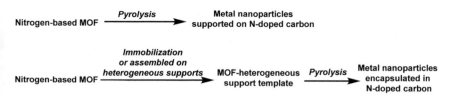

Scheme 6.17 Preparation of N-doped carbon supported nanoparticles by using MOFs as precursors

specific nitrogen containing MOF was transformed into graphitic N-doped carbon with high-density special basic sites, while the Co^{2+} ions were reduced to give Co-nanoparticles which were dispersed on or embedded in N-doped graphitic structures [30]. These Co-nanoparticles resulted in reusable catalysts for base-free transfer hydrogenation of nitriles to primary amines and secondary imines in good to excellent yields by using isopropanol (i-PrOH) as hydrogen source (Scheme 6.18) [30]. The progress and selectivity of this nitrile transfer hydrogenation reaction is highly dependent on the amount of isopropanol, which was used as both solvent and hydrogen

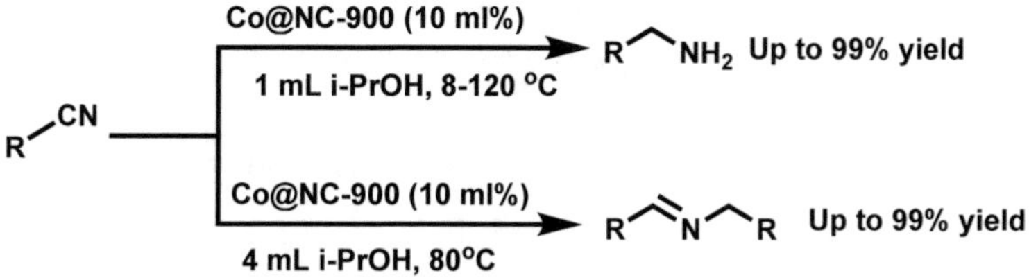

Scheme 6.18 Transfer hydrogenation of nitriles by using Co@NC-900 catalyst

Fig. 6.4 Preparation of cobalt nanoparticles and single atom-based catalysts by the pyrousls of Co-MOF on carbon

source. In 1 mL of isopropanol, the reaction gave selectively benzyl amine as the desired product. However, in 4 mL of isopropanol, the secondary imine was majorly formed, which is the intermediate or side product formed during the hydrogenation of nitriles. Thus, in 1 mL of isopropanol solvent Co@NC-900 works more efficiently than its reaction in 4 mL of isopropanol [30].

In 2017, Beller and coworkers [31] have reported the preparation of nitrogen doped graphitic shell encapsulated cobalt nanoparticles and single atoms by template synthesis of cobalt-DABCO-TPA MOF on carbon and subsequently pyrolysis at 800 °C under argon for 2 h (Co-DABCO-TPA@C-800) (Fig. 6.4). The resulting cobalt particles created a stable and reusable catalysts, which enabled the selective reductive amination for the synthesis of primary, secondary, tertiary and *N*-methylamines (>140 examples) [31]. Amines represent privileged compounds widely applicable in many science areas such as chemistry, biology, medicine, materials and energy [34–36]. For their synthesis, reductive amination represent sustainable and widely used methods in laboratories and industry [31].

Applying Co-DABCO-TPA@C-800 catalyst, the reductive amination reaction couples easily accessible carbonyl compounds (aldehydes, ketones) with ammonia, amines or nitro compounds in presence of molecular hydrogen under industrially viable and scalable conditions. The method offers cost-effective access to numerous amines, amino acid derivatives, and more complex drug targets (Schemes 6.19 and 6.20) [31]. By using this cobalt-based reductive amination protocol, selected existing drugs were prepared in good to excellent yields and the NH_2 moiety has also been introduced in functionalized and structurally complex compounds (Scheme 6.20) [31].

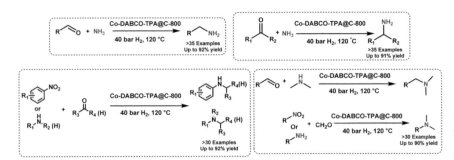

Scheme 6.19 Co-DBCO-TPA@C-800 catalyzed reductive aminations for synthesis various kinds of amines

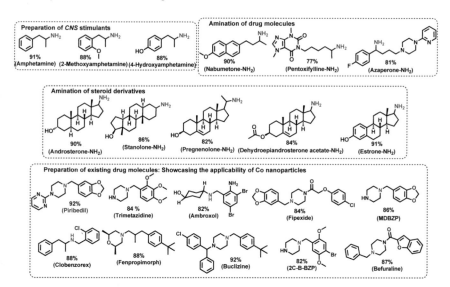

Scheme 6.20 Co-DBCO-TPA@C-800 synthesis of existing drug molecules and amination of life science and steroid derivatives

6.3   Supported Ni-Based Nanoparticles Surrounded by Nitrogen Doped Carbon Layers

Similar to iron and cobalt, a Ni-phenanthroline complex was also immobilized and pyrolyzed on SiO_2 support to obtain N-doped graphene surrounded Ni-nanoparticles [37]. The pyrolysis of Ni- phenanthroline on Aerosil OX 50 at 1000 °C under argon for 2 h resulted in the formation of intermetallic nickel silicide ($Ni–Si/NixOy–SiO_2$) core-shell nanoparticles distributed on the surface of the silica support, covered by N-doped graphene enriched with Si atoms (Fig. 6.5) [37].

This Ni–Si-based nanocatalyst enabled efficient and selective hydrogenation of nitroarenes, carbonyl compounds, nitriles, *N*-heteroarenes and unsaturated carbon–carbon bonds (Scheme 6.21) [37]. Advantageously, this Ni–Si-based nanomaterial creates a general hydrogenation catalysts alternative to Raney Ni [37].

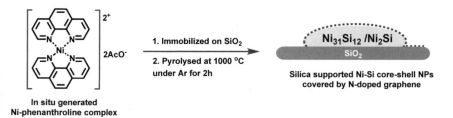

Fig. 6.5 Preparation of silica supported Ni–Si intermetallic core-shell NPs covered by N-doped graphene

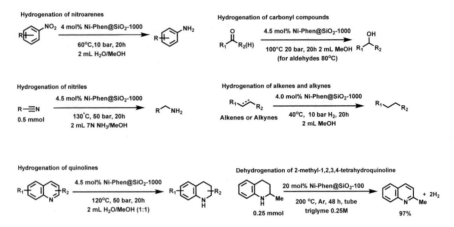

Scheme 6.21 Intermetallic Ni–Si-based catalysts for hydrogenation reactions

Recently, Kempe et al. [38] have reported alumina supported homogeneously distributed Ni-NPs embedded in N-doped carbon layer by the pyrolysis of Ni-salen complex on γ-Al_2O_3 (Fig. 6.6). The resulting Ni-nanoparticles constitute excellent reductive amination catalysts for the synthesis of primary amines from carbonyl compounds and aqueous ammonia in presence of molecular hydrogen under mild reaction conditions (Scheme 6.22) [38].

This Ni-nanocatalyst showed excellent reactivity to all kinds of aldehydes and ketones and as a result both linear and branched benzylic, heterocyclic and aliphatic primary amines were accessed in good to excellent yields (Scheme 6.22) [38].

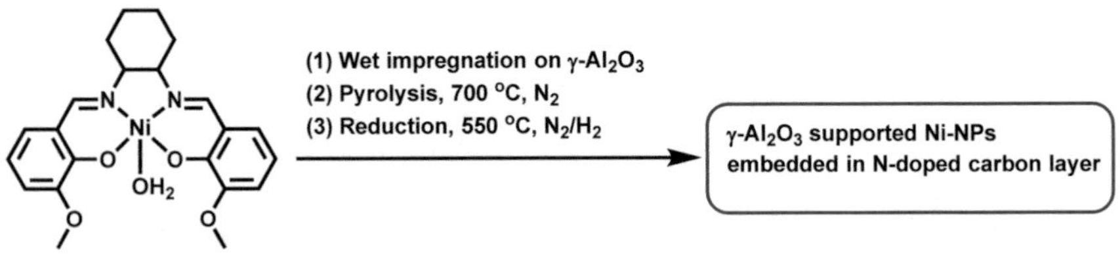

Fig. 6.6 Preparation of Ni-NPs embedded in N-doped carbon layer

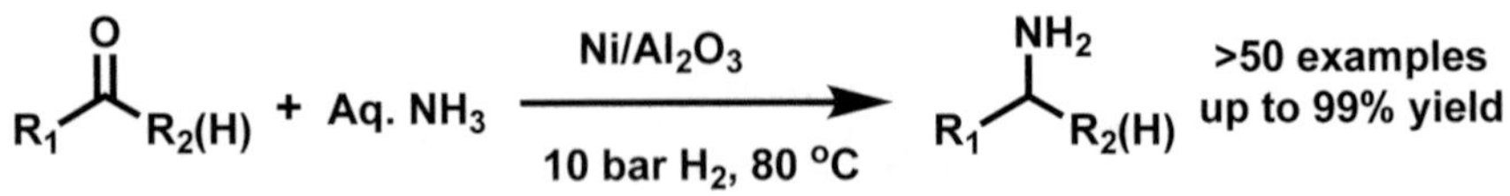

Scheme 6.22 Ni/Al_2O_3 catalyzed reductive amination for the synthesis primary amines

References

1. Jagadeesh RV, Surkus A-E, Junge H, Pohl M-M, Radnik J, Rabeah J, Huan H, Schünemann V, Brückner A, Beller M (2013) Nanoscale Fe$_2$O$_3$-based catalysts for selective hydrogenation of nitroarenes to anilines. Science 342:1073–1076. https://doi.org/10.1126/science.1242005
2. Jagadeesh RV, Stemmler T, Surkus A-E, Junge H, Junge K, Beller M (2015) Hydrogenation using iron oxide-based nanocatalysts for the synthesis of amines. Nat Protoc 10:548–557. https://doi.org/10.1038/nprot.2015.025
3. Ono N (2001) The nitro group in organic synthesis. Wiley-VCH, New York
4. Natte K, Neumann H, Jagadeesh RV, Beller M (2017) Convenient iron-catalyzed reductive aminations without hydrogen for selective synthesis of N-methylamines. Nat Commun 8:1344. https://doi.org/10.1038/s41467-017-01428-0
5. Jagadeesh RV, Junge H, Beller M (2014) Green synthesis of nitriles using non-noble metal oxides-based nanocatalysts. Nat Commun 5:4123. https://doi.org/10.1038/ncomms5123
6. Jagadeesh RV, Junge H, Beller M (2015) "Nanorust"-catalyzed benign oxidation of amines for selective synthesis of nitriles. Chemsuschem 8:92–96. https://doi.org/10.1002/cssc.201402613
7. Cui X, Li Y, Bachmann S, Scalone M, Surkus A-E, Junge K, Topf C, Beller M (2015) Synthesis and characterization of iron–nitrogen-doped graphene/core–shell catalysts: Efficient oxidative dehydrogenation of N-heterocycles. J Am Chem Soc 137(10652):10658. https://doi.org/10.1021/jacs.5b05674
8. Jaiswal G, Landge VG, Jagadeesan D, Balaraman E (2017) Iron-based nanocatalyst for the acceptorless dehydrogenation reactions. Nat Commun 88:2147. https://doi.org/10.1038/s41467-017-01603-3
9. Lefèvre M, Proietti E, Jaouen F, Dodelet J-P (2009) Iron-based catalysts with improved oxygen reduction activity in polymer electrolyte fuel cells. Science 324:71–74. https://doi.org/10.1126/science.1170051
10. Sahoo B, Kreyenschulte C, Agostini G, Lund H, Bachmann S, Scalone M, Junge K, Beller M (2018) A robust iron catalyst for the selective hydrogenation of substituted (iso)quinolones. Chem. Sci. 9:81348141. https://doi.org/10.1039/C8SC02744G
11. Sridharan V, Suryavanshi PA, Menéndez JC (2011) Advances in the chemistry of tetrahydro-quinolines. Chem Rev 111:7157–7259. https://doi.org/10.1021/cr100307m
12. Westerhaus FA, Jagadeesh RV, Wienhöfer G, Pohl M-M, Radnik J, Surkus A-E, Rabeah J, Junge K, Junge H, Nielsen M, Brückner A, Beller M (2013) Heterogenized cobalt oxide catalysts for nitroarene reduction by pyrolysis of molecularly defined complexes. Nat Chem 5:537–543. https://doi.org/10.1038/nchem.1645
13. Jagadeesh RV, Stemmler T, Surkus A-E, Bauer M, Pohl M-M, Radnik J, Junge K, Junge H, Brückner A, Beller M (2015) Nature protocols. Cobalt-based nanocatalysts for green oxidation and hydrogenation processes 10:916–926. https://doi.org/10.1038/nprot.2015.049
14. Jagadeesh RV, Junge H, Pohl M-M, Radnik J, Brückner A, Beller M (2013) Selective oxidation of alcohols to esters using heterogeneous Co$_3$O$_4$-N@C catalysts under mild conditions. J Am Chem Soc 135(29):10776–10782. https://doi.org/10.1021/ja403615c
15. Otera J (2003) Esterification: methods, reactions, and applications. Wiley-VCH, Weinheim
16. Fraser FF, Lihua Y, Ravikumar PC, Funk L, Shook BC (2010) Nitrile-containing pharmaceuticals: efficacious roles of the nitrile pharmacophore. J Med Chem 53:7902–7917. https://doi.org/10.1021/jm100762r
17. Anbarasan P, Schareina T, Beller M (2011) Recent developments and perspectives in palladium-catalyzed cyanation, of aryl halides: synthesis of benzonitriles. Chem Soc Rev 40:5049–5067. https://doi.org/10.1039/c1cs15004a

18. Martin A, Kalevaru VN (2010) Heterogeneously catalyzed ammoxidation: a valuable tool for one-step synthesis of nitriles. ChemCatChem 2:1504–1522. https://doi.org/10.1002/cctc.201 000173
19. Chen F, Surkus A-E, He L, Pohl M-M, Radnik J, Topf C, Junge K, Beller M (2015) Selective catalytic hydrogenation of heteroarenes with N-graphene-modified cobalt nanoparticles (Co3O4–Co/NGr@α-Al₂O₃). J Am Chem Soc 137:11718–11724. https://doi.org/10.1021/jacs.5b06496
20. Chen F, Topf C, Radnik J, Kreyenschulte C, Lund H, Schneider M, Surkus A-E, He L, Junge K, Beller M (2016) Stable and inert cobalt catalysts for highly selective and practical hydrogenation of C≡N and C=O bonds. J Am Chem Soc 138:8781–8788. https://doi.org/10.1021/jacs.6b03439
21. Chen F, Kreyenschulte C, Radnik J, Lund H, Surkus A-E, Junge K, Beller M (2017) Selective semihydrogenation of alkynes with N-graphitic-modified cobalt nanoparticles supported on silica. ACS Catal 7:1526–1532. https://doi.org/10.1021/acscatal.6b03140
22. Oger C, Balas L, Durand T, Galano J-M (2013) Are alkyne reductions chemo-, regio-, and stereoselective enough to provide pure (Z)-olefins in polyfunctionalized bioactive molecules. Chem Rev 113:1313–1350. https://doi.org/10.1021/cr3001753
23. Patai S (1964) The chemistry of alkenes. Interscience
24. Tang C, Surkus A-E, Chen F, Pohl M-M, Agostini G, Schneider M, Junge H, Beller M (2017) A stable nanocobalt catalyst with highly dispersed CoNx active sites for the selective dehydrogenation of formic acid. Angew Chem Int Ed 56:16616–16620. https://doi.org/10.1002/anie.201710766
25. Mellmann D, Sponholz P, Junge H, Beller M (2016) Formic acid as a hydrogen storage material-development of homogeneous catalysts for selective hydrogen release. Chem Soc Rev 45:3954–3988. https://doi.org/10.1039/C5CS00618J
26. Staffell I, Scamman D, Abad AV, Balcombe P, Dodds PE, Ekins P, Shah N, Ward KR (2019) The role of hydrogen and fuel cells in the global energy system. Energy Environ Sci 12:463–491. https://doi.org/10.1039/C8EE01157E
27. Han Y, Wang Z, Xu R, Zhang W, Chen W, Zheng L, Zhang J, Luo J, Wu K, Zhu Y, Chen C, Peng Q, Liu Q, Hu P, Wang D, Li Y (2018) Ordered porous nitrogen-doped carbon matrix with atomically dispersed cobalt sites as an efficient catalyst for dehydrogenation and transfer hydrogenation of N-heterocycles. Angew Chem Int Ed 57:11262–11266. https://doi.org/10.1002/anie.201805467
28. Dang S, Zhu Q-L (2017) Nanomaterials derived from metal–organic frameworks. Nat Rev Mater 3:17075. https://doi.org/10.1038/natrevmats.2017.75
29. Shen K, Chen X, Chen J, Li Y (2016) Development of MOF-derived carbon-based nano-materials for efficient catalysis. ACS Catal 6:5887–5903. https://doi.org/10.1021/acscatal.6b01222
30. Long J, Shen K, Li Y (2017) Bifunctional N-doped Co@C catalysts for base-free transfer hydrogenations of nitriles: controllable selectivity to primary amines vs imines. ACS Catal 7:275–284. https://doi.org/10.1021/acscatal.6b02327
31. Jagadeesh RV, Murugesan K, Alshammari AS, Neumann H, Pohl M-M, Radnik J, Beller M (2017) MOF-derived cobalt nanoparticles catalyze a general synthesis of amines. Science 358:326–332. https://doi.org/10.1126/science.aan6245
32. Corma A, Garcı´a H, Llabrés i Xamena FX (2010) Engineering metal organic frameworks for heterogeneous catalysis. Chem Rev 110:4606–4655. https://doi.org/10.1021/cr9003924
33. Furukawa H, Cordova KE, O'Keeffe M, Yaghi OM (2013) The chemistry and applications of metal-organic frameworks. Science 341:1230444. https://doi.org/10.1126/science.1230444
34. Lawrence SA (2004) Amines: synthesis, properties and applications. Cambridge University Press, Cambridge
35. Ricci A (2008) Amino group chemistry: from synthesis to the life sciences. Wiley-VCH, Weinheim
36. http://njardarson.lab.arizona.edu/sites/njardarson.lab.arizona.edu/files/Top200PharmaceuticalProductsRetailSales2015LowRes.pdf

37. Ryabchuk P, Agostini G, Pohl M-M, Lund H, Agapova A, Junge H, Junge K, Beller M (2018) Intermetallic nickel silicide nanocatalysts—a non-noble metal–based general hydrogenation catalyst. Sci Adv 4:eaat0761. https://doi.org/10.1126/sciadv.aat0761

38. Hahn G, Kunnas P, de Jonge N, Kempe R (2019) General synthesis of primary amines via reductive amination employing a reusable nickel catalyst. Nat Catal 2:71–77. https://doi.org/10.1038/s41929-018-0202-6

Chapter 7
Catalysis by Metal Nanoparticles Encapsulated Within Metal–Organic Frameworks

Amarajothi Dhakshinamoorthy and Hermenegildo Garcia

Abstract Supported metal nanoparticles (NPs) are a class of heterogeneous catalysts widely used to promote hydrogenations, chemical reductions, oxidations, and
coupling reactions. The metal–support interaction is crucial to determine not only
the catalytic activity of these materials, but also their stability. Since high catalytic
activity requires small size metal NPs, the most common deactivation mechanism
occurs by agglomeration of these NPs and increase of their particle size. One general
strategy to minimize particle growth in encapsulation inside porous hosts and MOFs
(MOFs) have been used for this purpose. This chapter describes the use of MOFs as
hosts to encapsulate metal NPs, covering the main properties of MOFs, encapsulation procedures and the most important reactions. Special attention has been paid to
describe the use of metal NPs supported on MOFs to promote coupling reactions,
since particularly for these type of reactions, MOFs have shown a clear superiority
respect to alternative supports. The final sections show some directions on the current
development of the field.

Keywords Heterogeneous catalysis · Porous supports · Metal–organic frameworks
as hosts · Coupling reactions · Alcohol oxidation

7.1 Introduction

Metal nanoparticles (NPs) exhibit general catalytic activity in a large variety of
reaction types including hydrogenation, aerobic oxidation, and cross-coupling reactions [1, 2]. It has been found that the activity of these metal NPs in catalysis is
due to the atoms located on the external surface having coordinatively unsaturated
positions [3]. In general, experimental and theoretical studies have shown that the

A. Dhakshinamoorthy
School of Chemistry, Madurai Kamaraj University, Madurai 625021, India
e-mail: admguru@gmail.com

H. Garcia (✉)
Departamento de Quimica, Instituto Universitario de Tecnologia Quimica (CSIC-UPV),
Universitat Politecnica de Valencia, Valencia 46022, Spain
e-mail: hgarcia@qim.upv.es

© Springer Nature Switzerland AG 2020   221
P. W. N. M. van Leeuwen and C. Claver (eds.), *Recent Advances in Nanoparticle Catalysis*,
Molecular Catalysis 1, https://doi.org/10.1007/978-3-030-45823-2_7

activity of surface metal atoms increases as its coordination number decreases from highly packed facets to less packed planes, edges, and particularly atoms at the vertex being particularly reactive [4, 5]. For these reasons, it is a general observation that the catalytic activity of metal NPs correlates well with the average particle size, decreasing in activity as the particle size increases [6–8].

One aspect related with the presence of coordinatively unsaturated metal atoms on the surface that, as just commented, are the active catalytic centers is the surface energy of the NPs. This surface energy indicates the stability that can be gained if these surface atoms would become fully coordinated. In this regard, the surface energy of small NPs is very large, rendering them very unstable. Due to this instability, the natural tendency of metal NPs is to undergo agglomeration to increase the percentage of saturated surface atoms, thereby reducing the surface energy. This tendency to grow makes small NPs difficult to stabilize, since they will aggregate spontaneously until dimensions of few micrometers are achieved. In view of the above considerations and the contradiction between catalytic activity of metal NPs and their stability, it is extremely important to develop strategies to stabilize metal NPs without affecting their catalytic activity.

One of the ways to minimize the growth of metal NPs in colloidal solutions is the use of ligands that by interacting with the surface atoms thwarts their growth [9]. However, the penalty of this methodology is that the ligands also block the catalytically most active centers of the metal NP and generally the gain in stability results in a decrease in activity. Scheme 7.1 illustrates the use of ligands to stabilize metal NPs in comparison with other alternative strategies.

Besides the use of ligands as stabilizers of colloid suspensions of metal NPs, other general strategies that combine stabilization of particle size with remarkable catalytic

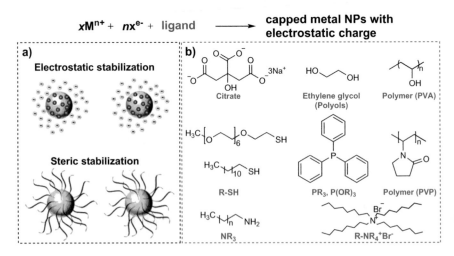

Scheme 7.1 Some of the commonly used ligands to stabilize colloidal metal NPs by electrostatic and steric stabilization. Reproduced with permission from Ref. [10] copyright 2018 Royal Society of Chemistry

Table 7.1 Comparison of the advantages and disadvantages of organic and inorganic supports stabilized NPs

	Advantages	Disadvantages
NPs@organic supports	Low cost/scalable Easy functionalization Biocompatible and biodegradable	Poor thermal stability Relatively low chemical stability Weak stabilization of NPs Poor pore size tunability
NPs@inorganic supports	High thermal and chemical stability Tunable pore size and shape Easy functionalization High surface area Compatible with high temperature and a wide range of pH values	Poor biodegradability Toxicity issues Adventitious catalytic sites

activity are the use of high surface area supports to deposit these metal NPs [11–13]. Among the supports that have been most frequently used to develop heterogeneous catalysts, metal oxides [14] are probably the most used, together with activated carbons [15] and organic polymers [16], either natural or synthetic. The advantages of metal oxides are that they can be obtained with large surface area frequently above 100 m^2/g and that they can establish a strong metal–support interaction due to the presence of surface hydroxyl groups. In the case of organic polymers, some of the synthetic ones do not frequently establish strong interactions with metal NPs, while natural biopolymers having hydroxyl groups, such as polysaccharides, are more suited to establish this interaction. However, a general drawback of organic substrates is their lack of stability under reaction conditions and this limitation has to be considered when selecting appropriate support. Table 7.1 shows a summary of main advantages and disadvantages of the different supports for metal NPs.

In addition to strong metal–support interaction, another approach that can result in stabilization of metal NPs is their immobilization inside a restricted space that due to the limited geometrical dimensions precludes particle size growth by mechanical reasons. It should be commented that spatial confinement inside pores and cavities is also compatible with the occurrence of strong metal–support interactions and in this way the two effects can operate simultaneously to increase the stability of metal NPs.

The present chapter describes the fundamentals, preparation procedures, characterization techniques and catalytic activity of selected metal NPs encapsulated inside the pores of MOFs which showed the best activity in each type of reactions. MOFs are one type of porous material that complements and exhibit unique features with respect to other porous materials including zeolites and mesoporous silicas. With respect to other porous materials, MOFs exhibit a large versatility in the design and synthesis, allowing the preparation of MOFs with any di- and polyvalent transition metal and with a large variety of organic linkers including polycarboxylic acids, nitrogen heterocycles and phosphorus-containing polytopic ligands.

In the next section, we will describe briefly the structure and composition of MOFs, paying particular attention to their stability, porosity and their ability to incorporate guests and particularly metal NPs. It should be commented that the support on which metal NPs are deposited or occluded can play other roles besides stabilization of metal NP size. In this way, it is well known in the area that samples having particles of the same size can exhibit contrasting catalytic activity and even selectivity as a function of the nature of the support. In some reaction mechanisms, it has been proposed that the support plays an active role not only by adsorbing reagents, but also by participating in some steps of the reaction mechanism. This active role of the support is particularly relevant in the present case since frequently MOFs have sites that have intrinsic catalytic activity, particularly as Lewis acids promoting cascade and oxidation reactions, due to the presence of coordinatively unsaturated positions at the metal nodes. As will be commented in other sections, the use of MOFs as support for metal NPs enables the development of multifunctional catalysts and more specifically catalysts having sites on the MOFs and sites on the occluded metal NPs. These types of multifunctional catalysts are especially relevant for the development of tandem reactions in where more than one elementary process occur simultaneously during the reaction.

7.2   MOFs Structure and Properties

MOFs are crystalline porous materials constituted by unit cells built of metallic nodes and multipodal rigid organic connectors. The nodes can be single metal ions, like Al^{3+} in MIL-53(Al) [17], or can be few metal atoms connected or not by oxygen or hydroxyl groups, such as the case of MIL-100(Fe) [18] that contains nodes of three Fe^{3+} ions connected to a central oxygen (Fe_3-μ-O). These metal nodes establish strong directional metal–ligand coordinative bonds with the organic linkers. Among them, aromatic di- or tricarboxylic acids are widely used to prepare a large variety of MOFs. The lattice and porosity derives from the directionality of the coordination bonds around the metal nodes and the geometry of the linker binding sites. Figure 7.1 illustrates the general structure of a MOF.

Fig. 7.1 Idealized structure of MOF and its components showing the directionality of the metal–ligand coordination bond

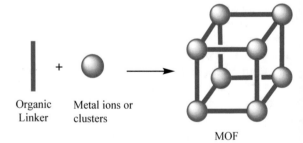

Organic Linker   Metal ions or clusters

MOF

Although by now there is a large number of MOFs that have been reported, only a handful of them have been widely employed, particularly in the field of heterogeneous catalysis [19–21]. Among the main reasons that justify the limited number of MOFs that have been most frequently preferred in catalysis, the major ones are the easy and reproducible preparation procedure from available precursors, but most importantly the chemical and physical stability of the framework. Generally, due to the nature of the metal–ligand bonds, MOFs are comparatively less robust than other purely inorganic porous materials, such as zeolites or porous aluminosilicates. In this way, particularly MOFs with divalent cations, such as Zn^{2+} and Cu^{2+}, are in general not very stable either thermally or in the presence of solvents or reagents. Thus, for instance, MOF-5 that is a zinc-terephthalate MOF becomes transformed into other materials, such as MOF-5' of undefined structure upon storage [22]. Similarly, $Cu_3(BTC)_2$ (BTC: 1,3,5-benzenetricarboxylate) cannot stand certain solvents and reagents such as amines or thiols [23]. For these reasons, there is a general believe that MOFs are not very stable porous materials, but there are certain MOF structure that exhibit a remarkable chemical and thermal stability that can be maintained unaltered during liquid-phase reactions.

Thus, in contrast to the case of MOFs of divalent cations [24], there are several MOFs that are extremely stable and enjoy a very robust lattice. Among them, MIL-101(Fe), MIL-100(Cr) and UiO-66(Zr) MOFs are remarkably stable and for these reasons they have been most frequently used as heterogeneous catalysts [25, 26]. Due to their robustness, these MOFs are also among the most used hosts to encapsulate metal NPs, immobilizing them and maintain a constant particle size under reaction condition. In the following paragraphs, we will comment briefly on the structure and properties of these robust MOFs.

Probably, one of the most robust MOF materials is UiO-66(Zr) (UiO: University of Oslo) the structure [27] of which is constituted by octahedral $Zr_6O_4(OH)_4$ metallic nodes that are coordinated with twelve terephthalate linkers forming super tetrahedral and super octahedral cages. Figure 7.2 shows the primary unit and the structure of UiO-66. This material has very large surface area typically above 1200 m^2/g with pore volume of 0.44 cm^3/g [26]. One important property of UiO-66(Zr) is that the material can be heated up to 400 °C without any change in the structure. Further heating causes dehydroxylation of the metallic nodes, but the process can still be

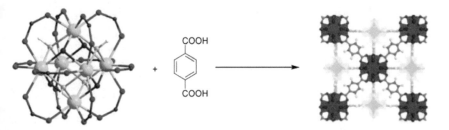

Fig. 7.2 Components and structure of UiO-66 MOF. Reproduced with permission from Ref. [28]. Copyright 2008 American Chemical Society

reversed by lowering the temperature under appropriate conditions. UiO-66(Zr) is also chemically stable in many solvents and reagents. The large cavities of dimensions about 10 nm allow the inclusion of guest, particularly metal NPs.

Another stable MOF also widely used as heterogeneous catalysts is MIL-100 (Scheme 7.2). This structure can be prepared with different trivalent cations, particularly with Cr^{3+} or Fe^{3+} ions. The metal nodes are trimeric M^{3+} cations connected to a central oxygen atom that is bonded to the three cations (μ_3-O) [29]. Each of the metal ions has an octahedral coordination with four positions occupied by two carboxylate groups of BTC, the other two positions around each metal ion are the central oxygen atom and an exchangeable ligand. In two thirds of the cases, the exchangeable ligand is a solvent molecule, such as DMF or water, and the other third is a charge compensating anion such as fluoride, chloride or hydroxide to reach the electroneutrality of the solid. This primary building unit defines two types of cages with pentagonal or hexagonal entrances with dimensions of about 2.4–2.9 nm.

MIL-101 is again very stable thermally, particularly in the case of Cr^{3+}. The corresponding MIL-101(Cr) being able to stand temperatures above 350 °C, without any change. Similar treatment temperature above 200 °C in MIL-100(Fe) causes a partial reduction of Fe(III) to Fe(II), but the structure is maintained and the reoxidation of Fe(II) to Fe(III) can take place reversible upon cooling. Scheme 7.2 shows the building units of MIL-100 MOF and its pore dimensions. MIL-101 is also another highly porous MOF with surface area frequently above 2500 m^2/g and pore volumes

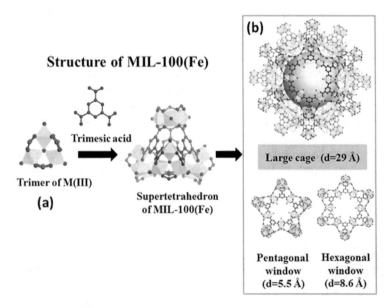

Scheme 7.2 (a) Oxo-centered trimers of iron octahedra; (b) combined with BTC linker leading to the MIL-100(Fe) framework with a zeolite topology of the MTN type. Reproduced with permission from Ref. [29]. Copyright 2016 American Chemical Society

Scheme 7.3 Some
properties of MOFs that are
relevant from the view point
of catalysis

Properties of MOFs

- ❖ High metal density
- ❖ Coordinatively unsaturated sites
- ❖ Thermal/chemical stability
- ❖ High surface area
- ❖ High pore volume
- ❖ Easy scalable
- ❖ Tunable pore volume
- ❖ Post-synthetic modification

Scheme 7.3 Some properties of MOFs that are relevant from the view point of catalysis

over 1 cm^3/g. The formula of MIL-101 is [M$_3$(O)X(BDC)$_3$(H$_2$O)$_2$] (BDC: 1,4-benzenedicarboxylate; X: OH or F), where M is a trivalent cation, typically Fe^{3+} or Cr^{3+}, X is anion and BDC is the organic linker [30].

The main properties of MOFs from the structural point of view are the large surface area and porosity. In fact, MOFs are among the materials with the lowest framework density, meaning that the mass in a certain volume for these materials is the smallest value so far achieved. This high percentage of empty volume is highly relevant for their use as hosts to incorporate inorganic guests and for the easy diffusion of substrates through the pores. Other important properties of MOFs are summarized in Scheme 7.3, and they include synthesis by design, wide range of possible chemical compositions, large percentage of transition metal and the possibility that MOFs offer to be modified by post-synthetic modification. These post-synthetic modifications frequently correspond to well established organic reactions occurring on the organic linker.

The next section will describe different preparation procedures to perform encapsulation of metal NPs, paying special attention to ensuring internal location of the resulting metal NPs.

7.3 Encapsulation Procedure

Metal NPs are typically obtained by reduction of convenient metal precursors, in particular metal complexes or metal salts. The first step in most of the synthesis consists, therefore, in the adsorption of metal precursor within the interior of the

MOF pore system. This adsorption can be carried out in the gas, liquid or solid phases. Scheme 7.4 summarizes some of the most widely employed encapsulation methods.

In some of the initial reports on the encapsulation of metal NPs inside MOF, Fischer and co-workers reported the gas-phase adsorption of volatile organometallic and metal carbonyl complexes such as Au(CO)Cl. Initially, the MOFs selected as hosts were, however, not the most stable ones, but the adsorption procedure was developed with the objective to ensure the internal location of the metal NP. For this reason, the MOF was initially submitted to a pre-treatment to eliminate adsorbed solvent molecules and water. After desorption of weakly adsorbed species, the pre-treated material was exposed to the presence of volatile organometallic complexes. Due to the large proportion between internal and external surface, adsorption in the gas phase in porous materials results preferentially in the internal location of the adsorbate. Subsequently, metal NPs are formed by decomposition of the organometallic complex by physical and chemical means, hydrogenation of the metal precursor being a simple procedure, compatible with preservation of the MOF structure. In that way, Pd NPs were incorporated inside MOF-5 [31].

Besides gas phase, adsorption can also be performed in liquid media. In one of the examples reported, Au NPs were prepared as colloidal suspension using PVP as capping and protecting agent and these small Au NPs were adsorbed inside the pores of MIL-101 without the need of additional treatment [32]. The mechanism of this incorporation of performed Au NPs inside MOF is unknown since the particle size of Au NPs is similar to the dimensions of the pores in MIL-101 and the presence of PVP as a shell stabilizing the colloidal solution should in principle disfavor diffusion of preformed Au NPs inside the pores.

One of the most common and probably general preparation procedures of metal NPs inside MOFs by liquid phase adsorption is the method known as double-solvent method [33]. This method is based on the adsorption of the minimum amount of a polar solvent, typically water containing a metallic salt precursor, in a large excess of highly apolar solvent in where the MOF is suspended. The rationale for the double-solvent method is the high polarity contrast between an apolar solvent in large excess and a polar solvent containing the metal salt precursor in small volume that can become incorporated completely inside the polar pores of the MOF if the volume

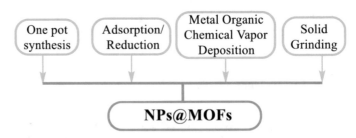

Scheme 7.4 Most widely used procedures for encapsulation of metal NPs inside MOFs

of this solvent is less than the internal pore volume. Subsequently, after adsorption of the precursor, the solid particles of the MOF have to be washed before proceeding to generation of the metal NP by chemical reduction.

Besides adsorption from the gas or liquid phases, formation of metal NPs inside MOFs has also been reported by adsorption in the solid phase. Thus, Haruta and co-workers have reported that grinding MOF-5 with a solid gold complex in a mortar is sufficient to produce the incorporation inside the MOF of the gold complex that subsequently can be reduced into Au NPs [34].

From the current state of the art, it can be summarized that there is a variety of procedures that yield metal NPs and even small metal oxide NPs for those metals like Zn and Ti that easily form metal oxides and that these procedures can be applied to a large variety of MOF structures, including those like Zr MOFs and MIL-100/101 families that produce remarkably stable catalysts [26, 35, 36].

One issue of concern when preparing guests inside porous materials, and particularly inside MOFs, is to provide convincing evidence of the internal location of the incorporated guest. This is always a very difficult task and probably the most convincing technique to address this issue is high-resolution transmission electron tomography. In this technique, a series of TEM images of the same MOF particle are taken with different submicrometric increment in the vertical axis. These images are later used to reconstruct a three-dimensional mapping of the crystallite, the process being analogous to confocal optical microscopy widely used to determine internalization of particles inside living cells, except that the resolution in the case of electron microscope is in the nanometer scale. Unfortunately, transmission electron tomography is not yet a routine technique, and therefore, it is not still widely available to characterize samples of metal NPs inside MOFs on a regular basis.

For this reason, other methods should be implemented to provide some support to the internal location of the metal NPs. In this context, conventional TEM, especially in the dark field mode, should determine the average particle size and its distribution. If the metal NPs are inside the pores their diameter has to be commensurate with the dimension of the pores. In other words, large particles should not be present and only particles with diameters smaller than the cavity size should be observed by TEM images.

Besides electron microscopy, surface area measurements and catalytic activity tests can also provide indirect support to the internal location of the metal NPs. Thus, isothermal gas adsorption after incorporation of metal NPs should determine a decrease in the internal surface area and pore volume. Similarly by testing the catalytic activity of the materials for substrates with similar chemical reactivity, but different molecular dimensions, smaller and larger than the pores of MOFs, the contrasting activity of the metal NPs inside the MOF should reflect that only those substrates that can diffuse inside the pores undergo chemical reaction, while those others that cannot access the internal pores would not undergo chemical transformation under same conditions.

The following section will describe selected examples of a variety of reactions with the aim to illustrate the advantage and the potential of the use of metal NPs encapsulated within MOFs (metal NPs/MOFs) as heterogeneous catalysts. The reader is also referred to the existing literature for a complete coverage on this field.

7.4 Catalysis by Metal NPs@MOFs

7.4.1 Suzuki–Miyaura Cross-Coupling

C-C bond forming reactions are important because their application as synthetic routes in organic transformations. One of these reactions that has become quite useful due to the high yields and mild conditions to form asymmetrically substituted biaryls is the Suzuki–Miyaura cross-coupling of aryl halides and arylboronic acids catalyzed by Pd NPs.

In one of the seminal contributions using metal NPs/MOFs as heterogeneous catalysts, Pd NPs were supported over a highly robust MOF, namely MIL-101(Cr), and the resulting Pd/MIL-101(Cr) was used as a heterogeneous catalyst for the Suzuki–Miyaura cross-coupling reaction in aqueous medium [37]. Characterization of the material showed that Pd NPs were distributed homogenously throughout the MIL-101(Cr) particle. The mean diameter of Pd NPs was 1.9 ± 0.7 nm as evidenced by TEM measurements. Further, powder XRD revealed that the crystalline nature of MIL-101(Cr) was maintained during the deposition of Pd NPs. XPS analysis indicated the existence of Pd in its metallic form. The Suzuki–Miyaura cross-coupling reaction between 4-chloroanisole and phenylboronic acid using Pd/MIL-101(Cr) reached 82% yield after 6 h, with NaOMe as the base at 80 °C under inert atmosphere (Scheme 7.5). It is interesting to note that aryl chlorides are among the less reactive substrates. To put the activity of Pd/MIL-101(Cr) into context, analogous Pd catalysts such as Pd/C and Pd/ZIF-8 afforded only 35 and 16% yields, respectively under identical conditions. This comparison clearly establishes the superior performance of MIL-101(Cr) as host and support in promoting this reaction. The enhanced activity of Pd/MIL-101(Cr) compared to Pd/ZIF-8 was attributed to the larger surface area and pore size of MIL-101(Cr), thus ensuring high dispersion of Pd NPs and facilitating diffusion of reactants and products throughout the pores. A series of substrates were tested to determine the scope of Pd/MIL-101(Cr) for this C-C coupling, observing 81–97% yields for all the substrates tested under identical conditions.

In other example, highly dispersed Pd NPs were encapsulated in the cages of MIL-101 using a "double solvent" method coupled with subsequent reduction with

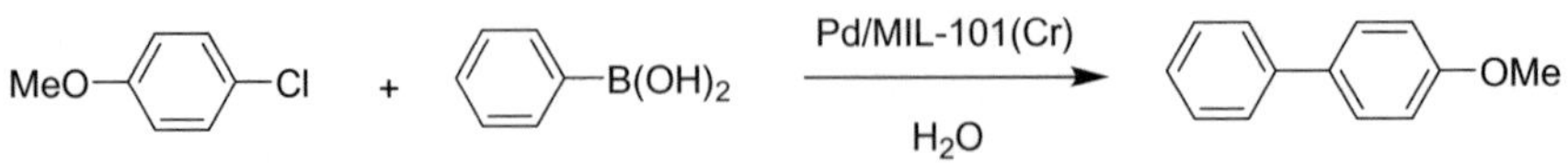

Scheme 7.5 Suzuki–Miyaura cross-coupling reaction of highly deactivated 4-chloroanisole catalyzed by Pd/MIL-101(Cr)

$NaBH_4$ (Scheme 7.6) [38]. As commented earlier (Sect. 7.3), this is one of the effective strategies employed in recent times to precisely deposit metal NPs within the pores of MOF, avoiding effectively the aggregation of Pd NPs on the external surface of the material. Powder XRD confirmed the robust nature of MIL-101(Cr) structure during Pd loading, while XPS indicated the oxidation state of Pd as zero. The average size of the Pd NPs was 2.4 nm, meaning that the NPs can fit within the mesoporous cavities of MIL-101(Cr). Interestingly, the activity of Pd/MIL-101(Cr) was tested for the reaction between bromobenzene and phenylboronic acid using potassium carbonate as base in water–ethanol mixture at room temperature reaching 99% yield. The activity of this Pd/MIL-101(Cr) was also tested to prepare a series of other biphenyl derivatives, achieving 90–99% yields under these conditions. ICP analysis showed the absence of Pd leaching to the solution and the hot filtration test showed heterogeneity of the reaction. The catalyst was reused for five cycles with no noticeable decay in its activity and maintaining the crystal structure of MIL-101 (Cr).

One of the open issues in metal NPs encapsulated within MOFs is to further tune the polarity and properties of the MOF cavity to enhance the catalytic activity and stability of occluded metal NPs. This tuning can be simply achieved by using substituted organic linkers, where the substitution can interact with the metal NPs.

Pd NPs have been immobilized over an amino-functionalized MIL-101(Cr) MOF to obtain Pd/MIL-101(Cr)-NH$_2$ [39]. This catalyst was characterized by SEM and TEM images, observing no changes in the morphology of MOF crystals during the deposition of Pd. Powder XRD showed that the structural integrity of MIL-101(Cr)-NH$_2$ was not altered during the deposition of Pd NPs. The metallic Pd NPs were homogeneously distributed throughout MIL-101(Cr)-NH$_2$ host with the average particle size of Pd being 2.6 nm. The activity of Pd/MIL-101(Cr)-NH$_2$ was tested for the cross-coupling reaction between phenylboronic acid and bromobenzene with cesium carbonate as base in water at room temperature, reaching 99% biphenyl yield. Interestingly, a quantitative yield was also observed for the reaction between pinacol phenylboronate and bromobenzene under identical conditions. The

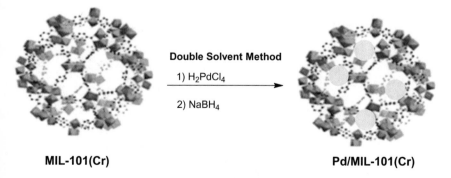

Double Solvent Method

1) H_2PdCl_4

2) $NaBH_4$

MIL-101(Cr)   **Pd/MIL-101(Cr)**

Scheme 7.6 Preparation of Pd/MIL-101(Cr) by double solvent method, incorporating first the Pd^{2+} salt and subsequent reduction to form Pd(0). Reproduced with permission from Ref. [38] Copyright 2014 Royal Society of Chemistry

Pd leaching was found to be 0.17 ppm. Excellent yields were also achieved for a wide variety of substrates including electron donating and withdrawing substituents, large dimension aryl halides, heteroaryl halides and chloroarenes under remarkably mild conditions that include the use of water as solvent, aerobic conditions and room temperature (Scheme 7.7). Furthermore, the solid catalyst was recycled at least ten times without any change in its catalytic performance.

One point to be addressed is whether or not the internal location of metal NPs is beneficial from the catalytic point of view, increasing or not the activity. In this regard, Cao and co-workers developed a simple direct anionic exchange method to introduce Pd in a MOF, followed by subsequent chemical reduction with NaBH$_4$ to deposit Pd NPs on amino-functionalized MIL-53(Al)-NH$_2$ [40]. The structural integrity of MIL-53(Al)-NH$_2$ was retained during the loading of Pd NPs. TEM images of the resulting Pd/MIL-53(Al)-NH$_2$ showed that the mean diameter of Pd NPs was 3.12 nm and mostly found on the external surface of MIL-53(Al)-NH$_2$. The catalytic performance of Pd/MIL-53(Al)-NH$_2$ was investigated in the Suzuki–Miyaura cross-coupling reaction of bromobenzene with phenylboronic acid using Na$_2$CO$_3$ as base in water–ethanol mixture at 40 °C achieving 94% yield with a turnover number (TON) of 198. Under identical conditions, the reaction of chlorobenzene and phenylboronic acid afforded only trace yield of biphenyl [40]. Further, the activity of Pd/MIL-53(Al) and Pd/C was found to be 45% (TON: 98) and 37% (TON: 88) under identical conditions [40]. A series of aryl halides were coupled with phenylboronic acid to obtain their respective biphenyl derivatives with yields ranging between 87 and 97% under identical conditions. The catalyst was reused for five cycles with no appreciable

Scheme 7.7 Scope of Pd/MIL-101(Cr)-NH$_2$ as Suzuki–Miyaura catalyst for the coupling of various aryl halides and phenylbornonic acid

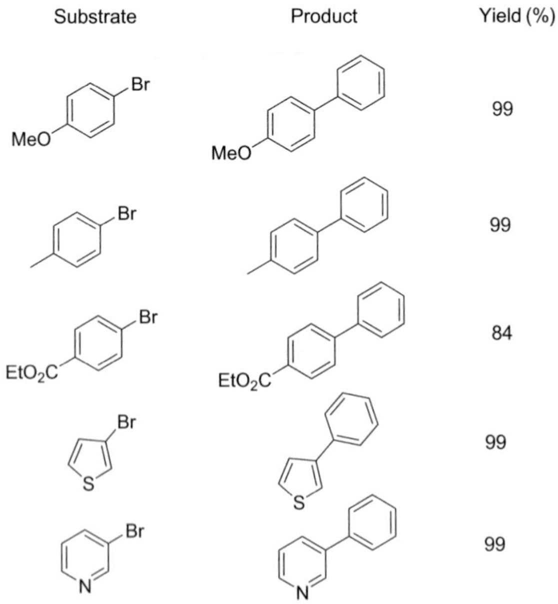

Substrate	Product	Yield (%)
		99
		99
		84
		99
		99

change in its activity. The reaction was found to be heterogeneous in nature with the Pd leaching of 0.3 ppm as evidenced by ICP-AES analysis. It should be commented that the performance of this Pd/MIL-53(Al)-NH$_2$ catalyst is remarkable in spite that the Pd NPs are mostly on the external surface. In addition, the present study with MIL-53(Al)-NH$_2$ shows the positive influence of the amino group on the activity. In some other examples, the most common observation is that the presence of amino groups interacting with the metal NP increases catalyst stability by minimizing the metal NPs growth.

Pd NPs have been also supported on UiO-66-NH$_2$ MOF and its catalytic performance has been examined in Suzuki–Miyaura cross-coupling reaction [41]. Although TEM images indicated uniform distribution of Pd NPs, the mean diameter of Pd NPs was 5.28 ± 0.5 nm, and therefore, they seem to be too large to be accommodated within UiO-66-NH$_2$ cavities. XPS analysis revealed the existence of metallic Pd in the as-prepared catalyst. The activity of this solid was tested in the Suzuki–Miyaura cross-coupling of phenylboronic acid and 4-iodoanisole, observing a turnover frequency (TOF) value of $2190\,h^{-1}$. A variety of biphenyl derivatives were prepared using Pd/UiO-66-NH$_2$ as catalyst under identical conditions with the yields ranging between 80 and 95%. Reusability experiments indicated identical activity for five consecutive cycles. Furthermore, powder XRD patterns of the five times used solid indicated that the structural integrity of UiO-66-NH$_2$ was not modified during repeated cycles. The Pd leaching was 4% in the first two cycles, but no further Pd leaching occurred in the subsequent uses.

Pd NPs (5 nm) were supported on nanoscale ScBTC NMOFs (NMOF: nanoscale MOF) by using a microwave-assisted impregnation process and the activity of Pd/N-ScBTC was tested in the Suzuki–Miyaura cross-coupling reaction between aryl/heteroaryl halides and arylboronic acids. The corresponding biphenyl was formed in high yields (75–97%) [42]. A commercial Pd/C catalyst and bulk ScBTC MOFs-supported Pd catalysts were also prepared and included in the study, selecting the coupling of aryl chloride and phenylboronic acid under identical conditions as probe reaction. The order of reactivity observed was as Pd/C < Pd-MOFs < Pd-NMOFs showing the benefits of reducing the MOF particle size to nanoscale on the catalytic activity. Pd/N-ScBTC was reused for five cycles with no decay in its activity.

7.4.2 Ullmann Coupling

For the synthesis of symmetrically substituted biaryls, the Ullmann homo coupling of aryl halides may present the advantage of avoiding the need of arylboronic acids and their derivatives (Scheme 7.8). In another example of the use of metal NPs/MOFs as heterogeneous catalysts, the performance of Pd/MIL-101(Cr) was examined in Ullmann reaction of aryl chlorides (Scheme 7.8) [37]. As one illustrative example, the homocoupling of 4-chloroanisole using Pd/MIL-101(Cr) with NaOMe as a base under inert atmosphere in water at 80 °C resulted in 97% yield of

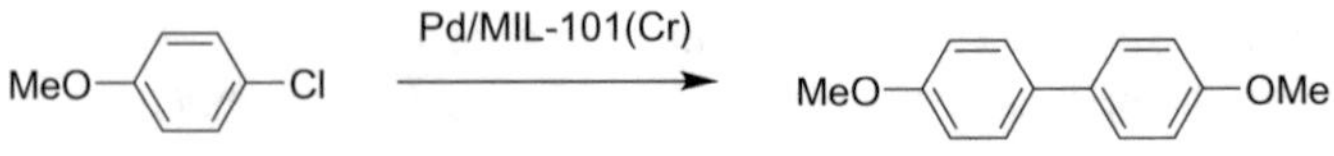

Scheme 7.8 Pd/MIL-101(Cr) catalyzed Ullmann coupling of 4-chloroanisole

4,4'-dimethoxybiphenyl. In contrast, the yield was very poor (3%) with Pd^{II}/MIL-101(Cr) as catalyst under identical conditions, thus revealing the role of metallic Pd NPs as active sites. A series of aryl chlorides were coupled under identical conditions with the yields ranging between 95–97%. The Pd/MIL-101(Cr) catalyst was reused five times without any loss in its efficiency in the Ullmann coupling of 4-chloroanisole. The crystalline structure of the catalyst was mostly retained after five catalytic cycles as evidenced by powder XRD. A TEM image of the five times reused catalyst showed that the mean diameter of Pd NPs is retained compared to the fresh sample.

7.4.3   Sonogashira Cross-Coupling

Other type of Pd-catalyzed C-C bond forming reaction is the Sonogashira cross-coupling between aryl halides and terminal alkynes (Scheme 7.9). There are also studies showing the superior activity of Pd NPs encapsulated with MOFs as heterogeneous catalysts for this reaction. Thus, Pd NPs were deposited at room temperature on MOF-5 (Pd/MOF-5) and its catalytic activity was tested in the Sonogashira coupling [43]. Powder XRD indicated that the crystalline nature of MOF-5 was retained during the loading of Pd NPs. SEM images showed identical morphology of MOF-5 before and after Pd NPs loading. TEM images indicated that the mean diameter of Pd NPs was around 3–6 nm, some of them probably too large to fit with MOF-5 cavities. Catalytic data have shown that Pd/MOF-5 is able to promote the Sonogashira coupling between iodobenzene and phenylacetylene using potassium phosphate as a base at 80 °C in methanol with 98% yield (Scheme 7.9). The catalytic reaction was found to be heterogeneous in nature, and the leaching of Pd was 2.1 ppm. Pd/MOF-5 was reused for five cycles, observing a gradual decrease in the yield. Thus, the yield of the coupling product was 98, 94, 60, 61 and 49% in the 1st, 2nd, 3rd, 4th and 5th cycles, respectively. These results clearly indicate that the selection of a highly robust MOF is crucial to achieve high activity and catalyst stability in the coupling reactions, since many of them require the need of a strong base that can be highly detrimental for MOF stability. Analysis by XPS of the five

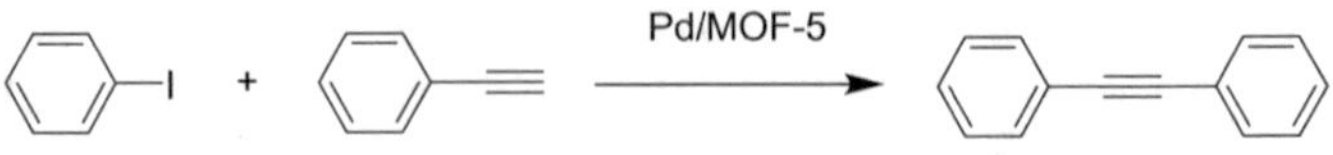

Scheme 7.9 Pd/MOF-5 catalyzed Sonogashira coupling reaction between iodobenzene and phenylacetylene

times used Pd/MOF-5 indicated the presence of Pd(II), thus showing the oxidation of Pd(0) to Pd(II) under the experimental conditions. Further, TEM analysis of the reused catalyst clearly showed agglomeration of Pd NPs, thus explaining one of the reasons of catalyst deactivation. In spite of this lack of stability, Pd/MOF-5 was used as heterogeneous catalyst for a wide range of substrates showing acceptable yields. Therefore, it appears that one of the major limitations of this catalyst is the inability of MOF-5 to preserve the small Pd NP size as Pd undergoes oxidation and agglomeration. It could be that the poor framework stability of MOF-5 is at the origin of the evolution of Pd NPs.

7.4.4 Heck Cross-Coupling

Together with the Suzuki–Miyaura and Sonogashira cross-couplings, the Heck reaction between a vinylic reagent and aryl halide catalyzed by Pd has become one of the most important C–C bond forming reactions in modern organic synthesis (Scheme 7.10). Pd NPs encapsulated within MOFs can also promote this reaction. Thus, highly dispersed Pd NPs with the average diameter of 3.2 nm were supported on NH_2-BDC/BDC mixed-linker MIL-53(Al) [44]. XPS analysis revealed the existence of metallic Pd, the presence of Pd(II) being undetectable. Among the various catalysts prepared with different percentages of NH_2-BDC, Pd NPs deposited over MIX-MIL-53(Al) (MIX: mixed linker) with 50% amino-functionalized linker showed the highest activity to promote the Heck cross-coupling reactions. The activity of Pd/MIX-MIL-53(Al)-NH_2-50 was studied in the Heck cross-coupling reaction between styrene and bromobenzene (Scheme 7.10) in Et_3N as base in DMF at 120 °C, reaching 93% yield with a TOF value of 310 h^{-1}. In contrast, the activities of other catalysts like Pd/MIX-MIL-53(Al)-NH_2-10 and Pd/MIX-MIL-53(Al)-NH_2-90 under identical conditions were 76 and 86% yields with TOF values of 254 and 287 h^{-1}. On other hand, the product yields of single linker Pd/MIL-53(Al)-NH_2 and Pd/MIL-53(Al) were 81 and 26% yields, respectively, under similar conditions. Further, Pd/C as catalyst afforded under identical conditions 49% yield which is two-fold lower than the activity observed with the optimal MIX MOF catalyst. A series of substrates with substituents of different electronic properties were tested under identical conditions observing 81–97% yields of the coupling product. ICP analysis indicated 0.1 ppm of Pd leaching to the solvent and the hot filtration test confirmed the heterogeneity of the catalysis. Reusability experiments were performed for the coupling between

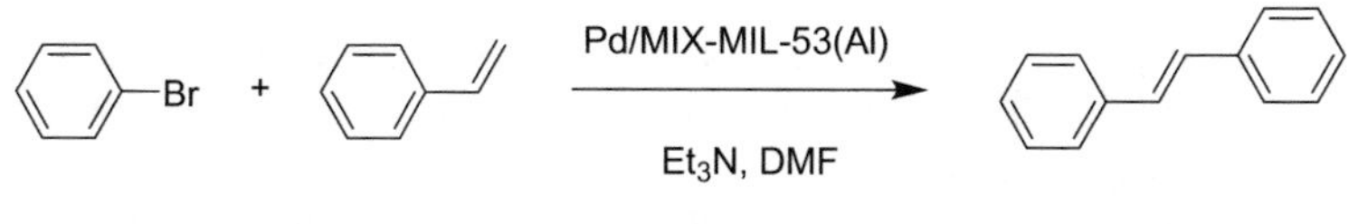

Scheme 7.10 Heck cross-coupling reaction between bromobenzene and styrene catalyzed by Pd/MIX-MIL-53(Al) catalyst

4-methoxy-1-bromobenzene and styrene, observing that the activity of Pd/MIX-MIL-53(Al)-NH$_2$ is mostly maintained for five cycles, the yield decreasing slightly from 81 to 76%. The TEM image of the five times reused catalyst revealed that the mean diameter of the Pd NPs was 3.4 nm which is identical to that of the fresh catalyst.

Pd/MIL-101 prepared by double-solvent method has also been reported as heterogeneous catalyst for the Heck cross-coupling reaction [38] between styrene and iodobenzene in DMF as solvent under air atmosphere using potassium carbonate as base, reaching a yield of 98% after 2 h. The catalyst also exhibited wide substrate scope demonstrating a very high activity and practical applicability. The solid catalyst was also recovered and reused for five cycles with no appreciable decay in its activity. These results clearly indicate that the entrapped Pd NPs within the pores of structurally robust MIL-101 remain highly active and that the preparation method also plays a vital role in determining the activity and stability of a solid catalyst.

The Brönsted basicity in activated MOF-808 (denoted as MOF-808a) was used as support for the deposition of Pd NPs to obtain Pd/MOF-808a that was used as catalyst for the Heck cross-coupling between iodobenzene and styrene under base-free conditions [45]. The existence of Brönsted basicity in MOF-808a was confirmed by CO$_2$-TPD-MS, in situ DRIFTS, and acid–base titration. Pd NPs were highly dispersed over the basic MOF-808a solid with a diameter of 4 ± 2 nm, indicating that the Pd NPs are mainly located on the external surface since the pore dimensions of MOF-808a are smaller than 4 nm. Pd/MOF-808a as catalyst without any additional base afforded in DMF at 130 °C 93.6% yield of diphenylethene. In contrast, no catalytic activity was observed under identical conditions for MOF-808a due to the lack of Pd NPs as active sites. Pd/MIL-101 and Pd/UiO-66 gave under identical conditions without any base 12 and 10% yields of stilbene. These results clearly establish the role of Brönsted basicity of MOF-808a cooperating to the Heck reaction. Recycling experiment indicated that the yield of the Heck product decreases slightly after the first run and then the yield remained identical up to 10th cycle with a value of 84%, which is not too far from the initial activity of the fresh material. Powder XRD patterns revealed that Pd/MOF-808a solid maintains a good crystallinity with no obvious deterioration of the structural integrity after the 10th cycle. Overall, the catalytic activity of Pd/MOF-808a is promising showing how the basic sites installed within the framework of MOF can contribute to C–C bond forming reaction.

7.5 Alcohol Oxidation

Besides C–C bond formation, oxidation and reduction reactions are useful reactions in organic synthesis. With respect to oxidations, one of the targets is to use oxygen and air as oxidizing reagent.

Regarding aerobic oxidations, Au NPs have shown to exhibit catalytic activity to promote reactions using molecular oxygen. In this context, Au NPs have been incorporated within MOFs and the resulting materials used as heterogeneous catalysts for aerobic oxidations. In particular, Au NPs/MOF can catalyze the aerobic alcohol oxidation.

In one of the seminal works, Au NPs were deposited by a colloidal method with polyvinylpyrrolidone (PVP) as protecting agent on the pores of MIL-101(Cr) to obtain Au/MIL-101(Cr) and its catalytic performance was assessed in the aerobic oxidation of alcohols [32]. Characterization data indicated that Au NPs are uniformly distributed over MIL-101 with the mean diameter of Au NPs of 2.3 ± 1.1 nm. Powder XRD indicated that the crystallinity of the MOF has been preserved during the loading of Au NPs in agreement with the robust nature of MIL-101(Cr). The aerobic oxidation of benzyl alcohol using Au/MIL-101(Cr) as solid catalyst at 80 °C under base-free conditions afforded 99% conversion with quantitative selectivity to benzaldehyde in toluene. Under base-free condition, Au/MOF-5 catalyst gave 31% yield of benzaldehyde at about 69% conversion in the oxidation of benzyl alcohol. This superior activity of Au/MIL-101 was due to the high dispersion of Au NPs and to the nature of support. Reusability experiments were performed with 4-methoxybenzyl alcohol and the conversion was steadily of 99% after six cycles. TEM images of the reused catalyst showed identical particle size of Au NPs than the fresh material, thus proving the ability of MIL-101 to preserve Au NPs without agglomeration. Hot filtration experiment confirmed the heterogeneity of the process. Notably, the TOF value for the conversion of 1-phenylethanol to acetophenone at 160°C using Au/Ga$_3$Al$_3$O$_9$ was 25,000 h^{-1} [46], whereas the TOF value achieved with Au/MIL-101 catalyst was 29,300 h^{-1} under similar conditions (Scheme 7.11). A series of alcohols including benzyl alcohol with electron donating or electron withdrawing groups, secondary alcohols, aliphatic alcohols, acyclic and heterocyclic alcohols were successfully oxidized to their respective carbonylic compounds in very high conversions and selectivities under identical conditions. In a subsequent study on the influence of amino substituent on the linker, it was found that Au/MIL-101 (Cr)-NH$_2$ exhibits similar activity with an even improved stability for twenty cycles, a fact that was attributed to the stabilization effect of –NH$_2$ groups on Au NPs [47].

Au NPs have also been encapsulated within the pores of UiO-66 by employing double-solvent method to get Au/UiO-66 solid [48]. As expected, one of the advantages of this method is the internalization of Au NPs within the pores of the solid matrix. Characterization data indicated that Au NPs are present within the UiO-66 pores and no NPs were seen on the external surface. The average size of Au NPs was 5 nm. The activity of Au/UiO-66 was evaluated in the aerobic oxidation of benzyl alcohol at 80 °C using potassium carbonate in water, observing 54% conversion with complete selectivity to benzaldehyde (Scheme 7.12). Under identical conditions,

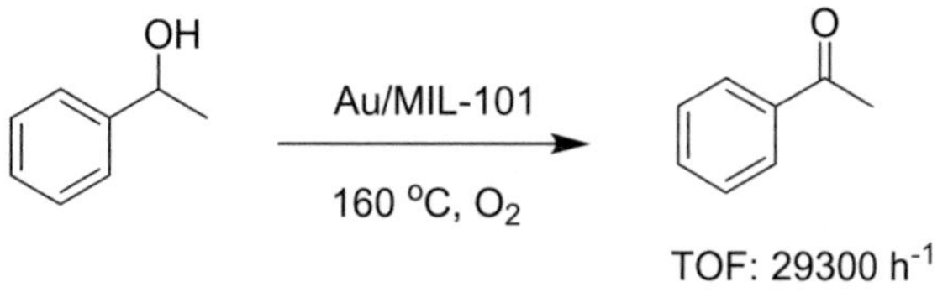

Scheme 7.11 Catalytic activity of Au/MIL-101 in the aerobic oxidation of 1-phenylethanol

OH → Au/MIL-101, 160 °C, O$_2$ → O

TOF: 29300 h^{-1}

Scheme 7.12 Aerobic oxidation of benzyl alcohol catalyzed by Au/UiO-66

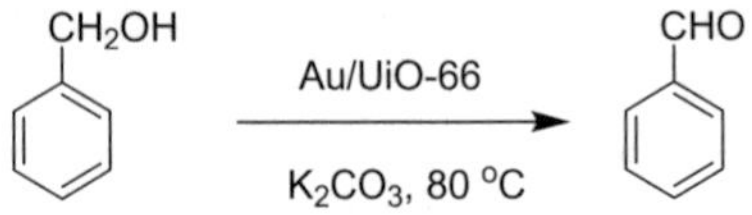

$$\text{CH}_2\text{OH} \xrightarrow[\text{K}_2\text{CO}_3,\ 80\ ^\circ\text{C}]{\text{Au/UiO-66}} \text{CHO}$$

Au^{3+}/UiO-66 afforded 3% conversion of benzyl alcohol which clearly illustrates the superior performance of Au NPs in promoting this aerobic oxidation reaction. The catalyst was reused for at least eight times without any decrease in its activity or leaching of active sites, thus showing high catalyst stability.

Garcia and co-workers have also shown that Au/UiO-66 can be used as heterogeneous solid catalyst for the aerobic oxidation of benzyl alcohol as a base-free catalyst [49]. Powder XRD analysis confirmed that the crystallinity of UiO-66 has been preserved during the deposition of Au NPs in accordance with the robustness of this structure. XPS analysis showed the metallic nature of Au NPs. The average Au NPs size was 7 nm in Au/UiO-66 catalyst and Au NPs were homogeneously dispersed along the UiO-66 host. Importantly, transmission electron tomography provides a reconstructed 3D image of one of the Au/UiO-66 particle showing that the Au NPs are incorporated inside the MOF crystallite. Au/UiO-66 exhibited 94% benzyl alcohol conversion with complete selectivity to benzaldehyde at 100 °C in toluene, while the conversion was 83% using air as oxidizing agent under similar conditions, although at longer time. The catalyst was reused twice with negligible Au leaching (0.18%). Powder XRD indicated no loss of crystallinity.

Pd NPs were supported on DUT-67(Zr) MOF (Pd/DUT-67; DUT: Dresden University of Technology) with the average particle size of Pd being 12–17 nm. The catalytic activity of Pd/DUT-67 was studied in the aerobic oxidation of alcohols in water and potassium carbonate as base [50]. Among the various conditions employed for the oxidation of benzyl alcohol in water, Pd/DUT-67 exhibited complete conversion and selectivity at 100 °C. Various benzylic alcohols were also converted to their respective aldehydes in moderate to high yields under identical conditions. In contrast, the activity of Pd/DUT-67 in the oxidation of 1-phenylethanol was only 15% at 100 °C in water. Although TEM image of the ten times reused catalyst showed that the material was mostly destroyed, the catalytic activity remained intact up to ten cycles. Therefore, it remains to be clarified what is exactly the role of porosity and crystallinity in the activity of this material.

MIL-88B-NH$_2$ was synthesised by using iron(III) chloride hexahydrate and 2-aminoterephthalic acid (NH$_2$-BDC) as organic linker. This MOF exhibited high surface area and was used as support for loading of metal NPs. In another study, Pd NPs were loaded within the pores of the flexible and robust MIL-88B-NH$_2$. The resulting Pd@MOF material was coated with a layer of mesoporous silica to obtain Pd@MIL-88B-NH$_2$@nano-SiO$_2$ (Scheme 7.13). The catalytic activity of Pd@MIL-88B-NH$_2$@nano-SiO$_2$ was tested in the aerobic oxidation of secondary alcohols [51]. Powder XRD pattern did not show any structural change upon loading metal NPs on the MOF. The metallic Pd NPs were uniformly distributed in the MOF with an average

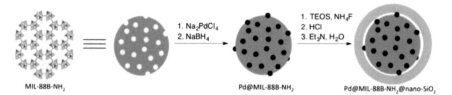

Scheme 7.13 Preparation procedure of Pd@MIL-88B-NH$_2$@nano-SiO$_2$. Reproduced with permission from ref [51]. Copyright 2015 American Chemical Society

particle size of 2–3 nm as indicated by TEM images. Pd@MIL-88B-NH$_2$@nano-SiO$_2$ promoted efficiently the aerobic oxidation of 1-phenylethanol to acetophenone in 98% conversion at 150 °C in xylene as solvent. Oxidation of secondary alcohols is typically more difficult than that of primary alcohol due to steric reasons and typically it requires more active catalysts and harsher reaction conditions. Catalytic activity data have shown that Pd@MIL-88B-NH$_2$@nano-SiO$_2$ was also highly efficient for many other substrates, without the formation of by-products. Pd@MIL-88B-NH$_2$@nano-SiO$_2$ was also used to prepare a catalytic teabag which enabled easy recovery and reuse. Reusability experiments indicated, however, a significant gradual deactivation from the 1st to 5th cycle from 85 to 38% conversion. This deactivation was attributed to diffusion limitations and the blocking of the pores that impede of interaction of substrate and active sites. In sharp contrast of the deactivation of the catalyst in batch mode, a continuous flow process was developed for Pd@MIL-88B-NH$_2$@nano-SiO$_2$. Under these conditions, Pd@MIL-88B-NH$_2$@nano-SiO$_2$ was used for at least 7 days without any deactivation of catalyst with a TOF value of 5 h^{-1}. This enhanced stability of continuous flow process is a convincing proof that the stability of MOF catalyst can be improved by the protecting effect of the silica matrix without imposing diffusion limitations.

Neocuproine ligand was attached to Fe-MIL-101-NH$_2$ by forming an amide (CO-NH) bond as shown in Scheme 7.14 [36]. The functionalized Fe-MIL-101-NH$_2$_Neo was used as solid support for the stabilization of Pd-Ce NPs (Pd-Ce/Fe-MIL-101-NH$_2$_Neo). TEM images showed that the average particle size of Pd-Ce NPs was around 5 nm with a homogeneous dispersion. The catalytic performance of Pd-Ce/Fe-MIL-101-NH$_2$_Neo was examined in the selective oxidation of glycerol to dihydroxyacetone (DHA) reaching 55% yield. In comparison, Pd/Fe-MIL-101-NH$_2$_Neo, Ce/Fe-MIL-101-NH$_2$_Neo and Au/Fe-MIL-101-NH$_2$_Neo were prepared and used for glycerol oxidation to DHA under identical conditions, achieving 17, 4 and 10% yields, respectively. ICP analysis indicated that the leaching of Pd and Ce was less than 2.7 and 3.5 ppm, respectively. The catalyst was reused for three cycles with identical activity.

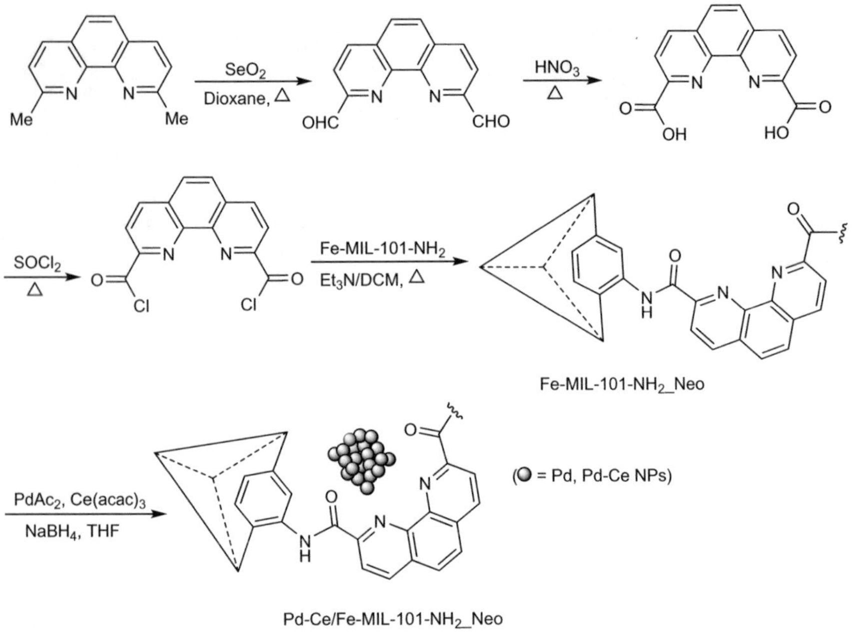

Scheme 7.14 Synthesis of Pd-Ce/Fe-MIL-101-NH$_2$_Neo solid catalyst

7.6 Hydrocarbon Oxidation

Colloidal Au and Pd NPs were incorporated on MIL-101 to obtain Au-Pd/MIL-101 solid catalyst [52]. Powder XRD patterns of the parent material and after the loading of Au and Pd NPs were identical, thus showing the robust nature of MIL-101. Further, TEM analysis revealed that the Au-Pd NPs were evenly distributed with the average particle size of 2.40 ± 0.63 nm. Au and Pd were on the MIL-101 support mostly in the form of bimetallic alloys. Au-Pd/MIL-101 efficiently promoted the liquid-phase aerobic oxidation of cyclohexane with the cyclohexane conversion higher than 40% with around 85% selectivity to cyclohexanone and cyclohexanol (TOF: 19000 h^{-1}) under mild and solvent-free conditions. Furthermore, the Au–Pd alloy catalyst showed higher reactivity than their pure single metal counterparts and also than the Au and Pd physical mixture. In addition, Au-Pd/MIL-101 catalyst maintained its catalytic activity for four catalytic cycles without any decay in the conversion of cyclohexane and the selectivity of cyclohexanone and cyclohexanol (85%). TEM images of the four times used catalyst showed a minimum agglomeration of NPs with the average size of 2.59 ± 0.51 nm. This small increment in the average particle size did not influence the catalytic activity. The high activity and selectivity observed for Au-Pd/MIL-101 in cyclohexane aerobic oxidation was due to the synergic effect of bimetallic Au-Pd NPs. In addition, Au-Pd/MIL-101 was active and selective in the aerobic oxidation of a range of saturated primary and

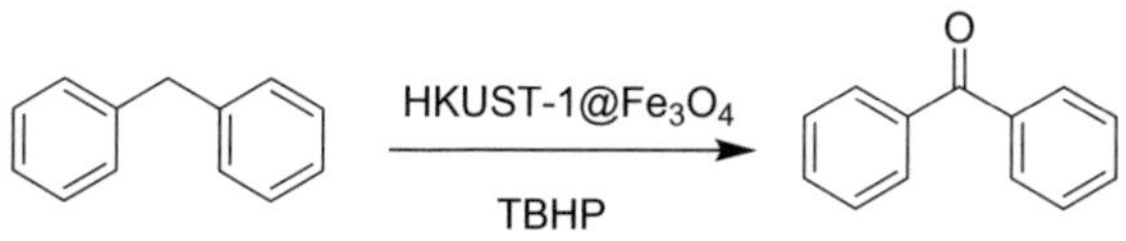

Scheme 7.15 Oxidation of diphenylmethane to benzophenone by TBHP catalyzed by HKUST-1@Fe$_3$O$_4$ catalyst

secondary C–H bonds including oxidation of cyclooctane, tetralin, diphenylmethane, ethylbenzene and toluene.

HKUST-1@Fe$_3$O$_4$ (HKUST: Hong Kong University of Science and Technology) solid catalyst was prepared by chemically bonding core NPs of growing HKUST-1 thin layers with Fe$_3$O$_4$ nanospheres connected by carboxyl groups [53]. The activity of HKUST-1@Fe$_3$O$_4$ was studied in the benzylic oxidation of diphenylmethane to benzophenone. The carboxyl functionalized Fe$_3$O$_4$ cores were about 20 nm diameter. Fe$_3$O$_4$ NPs were encapsulated by a HKUST-1 shell through in situ generation of HKUST-1 frameworks by the addition of BTC and Cu(OAc)$_2$ into the mixture. Oxidation of diphenylmethane (Scheme 7.15) in benzonitrile using TBHP as an oxidant at 80 °C was catalyzed by HKUST-1@Fe$_3$O$_4$ catalyst, achieving 94.7% conversion with 95.2% selectivity. Under identical conditions, control experiments were performed independently with Fe$_3$O$_4$ and HKUST-1 in the oxidation of diphenylmethane, attaining conversions of 56.4 and 86.5%, respectively. These values are significantly lower than those of the core-shell composite showing the synergy between the two components. The catalytic activity of HKUST-1@Fe$_3$O$_4$ was sustained for three cycles without any notable decay in its activity. Powder XRD pattern of the recovered HKUST-1@Fe$_3$O$_4$ sample showed identical pattern with that of the fresh material. This solid catalyst was further tested with other substrates, and a series of ketones with aryl side chain groups were synthesized with conversion ranging from 94.7 to 99% and selectivity from 95.2 to 99%.

7.7   Tandem Reactions

Pd/MOF-808a has also been used as heterogeneous solid catalyst for a tandem reaction [45]. In the first step, Pd NPs were involved in the activation of molecular oxygen, while the presence of a base in the framework facilitates deprotonation of the alcohol, thus, resulting in the efficient aerobic oxidation of benzyl alcohol. In the second step, benzaldehyde in methanol is transformed to methyl benzoate with the assistance of acid centers on Zr nodes or basic sites in MOF-808a. In this way, Pd/MOF-808a showed a complete conversion of benzyl alcohol with 75% selectivity to methyl benzoate at 80 °C [45]. In contrast, Pd/UiO-66 catalyst without any basic site gave benzaldehyde as the main product in 77% yield and 22% of methyl benzoate (Scheme 7.16). These experiments illustrate that the acid–base properties

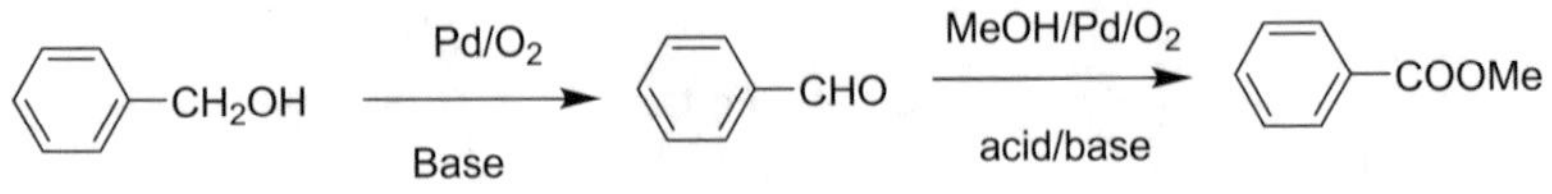

Catalyst	Conversion (%)	Yield (%)	
		C_6H_5CHO	C_6H_5COOMe
Pd/MOF-808a	>99	23	75
Pd/MOF-808a-2nd use	98.6	22.9	76
Pd/MOF-808a-3rd use	96.1	21.5	75.2
Pd/MIL-101	89	33	50
Pd/UiO-66	95	77	18

Scheme 7.16 Catalytic activity of a series of Pd-containing MOF for the tandem reaction of benzyl alcohol to methyl benzoate

of Pd/MOF-808a makes this material to act as a bifunctional catalyst, promoting the tandem reaction. In comparison, Pd/MIL-101 exhibited 89% conversion and slightly higher yield of methyl benzoate (50%) than Pd/UiO-66, a fact that was attributed to the higher activity of Cr^{III} ions of Pd/MIL-101. Furthermore, the reusability test with Pd/MOF-808a catalyst showed that the conversion and selectivity are retained in three consecutive cycles and no decay in the activity was observed.

Alcohol oxidation to carboxylic acid and its subsequent esterification is another case of a tandem reaction. In one example of metal NPs encapsulated within robust MOFs as catalysts for tandem reactions, Au/UiO-66 was prepared by different methods and employed as heterogeneous catalysts for the oxidative valorization of furfuryl alcohol with methanol (Scheme 7.17) [54]. Among the various methods employed for the preparation of Au/UiO-66, the solid prepared using impregnation-reduction method (Au NPs size 3–5 nm) showed complete conversion of furfuryl alcohol with 100% selectivity to methyl 2-furoate.

Trying to valorise biomass derived products, a series of Au NPs supported on Au/UiO-66-X (X: H, NH_2, NO_2 and COOH) were prepared by impregnation-reduction method and employed for the oxidative condensation of furfural with acetaldehyde formed in situ by ethanol. The experimental results have shown that the nature of Au/UiO-66-X can efficiently alter the performance of the catalyst in the oxidative condensation and oxidative esterification. Among these catalysts, Au/UiO-66-COOH (Au NPs size 6–10 nm) exhibited the best activity, reaching higher selectivity (84%) to furan-2-acrolein through the oxidative condensation of furfural with ethanol. Both Au/UiO-66 and Au/UiO-66-COOH catalysts were reused for five cycles with no decay in their activity. It should be commented that the Au

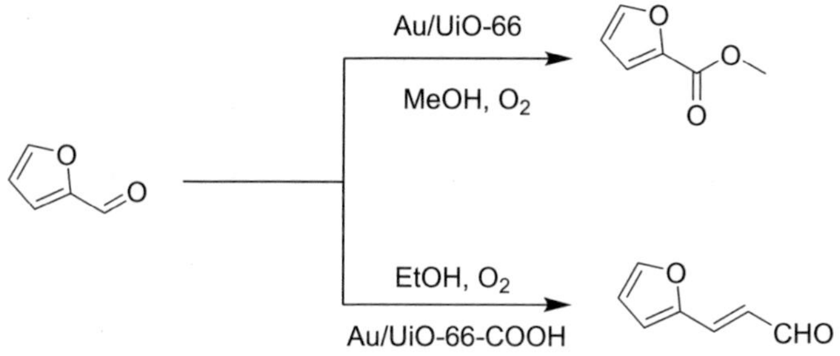

Scheme 7.17 Oxidative transformation of furfural with Au/UiO-66 and Au/UiO-66-COOH catalysts

particle size seems to be too large for the internal location of the NPs, and, therefore, it could still be room for improvement by decreasing the particle size.

7.8 CO Oxidation

A hybrid catalyst comprising of polymer-coated Ru NPs (Ru-PVP) was encapsulated in a porous ZIF-8 MOF [55]. The mean diameter of Ru NPs in the resulting Ru-PVP@ZIF-8 was measured to be 1.7 ± 0.4 nm by TEM images. This mean diameter of Ru NPs is similar to the Ru-PVP on carbon active (Ru-PVP/C) used for comparison as a control catalyst that was 1.5 ± 0.3 nm. The activity of Ru-PVP@ZIF-8 was assessed in the selective CO oxidation reaction in the presence of H_2. The CO oxidation reaction with Ru-PVP@ZIF-8 occurred at lower temperature (50% CO conversion (T_{50}) was determined to be 60 °C) than Ru-PVP/C (T_{50} at 97 °C) although both catalysts exhibit a similar chemisorption of CO around 13.0 cm^3 g(Ru)$^{-1}$. The activity of Ru-PVP@ZIF-8 was comparatively higher and also showed higher CO_2 selectivity than with carbon-supported Ru-PVP (Ru-PVP/C). These results clearly indicate the effective role played by the porous ZIF-8 MOF by affording a more suitable environment for the reaction with O_2 and CO gases.

7.9 Conclusions

The present chapter has shown that there are by now sufficient examples showing the advantages in terms of superior performance and stability of heterogeneous catalysts based on MOFs as hosts encapsulating metal NPs. Some of the most active heterogeneous catalysts reported so far for cross-coupling reactions and oxidations are based on incorporation of metal NPs inside robust MOFs, particularly of the MIL family.

The main reasons for this excellent catalytic performance are the small particle size of the incorporated metal NPs, commonly about 3 nm or below, the favorable diffusion of substrates and reagents to the active sites, due to the high porosity of the MOFs, and the remarkable stability of the metal NP/MOF composite due to the structural robustness of the host that thwarts agglomeration of embedded NPs.

In some cases, the occurrence of a synergy between the host and the metal NPs has been proposed. It has also been determined that the catalytic activity, selectivity and stability of the metal NP/MOF composite can be improved by the presence of substituents on the aromatic ring. These substituents can establish additional interactions with the metal NPs in the cavities, providing additional stability, while still leaving many atoms of the NP, mainly at places where no linkers are present, free to interact with substrates and reagents. Additionally, the presence of substituents on the linker can modify the polarity of the internal spaces and can tune the dimensions of the pores. It is clear that the influence of substituents present on the organic linkers and the influence of modification of the nodes are factors that deserve further study in order to determine the performance gain that can be reached by applying these strategies.

Although the number of examples of the use of metal NPs inside MOFs is likely to continue growing in the near future, by exploring new reactions as well as other hosts based on highly porous and robust MOFs, a vast potential of these metal NPs/MOF composite catalysts remains to be exploited for the promotion of tandem reactions. Tandem reactions in where two or more elementary reactions occur in a single step typically require the action of catalytic sites of different nature. Metal NPs encapsulated within MOFs can offer the combination of the catalytic activity of the metal NP with that of other sites located at the metal nodes or at the organic linker. Metal nodes having exchangeable positions are Lewis acid centers, while substituents on the linker can be either Brönsted acids or bases. The combination of all these sites on the encapsulated metal NP and at the MOF lattice can easily renders a multifunctional catalyst that could be suitable to promote tandem reactions.

Thus, metal NPs included in MOFs can be suitable solid catalysts, not only for those reactions promoted by metal NPs in other high surface area supports, but they can also act as multifunctional catalysts and in this way promote one-pot multiple reactions.

Acknowledgements AD thanks the University Grants Commission, New Delhi, for the award of an Assistant Professorship under its Faculty Recharge Programme. AD also thanks the Department of Science and Technology, India, for the financial support through Extra Mural Research Funding (EMR/2016/006500). Financial support by the Spanish Ministry of Economy and Competitiveness (Severo Ochoa and RTI2018-098237-B-C21) and Generalitat Valenciana (Prometeo 2017-083) is gratefully acknowledged.

References

1. Phan NTS, Van Der Sluys M, Jones CW (2006) On the nature of the active species in palladium catalyzed Mizoroki-Heck and Suzuki-Miyaura couplings—homogeneous or heterogeneous catalysis. A critical review. Adv Synth Catal 348:609–679
2. Astruc D, Lu F, Aranzaes JR (2005) Nanoparticles as recyclable catalysts: the frontier between homogeneous and heterogeneous catalysis. Angew Chem Int Ed 44:7852–7872
3. Tao AR, Habas S, Yang P (2008) Shape control of colloidal metal nanocrystals. Small 4:310–325
4. Viñes F, Gomes JRB, Illas F (2014) Understanding the reactivity of metallic nanoparticles: beyond the extended surface model for catalysis. Chem Soc Rev 43:4922–4939
5. Okumura M, Fujitani T, Huang J, Ishida T (2015) A career in catalysis: Masatake Haruta. ACS Catal 5:4699–4707
6. Roldan Cuenya B (2010) Synthesis and catalytic properties of metal nanoparticles: size, shape, support, composition, and oxidation state effects. Thin Solid Films 518:3127–3150
7. Suchomel P, Kvitek L, Prucek R, Panacek A, Halder A, Vajda S, Zboril R (2018) Simple size-controlled synthesis of Au nanoparticles and their size-dependent catalytic activity. Sci Rep 8:4589–4599
8. Taketoshi A, Haruta M (2014) Size- and structure-specificity in catalysis by gold clusters. Chem Lett 43:380–387
9. Narayanan R, El-Sayed MA (2005) Catalysis with transition metal nanoparticles in colloidal solution: nanoparticle shape dependence and stability. J Phys Chem B 109:12663–12676
10. Rossi LM, Fiorio JL, Garcia MAS, Ferraz CP (2018) The role and fate of capping ligands in colloidally prepared metal nanoparticle catalysts. Dalton Trans 47:5889–5915
11. Li G, Zhao S, Zhang Y, Tang Z (2018) Metal-organic frameworks encapsulating active nanoparticles as emerging composites for catalysis: recent progress and perspectives. Adv Mater 30:1800702
12. Dhiman M, Polshettiwar V (2018) Supported single atom and pseudo-single atom of metals as sustainable heterogeneous nanocatalysts. ChemCatChem 10:881–906
13. Dhakshinamoorthy A, Li Z, Garcia H (2018) Catalysis and photocatalysis by metal organic frameworks. Chem Soc Rev 47:8134–8172
14. Lia G, Tang Z (2014) Nanoscale 6:3995–4011
15. Munnik P, de Jongh PE, de Jong KP (2015) Recent developments in the synthesis of supported catalysts. Chem Rev 115:6687–6718
16. Gross E, Dean Toste F, Somorjai GA (2015) Polymer-encapsulated metallic nanoparticles as a bridge between homogeneous and heterogeneous catalysis. Catal Lett 145:126–138
17. Boutin A, Coudert F-X, Springuel-Huet M-A, Neimark AV, Férey G, Fuchs AH (2010) The behavior of flexible MIL-53(Al) upon CH_4 and CO_2 adsorption. J Phys Chem C 114:22237–22244
18. Horcajada P, Surblé S, Serre C, Hong D-Y, Seo Y-K, Chang J-S, Grenèche J-M, Margiolakid I, Férey G (2007) Synthesis and catalytic properties of MIL-100(Fe), an iron(iii) carboxylate with large pores. Chem Commun 2007:2820–2822
19. Hu Z, Zhao D (2017) Metal–organic frameworks with Lewis acidity: synthesis, characterization, and catalytic applications. CrystEngComm 19:4066–4081
20. Liu M, Wu J, Hou H (2019) Metal–organic framework (MOF)-based materials as heterogeneous catalysts for C–H bond activation. Chem Eur J 25:2935–2948
21. Wang C, An B, Lin W (2019) Metal-organic frameworks in solid-gas phase catalysis. ACS Catal 9:130–146
22. Ming Y, Kumar N, Siegel DJ (2017) Water adsorption and insertion in MOF-5. ACS Omega 2:4921–4928
23. Dhakshinamoorthy A, Alvaro M, Concepcion P, Garcia H (2011) Chemical instability of $Cu_3(BTC)2$ by reaction with thiols. Catal Commun 12:1018–1021
24. Li N, Xu J, Feng R, Hu T-L, Bu X-H (2016) Governing metal-organic frameworks towards high stability. Chem Commun 52:8501–8513

25. Kim D-W, Kim H-G, Cho D-H (2016) Catalytic performance of MIL-100 (Fe, Cr) and MIL-101 (Fe, Cr) in the isomerization of endo- to exo-dicyclopentadiene. Catal Commun 73:69–73
26. Dhakshinamoorthy A, Santiago-Portillo A, Asiri AM, Garcia H (2019) Engineering UiO-66 metal organic framework for heterogeneous catalysis. ChemCatChem 11:899–923
27. Bai Y, Dou Y, Xie L-H, Rutledge W, Li J-R, Zhou H-C (2016) Zr-based metal-organic frameworks: design, synthesis, structure, and applications. Chem Soc Rev 45:2327–2367
28. Cavka JH, Jakobsen S, Olsbye U, Guillou N, Lamberti C, Bordiga S, Lillerud KP (2008) A new zirconium inorganic building brick forming metal organic frameworks with exceptional stability. J Am Chem Soc 130:13850–13851
29. Patra S, Sene S, Mousty C, Serre C, Chausse A, Legrand L, Steunou N (2016) Design of laccase-metal organic framework-based bioelectrodes for biocatalytic oxygen reduction reaction. ACS Appl Mater Interfaces 8:20012–20022
30. Bhattacharjee S, Chen C, Ahn W-S (2014) Chromium terephthalate metal-organic framework MIL-101: synthesis, functionalization, and applications for adsorption and catalysis. RSC Adv 4:52500–52525
31. Rösler C, Fischer RA (2015) Metal–organic frameworks as hosts for nanoparticles. CrystEng-Comm 17:199–217
32. Liu H, Liu Y, Li Y, Tang Z, Jiang H (2010) Metal–organic framework supported gold nanoparticles as a highly active heterogeneous catalyst for aerobic oxidation of alcohols. J Phys Chem C 114:13362–13369
33. Aijaz A, Karkamkar A, Choi YJ, Tsumori N, Rönnebro E, Autrey T, Shioyama H, Xu Q (2012) Immobilizing highly catalytically active Pt nanoparticles inside the pores of metal-organic framework: a double solvents approach. J Am Chem Soc 134:13926–13929
34. Ishida T, Nagaoka M, Akita T, Haruta M (2008) Deposition of gold clusters on porous coordination polymers by solid grinding and their catalytic activity in aerobic oxidation of alcohols. Chem Eur J 14:8456–8460
35. Dhakshinamoorthy A, Asiri AM, Garcia H (2017) Metal organic frameworks as versatile hosts of Au nanoparticles in heterogeneous catalysis. ACS Catal 7:2896–2919
36. Li X, Kaizen Tjiptoputro A, Ding J, Min Xue J, Zhu Y (2017) Pd-Ce nanoparticles supported on functional Fe-MIL-101-NH$_2$: an efficient catalyst for selective glycerol oxidation. Catal Today 279:77–83
37. Yuan B, Pan Y, Li Y, Yin B, Jiang H (2010) A highly active heterogeneous palladium catalyst for the Suzuki-Miyaura and Ullmann coupling reactions of aryl chlorides in aqueous media. Angew Chem Int Ed 49:4054–4058
38. Shang N, Gao S, Zhou X, Feng C, Wang Z, Wang C (2014) Palladium nanoparticles encapsulated inside the pores of a metal–organic framework as a highly active catalyst for carbon–carbon cross-coupling. RSC Adv 4:54487–54493
39. Pascanu V, Yao Q, Bermejo Gomez A, Gustafsson M, Yun Y, Wan W, Samain L, Zou X, Martin-Matute B (2013) Sustainable catalysis: rational Pd loading on MIL-101Cr-NH$_2$ for more efficient and recyclable Suzuki-Miyaura reactions. Chem Eur J 19:17483–17493
40. Huang Y, Zheng Z, Liu T, Lü J, Lin Z, Li H, Cao R (2011) Palladium nanoparticles supported on amino functionalized metal-organic frameworks as highly active catalysts for the Suzuki-Miyaura cross-coupling reaction. Catal Commun 14:27–31
41. Kardanpour R, Tangestaninejad S, Mirkhani V, Moghadam M, Mohammadpoor-Baltork I, Khosropour AR, Zadehahmadi F (2014) Highly dispersed palladium nanoparticles supported on amino functionalized metal-organic frameworks as an efficient and reusable catalyst for Suzuki cross-coupling reaction. J Organomet Chem 761:127–133
42. Zhang L, Su Z, Jiang F, Zhou Y, Xu W, Hong M (2013) Catalytic palladium nanoparticles supported on nanoscale MOFs: a highly active catalyst for Suzuki-Miyaura cross-coupling reaction. Tetrahedron 69:9237–9244
43. Gao S, Zhao N, Shu M, Che S (2010) Palladium nanoparticles supported on MOF-5: a highly active catalyst for a ligand- and copper-free Sonogashira coupling reaction. Appl Catal A Gen 388:196–201

44. Huang Y, Gao S, Liu T, Lu J, Lin X, Li H, Cao R (2012) Palladium nanoparticles supported on mixed-linker metal-organic frameworks as highly active catalysts for Heck reactions. ChemPlusChem 77:106–112
45. Yan X, Wang K, Xu X, Wang S, Ning Q, Xiao W, Zhang N, Chen Z, Chen C (2018) Brønsted basicity in metal-organic framework-808 and its application in base-free catalysis. Inorg Chem 57:8033–8036
46. Su FZ, Liu YM, Wang LC, Cao Y, He HY, Fan KN (2008) Ga-Al mixed-oxide-supported gold nanoparticles with enhanced activity for aerobic alcohol oxidation. Angew Chem Int Ed 47:334–337
47. Santiago-Portillo A, Cabrero-Antonino M, Álvaro M, Navalón S, García H (2019) Tuning the microenvironment of gold nanoparticles encapsulated within MIL-101(Cr) for the selective oxidation of alcohols with O_2: influence of the amino terephthalate linker. Chem Eur J 25:9280–9286
48. Zhu J, Cheng Wang P, Lu M (2014) Selective oxidation of benzyl alcohol under solvent-free condition with gold nanoparticles encapsulated in metal-organic framework. Appl Catal A Gen 477:125–131
49. Leus K, Concepcion P, Vandichel M, Meledina M, Grirrane A, Esquivel D, Turner S, Poelman D, Waroquier M, Van Speybroeck V, Van Tendeloo G, Garcia H, Van Der Voort P (2015) Au@UiO-66: a base free oxidation catalyst. RSC Adv 5:22334–22342
50. Abedi S, Morsali A (2017) Improved activity of palladium nanoparticles using a sulfur-containing metal–organic framework as an efficient catalyst for selective aerobic oxidation in water. New J Chem 41:5846–5852
51. Pascanu V, Bermejo Gomez A, Ayats C, Eva Platero-Prats A, Carson F, Su J, Yao Q, Pericas MA, Zou X, Martín-Matute B (2015) Double-supported silica-metal–organic framework palladium nanocatalyst for the aerobic oxidation of alcohols under batch and continuous flow regimes. ACS Catal 5:472–479
52. Long J, Liu H, Wu S, Liao S, Li Y (2013) Selective oxidation of saturated hydrocarbons using Au–Pd alloy nanoparticles supported on metal-organic frameworks. ACS Catal 3:647–654
53. Chen Y, Huang X, Feng X, Li J, Huang Y, Zhao J, Guo Y, Dong X, Han R, Qi P, Han Y, Li H, Hu C, Wang B (2014) Facile fabrication of magnetically recyclable metal–organic framework nanocomposites for highly efficient and selective catalytic oxidation of benzylic C–H bonds. Chem Commun 50:8374–8377
54. Ning L, Liao S, Liu X, Guo P, Zhang Z, Zhang H, Tong X (2018) A regulatable oxidative valorization of furfural with aliphatic alcohols catalyzed by functionalized metal-organic frameworks-supported Au nanoparticles. J Catal 364:1–13
55. Sadakiyo M, Kon-no M, Sato K, Nagaoka K, Kasai H, Kato K, Yamauchi M (2014) Synthesis and catalytic application of PVP-coated Ru nanoparticles embedded in a porous metal–organic framework. Dalton Trans 43:11295–11298

Chapter 8
Earth-Abundant d-Block Metal Nanocatalysis for Coupling Reactions in Polyols

Marc Camats, Daniel Pla, and Montserrat Gómez

Abstract Green Chemistry concepts have directed chemists to conceive and develop sustainable procedures, from the starting materials choice through reaction and analysis conditions, including suitable engineering aspects, to the impact of products, comprising recycling, and waste management. Industrial processes in Fine and Pharmaceutical Chemistry sector have high E factors compared to oil and bulk chemicals industry. Thus, the development of catalytic methods leading to high added value products is crucial, as well as waste minimization through selective transformations. Catalysts from 3d metals, compared to "heavy" metals, are greener, although a combination of different approaches is needed for efficient and viable processes. In contrast to 4d and 5d metals, catalysis with earth-abundant metals is less developed, even less concerning nanocatalysts. Metal nanoparticles (MNPs), due to their unique electronic and structural properties, induce original reactivities allowing a plethora of transformations. Besides, solvents, present in most steps, represent a major economic and environmental concern. In addition, they can have a dramatic influence on the stabilization of MNP and hence, a huge impact on catalytic activity and recycling. This chapter gives a perspective on 3d metal-based nanocatalysts in polyols applied to couplings, reactions present in many methodologies to produce fine chemicals in a sustainable fashion.

Keywords 3d metals · Metal nanoparticles · Catalysis · Couplings · Polyols

Dedicated to Prof. Guillermo Muller for his thorough contributions in Organometallic Chemistry and Homogeneous Catalysis.

M. Camats · D. Pla · M. Gómez (✉)
Laboratoire Hétérochimie Fondamentale et Appliquée (UMR 5069), Université de Toulouse, CNRS, 118 Route de Narbonne, 31062 Toulouse Cedex 9, France
e-mail: gomez@chimie.ups-tlse.fr

D. Pla
e-mail: pla@lhfa.fr

8.1   Introduction

In the 1990s, a ground-breaking movement concerning the Earth preservation for future generations has appeared. In Chemistry, these actions led to the Green Chemistry concept [7, 128] articulated through the 12 Green Chemistry Principles [6]; shortly after, the Sandestin declaration established the bases of the Green Engineering Principles [8]. Notably, catalysis is one of the most important pillars of Green approaches [9]. Analyzing the evolution of metal-based catalysis, involving homogeneous and heterogeneous systems, it is evident to conclude that definitely the twentieth century has been the century of "heavy" metal-based catalysts, *i.e.*, catalytic processes concerning 4d and 5d metals (such as Mo, Ru, Rh, Ir, Pd, Pt, Au …), as internationally acknowledged by the Nobel Prizes in 2005 (Y. Chauvin, R. H. Grubbs and R. R. Schrock) and 2010 (R. F. Heck, E.-i. Negishi and A. Suzuki) for the development of metathesis processes (mainly Mo- and Ru-based catalysts) and Pd-catalyzed couplings in organic synthesis, respectively (www.nobelprize.org/prizes/chemistry/2005/summary; www.nobelprize.org/prizes/chemistry/2010/summary). Nevertheless, nature has exploited earth abundant 3d metals (Fe, Ni, Cu, Zn …), leading to metalloenzymes which exhibit a prominent specificity [16]. In coherence to a sustainable development, researches on the design of manufactured catalysts with first-row transition metals have exponentially grown since the 2000s (see the contributions collected in the special issue "First row metals and catalysis" of Chemical Reviews, 2019) [2, 4, 48, 50, 65, 79, 101, 123, 129] (Fig. 8.1).

On the other hand, solvents represent one of the major concerns for chemical transformations (employed in synthesis, extractions, purifications, analyses…), in particular volatile organic compounds which are generally toxic showing different levels of harmfulness and danger, submitted to severe regulations. Alternative solvents (water, ionic liquids and deep eutectic solvents, alcohols, supercritical fluids,

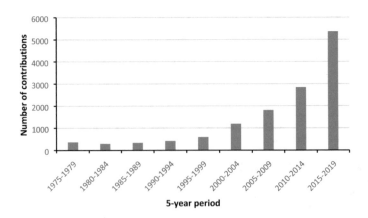

Fig. 8.1 Published contributions per 5-year period for first-row transition metals used in catalysis (data collected from SciFinder Database since 1975 up to July 2019)

renewable solvents, and combinations thereof ...) and solvent-free chemistry are useful and efficient approaches for many applications, needing investment and effort to adapt the current processes to the new reaction conditions [69]. In the frame of this chapter, we are interested in polyols, largely used in the industrial sector (highlighting the production of polymers such as polyurethanes, polyvinyl alcohol ...), because they present lower environmental impact than low-weight and volatile organic compounds, and they are particularly attractive for the synthesis and stabilization of nano-sized metal clusters.

Focusing on this type of nanomaterials, well-defined metal-based nanoparticles (objects showing dimensions in the range 1–100 nm) have known a huge expansion since the 1980s due to their distinctive properties, both physical and chemical, in comparison to molecular and bulk materials. Actually, this is a consequence of their electronic and structural features, making possible a vast number of applications [111, 90, 133], in particular in catalysis [1, 56, 89, 111, 112]. Even though metal nanoparticles (MNPs) have been largely applied in classical heterogeneous catalysis, with crucial participation in industrial processes mainly those related to oil area [137], nanocatalysis, which concerns MNPs dispersed in a solvent, has only been developed since the end of the last century. This exponential growth has benefited of the recent advances in characterization techniques (including operando approaches) which has permitted the design of catalysts at nanometric scale by means of controlling morphology and surface state of nanoobjects [5, 10, 63, 28]. Noticeably, reproducible synthetic methodologies are crucial for their further applications [107, 122]. The chemical strategies (commonly named bottom-up syntheses) often include solvents, from conventional organic compounds to alternative ones (showing a lower environmental impact), such as water [21, 110], ionic liquids [105], scCO$_2$, [32, 100, 134], and polyols [23, 40, 43]. It is important to mention that solvents may be involved in different aspects, for example, as reducing agents, stabilizers, or medium for trapping, nanocatalysts preserving their morphology during the catalytic transformation and thus facilitating their recycling. This is the case when polyols are present. In 1989, the polyol methodology was for the first time reported by Fiévet and coworkers where metal salts were reduced in ethylene glycol, obtaining well-defined MNPs [44, 45]. In this method, the polyol acts as solvent, reducing agent, and stabilizer when higher polyols are involved (e.g., polyphenols or polysaccharides), which prevents the agglomeration of MNPs in solution [36, 43].

These privileged physicochemical properties favor the preparation of tailor-made first-row transition metal nanocatalysts and derived nanocomposites in a controlled manner. Rational catalyst design is essential for the preparation of well-defined clusters and nanoparticles with optimal properties in terms of redox control, cooperative effects, and increased surface areas, all of them key factors for catalysis. Beyond the dual catalytic behavior related to both surface and reservoir of molecular species, nanocatalysis brings novel reaction manifolds due to the unique structural properties of MNPs, particularly their differential electronic structure as compared to bulk metals (e.g., facilitating single electron transfer processes via the Fermi level for electrons [42]).

In the present contribution, we focus on the recent advances in 3d metal nanocatalysts (for a recent review, see: [127]), involving polyols acting as stabilizer and/or reaction medium for coupling reactions, Carbon–Carbon and Carbon–Heteroatom bond formation processes, including multicomponent syntheses.

8.2 Carbon–Carbon Bond Formation

C–C cross-coupling reactions have been largely dominated by palladium-based catalyzed processes due to their efficiency and versatility. In particular, the ability of this metal to stabilize different kinds of species (complexes, nanoparticles, extended surfaces) and its relatively high robustness under many different reaction conditions has permitted the elucidation of the corresponding mechanisms [18, 120, 124]. From a sustainable chemistry point of view, the use of first-row transition metals is obviously preferred and a huge research has been developed in the last years (for instance, see the contributions published in the Accounts of Chemical Research special issue "Earth Abundant Metals in Homogeneous Catalysis," [27]).

8.2.1 Lewis Acid-Catalyzed Coupling Reactions

Heterocyclic motifs are present in a large variety of naturally occurring products along with industrial compounds [37, 59, 66, 102]. Multicomponent reactions represent an environmentally friendly approach to prepare polyfunctional compounds, in particular heterocycle derivatives, via one-pot processes involving three or more reactants, with high atom economy and easy implementation [19, 130, 138]. These transformations are often promoted by Lewis acids, which favor the kinetics directing the reaction pathway and in consequence improving the selectivity [51]. In this frame, Khurana and coworkers reported nickel nanoparticles (NiNPs) stabilized by polyethylene glycol (PEG-4000) and prepared by polyol-based methodology using ethylene glycol (EG) in the presence of $NaBH_4$, which were applied in the synthesis of spiropyrans [73], interesting materials particularly due to their unique molecular switch properties that can trigger structural isomerization under the effect of different external stimuli (light, mechanical stress, temperature...) [75, 83, 131]. They were synthesized by a multicomponent reaction, constituted of a tandem Knoevenagel-cyclo-condensation involving ninhydrin (or related cyclic dicarbonyl compounds), malonitrile, and dimedone (or related 1,3-dicarbonyl derivatives) (Scheme 8.1). The role of the nanocatalyst (mean size: ca. 7 nm determined by TEM) was evidenced by different control tests; in the absence of nickel, the reaction was much slower (some hours vs. some minutes) and the use of Ni powder (particle size <150 μm) led to moderate yields after 8 h of reaction. Ethylene glycol was a convenient solvent permitting a straightforward extraction of products by a biphasic system (using ethyl acetate as immiscible solvent with ethylene glycol), preserving the catalyst dispersed

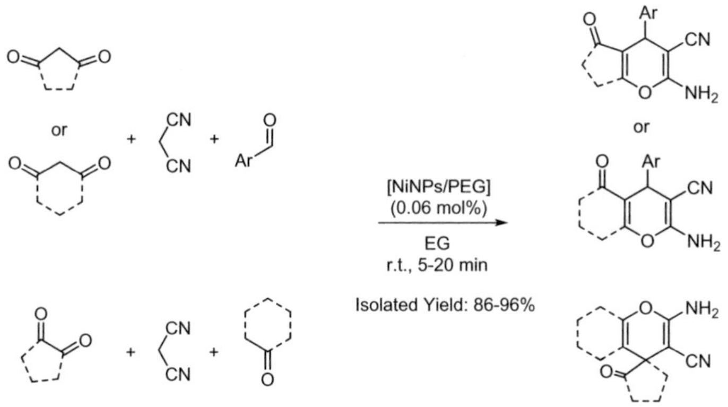
or
CN
CN
Ar
O
[NiNPs/PEG]
(0.06 mol%)
EG
r.t., 5-20 min
Isolated Yield: 86-96%
Ar
CN
NH2
or
Ar
CN
NH2
NH2
CN

Scheme 8.1 Synthesis of spiropyrans by a multicomponent reaction catalyzed by NiNPs and stabilized by PEG in ethylene glycol [73]

in the diol. NiNPs were efficiently reused up to 3 times. Authors postulated that nickel activates both carbonyl and nitrile groups.

The same authors previously reported Knoevenagel condensations of barbituric or Meldrum acid and aromatic aldehydes in EG, catalyzed by NiNPs and stabilized by PVP (polyvinylpyrrolidone) [71] and PEG [72], respectively (Scheme 8.2).

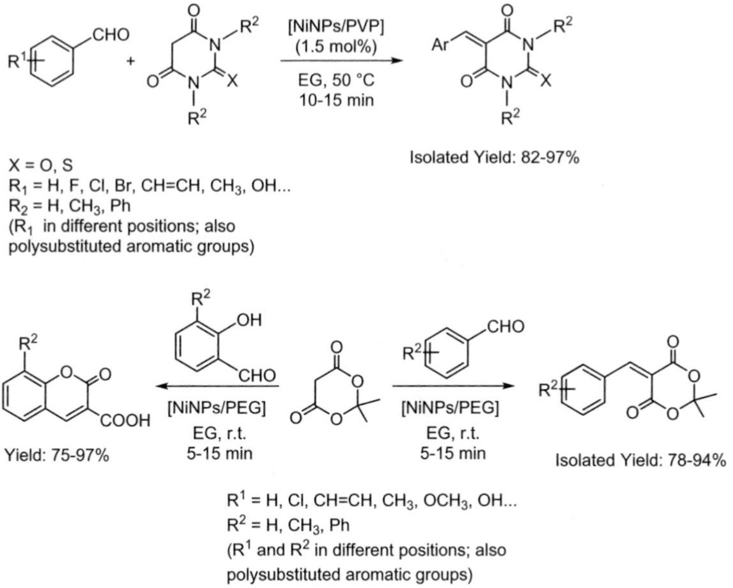
CHO
R1
R2
X
R2
[NiNPs/PVP]
(1.5 mol%)
EG, 50 °C
10-15 min
Ar
R2
X
R2
Isolated Yield: 82-97%
X = O, S
R1 = H, F, Cl, Br, CH=CH, CH3, OH...
R2 = H, CH3, Ph
(R1 in different positions; also
polysubstituted aromatic groups)
R2
OH
CHO
R2
COOH
Yield: 75-97%
[NiNPs/PEG]
EG, r.t.
5-15 min
CHO
R2
[NiNPs/PEG]
EG, r.t.
5-15 min
R2
Isolated Yield: 78-94%
R1 = H, Cl, CH=CH, CH3, OCH3, OH...
R2 = H, CH3, Ph
(R1 and R2 in different positions; also
polysubstituted aromatic groups)

Scheme 8.2 NiNPs stabilized by polymers catalyzed Knoevenagel condensations in ethylene glycol [71, 72]

Maleki and coworkers synthesized Fe_3O_4 nanoparticles stabilized by the biopolymer chitosan coming from chitin (polymer of N-acetylglucosamine) to be applied in three-component reactions yielding benzodiazepines [86] and benzimidazoloquinazolinones [84] (Scheme 8.3), which present biological and pharmacological activities [25, 58]. For both types of reactions, the catalyst was recycled up to four times preserving its catalytic performance, being ethanol the most appropriate solvent for these catalytic reactions. Authors proposed a similar pathway for both transformations.

In Fig. 8.2, the postulated mechanism for the synthesis of benzimidazolo[2,3-b]quinazolinones is illustrated. The reaction most likely starts by a Knoevenagel

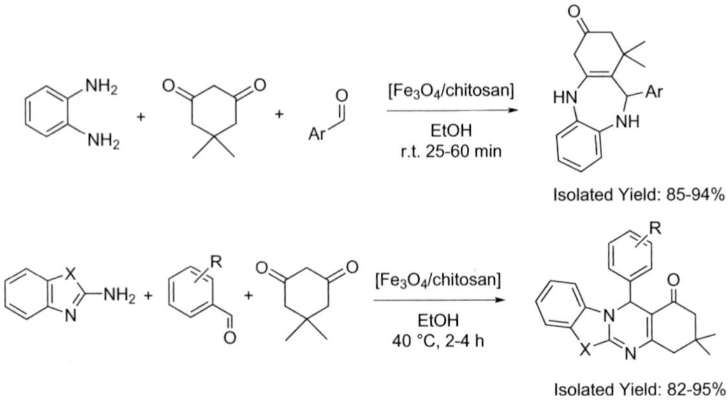

Scheme 8.3 Three-component reactions catalyzed by Fe_3O_4 nanoparticles stabilized by chitosan [84, 86]

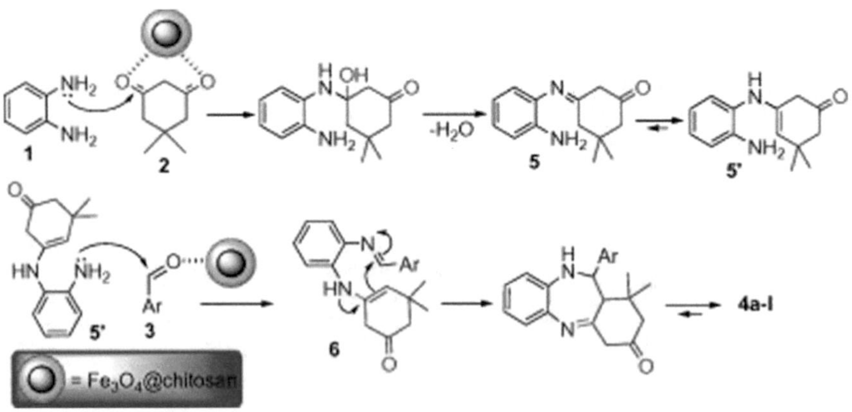

Fig. 8.2 Proposed mechanism for Fe-based catalyzed synthesis of benzimidazolo[2,3-b]quinazolinones. Reprinted from [84] with permission from Elsevier 2015, license no. 4639401066332

condensation between the aromatic aldehyde and dimedone; the resulting α,β-unsaturated ketone reacts with the 2-aminobenzimidazole or its corresponding thio-derivative via a Michael addition, giving an acyclic compound which undergoes a further intramolecular cyclization. Similar catalysts were also applied for the synthesis of 2-amino-4H-chromenes [108] and highly substituted pyridines [85].

Although the Lewis acid properties of Fe_3O_4 have proven to be useful for the synthesis of heterocycles, these properties seem to be enhanced in the presence of Cu(II) (Scheme 8.4) [87]. This catalytic system was reused 6 times with slight yield loss (first run: 98%; sixth run: 91%).

An interesting variant of this reactivity was reported by Mohammadi and Kassaee, showing that chromenylphosphonates could also be prepared in a similar manner in water with high yields and short reaction times (20–30 min), using functionalized magnetite with modified sulfo-chitosan (Scheme 8.5) [93].

These authors proposed a mechanism catalyzed by hydrogen–heteroatom (nitrogen or oxygen) interactions, where water is not an innocent solvent (Fig. 8.3) [93]. Similar behavior has been observed with magnetite functionalized with sulfo-PEG, Fe_3O_4@PEG-SO_3H [88].

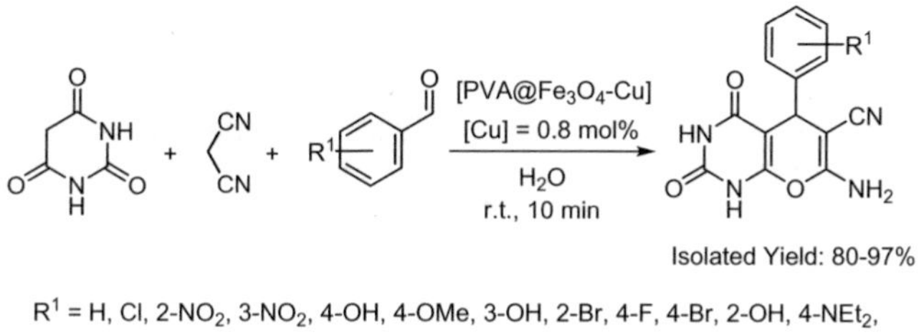
[PVA@Fe3O4-Cu]
[Cu] = 0.8 mol%
H2O
r.t., 10 min
Isolated Yield: 80-97%
R1 = H, Cl, 2-NO2, 3-NO2, 4-OH, 4-OMe, 3-OH, 2-Br, 4-F, 4-Br, 2-OH, 4-NEt2,

Scheme 8.4 Synthesis of pyrano[2,3-*d*]pyrimidines catalyzed by Fe_3O_4 modified with polyvinyl alcohol (PVA) and doped with Cu(II) salts [87]

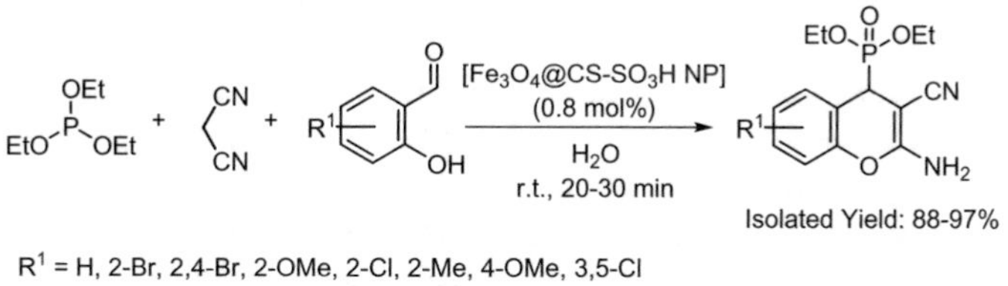
[Fe3O4@CS-SO3H NP]
(0.8 mol%)
H2O
r.t., 20-30 min
Isolated Yield: 88-97%
R1 = H, 2-Br, 2,4-Br, 2-OMe, 2-Cl, 2-Me, 4-OMe, 3,5-Cl

Scheme 8.5 Synthesis of 2-amino-4H-chromen-4-ylphosphonates catalyzed by Fe_3O_4@CS-SO_3H nanoparticles [93]

256 M. Camats et al.

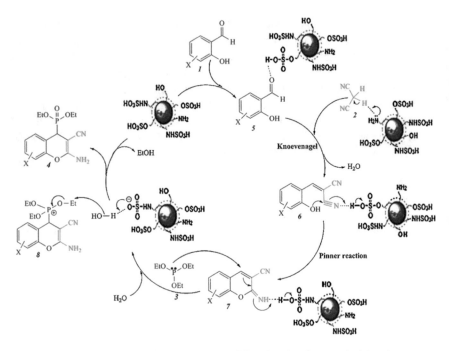

Fig. 8.3 Plausible mechanism for the synthesis of 2-amino-4H-chromen-4-ylphosphonates catalyzed by functionalized magnetite, Fe_3O_4@CS-SO_3H nanoparticles. Reprinted from [93], Copyright 2013, with permission from Elsevier, license no. 4640241387630

8.2.2 Cross-Coupling Reactions

The upbringing of d-block transition metals in catalyzed cross-coupling reactions is gaining ground against palladium thanks to their larger abundance than noble metals. The Sonogashira–Hagihara cross-coupling reaction enables the formation of a C–C bond between an sp^2-carbon halide (aryl or vinyl) and a terminal alkyne [117]. This transformation was initially described with a Pd(0) catalyst and a Cu(I) co-catalyst, but Cu alone [76, 80, 121] and other 3d transition metals have been reported as efficient catalysts for this transformation (Ni [15, 103, 104], Fe [67], Co [116]), fueling the important role of abundant metals in C–C bond forming reactions. Gaining a better understanding on the physicochemical processes leading to the formation of MNPs is key for the design of nanocatalysts with defined morphology [52, 97], notably when catalyst heterogenization with solid supports such as zeolites, titania, montmorillonite and carbonaceous materials is required. Robustness and catalytic activity of tailor-made nanostructured catalysts outperform in many cases classical ones [91].

C(sp)-C(sp²) Sonogashira cross-coupling Since synthesis of copper(I) acetylide, the first organometallic copper complex described in 1859 [20], the reactivity of copper toward the activation of terminal alkyne groups has yielded a number of interesting transformations for the synthesis of propargylic compounds [35]. The search of catalytic versions is key for exploiting the privileged reactivity of this metal from the point of view of sustainability [38]. Recent reports on the singular π-activation of alkynes leveraged by zero-valent CuNPs reveal the importance of this coordination in catalysis [33, 97].

In this context, the cross-coupling of terminal alkynes and acyl chlorides has been reported by Bhosale et al. using a Cu/Cu$_2$O catalyst prepared from Cu(OAc)$_2$ in one-step under microwave irradiation (3 min, 600 W) in 1,3-propanediol. The as-prepared catalytic material exhibited an irregular morphology that combines tubular domains and spherical particles ranging from 70 to 110 nm in size (Fig. 8.4) of a mixture of Cu(0) and Cu$_2$O as confirmed by high-resolution XPS analysis [17].

Taking into account the paramount importance that metal traces might have in catalysis, it is crucial to determine their contribution and potential cooperative effects. Firouzabadi et al. described the use of spherical paramagnetic Fe$_3$O$_4$ NPs (<30 nm in diameter) in ethylene glycol under ligand-free conditions for the Sonogashira–Hagihara reaction for the synthesis of (hetero)aryl alkynes from terminal alkynes and haloarenes (e.g., heteroaryl bromides, aryl iodides, Scheme 8.6) [46, 47]. In order to ascertain the role of Fe$_3$O$_4$ NPs as catalyst, the authors determined the traces of Pd, Ni, Cu, and Co present both in the reagents and solvent. Thus, the contamination concerning the above-mentioned metals was found to be 130 ppb Pd, 45 ppb Ni, 24 ppb Cu, and 21 ppb Co for K$_2$CO$_3$; as well as 20 ppb Pd, 33 ppb Cu, 27 ppb Ni,

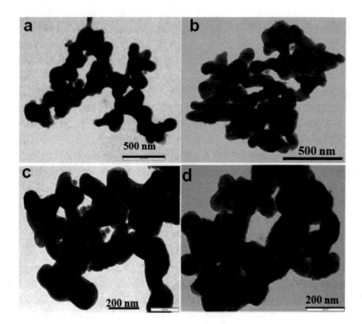

Fig. 8.4 TEM micrographs of Cu/Cu$_2$O NPs. Reproduced with permission from Royal Society of Chemistry 2014, license no. 4639380035278

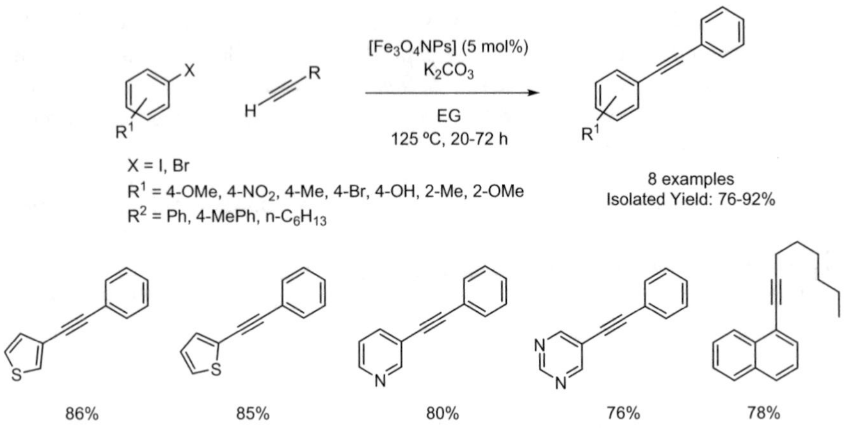

Scheme 8.6 Fe_3O_4 NPs catalyzed Sonogashira-Hagihara coupling reaction in EG [46]

and 24 ppb Co for ethylene glycol by ICP analyses. As catalyst controls, different metal salts [e.g., $Pd(OAc)_2$, CuCl, $NiCl_2$, $CoCl_2$] at concentrations ranging from 200 to 1500 ppb were tested, but the catalytic activity of Fe_3O_4 NPs at 5 mol% loadings proved to be superior [46]. The magnetic properties of Fe_3O_4 NPs allowed an easy catalyst recycling with only an overall 2% yield decrease after five consecutive runs.

An example of bimetallic nanocatalyst for Sonogashira cross-coupling reaction featuring PdCo NPs supported on graphene has been described by Dabiri and Vajargahy [31]. The synthesis of this nanocomposite was carried out following the polyol methodology. In particular, the co-reduction of $PdCl_2$ and $CoCl_2$ was performed in ethylene glycol, which acted as reducing and stabilizing agent for the immobilization of NPs on the 3D graphene support (Fig. 8.5). XPS analysis of the as-prepared nanocomposite revealed the presence of zero-valent Pd and Co (binding energies at 335.67 and 341.49 eV corresponding to Pd $3d_{5/2}$ and Pd $3d_{3/2}$; as well as 781.53 and 798.16 eV corresponding to Co $2p_{3/2}$ and Co $2p_{1/2}$), which match the literature values for PdCo alloys. Higher oxidation states, namely Pd(II), Co(II), and Co(III) were also detected, the latter arising from an oxidation on the NPs surface [31]. The prepared PdCo nanocomposite exhibited high catalytic activity for Sonogashira and Suzuki cross-coupling reactions of aryl halides with terminal alkynes and boronic acids, respectively, in water. Moreover, the catalyst was recycled up to seven times without loss in catalytic activity.

Aluminosilicate-based materials such as montmorillonite are largely used as catalyst supports. Liu et al. have recently described the co-immobilization of Cu and Pd on a montmorillonite-chitosan matrix. According to the authors, the resulting bimetallic CuPd nanocomposite features Pd coexisting in both Pd(0) and Pd(II) valence states, as well as Cu mainly in its Cu(II) as determined by XPS. However, the presence of Cu(0) and Cu(I) cannot be excluded relying solely on this technique. The as-prepared

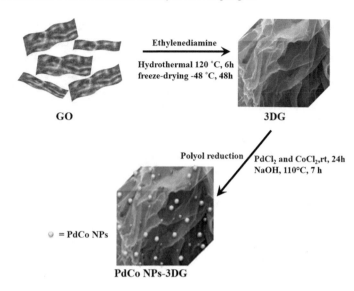

Fig. 8.5 Synthesis of PdCo NPs supported on graphene. Reproduced with permission from Ref. [31]. Copyright 2016 Wiley, license no. 4639270206261

nanocomposite containing spherical PdNPs below 3 nm in diameter catalyzed Sonogashira couplings of haloarenes and alkynes in 1,2-dimethoxyethane/H$_2$O, although Cu leaching resulted in an activity decrease upon recycling [81].

Alternatively, the stabilization of Pd at higher oxidation states could be achieved via the formation of stable Pd–N-heterocyclic carbene complexes. Yavuz et al. have recently reported Sonogashira cross-coupling reaction catalyzed by in situ generated Pd–1H-benzo[d]imidazolium complexes [from Pd(OAc)$_2$ and Cs$_2$CO$_3$], and CuNPs in PEG300 [132]. However, authors did not characterize the catalyst employed.

Suzuki–Miyaura-type cross-coupling During the last years, the controlled synthesis of NiNPs has drawn attention to several research groups, the main challenge being the exclusive preparation of zero-valent nickel nanomaterials due to the facile oxidation of Ni(0) species. Consequently, the lack of control of nickel oxidation in preformed nanoparticles is a persistent issue, as shown by several authors. Among them, one finds Tilley and coworkers in the synthesis of nickel nanocubes from Ni(acac)$_2$ under H$_2$ [78]; Hyeon's group in the preparation of NiNPs under thermal decomposition [99]; and Zarbin and coworkers in the synthesis of NiNPs following the polyol approach [98]. Chaudret and coworkers published an efficient method for the synthesis of nickel(0) nanorods, from [Ni(COD)$_2$] under hydrogen atmosphere [29]. More recently, zero-valent NiNPs were successfully prepared in neat glycerol, without exhibiting any oxide shell on their surface thanks to the low solubility of molecular oxygen in glycerol; these nanocatalysts were, in particular, highly efficient for the semi-hydrogenation of alkynes [106]. However, Ni-based nanoparticles have been scarcely applied in C–C cross-coupling reactions. Their efficiency is particularly remarkable when they are formed in situ, as proven by Lipshutz and coworkers using

NiNPs generated in water in the presence of appropriate surfactants (acting as stabilizers) and the Grignard reagent MeMgBr to activate the pre-catalyst ([$NiCl_2L_n$], where L is a (di)phosphine and $n = 1$ or 2), which were successfully applied in Suzuki–Miyaura couplings [62]. Simultaneously, Han and coworkers reported Ni-catalyzed carbonylative Suzuki-based reactions in PEG (Scheme 8.7) [135], process leading to biaryl ketones, motif present in different types of compounds (drugs, photosensitizers …). They compared the reactivity between preformed and in situ generated NiNPs, evidencing that those formed in situ were more active, probably because the latter are smaller, showing less aggregation that the preformed ones. Authors carried out control tests (with Hg and CS_2) in order to prove the nature of the catalytically active species (in the presence of these additives, the reaction did not work), concluding that the activity observed agrees with a surface-like reactivity.

Preformed NiNPs stabilized by triazole-modified chitosan were active nanocatalysts for Suzuki–Miyaura reactions between aryl halides and phenyl boronic acid derivatives (Scheme 8.8) [60].

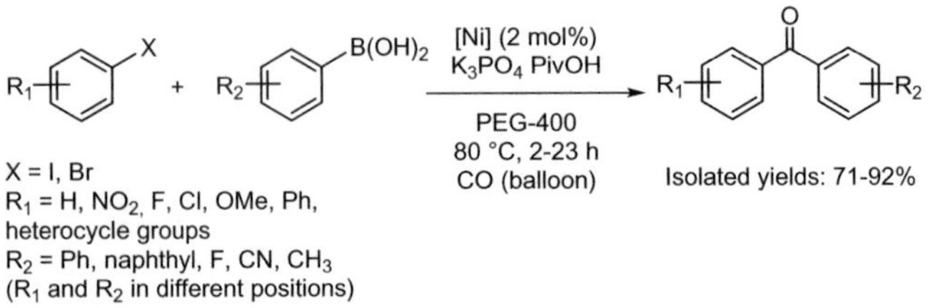
X
B(OH)2
R1
+
R2
[Ni] (2 mol%)
K3PO4 PivOH
PEG-400
80 °C, 2-23 h
CO (balloon)
O
R1
R2
X = I, Br
R1 = H, NO2, F, Cl, OMe, Ph,
heterocycle groups
R2 = Ph, naphthyl, F, CN, CH3
(R1 and R2 in different positions)
Isolated yields: 71-92%

Scheme 8.7 In situ generated NiNPs in PEG-400 applied in carbonylative Suzuki coupling reactions [135]

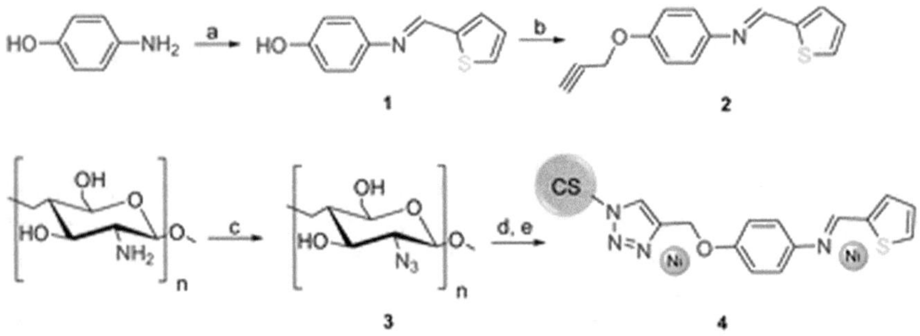
HO
NH2
a
b
N
S
O
1
2
OH
O
HO
NH2
n
c
N3
3
d, e
CS
N
N=N
Ni
4

Scheme 8.8 Synthetic path for the synthesis of NiNPs stabilized by modified chitosan: **a** thiophene-2-carbaldehyde, MeOH, reflux, 3 h; **b** propargyl bromide, K_2CO_3, acetone, 50 °C, 24 h; **c** triflyl azide, aqueous HCl · $NaHCO_3$, $CuSO_4 \cdot 5H_2O$, MeOH, r.t., 5 d; **d** CuI, **2**, DMF/THF (1 : 1), r.t., 72 h; **e** $NiCl_2$/EtOH solution, 12 h, r.t., then hydrazine hydrate, r.t., 2 h. New Journal of Chemistry by Center National de la Recherche Scientifique (France); Royal Society of Chemistry (Great Britain) Reproduced with permission of ROYAL SOCIETY OF CHEMISTRY in the format Book via Copyright Clearance Center, license no. 4639350882762 [60]

Heck–Mizoroki-type cross-coupling Since the pioneering and independent works from Gilman and Lichtenwalter [57], and Kharasch and Fields [70] in the 1930s and 1940s respectively, concerning the homocoupling reaction of Grignard reagents promoted by cobalt(II) salts, scarce works were carried out up to the 1990s ([22] and references therein). Cobalt complexes have proven their efficiency for the formation of $C(sp^2)$–$C(sp^2)$ bonds; particularly, Co-based catalysts are interesting for cross-couplings where alkyl halides are involved, because β-hydrogen elimination of alkyl intermediates is not favored in contrast to the analogous Pd and Ni organometallic species. In the frame of the present contribution, Co(II) anchored to chitosan-functionalized Fe_3O_4 NPs has found applications in Heck- and Sonogashira-type couplings [61]. Fe_3O_4 NPs containing chitosan were modified by reaction with methyl salicylate to give amide-phenol groups at their surface, which coordinate $CoCl_2$ (Scheme 8.9).

This represents an example of molecular-like catalytic reactivity using functionalized nanoparticles as support. For the Heck-type reaction, the catalyst was efficient for the coupling of aryl halides (chloro, bromo, iodo) with styrene or methyl acrylate using PEG as solvent (Scheme 8.10). In contrast to the non-functionalized Fe_3O_4 NPs, the functionalized ones were more active and could be recycled up to five times preserving their efficiency, without metal leaching. The Sonogashira-type coupling between aryl halides (bromo, iodo) and phenylacetylene derivatives gave moderate yields under harsher conditions than those used for the Heck couplings.

Kumada-type cross-coupling Kumada–Tamao–Corriu reaction, coupling between a Grignard reagent and an organic halide, was initially reported using Ni-based catalysts [30, 119]; other efficient systems such as those based on palladium and iron, have proven their efficacy, the main part of them involving molecular catalysts, but more recently copper, nickel, and palladium nanoparticles have been used as well ([64, 92] and references therein). Iron, representing the second more abundant metal in the Earth's crust, has found interesting applications in C–C couplings [3, 13, 96]. The main part of reported works assumes the contribution of molecular intermediate species in the catalytic transformation, but also nanoparticles have been identified for low oxidation states (zero-valent FeNPs should be more stable under catalytic conditions than organometallic Fe(0) complexes) and it has been postulated that they act as a reservoir of molecular species exhibiting higher oxidation state [11]. Bedford and coworkers studied the reaction of alkyl halides with aryl Grignard compounds catalyzed by in situ generated FeNPs from $FeCl_3$ in the presence of PEG-14000 (Scheme 8.11 and Fig. 8.6) [12]. Authors proved that the Grignard reagent acted as reducing agent. Preformed FeNPs/PEG (diameter in the range 7–13 nm, determined by TEM) afforded the same reactivity as those formed in situ.

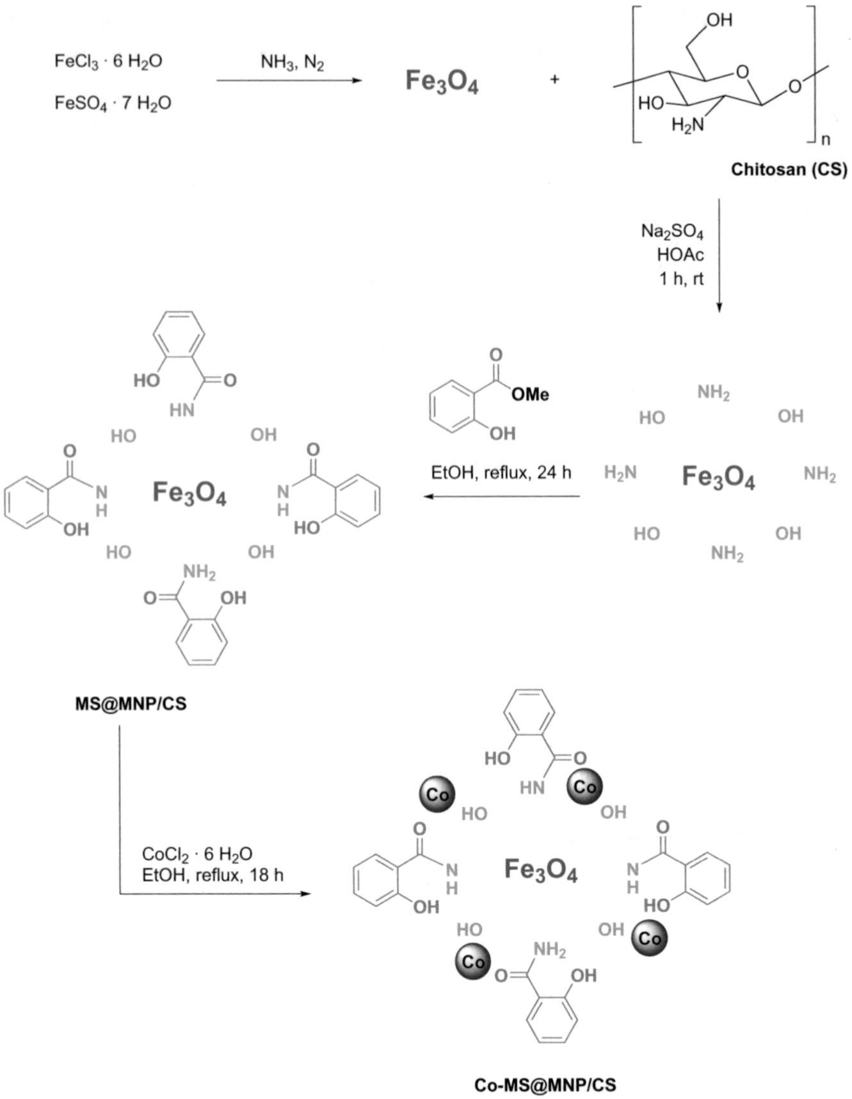

Scheme 8.9 Preparation of co-based nanocatalyst immobilized on Fe_3O_4 nanoparticles modified by chitosan. Green Chemistry by Royal Society of Chemistry (Great Britain) Reproduced with permission of ROYAL SOCIETY OF CHEMISTRY in the format Book via Copyright Clearance Center, license no. 4639360400827 [61]

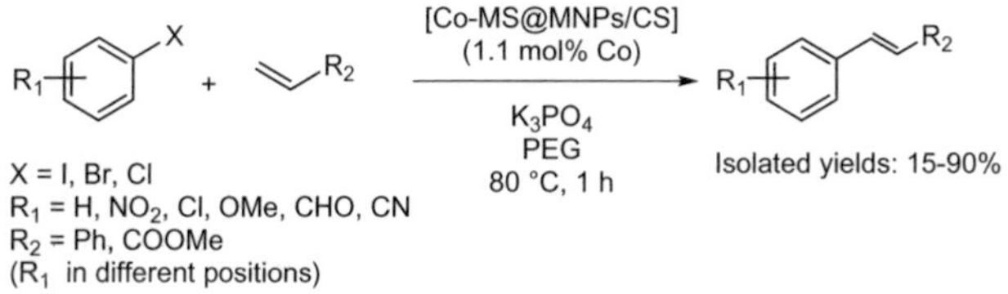

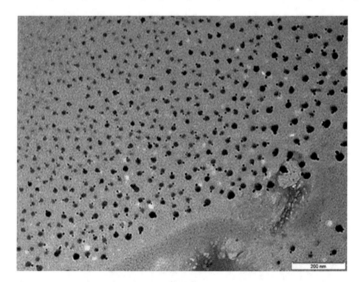

Scheme 8.10 Heck–Mizoroki-type couplings catalyzed by Co(II) supported on functionalized Fe_3O_4 nanoparticles (see Scheme 8.9) [61]

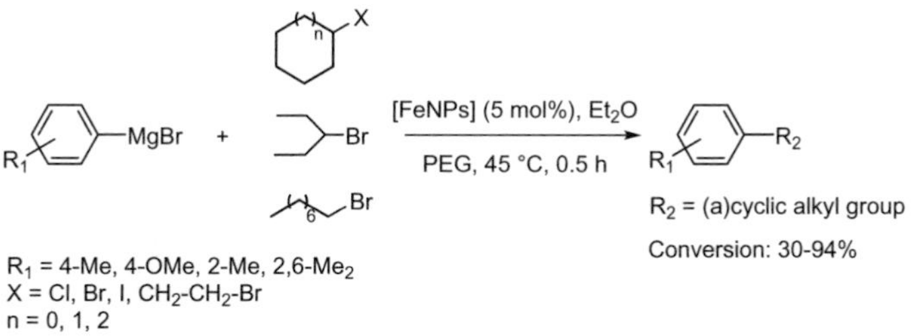

Scheme 8.11 $C(sp^2)$–$C(sp^3)$ cross-coupling catalyzed by FeNPs stabilized by PEG [12]

Fig. 8.6 TEM image corresponding to FeNPs generated in situ from $FeCl_3$ in the presence of 1,6-bis(diphenylphosphino)hexane with $4\text{-}MeC_6H_4MgBr$. Reproduced with permission of ROYAL SOCIETY OF CHEMISTRY in the format Book via Copyright Clearance Center, license no. 4640081165587 [12]

8.3   Carbon–Heteroatom Bond Formation

The activation of C–halogen bonds has been overwhelmingly used in the function-alization of arenes. Since the discovery of Ullmann's coupling in 1901, copper has been one of the most used earth-abundant metals in both C–C and C–heteroatom couplings (original paper of Ullmann: [125]; recent reviews: [14, 94, 109]). How-ever, many applications of this reactivity are especially limited due to the use of hazardous solvents such as DMF and DMSO. A lot of interest has been put into transforming this reactivity into greener approaches. In this context, considerable efforts have been recently developed in nanocatalytic systems capable of working in friendly environmental conditions [68, 127], polyols being alternative solvents of interest [41].

This strategy has been successfully applied in a wide range of C–N couplings using many different nitrogen-based reagents. Kidwai and coworkers applied preformed unsupported CuNPs in the N-arylation of aryl halides with anilines using PEG400 as solvent (Scheme 8.12) [74]. A large scope of aniline derivatives and also NH heterocycles was carried out.

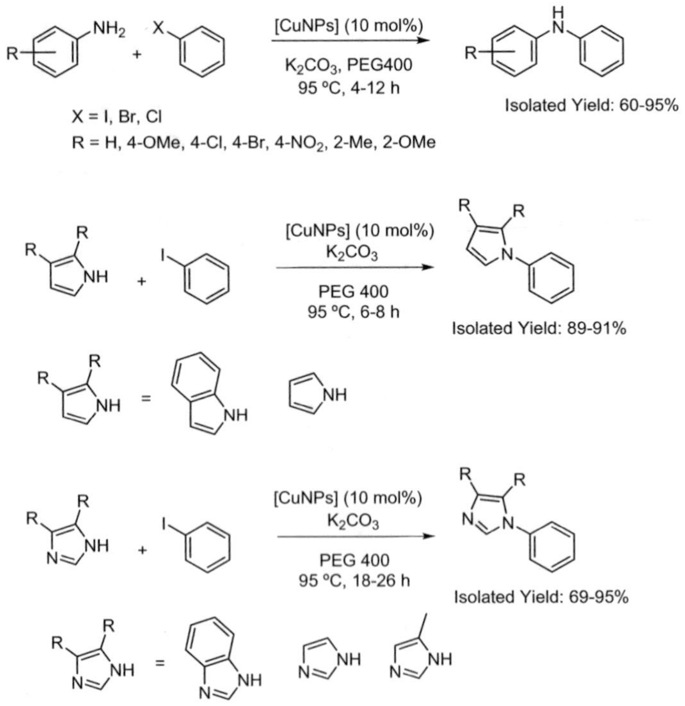

Scheme 8.12 CuNPs catalyzed C–N cross-couplings of anilines and NH heterocycles with aryl halides in PEG400 [74]

Graphene oxide functionalized with carboxamide groups (f-GO) was an efficient support to coordinate Cu(II) salts and further reducing them to give CuNPs immobilized on the solid. This CuNPs on f-GO were then treated with Fe_3O_4 nanoparticles leading to a magnetic catalytic material, which enhanced the catalyst recyclability (Fig. 8.7) [113].

The as-prepared heterogenized nanocatalyst was applied to Ullmann-type coupling for the synthesis of N-aryl amines using a deep eutectic solvent (choline chloride:glycerol $= 1:2$) (Scheme 8.13). Authors compared the efficiency of this catalytic system with other CuNPs, both supported and unsupported, concluding that their catalyst led to higher yields under smoother conditions. However, the recycling was moderate (up to 3 times without loss of activity).

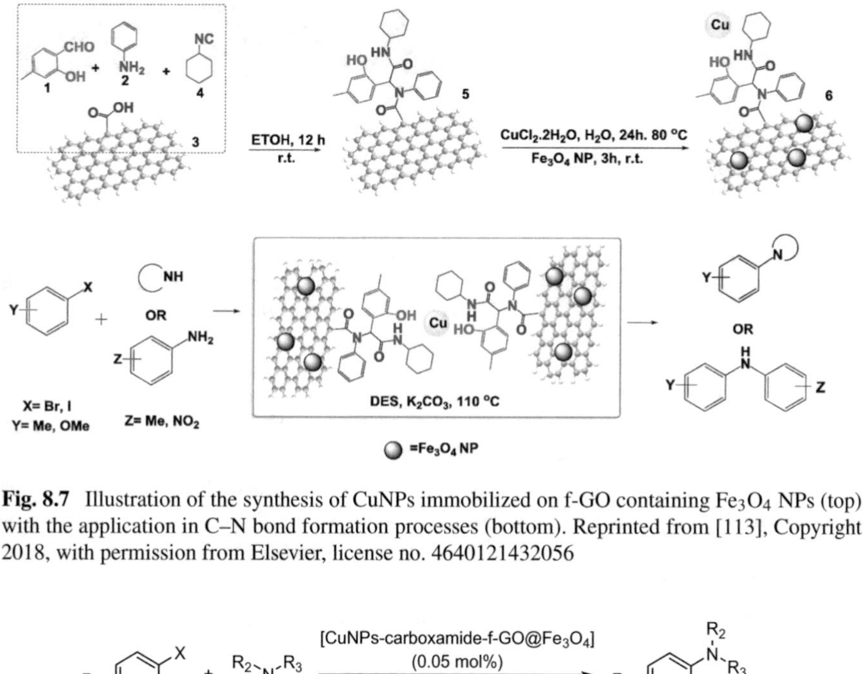

Fig. 8.7 Illustration of the synthesis of CuNPs immobilized on f-GO containing Fe_3O_4 NPs (top) with the application in C–N bond formation processes (bottom). Reprinted from [113], Copyright 2018, with permission from Elsevier, license no. 4640121432056

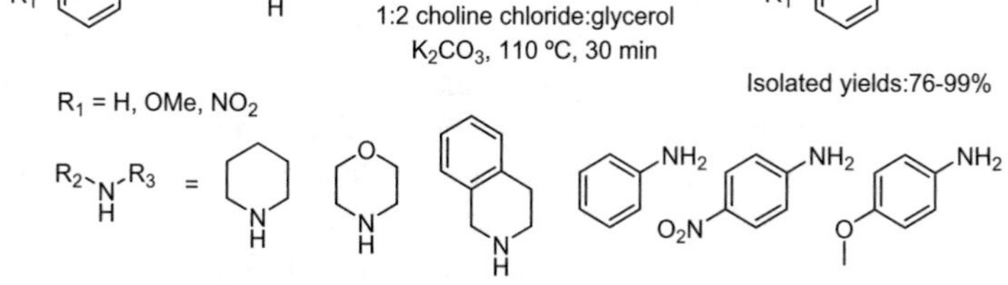

Scheme 8.13 CuNPs immobilized on a magnetic-modified graphene oxide (see Fig. 8.7) catalyzed C–N couplings of secondary amines and anilines [113]

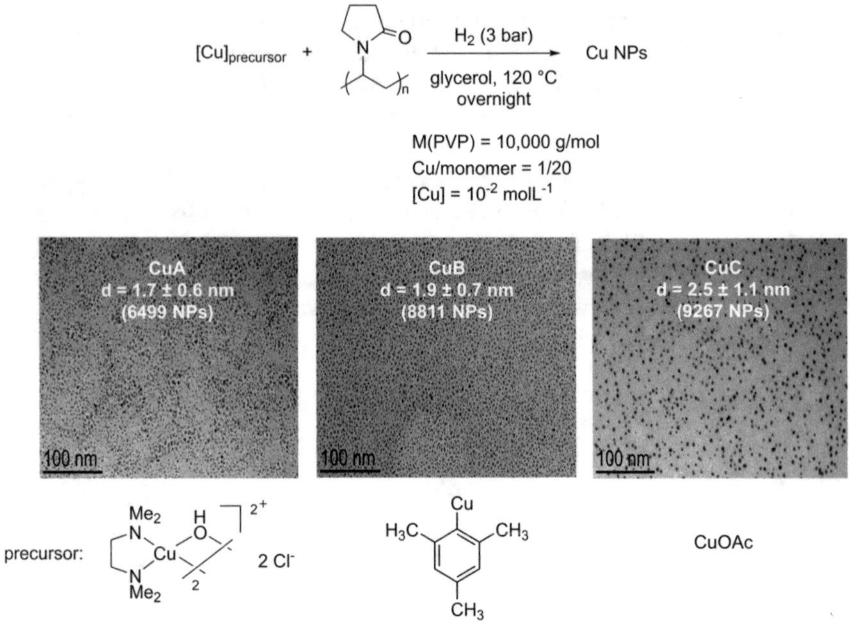

Fig. 8.8 Synthesis of CuNPs in glycerol from different copper precursors stabilized by the polymer PVP (top) and the corresponding TEM images of the different nanoparticles (bottom). Adapted with permission from [33], Copyright 2017 Wiley, license no. 4640130742799

The work of our group on the preparation of small CuNPs (mean diameter: 1.7–2.4 nm) in glycerol from the reduction of Cu(II) and Cu(I) precursors with PVP as stabilizer and under low pressure of H_2 (3 bar) avoided the formation of oxidized by-products coming from the solvent (Fig. 8.8). This approach represents the first report toward the synthesis of well-defined and stable CuNPs by a bottom-up strategy thanks to the low solubility of O_2 in this medium, circumventing the formation of oxide shells [33].

CuNPs nanoparticles were successfully applied in C–N couplings and in the synthesis of propargyl amines through different strategies, such as cross-dehydrogenative couplings and multicomponent reactions, both A^3 (aldehyde–alkyne–amine) and KA^2 (ketone–alkyne–amine) (Scheme 8.14). Authors carried out spectroscopic monitoring (UV-vis and FTIR analyses) concluding that the C–N coupling follows a surface reactivity, without formation of Cu(I) molecular species, like phenylethynylcopper(I), which would be poisoned by the presence of amines.

The selection of alternative aldehydes bearing heteroatoms in position 2 of the ring (e.g., 2-aminobenzaldehyde, 2-hydroxybenzaldehyde and 2-pyridinecarbaldehyde) provided a direct entry to the synthesis of heterocycles, namely indolizines, benzofurans and quinolines via a CuNP-catalyzed A^3-cycloisomerization tandem processes (Scheme 8.15).

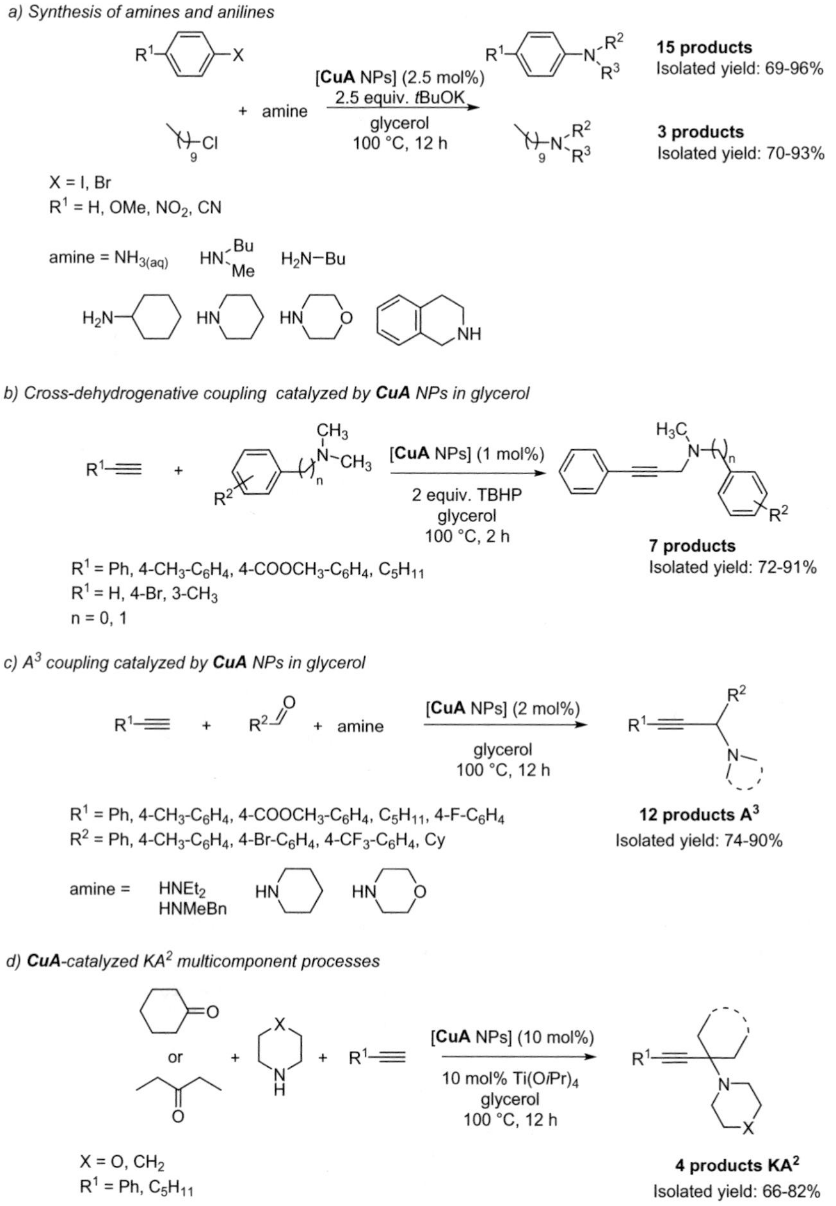

a) Synthesis of amines and anilines

X = I, Br
R^1 = H, OMe, NO$_2$, CN

amine = NH$_{3(aq)}$ HN$\overset{Bu}{\underset{Me}{}}$ H$_2$N–Bu

b) Cross-dehydrogenative coupling catalyzed by **CuA** NPs in glycerol

R^1 = Ph, 4-CH$_3$-C$_6$H$_4$, 4-COOCH$_3$-C$_6$H$_4$, C$_5$H$_{11}$
R^1 = H, 4-Br, 3-CH$_3$
n = 0, 1

c) A^3 coupling catalyzed by **CuA** NPs in glycerol

R^1 = Ph, 4-CH$_3$-C$_6$H$_4$, 4-COOCH$_3$-C$_6$H$_4$, C$_5$H$_{11}$, 4-F-C$_6$H$_4$
R^2 = Ph, 4-CH$_3$-C$_6$H$_4$, 4-Br-C$_6$H$_4$, 4-CF$_3$-C$_6$H$_4$, Cy

amine =   HNEt$_2$ HN(   )   HN(O)
          HNMeBn

12 products A^3
Isolated yield: 74-90%

d) **CuA**-catalyzed KA^2 multicomponent processes

X = O, CH$_2$
R^1 = Ph, C$_5$H$_{11}$

4 products KA^2
Isolated yield: 66-82%

Scheme 8.14 CuA nanoparticles catalyzed C–N couplings and multicomponent reactions [33] (for **CuA**, see Fig. 8.8)

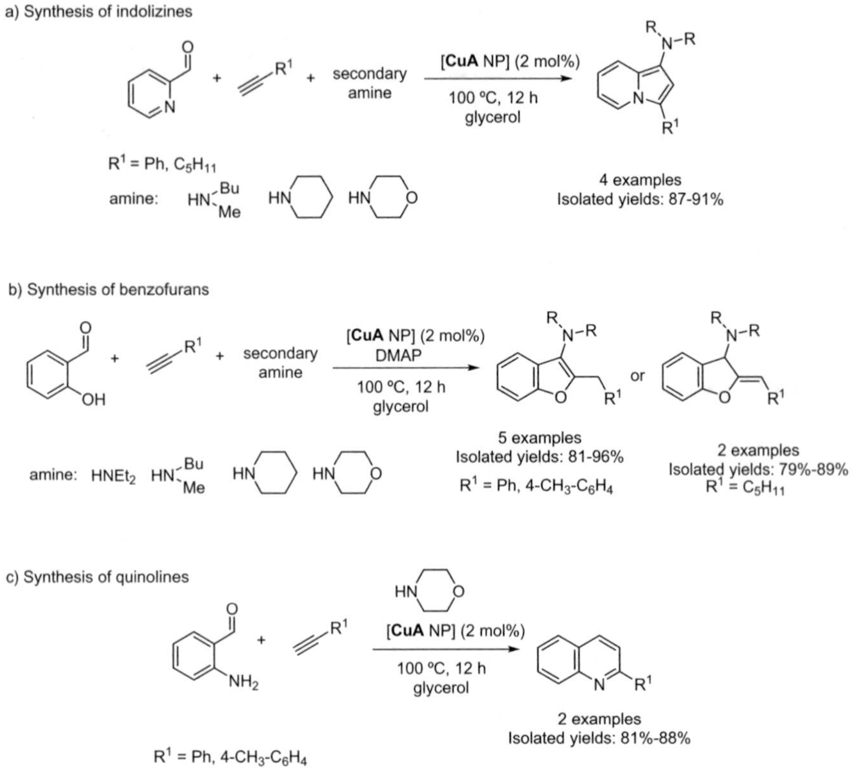

a) Synthesis of indolizines

R^1 = Ph, C$_5$H$_{11}$

amine: HN$\overset{Bu}{\underset{Me}{}}$ HN⟨⟩ HN⟨O⟩

4 examples
Isolated yields: 87-91%

b) Synthesis of benzofurans

amine: HNEt$_2$ HN$\overset{Bu}{\underset{Me}{}}$ HN⟨⟩ HN⟨O⟩

5 examples
Isolated yields: 81-96%
R^1 = Ph, 4-CH$_3$-C$_6$H$_4$

2 examples
Isolated yields: 79%-89%
R^1 = C$_5$H$_{11}$

c) Synthesis of quinolines

R^1 = Ph, 4-CH$_3$-C$_6$H$_4$

2 examples
Isolated yields: 81%-88%

Scheme 8.15 Synthesis of indolizines, benzofurans, and quinolines catalyzed by **CuA** nanoparticles in glycerol [33] (for **CuA**, see Fig. 8.8)

Moreover, Shah et al. have recently reported the A^3-coupling reaction for the synthesis of propargylamines and pyrrolo[1,2-a]quinolines using a heterogenized Cu catalyst (CuNPs@ZnO–polythiophene) at low catalyst loadings (Scheme 8.16)

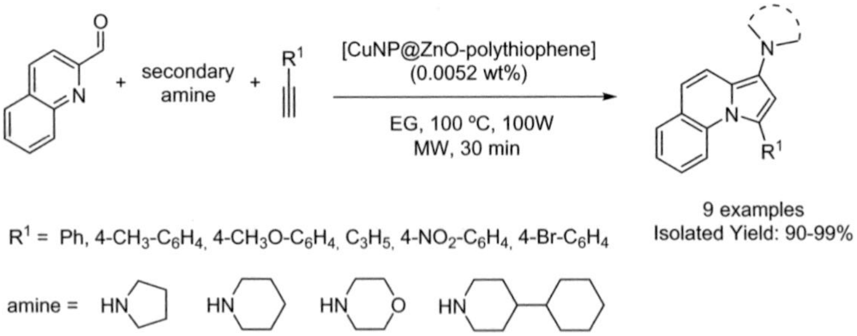

R^1 = Ph, 4-CH$_3$-C$_6$H$_4$, 4-CH$_3$O-C$_6$H$_4$, C$_3$H$_5$, 4-NO$_2$-C$_6$H$_4$, 4-Br-C$_6$H$_4$

9 examples
Isolated Yield: 90-99%

amine = HN⟨⟩ HN⟨⟩ HN⟨O⟩ HN⟨⟩⟨⟩

Scheme 8.16 CuNPs@ZnO-polythiophene catalyzed synthesis of pyrrolo[1,2-a]quinolines in EG under microwave irradiation [114]

[114]. Notably, the cyclization only takes place in an intramolecular fashion and no Cu leaching was detected in the EG reaction medium after catalyst filtration (ICP analyses).

Preformed Cu_2O nanoparticles (mean diameter: ca. 5 nm, Fig. 8.9) were applied in the coupling between iodoaryl derivatives and different nitrogen-based reagents, such as aryl and alkyl amines, but also aqueous ammonia, in glycerol [24]. This nanocatalyst was also efficient for the synthesis of thioethers through C–S couplings.

This catalytic system has also been applied for the activation of terminal alkyne groups toward azide–alkyne cycloaddition reaction (CuAAC) in glycerol (Scheme 8.17) [24].

Furthermore, Sharghi and Aberi reported the application of Cu_2O NPs in the synthesis of indazole derivatives through a three-component strategy in PEG300 (Scheme 8.18) [115].

Cu-based catalysts in polyol medium have also found interesting applications in C–S bond formation processes, as proved by our group using Cu_2O nanoparticles [24]. Primo, García, and coworkers reported the synthesis of Cu-based nanoparticles stabilized by chitosan, mainly constituted of Cu(0) but surrounded by a thin layer of copper oxides as proven by XPS [49]. This nanocatalyst was applied in the C–S coupling of aryl halides and thiophenol, being more active for iodo than bromo and chloro arenes, for the latter ones they were only active for aryl halides containing electron withdrawing substituents (Scheme 8.19). Authors observed that halide anions released to the medium during the catalytic reaction can act as poison for CuNPs, in addition to promoting metal leaching.

Generally speaking, C–O couplings are more challenging reactions mainly due to the limited functional compatibilities and the need of activated substrates [77, 118]. Besides, the use of polyol medium adds another difficulty because the solvent can compete with the substrate. In this frame, Biegi and Ghiasbeigi have recently reported the synthesis and full characterization of functionalized magnetite with isonicotinic acid with the aim of coordinating Cu(I) precursors (Fig. 8.10) [53]. The resulting catalytic material was efficiently applied to the synthesis of phenol and aniline derivatives (TOF up to 4494 h^{-1}), using a mixture of PEG and water as solvent (PEG:H_2O = 2:1) (Scheme 8.20). The catalytic phase was recycled up to 5 times with slight loss of yield, although the 30% copper loss reported by the authors after the fourth run was substantial (Cu content of catalyst before use: 57,037.8 ppm; after 4th run: 40,297.3 ppm).

8.4 Conclusions and Outlook

This chapter describes the use of nanocatalysts from Earth-abundant metals in polyol media applied in C–C and C–heteroatom coupling reactions. Despite the interesting reports mentioned in this contribution, nanocatalysis based on d-block transition metals in polyols is still in its infancy. This is an exponentially growing research

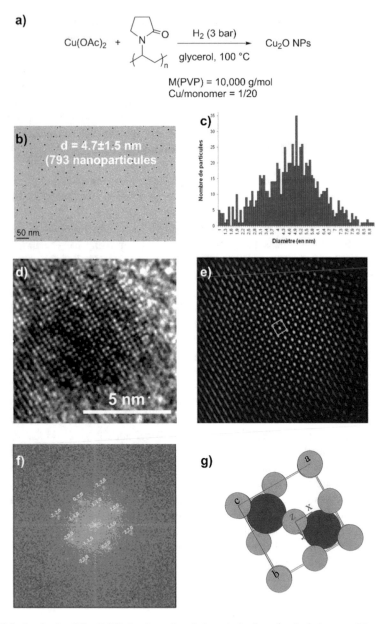

Fig. 8.9 Synthesis of Cu_2O NPs in glycerol and characterization after isolation at solid state and re-dispersion in glycerol. **a** Synthesis scheme; **b** TEM micrograph in glycerol; **c** size distribution histogram; **d** HR-TEM micrograph of one single particle of Cu_2O; **e** electronic diffraction spots by fast Fourier transform of a single particle; **f** filtered image showing the zone axis [0 0 1]; **g** cartoon of Cu_2O cubic structure. Adapted with permission from [24], Copyright 2014 Wiley, license no. 4640200545968

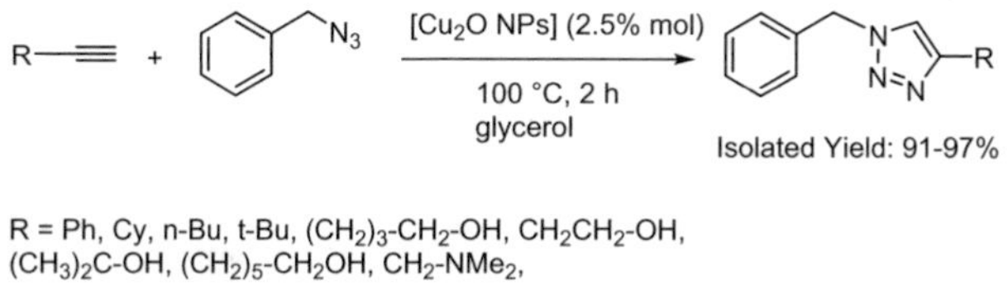

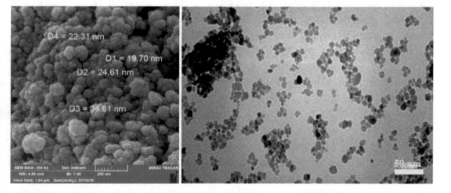

R = Ph, Cy, n-Bu, t-Bu, (CH₂)₃-CH₂-OH, CH₂CH₂-OH,
(CH₃)₂C-OH, (CH₂)₅-CH₂OH, CH₂-NMe₂,

Scheme 8.17 Cu_2O NPs catalyzed azide–alkyne cycloaddition in glycerol [24]

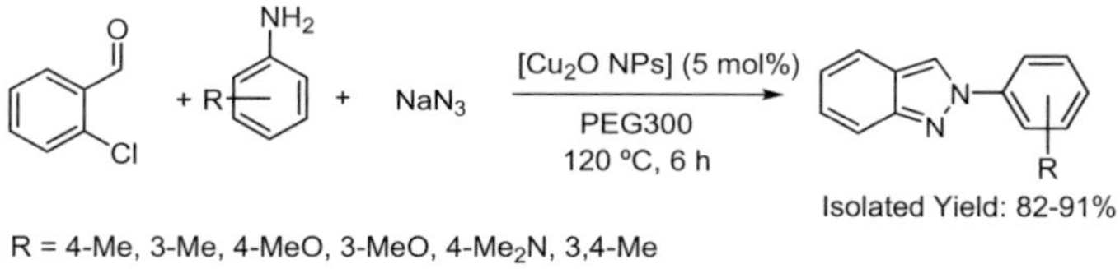

R = 4-Me, 3-Me, 4-MeO, 3-MeO, 4-Me₂N, 3,4-Me

Scheme 8.18 Cu_2O nanoparticles catalyzed the synthesis of $2H$-indazoles [115]

Scheme 8.19 CuNPs
stabilized by chitosan
applied in C–S couplings
[49] (CS stands for chitosan)

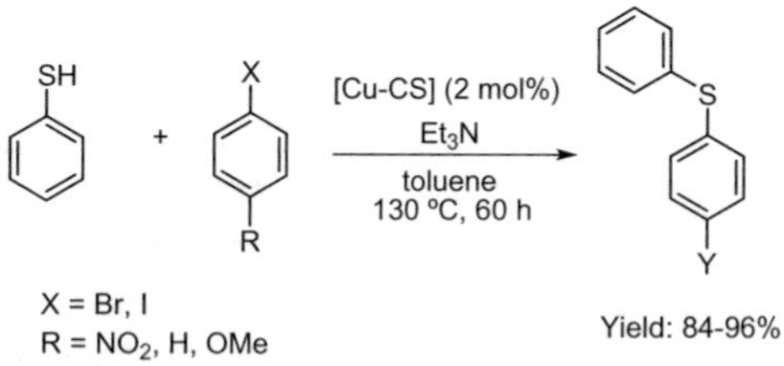

X = Br, I
R = NO₂, H, OMe

Yield: 84-96%

Fig. 8.10 SEM (left) and TEM (right) images of Cu(I) grafted to Fe_3O_4 modified by isonicotinic groups. Reprinted with permission from [53], Copyright 2019 Wiley, license no. 4640160235060

Scheme 8.20 Synthesis of
phenols catalyzed by
Cu-based catalyst supported
on modified Fe_3O_4 [53]

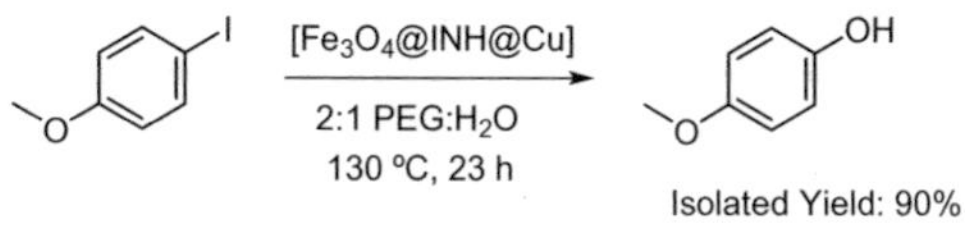

[Fe₃O₄@INH@Cu]

2:1 PEG:H₂O
130 °C, 23 h

Isolated Yield: 90%

field that exploits the non-innocent physicochemical properties of polyols as reducing, stabilizing, and dispersing agents in the quest for tailor-made nanocatalysts and nanocomposites with enhanced properties as compared to classical ones. In the framework of the development of greener and more sustainable processes, Cu, Ni, Co, and Fe nanocatalysts in polyol media represent key alternatives to overcome the dependence on the scarcity of noble metals in use currently for both academic and industrial purposes.

The Lewis acidity properties of 3d-transition metals confer them suitable properties as (co)catalysts to achieve new transformations by means of (i) Lewis acid base-adduct formation, thereby accelerating slow elementary steps, (ii) pKa of the reaction modulation (e.g., release of a Brønsted acid, proton transfer processes), or (iii) activating the catalyst precursors or off-cycle catalyst species by tuning its coordination sphere by anion abstraction. This chapter describes the reports on Ni, Cu, and Fe Lewis acid mediated transformations in polyol medium, but contributions from other abundant metals will surely appear in the literature in the years to come.

On the other hand, the abundance and redox properties of 3d-metal-oxide-based materials confer them large applicability as supports for catalysts. In particular, the magnetic properties of Fe_2O_3 and Fe_3O_4 as supports enable the recovery of the prepared composite materials. For instance, the use of such supports for the preparation of heterogenized Pd catalysts for C–C cross-coupling reactions and hydrogenations in polyol medium has been widely reported (for selected articles, see: [39, 54, 55, 82, 126, 136]). Other supports based on 3d-transition metals such as TiO_2 [95], CuO [26], and ZnO [114] have also been used for the same purpose. This heterogenization strategy efficiently enables the recoverability of the catalytic materials by magnetic separation, and also in some cases, the enhancement of TON is observed due to the synergy between catalyst and support. Furthermore, polymetallic systems merit further studies to exploit the cooperative effects between active metal centers in polyol medium [34]. The intrinsic properties of polyols in terms of favoring 3D organization via supramolecular interactions, their suitable oxidation potentials for the reduction of transition metal salts and organometallic complexes, as well as their dispersing abilities via solvation interactions, which often trigger an activity increase, confer them unique properties in nanocatalysis.

From a structural point of view, 3d-metals based-species present several oxidation states, often leading to paramagnetic intermediates of challenging elucidation. Given the specificity developed by nature in biocatalyzed transformations involving 3d-transition metals and the demonstrated efficiency of nanocatalysts discussed herein (e.g., Cu NPs for Sonogashira and C–heteroatom couplings, Ni NPs for Suzuki, Co NPs for Heck–Mizoroki, Fe NPs for Kumada-like couplings ...), the fundamental and applied research in this field foresees new reactivities and deep mechanistic insights taking advantage of the cutting-edge in operando techniques available nowadays.

References

1. Adil SF, Assal ME, Khan M, Al-Warthan A, Siddiqui MRH, Liz-Marzán LM (2015) Biogenic synthesis of metallic nanoparticles and prospects toward green chemistry. Dalton Trans 44(21):9709–9717. https://doi.org/10.1039/C4DT03222E
2. Ai W, Zhong R, Liu X, Liu Q (2019) Hydride Transfer reactions catalyzed by cobalt complexes. Chem Rev 119(4):2876–2953. https://doi.org/10.1021/acs.chemrev.8b00404
3. Alice W, Axel Jacobi von W (2013) Iron(0) nanoparticle catalysts in organic synthesis. Curr Org Chem 17(4):326–335. https://doi.org/10.2174/1385272811317040003
4. Alig L, Fritz M, Schneider S (2019) First-row transition metal (de)hydrogenation catalysis based on functional pincer ligands. Chem Rev 119(4):2681–2751. https://doi.org/10.1021/acs.chemrev.8b00555
5. An K, Alayoglu S, Ewers T, Somorjai GA (2012) Colloid chemistry of nanocatalysts: a molecular view. J Colloid Interface Sci 373(1):1–13. https://doi.org/10.1016/j.jcis.2011.10.082
6. Anastas P, Eghbali N (2010) Green chemistry: principles and practice. Chem Soc Rev 39(1):301–312. https://doi.org/10.1039/B918763B
7. Anastas PT, Warner JC (2000) Green chemistry. Oxford University Press, New York
8. Anastas PT, Zimmerman JB (2003) Peer reviewed: design through the 12 principles of green engineering. Environ Sci Technol 37(5):94A–101A. https://doi.org/10.1021/es032373g
9. Anastas PT, Kirchhoff MM, Williamson TC (2001) Catalysis as a foundational pillar of green chemistry. Appl Catal A 221(1):3–13. https://doi.org/10.1016/S0926-860X(01)00793-1
10. Astruc D (ed) (2008) Nanoparticles and catalysis. Wiley
11. Bedford RB (2015) How low does iron go? Chasing the active species in fe-catalyzed cross-coupling reactions. Acc Chem Res 48(5):1485–1493. https://doi.org/10.1021/acs.accounts.5b00042
12. Bedford RB, Betham M, Bruce DW, Davis SA, Frost RM, Hird M (2006) Iron nanoparticles in the coupling of alkyl halides with aryl Grignard reagents. Chem Commun 13:1398–1400. https://doi.org/10.1039/b601014h
13. Bedford RB, Brenner PB (2015) The development of iron catalysts for cross-coupling reactions. In: Bauer E (ed) Iron catalysis II. Springer International Publishing, Cham, pp 19–46
14. Beletskaya IP, Cheprakov AV (2004) Copper in cross-coupling reactions: the post-Ullmann chemistry. Coord Chem Rev 248(21):2337–2364. https://doi.org/10.1016/j.ccr.2004.09.014
15. Beletskaya IP, Latyshev GV, Tsvetkov AV, Lukashev NV (2003) The nickel-catalyzed Sonogashira-Hagihara reaction. Tetrahedron Lett 44(27):5011–5013. https://doi.org/10.1016/S0040-4039(03)01174-2
16. Beller M (2019) Introduction: first row metals and catalysis. Chem Rev 119(4):2089. https://doi.org/10.1021/acs.chemrev.9b00076
17. Bhosale MA, Sasaki T, Bhanage BM (2014) A facile and rapid route for the synthesis of Cu/Cu$_2$O nanoparticles and their application in the Sonogashira coupling reaction of acyl chlorides with terminal alkynes. Catal Sci Technol 4(12):4274–4280. https://doi.org/10.1039/C4CY00868E
18. Biffis A, Centomo P, Del Zotto A, Zecca M (2018) Pd metal catalysts for cross-couplings and related reactions in the 21st century: a critical review. Chem Rev 118(4):2249–2295. https://doi.org/10.1021/acs.chemrev.7b00443
19. Biggs-Houck JE, Younai A, Shaw JT (2010) Recent advances in multicomponent reactions for diversity-oriented synthesis. Curr Opin Chem Biol 14(3):371–382. https://doi.org/10.1016/j.cbpa.2010.03.003
20. Boettger R (1859) Ueber die Einwirkung des Leuchtgases auf verschiedene Salzsolutionen, insbesondere auf eine ammoniakalische Kupferchlorürlösung. Justus Liebigs Ann Chem 109(3):351–362. https://doi.org/10.1002/jlac.18591090318

21. Bulut S, Fei Z, Siankevich S, Zhang J, Yan N, Dyson PJ (2015) Aqueous-phase hydrogenation of alkenes and arenes: the growing role of nanoscale catalysts. Catal Today 247:96–103. https://doi.org/10.1016/j.cattod.2014.09.002
22. Cahiez G, Moyeux A (2010) Cobalt-catalyzed cross-coupling reactions. Chem Rev 110(3):1435–1462. https://doi.org/10.1021/cr9000786
23. Chahdoura F, Favier I, Gómez M (2014) Glycerol as suitable solvent for the synthesis of metallic species and catalysis. Chem Eur J 20(35):10884–10893. https://doi.org/10.1002/chem.201403534
24. Chahdoura F, Pradel C, Gómez M (2014) Copper(I) oxide nanoparticles in glycerol: a convenient catalyst for cross-coupling and azide-alkyne cycloaddition processes. ChemCatChem 6(10):2929–2936. https://doi.org/10.1002/cctc.201402214
25. Chakraborty S, Shah NH, Fishbein JC, Hosmane RS (2011) A novel transition state analog inhibitor of guanase based on azepinomycin ring structure: synthesis and biochemical assessment of enzyme inhibition. Bioorg Med Chem Lett 21(2):756–759. https://doi.org/10.1016/j.bmcl.2010.11.109
26. Chattopadhyay K, Dey R, Ranu BC (2009) Shape-dependent catalytic activity of copper oxide-supported Pd(0) nanoparticles for Suzuki and cyanation reactions. Tetrahedron Lett 50(26):3164–3167. https://doi.org/10.1016/j.tetlet.2009.01.027
27. Chirik P, Morris R (2015) Getting down to earth: the renaissance of catalysis with abundant metals. Acc Chem Res 48(9):2495. https://doi.org/10.1021/acs.accounts.5b00385
28. Corain B, Schmid G, Toshima N (eds) (2011) Metal Nanoclusters in Catalysis and Materials Science: The Issue of Size Control. Elsevier Science, Amsterdam
29. Cordente N, Respaud M, Senocq F, Casanove M-J, Amiens C, Chaudret B (2001) Synthesis and magnetic properties of nickel nanorods. Nano Lett 1(10):565–568. https://doi.org/10.1021/nl0100522
30. Corriu RJP, Masse JP (1972) Activation of grignard reagents by transition-metal complexes. A new and simple synthesis of trans-stilbenes and polyphenyls. J Chem Soc Chem Commun 3:144a. https://doi.org/10.1039/C3972000144A
31. Dabiri M, Vajargahy MP (2017) PdCo bimetallic nanoparticles supported on three-dimensional graphene as a highly active catalyst for sonogashira cross-coupling reaction. Appl Organomet Chem 31(4):e3594. https://doi.org/10.1002/aoc.3594
32. Dahl JA, Maddux BLS, Hutchison JE (2007) Toward greener nanosynthesis. Chem Rev 107(6):2228–2269. https://doi.org/10.1021/cr050943k
33. Dang-Bao T, Pradel C, Favier I, Gómez M (2017) Making copper(0) nanoparticles in glycerol: a straightforward synthesis for a multipurpose catalyst. Adv Synth Catal 359(16):2832–2846. https://doi.org/10.1002/adsc.201700535
34. Dang-Bao T, Pradel C, Favier I, Gómez M (2019) Bimetallic nanocatalysts in glycerol for applications in controlled synthesis. a structure–reactivity relationship study. ACS Appl. Nano Mater 2(2):1033–1044. https://doi.org/10.1021/acsanm.8b02316
35. Díez-González S (2016) Copper(I)–acetylides: access, structure, and relevance in catalysis. In: Pérez PJ (ed) Advances in Organometallic Chemistry, vol 66. Academic Press, pp 93–141
36. Dong H, Chen YC, Feldmann C (2015) Polyol synthesis of nanoparticles: status and options regarding metals, oxides, chalcogenides, and non-metal elements. Green Chem 17(8):4107–4132. https://doi.org/10.1039/C5GC00943J
37. Eicher T, Hauptmann S, Speicher A (2013) The chemistry of heterocycles: structures, reactions, synthesis, and applications. Wiley-VCH, Weinheim
38. Evano G, Blanchard N, Toumi M (2008) Copper-mediated coupling reactions and their applications in natural products and designed biomolecules synthesis. Chem Rev 108(8):3054–3131. https://doi.org/10.1021/cr8002505
39. Fakhri A, Naghipour A (2018) Organometallic polymer-functionalized Fe_3O_4 nanoparticles as a highly efficient and eco-friendly nanocatalyst for C-C bond formation. Trans Met Chem 43(5):463–472. https://doi.org/10.1007/s11243-018-0233-5
40. Favier I, Pla D, Gómez M (2018) Metal-based nanoparticles dispersed in glycerol: an efficient approach for catalysis. Catal Today 310:98–106. https://doi.org/10.1016/j.cattod.2017.06.026

41. Favier I, Pla D, Gómez M (2019) Palladium nanoparticles in polyols: synthesis, catalytic couplings and hydrogenations. Chem Rev. https://doi.org/10.1021/acs.chemrev.9b00204
42. Favier I, Toro M-L, Lecante P, Pla D, Gómez M (2018) Palladium-mediated radical homo-coupling reactions: a surface catalytic insight. Catal Sci Technol 8(18):4766–4773. https://doi.org/10.1039/C8CY00901E
43. Fiévet F, Ammar-Merah S, Brayner R, Chau F, Giraud M, Mammeri F, Viau G (2018) The polyol process: a unique method for easy access to metal nanoparticles with tailored sizes, shapes and compositions. Chem Soc Rev 47(14):5187–5233. https://doi.org/10.1039/C7CS00777A
44. Fievet F, Lagier JP, Blin B, Beaudoin B, Figlarz M (1989) Homogeneous and heterogeneous nucleations in the polyol process for the preparation of micron and submicron size metal particles. Solid State Ion 32–33:198–205. https://doi.org/10.1016/0167-2738(89)90222-1
45. Fievet F, Lagier JP, Figlarz M (2013) Preparing monodisperse metal powders in micrometer and submicrometer sizes by the polyol process. MRS Bull 14(12):29–34. https://doi.org/10.1557/S0883769400060930
46. Firouzabadi H, Iranpoor N, Gholinejad M, Hoseini J (2011) Magnetite (Fe$_3$O$_4$) nanoparticles-catalyzed sonogashira-hagihara reactions in ethylene glycol under ligand-free conditions. Adv Synth Catal 353(1):125–132. https://doi.org/10.1002/adsc.201000390
47. Firouzabadi H, Iranpoor N, Gholinejad M, Hoseini J (2011b) Magnetite (Fe$_3$O$_4$) nanoparticles-catalyzed Sonogashira-Hagihara reactions in ethylene glycol under ligand-free conditions. [Erratum to document cited in CA154:310261]. Adv Synth Catal 353(7):1027. https://doi.org/10.1002/adsc.201190017
48. Formenti D, Ferretti F, Scharnagl FK, Beller M (2019) Reduction of nitro compounds using 3d-non-noble metal catalysts. Chem Rev 119(4):2611–2680. https://doi.org/10.1021/acs.chemrev.8b00547
49. Frindy S, El Kadib A, Lahcini M, Primo A, Garcia H (2015) Copper nanoparticles stabilized in a porous chitosan aerogel as a heterogeneous catalyst for C-S cross-coupling. ChemCatChem 7(20):3307–3315. https://doi.org/10.1002/cctc.201500565
50. Gandeepan P, Müller T, Zell D, Cera G, Warratz S, Ackermann L (2019) 3d transition metals for C-H activation. Chem Rev 119(4):2192–2452. https://doi.org/10.1021/acs.chemrev.8b00507
51. Ganem B (2009) Strategies for innovation in multicomponent reaction design. Acc Chem Res 42(3):463–472. https://doi.org/10.1021/ar800214s
52. Gawande MB, Goswami A, Felpin F-X, Asefa T, Huang X, Silva R, Varma RS (2016) Cu and Cu-based nanoparticles: synthesis and applications in catalysis. Chem Rev 116(6):3722–3811. https://doi.org/10.1021/acs.chemrev.5b00482
53. Ghiasbeigi E, Soleiman-Beigi M (2019) Copper immobilized on isonicotinic acid hydrazide functionalized nano-magnetite as a novel recyclable catalyst for direct synthesis of phenols and anilines. Chem Select 4(12):3611–3619. https://doi.org/10.1002/slct.201803770
54. Gholinejad M, Zareh F, Najera C (2018) Nitro group reduction and Suzuki reaction catalysed by palladium supported on magnetic nanoparticles modified with carbon quantum dots generated from glycerol and urea. Appl Organomet Chem 32(1):e3984. https://doi.org/10.1002/aoc.3984
55. Ghorbani-Choghamarani A, Tahmasbi B, Moradi P (2016) Palladium-S-propyl-2-aminobenzothioate immobilized on Fe3O4 magnetic nanoparticles as catalyst for Suzuki and Heck reactions in water or poly(ethylene glycol). Appl Organomet Chem 30(6):422–430. https://doi.org/10.1002/aoc.3449
56. Ghosh Chaudhuri R, Paria S (2012) Core/shell nanoparticles: classes, properties, synthesis mechanisms, characterization, and applications. Chem Rev 112(4):2373–2433. https://doi.org/10.1021/cr100449n
57. Gilman H, Lichtenwalter M (1939) Relative reactivities of organometallic compounds. XXV. Coupling reaction with halides of group VIII metals. J Am Chem Soc 61(4):957–959. https://doi.org/10.1021/ja01873a056
58. Grasso S, Micale N, Monforte A-M, Monforte P, Polimeni S, Zappalà M (2000) Synthesis and in vitro antitumour activity evaluation of 1-aryl-1H,3H-thiazolo[4,3-b]quinazolines. Eur J Med Chem 35(12):1115–1119. https://doi.org/10.1016/S0223-5234(00)01195-8

59. Haji M (2016) Multicomponent reactions: a simple and efficient route to heterocyclic phosphonates. Beilstein J Org Chem 12:1269–1301. https://doi.org/10.3762/bjoc.12.121

60. Hajipour AR, Abolfathi P (2017) Novel triazole-modified chitosan@nickel nanoparticles: efficient and recoverable catalysts for Suzuki reaction. New J Chem 41(6):2386–2391. https://doi.org/10.1039/C6NJ03789E

61. Hajipour AR, Rezaei F, Khorsandi Z (2017) Pd/Cu-free Heck and Sonogashira cross-coupling reaction by Co nanoparticles immobilized on magnetic chitosan as reusable catalyst. Green Chem 19(5):1353–1361. https://doi.org/10.1039/C6GC03377F

62. Handa S, Slack ED, Lipshutz BH (2015) Nanonickel-catalyzed Suzuki–Miyaura cross-couplings in water. Angew Chem Int 54(41):11994–11998. https://doi.org/10.1002/anie.201505136

63. Heiz U, Landman U (eds) (2007) Nanocatalysis. Springer, Berlin

64. Heravi MM, Zadsirjan V, Hajiabbasi P, Hamidi H (2019) Advances in Kumada–Tamao–Corriu cross-coupling reaction: an update. Monatsh Chem 150(4):535–591. https://doi.org/10.1007/s00706-019-2364-6

65. Irrgang T, Kempe R (2019) 3d-metal catalyzed N- and C-alkylation reactions via borrowing hydrogen or hydrogen autotransfer. Chem Rev 119(4):2524–2549. https://doi.org/10.1021/acs.chemrev.8b00306

66. Joule JA, Mills K (2013) Heterocyclic chemistry at a glance, 2nd Edition. John Wiley & sons, Sussex

67. Kataria M, Pramanik S, Kaur N, Kumar M, Bhalla V (2016) Ferromagnetic α-Fe₂O₃ NPs: a potential catalyst in Sonogashira-Hagihara cross coupling and hetero-Diels–Alder reactions. Green Chem 18(6):1495–1505. https://doi.org/10.1039/C5GC02337H

68. Kaushik M, Moores A (2017) New trends in sustainable nanocatalysis: emerging use of earth abundant metals. Curr Opin Green Sustain Chem 7:39–45. https://doi.org/10.1016/j.cogsc.2017.07.002

69. Kerton F, Marriott R (2013) Alternative solvents for green chemistry, Edition 2. RSC Publishing

70. Kharasch MS, Fields EK (1941) Factors determining the course and mechanisms of Grignard reactions. IV. The effect of metallic halides on the reaction of aryl grignard reagents and organic halides 1. J Am Chem Soc 63(9):2316–2320. https://doi.org/10.1021/ja01854a006

71. Khurana JM, Vij K (2010) Nickel nanoparticles catalyzed Knoevenagel condensation of aromatic aldehydes with barbituric acids and 2-thiobarbituric acids. Catal Lett 138(1):104–110. https://doi.org/10.1007/s10562-010-0376-2

72. Khurana JM, Vij K (2011) Nickel nanoparticles catalyzed chemoselective Knoevenagel condensation of Meldrum's acid and tandem enol lactonizations via cascade cyclization sequence. Tetrahedron Lett 52(28):3666–3669. https://doi.org/10.1016/j.tetlet.2011.05.032

73. Khurana JM, Yadav S (2012) Highly monodispersed PEG-stabilized Ni nanoparticles: proficient catalyst for the synthesis of biologically important spiropyrans. Aust J Chem 65(3):314–319. https://doi.org/10.1071/CH11444

74. Kidwai M, Mishra NK, Bhardwaj S, Jahan A, Kumar A, Mozumdar S (2010) Cu nanoparticles in PEG: a new recyclable catalytic system for N-arylation of amines with aryl halides. ChemCatChem 2(10):1312–1317. https://doi.org/10.1002/cctc.201000062

75. Klajn R (2014) Spiropyran-based dynamic materials. Chem Soc Rev 43(1):148–184. https://doi.org/10.1039/C3CS60181A

76. Kotovshchikov YN, Latyshev GV, Lukashev NV, Beletskaya IP (2015) Alkynylation of steroids via Pd-free Sonogashira coupling. Org Biomol Chem 13(19):5542–5555. https://doi.org/10.1039/C5OB00559K

77. Kunz K, Scholz U, Ganzer D (2003) Renaissance of Ullmann and Goldberg reactions—progress in copper catalyzed C-N-C-O- and C-S-coupling. Synlett 2003(15):2428–2439. https://doi.org/10.1055/s-2003-42473

78. LaGrow AP, Ingham B, Cheong S, Williams GVM, Dotzler C, Toney MF, Tilley RD (2012) Synthesis, alignment, and magnetic properties of monodisperse nickel nanocubes. J Am Chem Soc 134(2):855–858. https://doi.org/10.1021/ja210209r

79. Langeslay RR, Kaphan DM, Marshall CL, Stair PC, Sattelberger AP, Delferro M (2019) Catalytic applications of vanadium: a mechanistic perspective. Chem Rev 119(4):2128–2191. https://doi.org/10.1021/acs.chemrev.8b00245

80. Liori AA, Stamatopoulos IK, Papastavrou AT, Pinaka A, Vougioukalakis GC (2018) A sustainable, user-friendly protocol for the Pd-free sonogashira coupling reaction. Eur J Org Chem 2018(44):6134–6139. https://doi.org/10.1002/ejoc.201800827

81. Liu Q, Xu M, Zhao J, Yang Z, Qi C, Zeng M, Wang B (2018) Microstructure and catalytic performances of chitosan intercalated montmorillonite supported palladium (0) and copper (II) catalysts for Sonogashira reactions. Int J Biol Macromol 113:1308–1315. https://doi.org/10.1016/j.ijbiomac.2018.03.066

82. Liu X, Zhao X, Lu M (2015) A highly water-dispersible and magnetically separable palladium catalyst based on functionalized poly(ethylene glycol)-supported iminophosphine for Suzuki-Miyaura coupling in water. Appl Organomet Chem 29(6):419–424. https://doi.org/10.1002/aoc.3308

83. Lukyanov BS, Lukyanova MB (2005) Spiropyrans: synthesis, properties, and application (review). Chem Heterocycl Compd 41(3):281–311. https://doi.org/10.1007/s10593-005-0148-x

84. Maleki A, Aghaei M, Ghamari N (2015) Synthesis of benzimidazolo[2,3-b]quinazolinone derivatives via a one-pot multicomponent reaction promoted by a chitosan-based composite magnetic nanocatalyst. Chem Lett 44(3):259–261. https://doi.org/10.1246/cl.141074

85. Maleki A, Jafari AA, Yousefi S, Eskandarpour V (2015) An efficient protocol for the one-pot multicomponent synthesis of polysubstituted pyridines by using a biopolymer-based magnetic nanocomposite. C R Chim 18(12):1307–1312. https://doi.org/10.1016/j.crci.2015.09.002

86. Maleki A, Kamalzare M (2014) An efficient synthesis of benzodiazepine derivatives via a one-pot, three-component reaction accelerated by a chitosan-supported superparamagnetic iron oxide nanocomposite. Tetrahedron Lett 55(50):6931–6934. https://doi.org/10.1016/j.tetlet.2014.10.120

87. Maleki A, Niksefat M, Rahimi J, Taheri-Ledari R (2019) Multicomponent synthesis of pyrano[2,3-d]pyrimidine derivatives via a direct one-pot strategy executed by novel designed cooperated Fe3O4@polyvinyl alcohol magnetic nanoparticles. Mater Today Chem 13:110–120. https://doi.org/10.1016/j.mtchem.2019.05.001

88. Maleki A, Zand P, Mohseni Z, Firouzi-Haji R (2018) Green composite nanostructure (Fe$_3$O$_4$@PEG-SO$_3$H): Preparation, characterization and catalytic performance in the efficient synthesis of β-amino carbonyl compounds at room temperature. Nano-Struct Nano-Objects 16:31–37. https://doi.org/10.1016/j.nanoso.2018.03.012

89. Martínez-Prieto LM, Chaudret B (2018) Organometallic ruthenium nanoparticles: synthesis, surface chemistry, and insights into ligand coordination. Acc Chem Res 51(2):376–384. https://doi.org/10.1021/acs.accounts.7b00378

90. Mei W, Wu Q (2018) Applications of Metal Nanoparticles in Medicine/Metal Nanoparticles as Anticancer Agents. In: Thota S, Crans DC (eds) Metal Nanoparticles: Synthesis and Applications in Pharmaceutical Sciences. Wiley-VCH Verlag GmbH & Co. KGaA, pp 169–190

91. Mitrofanov AY, Murashkina AV, Martín-García I, Alonso F, Beletskaya IP (2017) Formation of C-C, C–S and C–N bonds catalysed by supported copper nanoparticles. Catal Sci Technol 7(19):4401–4412. https://doi.org/10.1039/C7CY01343D

92. Moglie Y, Mascaró E, Nador F, Vitale C, Radivoy G (2008) Nanosized iron- or copper-catalyzed homocoupling of aryl, heteroaryl, benzyl, and alkenyl grignard reagents. Synth Commun 38(22):3861–3874. https://doi.org/10.1080/00397910802238726

93. Mohammadi R, Kassaee MZ (2013) Sulfochitosan encapsulated nano-Fe$_3$O$_4$ as an efficient and reusable magnetic catalyst for green synthesis of 2-amino-4H-chromen-4-yl phosphonates. J Mol Catal A: Chem 380:152–158. https://doi.org/10.1016/j.molcata.2013.09.027

94. Mondal S (2016) Recent advancement of Ullmann-type coupling reactions in the formation of C-C bond. ChemTexts 2(4):17. https://doi.org/10.1007/s40828-016-0036-2

95. Mondal P, Bhanja P, Khatun R, Bhaumik A, Das D, Manirul Islam S (2017) Palladium nanoparticles embedded on mesoporous TiO_2 material (Pd@MTiO$_2$) as an efficient heterogeneous catalyst for Suzuki-Coupling reactions in water medium. J Colloid Interface Sci 508:378–386. https://doi.org/10.1016/j.jcis.2017.08.046

96. Nakamura E, Hatakeyama T, Ito S, Ishizuka K, Ilies L, Nakamura M (2013) Iron-catalyzed cross-coupling reactions. Org React 1–210

97. Nasrollahzadeh M, Sajjadi M, Ghorbannezhad F, Sajadi SM (2018) A review on recent advances in the application of nanocatalysts in A3 coupling reactions. Chem Rec 18(10):1409–1473. https://doi.org/10.1002/tcr.201700100

98. Neiva EGC, Bergamini MF, Oliveira MM, Marcolino LH, Zarbin AJG (2014) PVP-capped nickel nanoparticles: synthesis, characterization and utilization as a glycerol electrosensor. Sens Actuator B Chem 196:574–581. https://doi.org/10.1016/j.snb.2014.02.041

99. Park J, Kang E, Son SU, Park HM, Lee MK, Kim J, Hyeon T (2005) Monodisperse nanoparticles of Ni and NiO: synthesis, characterization, self-assembled superlattices, and catalytic applications in the Suzuki coupling reaction. Adv Mater 17(4):429–434. https://doi.org/10.1002/adma.200400611

100. Patete JM, Peng X, Koenigsmann C, Xu Y, Karn B, Wong SS (2011) Viable methodologies for the synthesis of high-quality nanostructures. Green Chem 13(3):482–519. https://doi.org/10.1039/C0GC00516A

101. Peng J-B, Wu F-P, Wu X-F (2019) First-row transition-metal-catalyzed carbonylative transformations of carbon electrophiles. Chem Rev 119(4):2090–2127. https://doi.org/10.1021/acs.chemrev.8b00068

102. Pozharskii AF, Soldatenkov AT, Katritzky AR (2011) Heterocycles in industry and technology. In: Pozharskii AF, Soldatenkov AT, Katritzky AR (eds) Heterocycles in life and society, pp 209–246

103. Prabhu RN, Lakshmipraba J (2017) A nickel(II) thiosemicarbazonato complex: synthesis, structure, electrochemistry, and application in catalytic coupling of terminal alkynes with arylboronic acids. Transit Metal Chem 42(7):579–585. https://doi.org/10.1007/s11243-017-0162-8

104. Prabhu RN, Ramesh R (2016) Square-planar Ni(II) thiosemicarbazonato complex as an easily accessible and convenient catalyst for Sonogashira cross-coupling reaction. Tetrahedron Lett 57(44):4893–4897. https://doi.org/10.1016/j.tetlet.2016.09.049

105. Prechtl MHG (ed) (2017) Nanocatalysis in ionic liquids. Wiley-VCH, Weinheim

106. Reina A, Favier I, Pradel C, Gómez M (2018) Stable Zero-Valent nickel nanoparticles in glycerol: synthesis and applications in selective hydrogenations. Adv Synth Catal 360(18):3544–3552. https://doi.org/10.1002/adsc.201800786

107. Roucoux A, Schulz J, Patin H (2002) Reduced transition metal colloids: a novel family of reusable catalysts? Chem Rev 102(10):3757–3778. https://doi.org/10.1021/cr010350j

108. Safari J, Javadian L (2015) Ultrasound assisted green synthesis of 2-amino-4H-chromene derivatives catalyzed by Fe_3O_4-functionalized nanoparticles with chitosan as a novel and reusable magnetic catalyst. Ultrason Sonochem 22:341–348. https://doi.org/10.1016/j.ultsonch.2014.02.002

109. Sambiagio C, Marsden SP, Blacker AJ, McGowan PC (2014) Copper catalysed Ullmann type chemistry: from mechanistic aspects to modern development. Chem Soc Rev 43(10):3525–3550. https://doi.org/10.1039/C3CS60289C

110. San K, Shon Y-S (2018) Synthesis of Alkanethiolate-capped Metal nanoparticles using alkyl thiosulfate ligand precursors: a method to generate promising reagents for selective catalysis. Nanomaterials 8(5):346. https://doi.org/10.3390/nano8050346

111. Schmid G (ed) (2011) Nanoparticles: from theory to application. Wiley-VCH, Weinheim

112. Serp P, Philippot K (eds) (2012) Nanomaterials in Catalysis. Wiley-VCH, Weinheim

113. Shaabani A, Afshari R (2018) Magnetic Ugi-functionalized graphene oxide complexed with copper nanoparticles: Efficient catalyst toward Ullman coupling reaction in deep eutectic solvents. J Colloid Interface Sci 510:384–394. https://doi.org/10.1016/j.jcis.2017.09.089

114. Shah AP, Sharma AS, Jain S, Shimpi NG (2018) Microwave assisted one pot three component synthesis of propargylamines, tetra substituted propargylamines and pyrrolo[1,2-a]quinolines using CuNPs@ZnO-PTh as a heterogeneous catalyst. New J Chem 42(11):8724–8737. https://doi.org/10.1039/C8NJ00410B

115. Sharghi H, Aberi M (2014) Ligand-free copper(I) oxide nanoparticle catalyzed three-component synthesis of 2H-indazole derivatives from 2-halobenzaldehydes, amines and sodium azide in polyethylene glycol as a green solvent. Synlett 25(8):1111–1115. https://doi.org/10.1055/s-0033-1340979

116. Song J-Y, Zhou X, Song H, Liu Y, Zhao H-Y, Sun Z-Z, Chu W-Y (2018) Visible-light-assisted cobalt-2-(hydroxyimino)-1-phenylpropan-1-one complex catalyzed Pd/Cu-free Sonogashira-Hagihara cross-coupling reaction. ChemCatChem 10(4):758–762. https://doi.org/10.1002/cctc.201701253

117. Sonogashira K (2002) Development of Pd–Cu catalyzed cross-coupling of terminal acetylenes with sp2-carbon halides. J Organomet Chem 653(1):46–49. https://doi.org/10.1016/S0022-328X(02)01158-0

118. Swamy KCK, Kumar NNB, Balaraman E, Kumar KVPP (2009) Mitsunobu and related reactions: advances and applications. Chem Rev 109(6):2551–2651. https://doi.org/10.1021/cr800278z

119. Tamao K, Sumitani K, Kumada M (1972) Selective carbon-carbon bond formation by cross-coupling of Grignard reagents with organic halides. Catalysis by nickel-phosphine complexes. J Am Chem Soc 94(12):4374–4376. https://doi.org/10.1021/ja00767a075

120. Thomas AA, Denmark SE (2016) Pre-transmetalation intermediates in the Suzuki-Miyaura reaction revealed: the missing link. Science 352(6283):329–332. https://doi.org/10.1126/science.aad6981

121. Thomas AM, Sujatha A, Anilkumar G (2014) Recent advances and perspectives in copper-catalyzed Sonogashira coupling reactions. RSC Adv 4(42):21688–21698. https://doi.org/10.1039/C4RA02529F

122. Torimoto T, Kameyana T, Kuwabata S (2017) Top-down synthesis methods for nanoscale catalysts. In: Prechtl MHG (ed) Nanocatalysis in ionic liquids. Wiley-VCH, Weinheim, pp 171–205

123. Trammell R, Rajabimoghadam K, Garcia-Bosch I (2019) Copper-promoted functionalization of organic molecules: from biologically relevant Cu/O₂ model systems to organometallic transformations. Chem Rev 119(4):2954–3031. https://doi.org/10.1021/acs.chemrev.8b00368

124. Tsuji J (2004) Palladium reagents and catalysts: new perspectives for the 21st century. Wiley, West Sussex

125. Ullmann F, Bielecki J (1901) Ueber synthesen in der biphenylreihe. Ber Dtsch Chem Ges 34(2):2174–2185. https://doi.org/10.1002/cber.190103402141

126. Veisi H, Najafi S, Hemmati S (2018) Pd(II)/Pd(0) anchored to magnetic nanoparticles (Fe₃O₄) modified with biguanidine-chitosan polymer as a novel nanocatalyst for Suzuki-Miyaura coupling reactions. Int J Biol Macromol 113:186–194. https://doi.org/10.1016/j.ijbiomac.2018.02.120

127. Wang D, Astruc D (2017) The recent development of efficient Earth-abundant transition-metal nanocatalysts. Chem Soc Rev 46(3):816–854. https://doi.org/10.1039/C6CS00629A

128. Warner JC, Cannon AS, Dye KM (2004) Green chemistry. Environ Impact Assess Rev 24(7):775–799. https://doi.org/10.1016/j.eiar.2004.06.006

129. Wei D, Darcel C (2019) Iron catalysis in reduction and hydrometalation reactions. Chem Rev 119(4):2550–2610. https://doi.org/10.1021/acs.chemrev.8b00372

130. Wender PA (2014) Toward the ideal synthesis and molecular function through synthesis-informed design. Nat Prod Rep 31(4):433–440. https://doi.org/10.1039/C4NP00013G

131. Xia H, Xie K, Zou G (2017) Advances in spiropyrans/spirooxazines and applications based on fluorescence resonance energy transfer (FRET) with fluorescent materials. Molecules 22(12):2236

132. Yavuz K, Kuecuekbay H (2018) Efficient and green catalytic system incorporating new benzimidazolium salts for the Sonogashira cross-coupling reaction. Appl Organomet Chem 32(1):e3897. https://doi.org/10.1002/aoc.3897

133. Yonezawa T (2018) Application 78—preparation of metal nanoparticles and their application for materials. In: Naito M, Yokoyama T, Hosokawa K, Nogi K (eds) Nanoparticle technology handbook, 3rd edn. Elsevier, Amsterdam, pp 829–837

134. Zhang Y, Erkey C (2006) Preparation of supported metallic nanoparticles using supercritical fluids: a review. J Supercrit Fluids 38(2):252–267. https://doi.org/10.1016/j.supflu.2006.03.021

135. Zhong Y, Gong X, Zhu X, Ni Z, Wang H, Fu J, Han W (2014) In situ generated nickel nanoparticle-catalyzed carbonylative Suzuki reactions of aryl iodides with arylboronic acids at ambient CO pressure in poly(ethylene glycol). RSC Adv 4(108):63216–63220. https://doi.org/10.1039/C4RA10739J

136. Zhou L, Gao C, Xu W (2010) Robust Fe_3O_4/SiO_2-Pt/Au/Pd magnetic nanocatalysts with multifunctional hyperbranched polyglycerol amplifiers. Langmuir 26(13):11217–11225. https://doi.org/10.1021/la100556p

137. Zhou B, Hermans S, Somorjai GA (eds) (2004) Nanotechnology in catalysis. Kluwer/Plenum, New York

138. Zhu J, Bienaymé H (eds) (2006) Multicomponent reactions. Wiley, Weinheim

Chapter 9
Metal Nanoparticles for Hydrogen Isotope Exchange

A. Palazzolo, J. M. Asensio, D. Bouzouita, G. Pieters, S. Tricard, and B. Chaudret

Abstract Since the mid-1990s Hydrogen Isotope Exchange (HIE), consisting in the direct exchange of protium with its isotopes, has witnessed an enormous development (Atzrodt et al. in Angew Chem Int Ed, 57:3022–3047, 2018, [1], Atzrodt in Angew Chem Int Ed, 46:7744–7765, 2007, [2]). HIE reactions can nowadays be performed on a plethora of organic compounds by using both homogeneous and heterogeneous catalysis. Molecular catalysts remain the most commonly used due to their high reliability (Atzrodt et al. in Angew Chem Int Ed, 57:3022–3047, 2018, [1]). However, metallic nanoparticles have started attracting the attention of the scientific community (Asensio et al. in Chem Rev, 120:1042–1084, 2020, [3]) because of their interesting characteristics such as:

1. their reactivity in between homogeneous and heterogeneous catalysts,
2. the possibility to deeply influence their chemical properties by varying the stabilizing agent,
3. the non-negligible advantages of (generally) simple workup procedures.

In this chapter, we will give an overview of the recent advances in HIE. First, we will describe the main applications of protium isotopes. Then, we will briefly discuss the main advances in catalytic HIE reactions in both homogeneous and heterogeneous phase. Finally, we will summarize the examples of HIE catalyzed by metallic nanoparticles that have been described in the literature.

Keywords Hydrogen Isotope Exchange · Catalytic deuteration · C-H activation · H-H activation · Metal nanoparticles

A. Palazzolo · G. Pieters
SCBM, CEA, Université Paris Saclay, F-91191 Gif-sur-Yvette, France

J. M. Asensio (✉) · D. Bouzouita · S. Tricard · B. Chaudret
LPCNO, INSA, CNRS, Université de Toulouse, 135, Avenue de Rangueil, F-31077 Toulouse, France
e-mail: asensior@insa-toulouse.fr

© Springer Nature Switzerland AG 2020
P. W. N. M. van Leeuwen and C. Claver (eds.), *Recent Advances in Nanoparticle Catalysis*,
Molecular Catalysis 1, https://doi.org/10.1007/978-3-030-45823-2_9

9.1 Utilization of Protium Isotopes

Deuterium and tritium are nowadays employed in several research fields spanning drug development processes, material chemistry and fundamental mechanistic investigations [4]. Deuterium is a stable isotope of hydrogen that contains one proton, one electron and one neutron; thus, it is stated as both ^2H and D (IUPAC). Deuterium was discovered in 1931 by Urey et al. [5]. Its name was derived from the Greek *deuteros*, namely second [6]. Tritium is a radioactive isotope of hydrogen that contains one proton, one electron and two neutrons. Thus, it is stated as both ^3H and T (IUPAC). Firstly produced by Oliphant, Harteck and Rutherford bombarding deuterated inorganic salts with deuterium ions, [7] tritium has a half-life of 12.32 years and decays to ^3He emitting low-energy β^- particles [8]. One of the most important characteristics of hydrogen isotopes is their strong kinetic isotope effect (KIE). Primary KIE is defined as "the ratio between the kinetic constants of a chemical transformation when one of the atoms of a reactant is substituted by one of its isotopes" [9]. This phenomenon can be explained considering that "heavier" isotopes possess lower vibrational frequency and zero-point energy (ZPE) [10–12]. Although this explanation does not consider the influence of tunneling, it can be used as a valid approximation to explain several experimental observations. In the case of C-D and C-H bonds, there is an important energy difference at the ZPE which becomes almost inexistent at the transition state; such a difference is experimentally translated in a very important primary KIE (Fig. 9.1).

Moreover, different types of secondary KIE can emerge when a C-D bond breaking is not directly involved in the chemical transformation. Secondary KIEs often arise upon changing of hybridization or through the involvement of hyperconjugation. Nevertheless, they are much smaller in magnitude compared to primary KIE. The

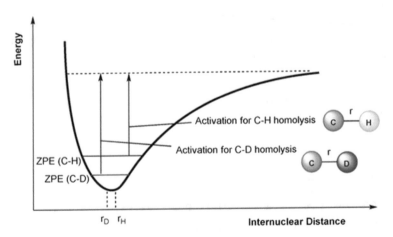

Fig. 9.1 Energetic profile for C-H and C-D bonds showing the difference in ZPE. Reprinted with permission from Ref. [4]. Copyright 2018 Wiley [4]

kinetic constant of a reaction at 25 °C that involves a C-H bond breaking is, theoret-
ically, around 6.5 times faster than the same reaction involving a C-D bond breaking
(Eq. 9.1).

$$\frac{k_H}{k_D} = e^{\frac{hc(v_H - v_D)}{2kT}} \tag{9.1}$$

Equation **9.1**, $k_H - k_D$ kinetic constant for C-H and C-D bonds, h Planck's
constant, c speed of the light, $v_H - v_D$ elongation frequencies for C-H and C-D
bonds, k Boltzmann's constant, T temperature.

One of the most classical (or common) applications of deuterated compounds is
their use as tools to understand reaction mechanisms. By observing the magnitudes
of isotope effects, it is possible to deduce which site might contribute to the chemical
mechanism of a reaction. Therefore, KIE can be used for:

1. determining the absolute rates in two parallel reactions,
2. distinguishing two chemical processes in competition experiments, when
   "labeled" and "unlabeled" compounds are placed in the same flask,
3. assessing the reactivity of two different C-H bonds within the same molecule.

These techniques are used to study inter/intramolecular organic transformations
as well as enzymatic reactions [13–15] and organometallic processes. Additionally,
deuterium can be incorporated into metabolized sites of bioactive compounds in order
to alter their pharmacokinetic profiles [16–19]. Metabolic inactivation of drugs is one
of the main reasons of discard during drug development processes, because it can
potentially lead to the production of toxic metabolites, to the inactivation of the drug
or to too low concentrations in blood. On the other hand, deuterated drugs may possess
enhanced pharmacokinetic properties thanks to lower metabolic rates. Recently, FDA
has allowed the first deuterated drug to reach the market [20]. Tetrabenazine is a
vesicular monoamine transporter 2 (VMAT2) inhibitor, and it is used for the treatment
of chorea associated with Huntington disease. The active metabolite, issue from the
reduction of the carbonyl moiety, is rapidly oxidized on the cathecolic methoxy
groups and thus excreted. The deuteration of these positions increases the half-life
of tetrabenazine and permits reduction of the daily dose (Fig. 9.2).

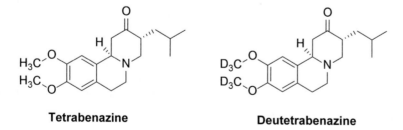

Tetrabenazine   **Deutetrabenazine**

Fig. 9.2 Chemical structures of tetrabenazine and deutetrabenazine

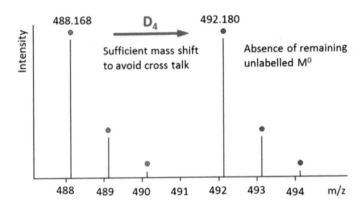

Fig. 9.3 Desired mass shift of deuterated internal standard to avoid signal overlapping [4]. Reprinted with permission from Ref. [4]. Copyright 2018 Wiley

Deuterated molecules are also routinely applied as internal standards for quantitative LC-MS/MS analysis. The precise quantification of a substance using MS may be difficult because of many factors (matrix effects, ion suppression). For this reason, the quantitative analysis of molecules in a complex matrix usually requires an internal standardization involving the use of stable isotope labeled internal standards (SILSs) [21−22]. The latter are particularly advantageous because they display the same chemical and physical properties of the analyte, but they possess different molecular weights. To be used as internal standard, a deuterated sample should have ideally the following specifications:

1. containing a negligible amount of unlabeled molecule (less than 1%),
2. possessing a deuterium content of 3–5 atoms to avoid signal superimposition (Fig. 9.3),
3. displaying an isotopic distribution as narrow as possible to increase the accuracy of the measurement.

In a common procedure, a known quantity of SILS is added to the biological sample containing the compound to be quantified. After purification, it is possible to calculate the initial quantity of the desired molecule by comparing the peaks of "labeled" standard and "unlabeled" molecule in the MS spectrum. SILS is also used to assess drug-drug interactions, to detect and quantify illegal drugs, [23−24] for anti-doping tests [25−26] and to test the presence of a variety of contaminants [27−28].

On the other hand, tritium-labeled compounds are widely applied during early drug development processes. For example, tritiated molecules are used in radioligand binding assays, which measure the interaction between two molecules, generally a ligand and a target. Such a use is due to two main factors:

1. tritium's high specific activity,
2. the fact that tritium labeling does not modify the interaction between the molecule of interest and its target (compared to other probing techniques).

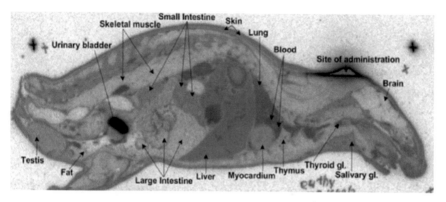

Fig. 9.4 Quantitative whole-body autoradiography (QWBA) showing the site of administration of a radioactive drug [4]. Reprinted with permission from Ref. [4]. Copyright 2018 Wiley

Receptor affinity can be easily measured by saturation experiments which provide the half maximal inhibitory concentration (IC_{50}) of the unlabeled molecule [29]. Tritiated compounds can also be used in later phases of drug development process such as adsorption, distribution, metabolism and excretion (ADME) studies. Among them, ex vivo imaging techniques such as whole-body autoradiography (WBA) can be used to determine the in situ localization of radiolabeled xenobiotics in laboratory animals (Fig. 9.4) [30].

9.2   Current Approach to the Synthesis of Deuterated and Tritiated Molecules

The synthesis of molecules labeled with hydrogen isotopes can be achieved throughout two pathways. First, one can envisage the use of isotopically labeled building blocks for the synthesis of the desired molecules. This approach is the most reliable one because of the high control on the site of the isotope incorporation and the high level of isotopic enrichment achievable. However, the necessity of reworking a synthetic pathway in order to incorporate a labeled fragment may be difficult, especially for complicated compounds such as biomolecules. Moreover, in the case of tritiations, high amounts of radioactive wastes are produced. In this context, direct exchange of hydrogen with its isotopes represents a fascinating alternative to classical synthesis. In fact, Hydrogen Isotope Exchange (HIE) is easier, cheaper and leads to less radioactive wastes because it can be done directly on the compound of interest (late stage functionalization) [31]. On the other hand, HIE methods are still under development, and not all functionalities are nowadays compatible. In the example reported in Fig. 9.5, [32] deuterated paroxetine was prepared through a five-step synthesis with an overall yield of 20%. The same molecule could be deuterated (on

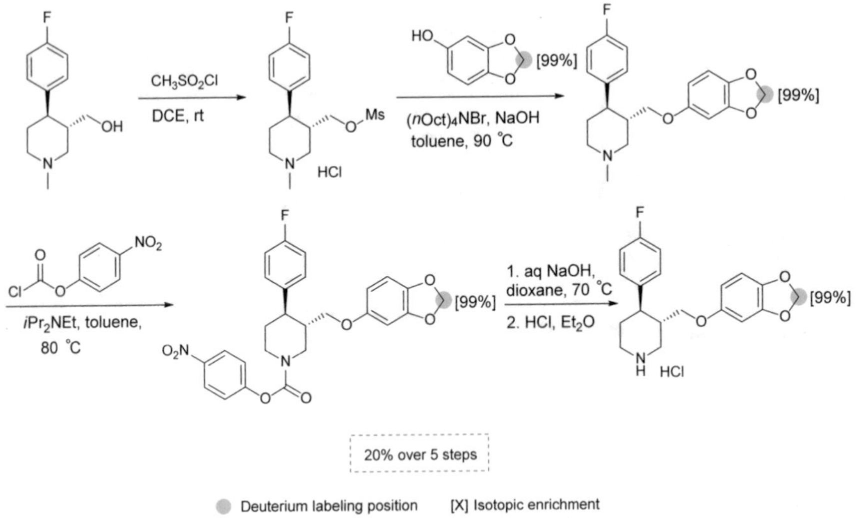

Deuterium labeling position   [X] Isotopic enrichment

Fig. 9.5 Example of the synthesis of deuterated paroxetine [32]

other positions) with high isotopic enrichments in one step using HIE catalyzed by RuNPs (see—Fig. 9.17).

As stated in the introduction, molecular catalysts such as Crabtree's type iridium complexes are among the most widely employed systems for catalytic HIE reactions. Initially used as hydrogenation catalysts, [33] Ir(I) complexes were firstly explored in the context of HIE by Heys [34] in 1992. In this work, the labeling of different compounds containing functional groups such as amides or esters was examined. The hypothesized reaction mechanism envisages the initial activation of the Ir(I) to Ir(III) upon the oxidative addition of D_2 and further release of the cyclooctadiene ligand (Fig. 9.6). The rate-limiting step is the C-H activation which possesses a high energy barrier and leads to the formation of a metallacycle intermediate (**9-10**).

The second generation of iridium catalysts started in 2008, when Kerr reported a bulky cationic Ir(I) pre-catalyst able to promote ortho-HIE to different directing groups (Fig. 9.7). The bulkiness as well as the electron richness of the ligands played a fundamental role avoiding the formation of inactive iridium clusters and favoring both oxidative addition and reductive elimination [35]. Extensive theoretical investigations confirmed that the rate-limiting step of the reaction is the C-H activation.

In an extension of this work, Atzrodt showed that Kerr's catalyst can be successfully applied to the ortho-directed labeling of various heterocycles such as pyrimidine, imidazole, oxazole, thiazole and their benzofused analogs [36]. Recently, several efforts have been done in the direction of C(sp³)-H activation [37]. Kerr showed that iridium-catalyzed HIEs can be easily achieved on aliphatic C-H bonds using mild conditions. Interestingly, the investigated catalyst tolerated the use of diverse DGs including pyridine, pyrimidine, quinoline, thiazole and benzothiazole.

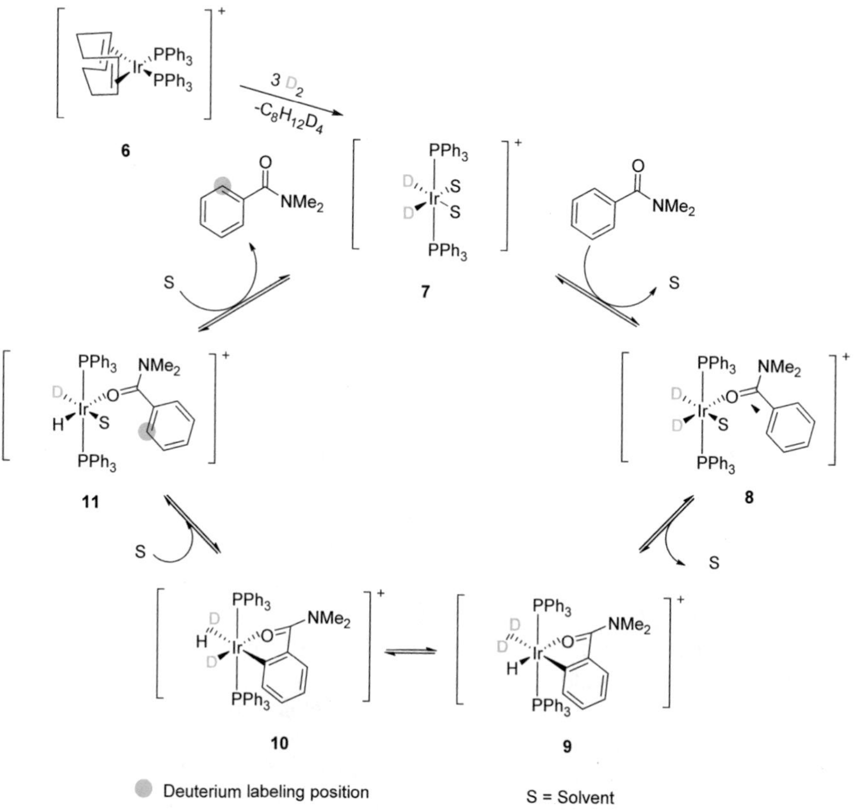

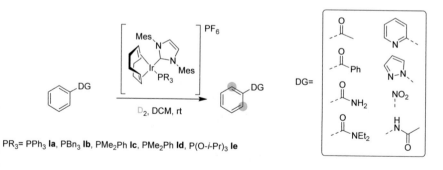

Fig. 9.6 Reaction mechanism proposed by Heys and co-workers for the Ir(III) catalyzed HIE of amides and esters [34]

Fig. 9.7 Supported directing group by the first generation of Kerr catalysts [35]

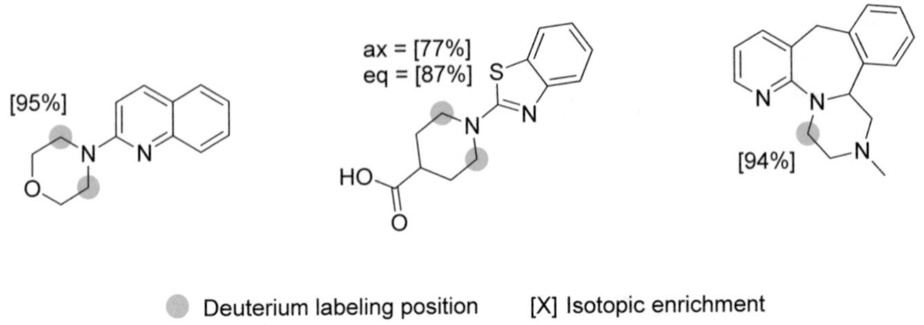

ax = [77%]
eq = [87%]

[95%]

[94%]

HO

● Deuterium labeling position   [X] Isotopic enrichment

Fig. 9.8 Selected examples from Kerr and co-workers' labeling of aliphatic amines directed by nitrogen-containing heterocycles [38]

Cyclic aliphatic moieties such as morpholines, piperidines and piperazines can be labeled under mild conditions and with very low catalytic loadings (Fig. 9.8) [38].

Although less efficient than cationic catalysts, [39] neutral iridium catalysts have also been widely studied in HIE. In 2015, Kerr described a well-designed neutral iridium(I) complex bearing a chlorine substituent which allowed the labeling of primary sulfonamides [40]. In addition, the different reactivity of neutral and less hindered Ir(I) species has been confirmed by its use in the selective labeling of highly reactive moieties such as aldehydes [41].

Along with iridium, ruthenium possesses a chemical reactivity which makes it one of the most studied metals in HIE. For example, ruthenium is efficient for the labeling of molecules that lack strongly coordinating functional groups. Leitner and co-workers reported the deuterium labeling of benzene derivatives and heteroaromatics under mild conditions using Ru(II) pincer complexes and D_2O as isotopic source (Fig. 9.9). Theoretical calculations revealed that the isotopic uptake is governed by steric effects [42]. Recently, Szymczak and co-workers reported

PtBu$_2$
H$_2$
N–Ru–H
H
PtBu$_2$

(1 mol%)

D$_2$O, hexane, 50 °C

[5%]
[5%]
[84%]
[28%]

● Deuterium labeling position   [X] Isotopic enrichment

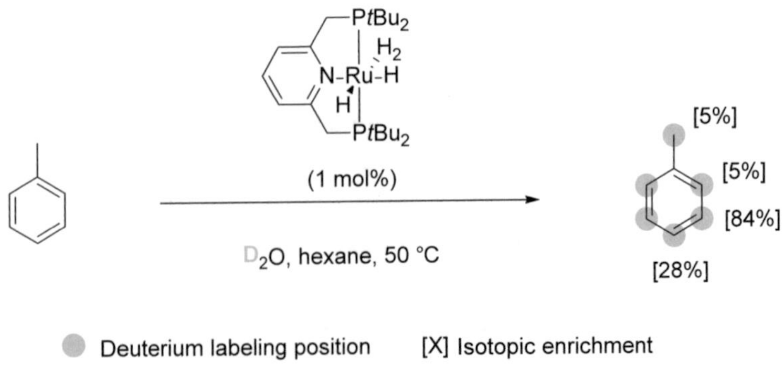

Fig. 9.9 Leitner and co-workers' deuteration of benzene derivatives in the absence of a directing group. The regioselectivity is ruled by steric effects [42]

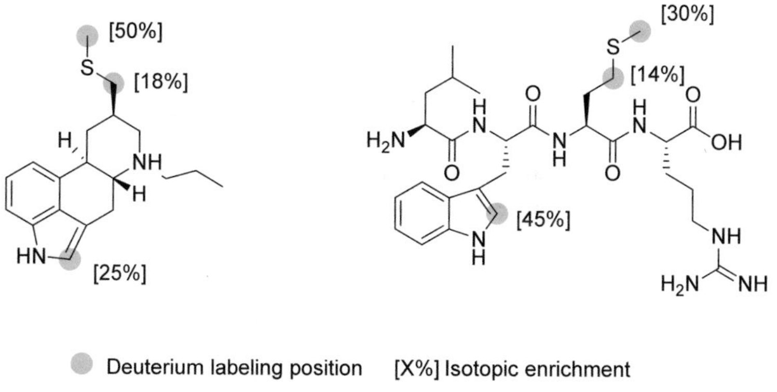

Deuterium labeling position   [X%] Isotopic enrichment

Fig. 9.10 Selected examples from Pieters et al. of deuteration of thioethers with Ru/C showing the high complexity of the substrates which can be labeled with this method [46]

H/D exchange catalyzed by an electron-poor Ru(II) cationic catalyst, which allows complete stereoretentive labeling of chiral amines using D_2O as isotopic source [43].

On the other hand, heterogeneous ruthenium catalysts are known to promote HIE on compounds containing hydroxyl groups. For instance, Sajiki reported a catalytic reaction involving Ru/C and D_2O as isotopic source under hydrogen atmosphere for the efficient labeling of linear, branched, cyclic, primary and secondary alcohols [44]. The same group reported subsequently the labeling of protected sugars with the same catalytic system [45]. Recently, Pieters and co-workers showed that Ru/C is capable of performing HIE directed by thioethers (Fig. 9.10) despite the fact that they are known to efficiently poison heterogeneous catalysts. By increasing the catalytic loading for this reaction, it was possible to label very complex molecules including peptides and drugs [46].

Palladium-based catalysts have also been widely explored in HIE chemistry. Homogeneous Pd catalysts are not much used because of the high stability of alkyl-palladated intermediates. Nevertheless, some examples of HIE catalyzed by Pd complexes have been reported. Reaction mechanisms involve the formation of cyclopalladated species, which are further hydrolyzed with acids [47] or D_2/T_2 [48]. On the other hand, heterogeneous Pd species have been largely employed in the field of HIE. In 2005, Sajiki et al. reported an interesting Pd/C catalyzed HIE protocol for the labeling of phenylalanine employing D_2O under H_2 atmosphere (1 bar). The reaction is highly selective for the benzylic position [49]. The same catalytic system also allowed the labeling of nucleobases derivatives with high isotopic enrichments but in harsh conditions (160–180 °C in a sealed tube), which limited its employability to very simple substrates [50]. Recent efforts on transition metal catalyzed C-H activation go strongly in the direction of the use of earth-abundant metals because of their lower cost and higher availability. In this field, Chirik described the first iron catalyzed synthesis of labeled drugs using D_2 or T_2 as isotopic source (Fig. 9.11). In contrast with the typical ortho-directed HIE of most transition metals, the iron catalyst activates the more electron-poor and accessible C-H bonds. Despite its high

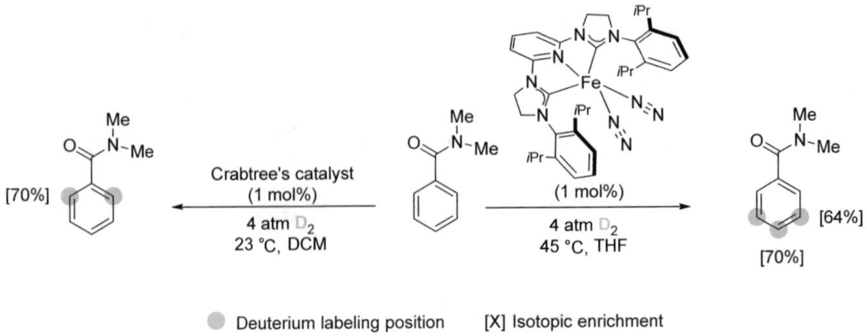

Deuterium labeling position   [X] Isotopic enrichment

Fig. 9.11 Difference in regioselectivity between Crabtree and Chirik's iron catalysts [51]

sensitivity to oxygen, moisture and protic functionalities, this catalyst allows efficient tritiation of pharmaceuticals employing very low pressures of the isotopic gas source [51].

The same group recently developed cobalt and nickel diimine complexes for the labeling of benzylic sites and various azines and diazines [52–54]. However, with nickel, the formation of nickel nanoparticles or clusters under D_2 atmosphere cannot be excluded.

9.3 Nanoparticles for HIE

Recently, metallic nanoparticles (NPs) have become a prominent tool for the activation of C-H bonds, [55] yet their use in HIE chemistry has just recently been described. The first example of H/D exchange catalyzed by metal nanoparticles was reported by Ott et al. in 2005 [56]. In this work, Ir nanoparticles with a size of 2.1 ± 0.6 nm were prepared in ionic liquids. These NPs catalyzed the H-D exchange on 1-butyl-3-methylimidazolium, not only on the imidazolium cation, but also on the alkyl chain. The high deuteration percentage at position 2 of the cycle was explained by the coordination of imidazolium on the surface of the NPs. This work opened the field to the stabilization of NPs by "N-Heterocyclic Carbene" (NHC) ligands. Later on, Sullivan et al. synthesized DMAP-stabilized Pd nanoparticles with a mean diameter of 3.4 ± 0.5 nm via reduction of Na_2PdCl_4 in water by sodium borohydride, which were used in the deuteration of pyridines (Fig. 9.12) [57]. Control experiments revealed that the starting Pd(II) was not active in H/D exchange reaction. After supporting the Pd NPs on thiol-modified carbon nanotubes, an enhancement in their catalytic activity in H/D exchange on various pyridines was observed. However, low deuteration degrees were obtained on substrate carrying hydroxyl groups. In addition, deuteration degree was higher in the ortho positions in respect of the hydroxyl group, which suggests that coordination of 4-hydroxypyridyl through the O atom to the surface of the NPs atom

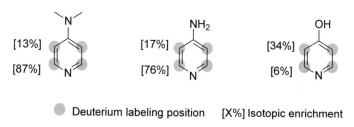

Deuterium labeling position   [X%] Isotopic enrichment

Fig. 9.12 Molecules deuterated by Sullivan et al. through PdNP catalysis [57]

took place preferentially. Shapley also reported the deuteration of various nitrogen-containing heterocycles catalyzed by Pd NPs stabilized by polyvinylpyrrolidone (PVP). The latter are prepared by thermal decomposition of $Pd(OAc)_2$ and possess a mean diameter of 4.5 nm. These catalysts were stable colloidal dispersions for months and promoted the α deuteration of various N-containing heterocycles using D_2O as isotopic source [58].

A different approach for the preparation of metallic NPs consists in the controlled decomposition of organometallic precursors under mild conditions. This "organometallic" approach has been widely employed by our research groups to synthesize Ru NPs. For example, [Ru(COD)(COT)] can be decomposed to yield Ru NPs with a mean size of 1.1 nm in the presence of a stabilizing agents such as polymeric polyvinylpyrrolidone (PVP) and hydrogen (Fig. 9.13) [59].

As this approach limits the production of surface contaminants, it grants a higher control on the surface species and a potentially higher catalytic activity. In 2005, Pery et al. investigated the presence of mobile hydrides at the surface of Ru NPs/HDA (HDA = hexadecylamine) using solid state ^2H MAS NMR. Surface hydrides were rapidly exchanged by deuterides putting the Ru NPs/HDA under deuterium gas atmosphere (1 bar). In this work, the authors observed by ^2H NMR analysis that Ru NPs promote H/D exchange at the HDA ligands at low D_2 pressures and low temperatures (Fig. 9.14). Thus, this work constituted the proof of concept that Ru NPs can be used as catalysts to perform HIE reactions under mild conditions [60].

Breso-Femenia et al. studied the deuteration of phosphorus compounds by Ru/PVP NPs and more precisely on three different phosphines: triphenylphosphine,

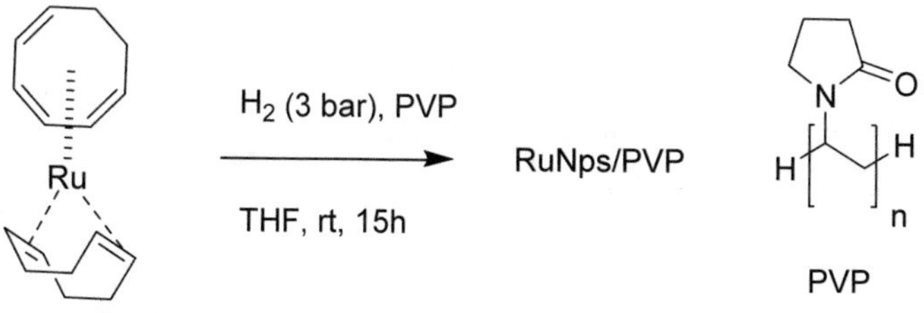

H_2 (3 bar), PVP
$\longrightarrow$ RuNps/PVP
THF, rt, 15h

PVP

Fig. 9.13 Synthesis of Ru NPs/PVP by decomposition of organometallic [Ru(COD)(COT)] [59]

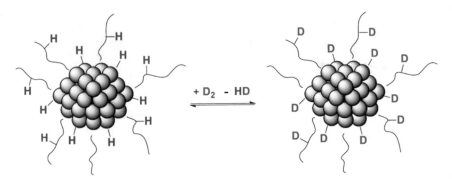

Fig. 9.14 H/D exchange promoted by RuNPs/HDA in the presence of D_2 gas [60]

triphenylphosphine oxide and triphenyl phosphite [61]. In the case of triphenylphosphine, a selective deuteration of the ortho position of the aromatic ring was observed with an incorporation of 1-6 deuterium atoms depending on the reaction time without detection of products from the aromatic ring reduction. Nevertheless, the Ru/PVP NPs were not able to deuterate the aliphatic groups of phosphines. Concerning triphenylphosphine oxide, an incorporation of deuterium was observed under the same conditions but with a reduction of the aromatic rings even at low temperatures, which can be explained by a π-coordination of the substrate through the aromatic ring. Then, for triphenyl phosphite, no deuteration took place under the same conditions. The authors proposed that the presence of O increases the distance between the surface of the nanoparticles and the aromatic ring, which disadvantages the H/D exchange (Fig. 9.15).

Later on, Pieters et al. demonstrated that Ru NPs/PVP were indeed able to deuterate diverse nitrogen-containing aromatic and aliphatic compounds using D_2 as isotopic source [62]. Thus, activation of either $C(sp^3)$- or $C(sp^2)$-H bonds next to a nitrogen atom under very mild conditions and with high regioselectivity was

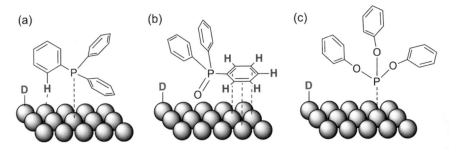

Fig. 9.15 Ru@PVP NPs catalyze H/D exchange on phosphines. Selectivity depends on the coordination mode of the ligand. **a** PPh$_3$ is deuterated in the ortho of the phenyl substituent. **b** OPPh$_3$ is not able to coordinate through the P atom. Thus, π-coordination leads to the reduction of the phenyl substituents. **c** In the case of P(OPh)$_3$, the distance between the ligand and the NPs surface inhibits the deuteration of the phosphine [3]

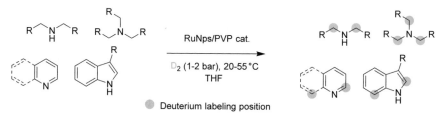

Fig. 9.16 Common nitrogen heterocyclic scaffolds which can be labeled under mild conditions using Ru NPs/PVP catalysis and deuterium gas as isotopic source [62]

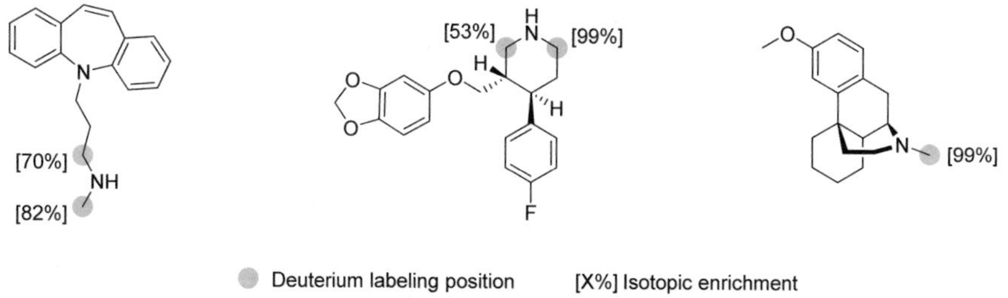

Fig. 9.17 Selected examples from the work Pieters et al. using Ru NPs/PVP and showing the high molecular complexity of the drugs successfully labeled [62]

achieved for several molecules such as pyridines, quinolines, indoles and alkyl amines (Fig. 9.16). Remarkably, RuNps/PVP catalysis granted access to a series of complex labeled drugs with high deuterium incorporation (Fig. 9.17).

Thereafter, Ru NPs/PVP were used to catalyze the deuteration of amino acids and peptides in D_2O. This work constitutes the first general method permitting stereo-retentive C-H deuteration [63]. The reaction is very regioselective to the α position of the amino group of the amino acids, and it does not require any protection of the carboxylic acid moiety. High deuterium uptake was observed for amino acids containing aliphatic, amido, amino and hydroxyl side chains, with the latter being stereoretentively labeled as well. Moreover, this method was successfully applied to biologically relevant di-, tri- and tetra-peptides (Fig. 9.18).

DFT calculations confirmed that the less energetic pathway starts with the coordination of the amine to the nanoparticle, followed by a C-H activation through oxidative addition onto a Ru surface atom [63]. The C-H bond breaking is the rate-limiting step, which proceeds via a 4-membered dimetallacycle intermediate (Fig. 9.19). The formation of this intermediate explains just partially the chiral outcome of the investigated transformation. In fact, the stereoretentivity needs also to be attributed to the fact that the H/D exchange is happening at the surface of the nanoparticle thanks to the high mobility of deuteride species. It is noteworthy to say that molecular catalysts can generally form only a monometallacycle; thus, their effectiveness in the C-H activation process strongly relies on a defined geometry. On the other hand,

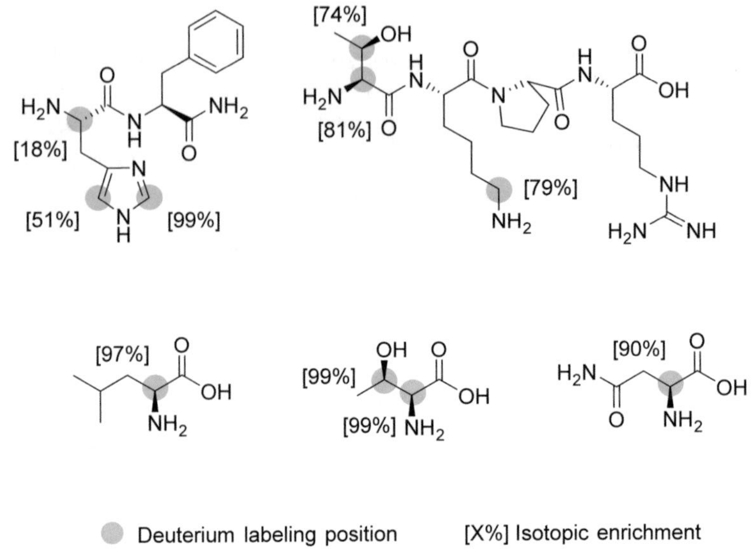

Fig. 9.18 Selected examples from Taglang et al. deuteration of aminoacids and peptides using RuNPs/PVP [63]

given the polyatomic nature of NPs surface, a larger diversity of key intermediates can be formed when they are used as catalysts for C-H bond activations.

One of the advantages of NHCs is that it is possible to play with the substituents and thus modulate their solubility. Water-soluble Ru NPs stabilized by NHC ligands functionalized with a sulfonate group were synthesized by Martínez-Prieto et al. in 2017 [64]. The selective H/D exchange on L-lysine at different pHs was studied (Fig. 9.20). At a pH of 10.4, two positions are selectively deuterated (α and ε) with low deuteration in the γ position (12.5%). The reaction efficiency decreases when reducing the pH value, with virtually no reactivity at a pH of 2.2. This loss of reactivity is understandable since under these conditions, the NH$_2$ groups are protonated, which disadvantages the coordination of the substrate on the surface of the NPs. On the contrary, at basic pH (13.4), a higher deuterium incorporation is observed. This incorporation is facilitated by the coordination of the two amine groups on the surface of the NPs and thus leads to an almost complete deuteration of positions α, β and γ (99%, 98.5% and 89.5%, respectively). The coordination of the substrate on the surface of the NPs was demonstrated by a chemical shift perturbation-nuclear magnetic resonance (CSP-NMR) study.

Ru NPs were also used by Bhatia et al. in H/D exchange of aminoacids by electrocatalysis [65]. The authors used Ru NPs supported on activated carbon (RuNPs/ACC). The described method allowed the deuteration of amines and alcohols in D$_2$O with a power supply. The platinum anode acts as a counter electrode, and the reaction takes place on the cathode made up of RuNPs/ACC. A better incorporation

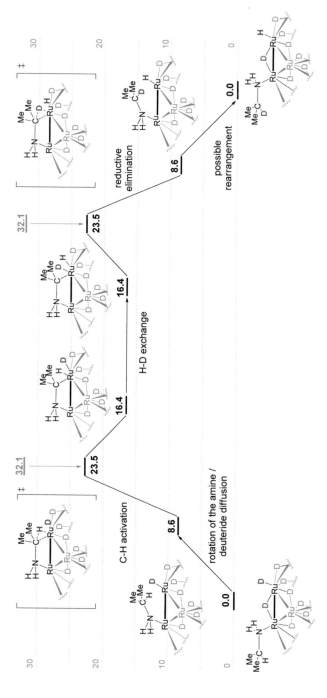

Fig. 9.19 DFT calculations showing the key 4-membered di ruthenacycle intermediates [63]

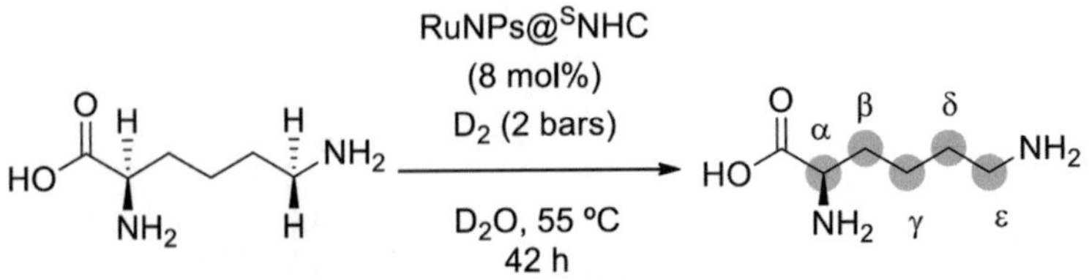

pH	α (%)	ε (%)	γ (%)	β (%)	δ (%)
2.2	6	2	-	-	-
6.9	95	70	-	-	-
8.4	97	92	-	-	-
10.4	99	98.5	12.5	-	-
11.0	99	98.5	45	-	-
13.2	99	98.5	89.5	10	-
13.8	76	98	31.5	-	-

Fig. 9.20 Activities of RuNPs@NHC on enantiospecific deuteration of L-lysine as a function of Ph [64]

of deuterium was observed on alcohols at the OH group's α position compared to amines. A mechanism similar to that proposed by Taglang et al. was presented.

The influence of introduction of Pt atoms at the surface of Ru NPs was studied by Bouzouita et al. [66]. In this work, the authors prepared several RuPt NPs (Fig. 9.21) and studied the influence of the Pt precursor on the surface composition and therefore on the catalytic reactivity of the NPs. The exchange of Ru atoms by Pt led to a decrease in the reaction rate of the H/D exchange in the α position of L-lysine

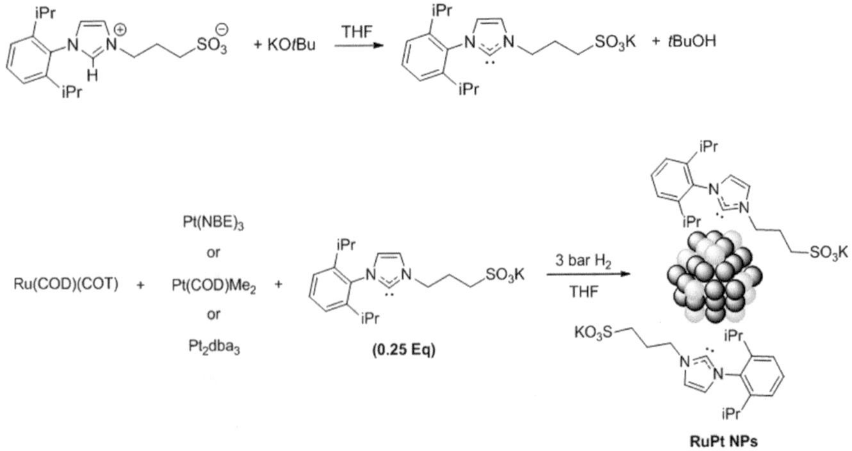

Fig. 9.21 Water-soluble Ru-Pt nanoparticles synthesis [66]. Reprinted with permission from Ref. [66]. Copyright 2019 Royal Society of Chemistry

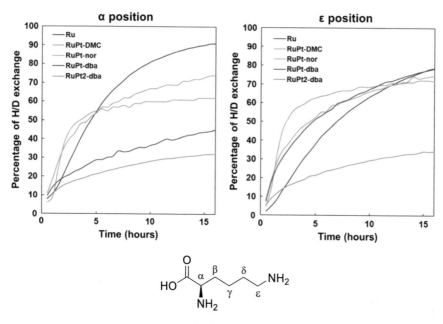

Fig. 9.22 Kinetic study of H/D exchange with different catalysts on the α and ε positions of L-lysine [66]. Reprinted with permission from Ref. [66]. Copyright 2019 Royal Society of Chemistry

without significant modification of the reactivity on the ε position (Fig. 9.22). To explain these effects, the authors proposed that a chelate effect involving the amine and acid groups of L-lysine would result in stronger adsorption of the groups near Cα at the surface of the RuPt. Thus, introduction of Pt at the surface would result in stronger coordination of carboxylate with a concomitant decrease in catalytic activity toward deuteration at Cα.

Palazzolo et al. have recently used Ru NPs@NHC to ameliorate HIE reactions of nucleobases derivatives. Indeed, in this case, Ru NPs/PVP were efficiently used in the deuteration of several biomolecules (nucleosides, nucleotides, xanthines) and drugs, but they were less efficient when more challenging conditions had to be adopted in the case of tritiation of drugs or deuteration of high molecular weight oligonucleotides. Indeed, the use of a more organosoluble stabilizing agent such as the carbene "ICy" (N,N-dicyclohexylimidazol-2-ylidene) deeply modified the reactivity of Ru NPs, which became more efficient toward C(sp^2)-H activation. The latter permitted the high specific activity in drugs tritiation in organic solvents. On the other hand, in the case of the deuteration of oligonucleotides, the use of water-soluble carbene "PriPr" [3-(2,6-diisopropylphenyl)-1-(3-sulfonatopropyl)-1H-imidazol-3-ium-2-ylidene] permitted the synthesis of an oligonucleotide biomolecule which could be used as internal standard for quantification in mass spectrometry (Fig. 9.23) [67].

The reactivity of Ru NPs in HIE reactions can be applied to light alkanes that do not contain directing groups. Rothermel, et al. prepared Ru NPs stabilized by

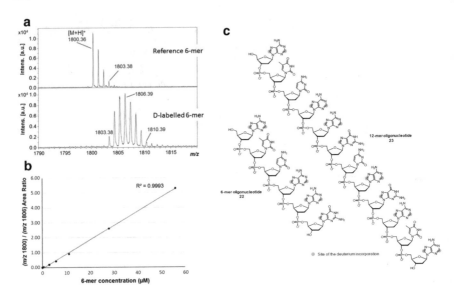

Fig. 9.23 **a** MALDI-TOF mass spectra of native and deuterium-labeled 6-mer oligonucleotide. Non-overlapping isotope massifs were observed after D-labeling. **b** Calibration curve obtained for the native 6-mer from 0.56 to 56 μM, using a deuterium-labeled 6-mer concentration set at a constant value of 280 μM (overall concentration). Most intense isotopes were used for native and D-labeled species (m/z 1800.36 and m/z 1806.39, respectively). **c** Structures of the labeled oligonucleotides [67]. Reprinted with permission from Ref. [67]. Copyright 2019 Wiley

bis(diphenylphosphino)butane (dppb) ligands, which were used in the deuteration of cyclohexane and cyclopentane [68]. The reactions were carried out in the neat substrate (1 or 2 mL) using gaseous D_2 as isotopic source (6 bar) at 60 °C. H/D exchange was much higher for cyclopentane (See Scheme 9), and the uptake of up to 4 D atoms was observed in this case, whereas cyclohexane only incorporated 1 D atom (Fig. 9.24). The reason of such a different reactivity was difficult to determine, but several hypotheses were considered. First, it was proven that the presence of ligands at the surface was not enough to explain the difference in reactivity. Then, thermodynamic explanations were discarded as cyclohexane and cyclopentane have similar bond cleavage energies (400 kJ/mol and 395–403 kJ/mol, respectively). Thus, the authors concluded that reactivity came from a specific recognition of the Ru surface for cyclopentane, the origin of which has not yet been determined.

9.4 Conclusions and Perspectives

HIE has attracted the attention of the catalysis community as deuterated compounds possess interesting applications in several fields. Although catalytic HIE has been traditionally performed by homogeneous complexes, recent studies have proved that

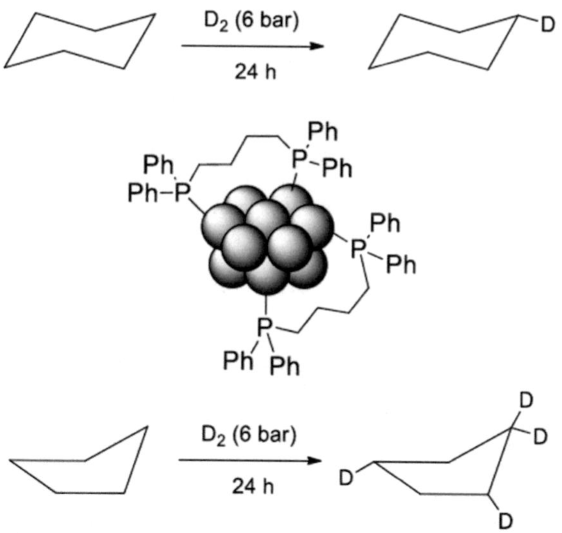

Fig. 9.24 Higher deuterium incorporation for cyclopentane than for cyclohexane with Ru NPs stabilized by dppb under mild conditions [68]

metallic NPs stabilized by polymers or ligands are powerful catalysts that permit us to selectively obtain high degrees of deuteration in a wide variety of substrates with high interest. However, until nowadays, the efforts have been focused on Ru NPs, which have demonstrated to be very active catalysts under mild conditions. Mechanistic studies pointed out that the reaction elapses through the σ-bond activation of the C-H bond by surface Ru. Nevertheless, there are no examples in the literature of NPs composed of other metals that have been traditionally used in HIE such as Ir, Pd and Ni. In addition, different reaction mechanisms may be involved in the reaction when changing the nature of the active metal (i.e., σ-bond metathesis or heterolytic cleavage), which surely will influence reaction selectivity. Thus, we believe that new efforts in HIE will be focused on exploring new compositions of metallic NPs, which may permit us to modulate the selectivity of the reaction and to broaden the scope of substrates. In any case, thanks to the level reached today on the synthesis and overall comprehension of HIE catalyzed by metallic NPs, it is possible to solve historical issues of HIE such as high molecular complexity or deuteration of compounds containing poisoning moieties (i.e., sulfur-containing compounds). These advantages will surely attract the attention of researchers in the field to the utilization of metallic NPs as catalysts for HIE reactions in solution.

References

1. Atzrodt J, Derdau V, Kerr WJ, Reid M (2018) C-H functionalisation for hydrogen isotope exchange. Angew Chem Int Ed 57:3022–3047
2. Atzrodt J, Derdau V, Fey T, Zimmermann J (2007) The renaissance of H/D exchange. Angew Chem Int Ed Engl 46:7744–7765

3. Asensio JM, Bouzouita D, van Leeuwen PWNM, Chaudret B (2020) σ-H–H, σ-C–H, and σ-Si–H bond activation catalyzed by metal nanoparticles. Chem Rev 120:1042–1084. https://doi.org/10.1021/acs.chemrev.1029b00368
4. Atzrodt J, Derdau V, Kerr WJ, Reid M (2018) Deuterium- and tritium-labelled compounds: applications in the life sciences. Angew Chem Int Ed 57:1758–1784
5. Urey HC, Brickwedde FG, Murphy GM (1932) A hydrogen isotope of mass 2. Phys Rev 39:164–165
6. O'Leary D (2012) The deeds to deuterium. Nat Chem 4:236
7. Oliphant ML, Harteck P (1934) Rutherford transmutation effects observed with heavy hydrogen. Nature 133:413
8. Lucas LL, Unterweger MP (2000) Comprehensive review and critical evaluation of the half-life of tritium. J Res Natl Inst Stand Technol 105:541–549
9. Atkins PW, De Paula J (2014) Atkins' physical chemistry
10. Wiberg KB (1955) The deuterium isotope effect. Chem Rev 55:713–743
11. Westheimer FH (1961) The magnitude of the primary kinetic isotope effect for compounds of hydrogen and deuterium. Chem Rev 61:265–273
12. Klinman JP (2010) A new model for the origin of kinetic hydrogen isotope effects. J Phys Org Chem 23:606–612
13. Nelson SD, Trager WF (2003) The use of deuterium isotope effects to probe the active site properties, mechanism of cytochrome P450-catalyzed reactions, and mechanisms of metabolically dependent toxicity. Drug Metab Dispos 31:1481–1498
14. Guengerich FP (2001) Common and uncommon cytochrome P450 reactions related to metabolism and chemical toxicity. Chem Res Toxicol 14:611–650
15. Chowdhury G, Calcutt MW, Nagy LD, Guengerich FP (2012) Oxidation of methyl and ethyl nitrosamines by cytochrome P450 2E1 and 2B1. Biochemistry 51:9995–10007
16. Howland RH (2015) Aspergillus, angiogenesis, and obesity: the story behind beloranib. J Psychosoc Nurs Ment Health Serv 53:13–16
17. Katsnelson A (2013) Heavy drugs draw heavy interest from pharma backers. Nat Med 19:656
18. Sanderson K (2009) Big interest in heavy drugs. Nature 458:269
19. Yarnell A, Wang L (2009) Student affiliates answer presidential video challenge. Chem Eng News 87:48
20. Mullard A (2017) FDA approves first deuterated drug. Nat Rev Drug Discov 16:305
21. Stokvis E, Rosing H, Beijnen JH (2005) Stable isotopically labeled internal standards in quantitative bioanalysis using liquid chromatography/mass spectrometry: necessity or not? Rapid Commun Mass Spectrom 19:401–407
22. Hewavitharana AK (2011) Matrix matching in liquid chromatography-mass spectrometry with stable isotope labelled internal standards-Is it necessary? J Chromatogr A 1218:359–361
23. Berg T, Karlsen M, Oeiestad AML, Johansen JE, Liu H, Strand DH (2014) Evaluation of 13C- and 2H-labeled internal standards for the determination of amphetamines in biological samples, by reversed-phase ultra-high performance liquid chromatography-tandem mass spectrometry. J Chromatogr A 1344:83–90
24. Metcalfe C, Tindale K, Li H, Rodayan A, Yargeau V (2010) Illicit drugs in Canadian municipal wastewater and estimates of community drug use. Environ Pollut 158:3179–3185
25. Piper T, Emery C, Saugy M (2011) Recent developments in the use of isotope ratio mass spectrometry in sports drug testing. Anal Bioanal Chem 401:433–447
26. Piper T, Thomas A, Thevis M, Saugy M (2012) Investigations on hydrogen isotope ratios of endogenous urinary steroids: reference-population-based thresholds and proof-of-concept. Drug Test Anal 4:717–727
27. Tran NH, Hu J, Ong SL (2013) Simultaneous determination of PPCPs, EDCs, and artificial sweeteners in environmental water samples using a single-step SPE coupled with HPLC-MS/MS and isotope dilution. Talanta 113:82–92
28. Benijts T, Dams R, Lambert W, De Leenheer A (2004) Countering matrix effects in environmental liquid chromatography-electrospray ionization tandem mass spectrometry water analysis for endocrine disrupting chemicals. J Chromatogr A 1029:153–159

29. Maguire JJ, Kuc RE, Davenport AP (2012) Radioligand binding assays and their analysis. Methods Mol Biol 897:31–77
30. Harrell AW, Sychterz C, Ho MY, Weber A, Valko K, Negash K (2015) Interrogating the relationship between rat in vivo tissue distribution and drug property data for > 200 structurally unrelated molecules. Pharmacol Res Perspect 3. e00173/00171-e00173/00112
31. Voges R, Heys JR, Moenius T (2009) Preparation of compounds labeled with tritium and carbon-14
32. Uttamsingh V, Gallegos R, Liu JF, Harbeson SL, Bridson GW, Cheng C, Wells DS, Graham PB, Zelle R, Tung R (2015) Altering metabolic profiles of drugs by precision deuteration: reducing mechanism-based inhibition of CYP2D6 by paroxetine. J Pharmacol Exp Ther 354:43–54
33. Crabtree R (1979) Iridium compounds in catalysis. Acc Chem Res 12:331–337
34. Heys R (1992) Investigation of iridium hydride complex [IrH2(Me2CO)2(PPh3)2]BF4 as a catalyst of hydrogen isotope exchange of substrates in solution. J Chem Soc Chem Commun, 680–681
35. Brown JA, Irvine S, Kennedy AR, Kerr WJ, Andersson S, Nilsson GN (2008) Highly active iridium(I) complexes for catalytic hydrogen isotope exchange. Chem Commun, 1115–1117
36. Atzrodt J, Derdau V, Kerr WJ, Reid M, Rojahn P, Weck R (2015) Expanded applicability of iridium(I) NHC/phosphine catalysts in hydrogen isotope exchange processes with pharmaceutically-relevant heterocycles. Tetrahedron 71:1924–1929
37. Valero M, Weck R, Guessregen S, Atzrodt J, Derdau V (2018) Highly selective directed iridium-catalyzed hydrogen isotope exchange reactions of aliphatic amides. Angew Chem Int Ed 57:8159–8163
38. Kerr WJ, Mudd RJ, Reid M, Atzrodt J, Derdau V (2018) Iridium-catalyzed Csp3-H activation for mild and selective hydrogen isotope exchange. ACS Catal 8:10895–10900
39. Cochrane AR, Irvine S, Kerr WJ, Reid M, Andersson S, Nilsson GN (2013) Application of neutral iridium(I) N-heterocyclic carbene complexes in ortho-directed hydrogen isotope exchange. J Labell Compd Radiopharm 56:451–454
40. Kerr WJ, Reid M, Tuttle T (2015) Iridium-catalyzed C-H activation and deuteration of primary sulfonamides: an experimental and computational study. ACS Catal 5:402–410
41. Kerr WJ, Reid M, Tuttle T (2017) Iridium-catalyzed formyl-selective deuteration of aldehydes. Angew Chem Int Ed 56:7808–7812
42. Prechtl MHG, Hoelscher M, Ben-David Y, Theyssen N, Loschen R, Milstein D, Leitner W (2007) H/D exchange at aromatic and heteroaromatic hydrocarbons using D2O as the deuterium source and ruthenium dihydrogen complexes as the catalyst. Angew Chem Int Ed 46:2269–2272
43. Hale LVA, Szymczak NK (2016) Stereoretentive deuteration of α-chiral amines with D_2O. J Am Chem Soc 138:13489–13492
44. Maegawa T, Fujiwara Y, Inagaki Y, Monguchi Y, Sajiki H (2008) A convenient and effective method for the regioselective deuteration of alcohols. Adv Synth Catal 350:2215–2218
45. Fujiwara Y, Iwata H, Sawama Y, Monguchi Y, Sajiki H (2010) Method for regio-, chemo- and stereoselective deuterium labeling of sugars based on ruthenium-catalyzed C-H bond activation. Chem Commun 46:4977–4979
46. Gao L, Perato S, Garcia-Argote S, Taglang C, Martinez-Prieto LM, Chollet C, Buisson D-A, Dauvois V, Lesot P, Chaudret B et al (2018) Ruthenium-catalyzed hydrogen isotope exchange of C(sp3)-H bonds directed by a sulfur atom. Chem Commun 54:2986–2989
47. Ma S, Villa G, Thuy-Boun PS, Homs A, Yu J-Q (2014) Palladium-catalyzed ortho-selective C-H deuteration of arenes: evidence for superior reactivity of weakly coordinated palladacycles. Angew Chem Int Ed 53:734–737
48. Yang H, Dormer PG, Rivera NR, Hoover AJ (2018) Palladium(II)-mediated C-H tritiation of complex pharmaceuticals. Angew Chem Int Ed 57:1883–1887
49. Maegawa T, Akashi A, Esaki H, Aoki F, Sajiki H, Hirota K (2005) Efficient and selective deuteration of phenylalanine derivatives catalyzed by Pd/C. Synlett, 845–847
50. Sajiki H, Esaki H, Aoki F, Maegawa T, Hirota K (2005) Palladium-catalyzed base-selective H-D exchange reaction of nucleosides in deuterium oxide. Synlett, 1385–1388

51. Yu PR, Hesk D, Rivera N, Pelczer I, Chirik PJ (2016) Nature. Nature Publishing Group vol 529, pp 195–199
52. Palmer WN, Chirik PJ (2017) Cobalt-catalyzed stereoretentive hydrogen isotope exchange of C(sp3)-H bonds. ACS Catal 7:5674–5678
53. Yang H, Zarate C, Palmer WN, Rivera N, Hesk D, Chirik PJ (2018) Site-selective nickel-catalyzed hydrogen isotope exchange in N-heterocycles and Its application to the tritiation of pharmaceuticals. ACS Catal 8:10210–10218
54. Zarate C, Yang H, Bezdek MJ, Hesk D, Chirik PJ (2019) Ni(I)-X complexes bearing a bulky α-diimine ligand: synthesis, structure, and superior catalytic performance in the hydrogen isotope exchange in pharmaceuticals. J Am Chem Soc 141:5034–5044
55. Pla D, Gomez M (2016) Metal and metal oxide nanoparticles: a lever for C-H functionalization. ACS Catal 6:3537–3552
56. Ott LS, Cline ML, Deetlefs M, Seddon KR, Finke RG (2005) Nanoclusters in ionic liquids: evidence for N-Heterocyclic carbene formation from imidazolium-based ionic liquids detected by ^{2}H NMR. J Am Chem Soc 127:5758–5759
57. Sullivan JA, Flanagan KA, Hain H (2008) Selective H-D exchange catalyzed by aqueous phase and immobilized Pd nanoparticles. Catal Today 139:154–160
58. Guy KA, Shapley JR (2009) H-D exchange between N-heterocyclic compounds and D^2O with a Pd/PVP colloid catalyst. Organometallics 28:4020–4027
59. Pan C, Pelzer K, Philippot K, Chaudret B, Dassenoy F, Lecante P, Casanove M-J (2001) Ligand-stabilized ruthenium nanoparticles: synthesis, organization, and dynamics. J Am Chem Soc 123:7584–7593
60. Pery T, Pelzer K, Buntkowsky G, Philippot K, Limbach H-H, Chaudret B (2005) Direct NMR evidence for the presence of mobile surface hydrides on ruthenium nanoparticles. Chem Phys Chem 6:605–607
61. Breso-Femenia E, Godard C, Claver C, Chaudret B, Castillón S (2015) Selective catalytic deuteration of phosphorus ligands using ruthenium nanoparticles: a new approach to gain information on ligand coordination. Chem Commun 51:16342–16345
62. Pieters G, Taglang C, Bonnefille E, Gutmann T, Puente C, Berthet J-C, Dugave C, Chaudret B, Rousseau B (2014) Regioselective and stereospecific deuteration of bioactive aza compounds by the use of ruthenium nanoparticles. Angew Chem Int Ed 53:230–234
63. Taglang C, Martinez-Prieto LM, del Rosal I, Maron L, Poteau R, Philippot K, Chaudret B, Perato S, Sam Lone A, Puente C et al. (2015) Enantiospecific C-H activation using ruthenium nanocatalysts. Angew Chem Int Ed 54:10474–10477
64. Martínez-Prieto LM, Baquero EA, Pieters G, Flores JC, de Jesús E, Nayral C, Delpech F, van Leeuwen PWNM, Lippens G, Chaudret B (2017) Monitoring of nanoparticle reactivity in solution: interaction of l-lysine and Ru nanoparticles probed by chemical shift perturbation parallels regioselective H/D exchange. Chem Commun 53:5850–5853
65. Bhatia S, Spahlinger G, Boukhumseen N, Boll Q, Li Z, Jackson JE (2016) Stereoretentive H/D exchange via an electroactivated heterogeneous catalyst at sp^3 C-H sites bearing amines or alcohols. Eur J Org Chem 2016:4230–4235
66. Bouzouita D, Lippens G, Baquero EA, Fazzini PF, Pieters G, Coppel Y, Lecante P, Tricard S, Martinez-Prieto LM, Chaudret B (2019) Tuning the catalytic activity and selectivity of water-soluble bimetallic RuPt nanoparticles by modifying their surface metal distribution. Nanoscale 11:16544–16552
67. Palazzolo A, Feuillastre S, Pfeifer V, Garcia-Argote S, Bouzouita D, Tricard S, Chollet C, Marcon E, Buisson D-A, Cholet S et al. (2019) Efficient access to deuterated and tritiated nucleobase pharmaceuticals and oligonucleotides using hydrogen-isotope exchange. Angew Chem Int Ed 58:4891–4895
68. Rothermel N, Bouzouita D, Roether T, de Rosal I, Tricard S, Poteau R, Gutmann T, Chaudret B, Limbach H-H, Buntkowsky G (2018) Surprising differences of alkane C-H activation catalyzed by ruthenium nanoparticles: complex surface-substrate recognition? Chem Cat Chem 10:4243–4247

Chapter 10
Progress in the Selective Semi-hydrogenation of Alkynes by Nanocatalysis

Jorge A. Delgado and Cyril Godard

Abstract The present chapter examines the literature related to the semi-hydrogenation of alkynes catalysed by systems obtained from colloidal synthetic methodologies. The strategies described for the enhancement of the alkene selectivity are classified according to its mode of operation. This scheme permits to fragment the complexity involved in this transformation into individual phenomena such as the effect of the particle-size, stabilizer, second metal, additives, support, and among other things. Special consideration is given to insights gained by combination of inputs from experiments and density functional theory (DFT). This chapter aims at providing the reader with an overview of the potential and outlook of the utilization of colloidal methodologies for the preparation of well-defined heterogeneous catalysts for both fundamental understanding and industrial applications.

Keywords Alkyne hydrogenation · Modified palladium · Semi-hydrogenation · Subsurface hydride · Single Atom Catalysts

10.1 Introduction

Without any doubt, modern tools in heterogeneous catalysis have permitted the development of not only profitable but also environmentally sustainable chemical processes [1, 2]. In this framework, hydrogenation reactions are possibly one of the most common catalytic transformations at industrial level; particularly, the semi-hydrogenation of alkynes has been a focus of attention in the last decade due to its relevance in the manufacturing of fine chemical and polymers [3–5]. In this context, the development of chemo- and stereo-selective catalysts has been identified as essential in order to enhance the process productivity, thus preventing side reactions such as

J. A. Delgado
Centre Tecnològic de la Química, Marcel lí Domingo s/n, Campus Sescelades,
43007 Tarragona, Spain

C. Godard (✉)
Departament de Química Física i Inorgánica, Universitat Rovira i Virgili,
Marcel lí Domingo s/n, Campus Sescelades, 43007 Tarragona, Spain
e-mail: cyril.godard@urv.cat

© Springer Nature Switzerland AG 2020
P. W. N. M. van Leeuwen and C. Claver (eds.), *Recent Advances in Nanoparticle Catalysis*,
Molecular Catalysis 1, https://doi.org/10.1007/978-3-030-45823-2_10

over-hydrogenation or oligomerization of the alkyne substrate [5, 6]. When referred to catalysts employed in the semi-hydrogenation of alkynes, a wide variety of transition metals has been employed, covering a wide range of particle sizes, most of them classified as either metal nanoparticles (M-NPs), metal nanoclusters (M-Clusters) and single-atom catalysts (SACs) (Scheme 10.1) [7].

Metals that easily activate the hydrogen molecule (e.g. Pt, Ni) normally display a high over-hydrogenation activity and suffer of poor control of the alkene selectivity. Conversely, metals with low hydrogen activation activity (e.g. Au, Ag) frequently display good alkene selectivities but require harsh reaction conditions in terms of both temperature and pressure [8]. Palladium appears as the metal that apparently meets the right balance towards hydrogen activation and outstanding activity under mild conditions, but its control of the selectivity when approaching full alkyne conversion is extremely poor [9].

In terms of catalyst preparation, solid supports such as oxides (SiO_2, Al_2O_3, TiO_2), $CaCO_3$ and carbon-based materials (carbon nanofibres or nanotubes, activated C, graphite) have been classically employed for the dispersion of the metallic palladium phase [10]. Although impregnation methodologies have dominated the preparation of heterogeneous catalysts, control of the NPs size or their distribution onto the support are common issues. In contrast, the use of colloidal methods provides access to well-defined M-NPs with accurate control of their size, shape and structure at the nanoscale [11]. Even though sophisticated synthetic methods prevent the scaling-up for productive applications, the possibility to prepare series of materials with well-defined properties has provided the scientific community with model catalysts that are ideal for fundamental studies and have contributed enormously to the understanding of the phenomena involved in the semi-hydrogenation of alkynes [12]. In addition, one of the recent trends in research on heterogeneous catalysis consists in the use of small metal clusters (M-clusters) and single-atom catalysts (SACs) [13]. Their special

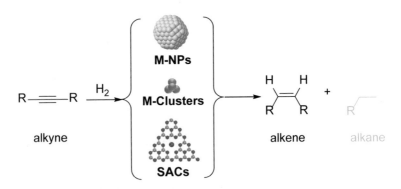

M-NPs: Pd, Ni, Cu, Ag, Au, Fe, Ru, Pt..
M-Clusters: Pd_2, Cu_{11}, Ni_6
SACs: Pd, Ni, Ru

Scheme 10.1 Catalysts employed in the semi-hydrogenation of alkynes characteristics linked to their sub-nanometre scale provide a typical performance in terms of activity, selectivity and stability, relevant for challenging reactions such as the semi-hydrogenation of acetylene [14, 15].

In the following sections, the literature on the selective hydrogenation of alkynes catalysed by metal nanoparticles, metal clusters and single atom catalysts is analysed in a comprehensive but not exhaustive manner, with a focus on materials prepared by colloidal methodologies. This chapter aims at describing the potential and outlook of the utilization of colloidal chemistry for the preparation of heterogeneous catalysts for both fundamental understanding and industrial applications. For this purpose, special attention is given to insights gained by combination of experimental with DFT input. The scope of our analysis is limited to the systems that employ hydrogen gas as the reducing agent in combination with a metallic phase. However, metal-free catalytic systems [16], or the use of alternative reducing agents has also provided promising results in the selective hydrogenation of alkynes [17–22].

10.1.1 General Reaction Mechanism

A general reaction mechanism for the hydrogenation of alkynes is depicted in Scheme 10.2. Possibly, the most widely accepted mechanism for this transformation consists in the two-step hydrogenation where the alkyne is first transformed into the alkene and subsequently to the alkane (path a). This process occurs with the desorption and re-adsorption of the alkene intermediate. If the two-step hydrogenation occurs in a consecutive manner, without desorption of the alkene from the metal surface, path b is taking place. For the case of light alkynes as acetylene, the oligomerization into hydrocarbons of C_4-C_{32} might become an important path (up to 20–40% selectivity), which is strongly favoured by the reaction conditions [23–26]). Intrinsic properties of the catalyst such as the metallic phase, as well as the process conditions, might influence the relevance of the reaction path for hydrogenation.

Scheme 10.2 Possible reaction paths during the hydrogenation of alkynes

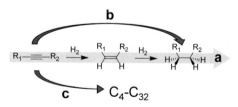

Scheme 10.3 Representation
of **a** thermodynamic and
b mechanistic selectivity in
the hydrogenation of alkynes

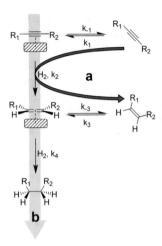

10.1.2   Types of Selectivity: Mechanistic Versus Thermodynamic

Scheme 10.3 depicts the two types of selectivity distinguished in the partial hydrogenation of alkynes: the thermodynamic and mechanistic selectivity [27]. The thermodynamic selectivity relies on the preferential adsorption of the alkyne in the presence of the alkene product ($k_1/k_{-1} \gg k_3/k_{-3}$). Such a favoured adsorption of the alkyne (defined in terms of the adsorption energies) displaces the produced alkene from the metal surface, thus preventing its over-hydrogenation. In contrast, the mechanistic selectivity becomes relevant when the hydrogenation of the alkyne is kinetically favoured compared to the alkene hydrogenation ($k_2 \gg k_4$) [28]. Since for most of the transition metals, the rate of alkene hydrogenation is greater than that of alkyne hydrogenation [29]; the selectivity observed in this reaction is frequently regarded as the thermodynamic one.

10.1.3   Approaches for the Design of Selective Catalysts

If thermodynamic factors at the origin of the selectivity in the semi-hydrogenation of alkynes, its modulation depends on delicate structural and electronic features of the catalysts. Tailoring such features are nowadays more accessible than ever thanks to the advances in nanoscience and the synthetic methodologies for the preparation of well-defined metal nanoparticles [30, 31]. In this context, unsupported nanoparticles offer exceptional opportunities as model catalysts, which in addition make them ideal candidates to shed light on the role of the support during catalysis [32].

Parameters intrinsic to the colloidal chemistry by itself such as the size, shape and composition of the metal nanoparticles, as well as the capping agent, the support combined with parameters intrinsic to the hydrogenation reaction (e.g. conditions, additives), make of the semi-hydrogenation reaction a complex multivariable equation. In the following sections, a comprehensive analysis of the principal structural, electronic and external effects that affect the catalytic performance in the semi-hydrogenation of alkynes is presented.

10.2 Particle-Size Effect

The effect of the metal dispersion and the particle size was early recognized of paramount importance in heterogeneous catalysis [33]. In the context of the semi-hydrogenation reaction, Kiwi-Minsker et al. reported the particle-size effect of unsupported PdNPs with diameters between 6 and 14 nm in the semi-hydrogenation of 1-hexyne (Fig. 10.1) [34]. Similar TOF values were observed in the range 6–11 nm while a 15fold-TOF increase was detected for larger nanoparticles. Differently, the selectivity towards 1-hexene was size independent. Similar behaviours in selectivity and TOF were observed in the semi-hydrogenation of 2-methyl-3-butyn-2-ol (MBY) catalysed by PdNPs of variable size [35]. In terms of the rationalization of the size sensitivity, in a complementary study, the authors associated the size effect with the fraction of Pd_{plane} sites which predominates in larger NPs; therefore, considering such sites as the most active for the hydrogenation of alkynes [7]. Works published by other authors agree in the observed structure sensitivity in this reaction [36].

For the case of structure sensitivity in the acetylene semi-hydrogenation, PdNPs in the range of 8–13 nm supported on Pd/CNF/SMF (Carbon nanofibres/sintered metal fibres) were evaluated [37]. Analogous to the trends observed for liquid substrates, TOF values were found to increase with the size of the PdNPs up to 11 nm, while

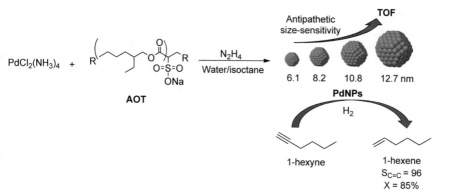

Fig. 10.1 Particle-size effect of PdNPs in the semi-hydrogenation of 1-hexyne

larger NPs displayed constant values. The observed behaviour below 11 nm was attributed to be "geometric" in nature due to the size-sensitive formation of a Pd–C_x phase [38]. Such carbon deposits can block Pd surface sites, thus affecting the activity. In a complementary study, the same authors reported that the decrease in PdNPs size from 8 to 2 nm resulted in a decrease in activity, and increase in selectivity to the ethylene product [39]. This time, the carbon deposits, important for smaller NPs, were responsible for site isolation thus improving the selectivity by a geometric effect.

Regarding the shape effect in catalysis, Kiwi-Minsker et al. reported the selective hydrogenation of 2-methyl-3-butyn-2-ol (MBY) employing PVP-stabilized Pd nanocubes (6 and 18 nm), cuboctahedrons (5.5 nm) and octahedrons (31 nm) [40]. Founded on the observed reactivities, two types of active sites were proposed for this reaction: those on the planes where the semi-hydrogenation of MBY to 2-methyl-3-buten-2-ol (MBE) takes place and those at the edges which are responsible for over-hydrogenation (Fig. 10.2). According to this criteria, the alkene product is adsorbed more strongly at low coordination sites, thus resulting in over-hydrogenation. Differently, a similar study carried out on acetylene revealed that the shape of NPs does not influence the selectivity, while the activity decreased in the order $Pd_{octahedrons}$ > $Pd_{cuboctahedrons}$ > Pd_{cubes} [25]. Considering the divergent observations regarding the size and shape effects, it can only be concluded that these effects cannot be generalized. In addition, due to the intimate relationship between size and shape, studies in this area need to consider their interplay.

Decreasing the particle size to the limit of the nanoscale, we can find firstly small metal clusters and ultimately single atoms. One of the recent focuses in research on heterogeneous catalysis is the use of single-atom catalysts (SACs) [13, 41]. In the recent literature, examples of Pd [14, 42–44], Ni [45], Au [46], Cu [47] and Ru [48]

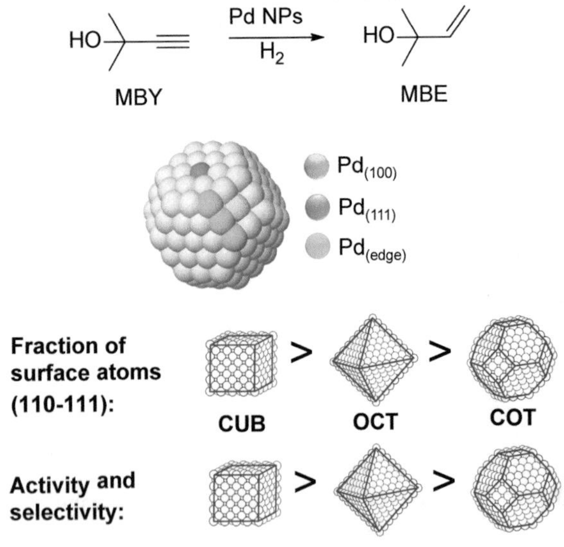

Fig. 10.2 Effect of the NPs shape in the semi-hydrogenation of 2-methyl-3-butyn-2-ol (MBY). Reproduced with permission from Ref. [15]. Copyright 2017 Royal Society of Chemistry

single-atom catalysts applied in the selective hydrogenation of alkynes were reported. For instance, Huang et al. reported the enhancement of both selectivity and coking-resistance using a single-atom Pd_1/C_3N_4 catalyst for acetylene hydrogenation in excess ethylene [14]. According to the authors, analysis of this material by in situ X-ray photoemission spectroscopy revealed that the considerable charge transfer from the Pd NPs to g-C_3N_4 likely plays an important role in the catalytic performance enhancement. Among the different methodologies employed for the preparation of single-atom catalysts, the use of metal-organic frameworks can be highlighted for the isolation/encapsulation of the corresponding metal species [45, 48]. More recently, the same authors reported the preparation of an atomically dispersed copper (Cu) catalyst supported on a defective nanodiamond-graphene (Cu_1/ND@G). This material exhibited excellent catalytic performance for the selective conversion of acetylene to ethylene (98% of ethylene selectivity at 95% conversion and stability for more than 60 h at 200 °C) [47]. According to the authors, this exceptional catalytic performance was due to the unique bonding structure and electronic properties of Cu atoms on Cu_1/ND@G which facilitate the acetylene activation and ethylene desorption. Although the following catalytic system does not fall into the definition of SACs (formally it is an immobilized molecular catalyst), its relevance in the field makes its inclusion in this analysis necessary. Li et al. reported the selective hydrogenation of acetylene catalysed by a cationic nickel atom confined in zeolite [49]. The catalyst consisted of a four-coordinated cationic nickel(II) confined in chabazite zeolite, (Ni@CHA). This material catalysed efficiently the selective hydrogenation of acetylene with extremely high selectivity towards ethylene (97% at full conversion and 180 °C). The chabazite framework is proposed as an inorganic ligand of the nickel centre, which also contributes to the heterolytic hydrogen dissociation by the local electrostatic field within the zeolite cage.

The application of small metal clusters in catalysis has also attracted the attention of the scientific community in the last decades [50]. In terms of fundamental understanding, Abdollahi and Farmanzadeh studied the reactivity of small Pd [51], Cu [52] and Ni [53] metal clusters (composed of 2–15 atoms) as catalysts for the semi-hydrogenation of acetylene in ethylene-rich mixtures using DFT. Differences in the activation energy barriers and adsorption energies of acetylene and ethylene revealed Pd_2, Cu_{11} and Ni_6 as potential candidates with optimal selectivity for this reaction.

10.3   Effect of a Second Phase

The introduction of a second phase, generally a metal, to dilute the active phase is a widely applied strategy for enhancing the catalyst performance in the semi-hydrogenation of alkynes. Traditionally, the outcome of such a dilution is classified according to electronic or geometric effects. The geometric effect refers to the formation of ensembles of specific character or to the blockage of unselective sites thus enhancing the selectivity towards the alkene. In contrast, the electronic effect

Fig. 10.3 Metal distribution
in several bimetallic arrays:
a cluster-in-cluster, b alloy
with random metal
distribution, c ordered alloy
and d core@shell

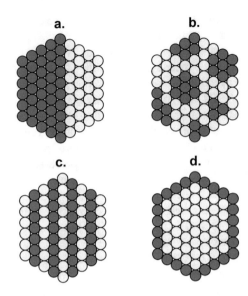

stands for the disruption of the electronic properties of the active metal in order
to promote the thermodynamic selectivity or to limit the occurrence of subsurface
species, generally associated to over-hydrogenation issues.

For a bimetallic formulation, different metal distributions might arise depending
on the methodology used for the synthesis. Figure 10.3 displays the cross-sections
of some of the most common bimetallic arrays in order of structural complexity
[54]. Such a metal distribution within the nanocluster will determine not only the
exposition degree and the type of active sites, (e.g. geometric isolation), but also
the electronic structure of the resulted metal blend which in turn might impact the
adsorption properties against substrate and products [55, 56].

In the following paragraphs, a brief review of some of the most representative
families of bimetallic formulations is presented with focus on the identification of
the electronic or structural descriptors that promote the selectivity. The analysis
starts with palladium-based bimetallic formulations (e.g. PdAg, PdCu, PdAu) passing
through other formulations (e.g. FeAu, RhNi, NiAu), metal borides (e.g. Ni_2B),
phosphides (Ni_2P) and nitrides (Ni_3N), and finally, a section dedicated to acetylene
semi-hydrogenation.

10.3.1 Palladium-Based Formulations

Possibly, the most studied bimetallic formulation for the semi-hydrogenation of
alkynes with extended industrial application is the PdAg catalysts. Kiwi-Minsker
et al. investigated the effect of a second metal in the semi-hydrogenation of dehy-
droisophytol (DIP) using Pd and bimetallic Pd-Ag and Pd-Cu NPs [57]. Bimetallic

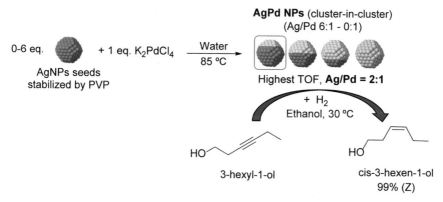

Fig. 10.4 Synthesis of bimetallic PdAg NPs and their evaluation in the semi-hydrogenation of 3-hexyl-1-ol

NPs with variable Pd/M ratios (1.5–5) were prepared by a two-step process using PVP-stabilized PdNPs as seeds. Characterization indicated the formation of different bimetallic structures: Pd@Ag core@shell and PdCu-mixed alloys. The incorporation of the second metal resulted in an increase of the alkene selectivity from 91 up to 97% (at 99% conversion) which was credited to the geometric effect of the dilution of the Pd phase by Ag or Cu and modification of the electronic properties of Pd.

Calver et al. reported the preparation of AgPd nanoparticles via galvanic exchange reactions between AgNPs seeds and K_2PdCl_4 (Ag/Pd = 0:1–6:1) and their evaluation in the semi-hydrogenation of 3-hexyl-1-ol (Fig. 10.4) [58]. From the tested series, Ag/Pd ratio of 2:1 exhibited the highest TOFs, while all the catalysts provided high selectivity for Z-3-hexen-1-ol (>99%). EXAFS analysis suggested the formation of cluster-in-cluster structures.

Possibly, one of the most remarkable contributions regarding PdAg formulations was delivered by Mitsudome et al. who reported the preparation core@shell Pd@Ag NPs of *ca.* 26 nm with variable Ag content (9–33 wt% Ag, Fig. 10.5) [59]. The Pd@Ag catalyst containing 17 wt% Ag provided excellent alkene selectivity in the semi-hydrogenation of 1-octyne (>99% selectivity at >99% conversion at r.t and 1 bar H_2). Outstandingly, no over-hydrogenation was registered even after extended reaction times. In contrast, over-hydrogenation issues were detected on catalysts provided with lower silver loading (<16.7%). The excellent control of the alkene selectivity by Pd@Ag was attributed to the cooperative interaction between Pd and Ag, where the Pd core contributed to the activation of hydrogen, while Ag served as the platform for the hydrogenation reaction. Due to the restriction of Pd inside the NP core, any risk of unselective process was prevented.

At the end of the last century, Bronstein et al. reported the evaluation of PdAu, PdPt and PdZn nanoparticles formed in block polystyrene poly-4-vinylpyridine (PS-b-P4VP) micelles in the semi-hydrogenation of dehydrolinalool (DHL) [55, 60]. Excellent alkene selectivity was obtained for all the catalysts (99.8% at full conversion) attributed to a dual promotion: on the one hand, by modification of the NPs

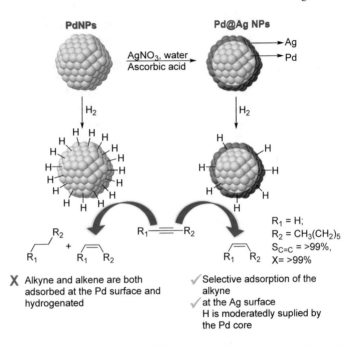

Fig. 10.5 Synthesis of Pd@Ag core@shell NPs and semi-hydrogenation of alkynes. Reproduced with permission from Ref. [15]. Copyright 2017 Royal Society of Chemistry

surface with pyridine units from the co-polymer and on the other hand, the geometric and electronic effects induced by the second metal.

The application of bimetallic PdCu systems in the selective hydrogenation of alkynes has increased in recent years. Recently, Godard et al. reported a facile and straightforward methodology for the preparation of either monometallic (Cu, Ni and Pd) or bimetallic nanocatalysts (NiCu and PdCu) stabilized by an N-heterocyclic carbene ligand [61, 62]. Both colloidal and supported nanoparticles (NPs) on carbon nanotubes (CNTs) were prepared via a one-pot synthesis and subsequently tested in the semi-hydrogenation of alkynes and alkynols. From the series of catalysts, the bimetallic PdCu/CNTs demonstrated superior performance for the case of aliphatic substrates such as 1-octyne and 4-octyne when compared with commercial references (Lindlar and Pd Nanoselect) [62]. This result was attributed to the electronic and geometric promotion of copper in the moderation of the reactivity of the active Pd phase being the equimolar composition of 50Pd50Cu the optimal in terms of activity and selectivity. Similarly, Buxaderas et al. recently reported a one-pot synthetic approach for the preparation of bimetallic PdCu NPs supported on mesostructured silica (CuPd@MCM-48) [63]. This material demonstrated to be active in the semi-hydrogenation of alkynes and highly chemoselective towards the alkyne functionality under 1 bar H_2 and 110 °C (Fig. 10.6).

Although less common than the previously commented formulations, other bimetallic systems based on Pd resulted of interest. Shi et al. reported the synthesis

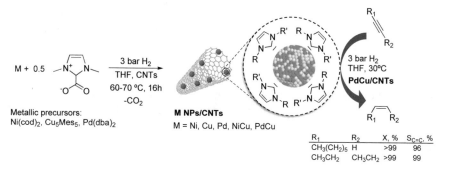

Fig. 10.6 Synthesis of CNTs supported monometallic (Ni, Cu, Pd) and bimetallic (NiCu and PdCu) NPs and their application in the hydrogenation of alkynes

of PdRhP nanoparticles by co-reduction of Pd(acac)$_2$ and Rh(acac)$_3$, employing trioctylphosphine (TOP) as the stabilizer [64]. These NPs were evaluated in the semi-hydrogenation of ethynylbenzene under mild conditions, evidencing an alkene selectivity of 92% at full conversion. Once more, the improved alkene selectivity was explained in terms of geometric and electronic effect displayed by Rh and P over the Pd phase.

While the design of bimetallic catalysts was focused over the past decades on controlling the nature of active sites, more recently, intermetallic compounds have arisen as attractive candidates for drastically changing the electronic and geometric states of active metals. For palladium-based catalysts, systems such as PdZn [65–67], PdGa [68] and PdIn [43, 69] were reported. Other intermetallic formulations include NiGa [70], AlFe [71], CoMnGe, CoFeGe [72] and AlCo [73]. The formation of isolated active ensembles in combination with electronic modifications which favour the desorption of the olefinic products are generally the justification of the performance enhancement observed with polymetallic catalysts in comparison with the pure active phases.

Another type of palladium-based catalysts that is worth mentioning is doped catalysts. For instance, the modification of palladium nanoparticles at interstitial sites with boron atoms was reported by Tsang et al. [74]. The doping process was carried out by treatment of a Pd/C catalyst with borane tetrahydrofuran. DFT calculations indicated that the presence of subsurface boron atoms altered the adsorption properties of the palladium surface atoms thus enhancing the alkene selectivity when compared with the initial Pd catalyst. In addition, the formation of subsurface hydrides often regarded as responsible of over-hydrogenation issues is prevented by the presence of boron at the interstices [75].

In recent years, the utilization of palladium sulphide was also reported as selective catalyst for the semi-hydrogenation of alkynes [76–78]. For instance, Anderson et al. reported the excellent performance displayed by the Pd$_4$S phase of palladium sulphide in the hydrogenation of acetylene in ethylene-rich mixtures (80% ethylene sel. at full conversion under 18 bar) [78]. The high selectivity was related to the crystal structure of Pd$_4$S with the unique spatial arrangement providing Pd atoms isolation. Other approaches attempted the functionalization or doping of the palladium surface with sulphur atoms. For instance, Perez-Ramirez et al. reported the direct formation of a nanostructured Pd_3S phase by a simple treatment of palladium nanoparticles supported on graphitic carbon nitride with aqueous sodium sulphide [77]. This material exhibited unparalleled performance in the semi-hydrogenation of alkynes in liquid phase. According to the authors, sulphur displayed a multifunctional role in both the isolation of palladium trimers and weakening the binding of organic intermediates.

10.3.2 Other Non-Pd-Based Formulations

De Vries et al. can be considered as one of the pioneers in applying Fe NPs in the semi-hydrogenation of 1-octyne [79, 80]. As a mechanism to prevent the formation of 1-octyne at early stages of the reaction (mechanistic selectivity) observed in the Fe NPs, the authors opted for the preparation of Fe bimetallic with Ni, Co and Cu (1:1) as well as all monometallic NPs [80]. The series displayed the following activity order, Fe < Co < Ni, and the alkene selectivities obtained with the bimetallic systems were intermediate compared to those obtained with the monometallic catalysts. Very recently, Gregory et al. reported the facile preparation of FeNPs by reduction of iron(II) acetylacetonate with diisobutylaluminum hydride (DIBAL-H) in THF in the absence of stabilizer [81]. The formed nanoparticles and nanoclusters enabled the synthesis of various Z-alkenes in high yields and with high stereocontrol under very mild conditions (1–3 bar H_2, 30 °C).

Moores et al. reported the application of iron@iron oxide core@shell nanoparticles of ca. 50 nm (Fe@FeO NPs) in the hydrogenation of alkynes in aqueous media [82, 83]. Although the alkene selectivity of these NPs was not remarkable (6% at 88% of conversion), they demonstrated to be stable and magnetically recoverable. The authors also suggested that the presence of an oxide shell did not block the activity of the NPs but still provided protection against oxidation by oxygen or water. According to other studies on iron-based catalysts [18, 79, 80, 84, 85], relevant alkene selectivities were only accessible in the presence of additives or promoting stabilizers.

Liu et al. reported the preparation of a layered double hydroxide-derived NiCu alloy (NiCu/MMO) and its evaluation in the semi-hydrogenation of acetylene. The bimetallic catalyst demonstrated improved alkene selectivity and longer stability when compared to the monometallic Ni catalyst [86]. Very recently, Chandler et al. reported a solution-phase synthesis for Ni and bimetallic NiAu NPs and its deposition on alumina [87]. Gold was added to the initial Ni NPs via galvanic displacement of Ni in organic solution in the presence of oleylamine as capping agent. In the hydrogenation of 1-hexyne, the alumina-supported NiAu catalysts provided intermediate activity and selectivity when compared to the monometallic Ni and Au catalysts.

10.3.3 Borides, Phosphides and Nitrides

In a series of early contributions, Brown et al. reported the preparation of nickel boride Ni_2B colloids and its application in the semi-hydrogenation of alkynes [88, 89]. The NPs were prepared by chemical reduction of nickel salts using sodium borohydride in ethanol or water. Interestingly, differences in the reactivity were detected as a function of the reaction solvent, and those prepared in ethanol exhibited the highest alkene selectivity in the semi-hydrogenation of 3-hexyne in the presence of quinoline (99% selectivity towards Z-3-hexene, Fig. 10.7) [89].

Later, the same authors reported the preparation and testing of Pd, Pt and Rh colloids by an analogous synthetic approach [90]. Based on the monitoring of the reaction, the authors proposed that the reactivity of the Pd catalyst was governed by thermodynamic selectivity while that of the Ni catalyst was attributed to a mechanistic selectivity [91].

Corma et al. pioneered the application of nickel phosphide in the semi-hydrogenation of alkynes. NiP colloids with variable Ni/P ratio were prepared by reaction of Ni NPs with P_4, followed by an annealing step at 220 °C [3, 4]. Ni_2P NPs revealed to hydrogenate a series of terminal and internal alkenes with moderate to good selectivities (Fig. 10.8). It is noteworthy that NPs prepared with lower phosphorus content ($Ni_{3.5}P$) evidenced poor control of the alkene selectivity, possibly because of the lack of dilution of the highly active nickel phase.

More recently, Perez-Ramirez et al. reported the preparation of two nickel phosphides, namely Ni_2P and Ni_5P_4 and their assessment in the semi-hydrogenation of 1-hexyne and 2-methyl-3-butyn-2-ol [92]. The phosphides exhibited a higher rate and selectivity than unmodified nickel catalysts. Higher activity and lower selectivity were observed when the alkynol was used as the substrate. According to DFT studies, this phenomenon could be attributed to differences in the product desorption related to the presence of the hydroxyl group. Very recently, the synthesis of N-doped carbon supported nickel nitride Ni_3N nanorods and their application in the alkyne semi-hydrogenation were reported [95]. Strong reaction conditions were required

Fig. 10.7 Preparation of P2-Ni catalyst and its application in the semi-hydrogenation of alkynes

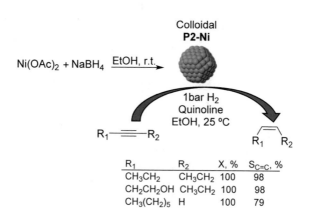

R_1	R_2	X, %	$S_{C=C}$, %
CH_3CH_2	CH_3CH_2	100	98
CH_2CH_2OH	CH_3CH_2	100	98
$CH_3(CH_2)_5$	H	100	79

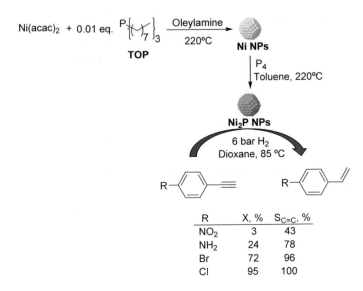

Ni(acac)$_2$ + 0.01 eq. TOP $\xrightarrow[220°C]{\text{Oleylamine}}$ Ni NPs

$\xrightarrow[\text{Toluene, 220°C}]{P_4}$ Ni$_2$P NPs

$\xrightarrow[\text{Dioxane, 85 °C}]{6 \text{ bar } H_2}$

R	X, %	$S_{C=C}$, %
NO$_2$	3	43
NH$_2$	24	78
Br	72	96
Cl	95	100

Fig. 10.8 Preparation of Ni$_2$P NPs and semi-hydrogenation of substituted phenylacetylenes

(20 bar H$_2$, 100 °C), and moderate to good alkene selectivities were observed. DFT calculations suggested that the relatively low adsorption energy of the alkene on Ni$_3$N (vs. pure Ni) might contribute to the high selectivity for alkene.

Similarly, Zhao et al. reported the preparation of palladium phosphide NPs by chemical reduction of Pd(acac)$_2$ in the presence of triphenylphosphine and trioctylphosphine as phosphorus source [93]. ICP and XRD analysis suggested the formation Pd$_{75}$P$_{25}$ NPs with an alloyed structure. The NPs evidenced high activity but addition of quinoline was necessary to prevent excessive over-hydrogenation. Higher alkene selectivities were reported by the same author using a trimetallic palladium copper nickel phosphide catalyst (Pd-Cu-Ni-P) [94]. The authors attributed this result to synergistic effects of the electron deficiency of Pd and the amorphous phase.

For the semi-hydrogenation of acetylene in ethylene-rich mixtures, Williams et al. studied the application of PdAg and PdAu/SiO$_2$ bimetallic catalysts prepared by electroless deposition. This synthetic method favoured the formation of core@shell NPs [96]. NPs with different degree of M-coverage (M = Ag or Au) were prepared by controlling the stoichiometry of the Pd and Ag/Au solutions. In catalysis, the selectivity towards ethylene was enhanced with the Au or Au coverage. Interestingly, similar activity and selectivity trends were observed for both Ag- and Au– Pd/SiO$_2$ thus suggesting that the bimetallic promotion for these particular catalysts was geometric and not electronic in nature. In this line, the bimetallic PdCu/CNTs catalyst reported by Godard et al. was also tested in the semi-hydrogenation of acetylene in ethylene-rich mixtures [62]. Excellent stability was observed even after 40 h of reaction with moderate to good ethylene selectivity (ca. 60%). The performance of the bimetallic catalyst was superior to that observed for a commercial pure Pd nanocatalyst (Pd/Al$_2$O$_3$, Strem Chemical), which exhibited poor ethylene selectivities along the test (<35%) with a fast deactivation during the first 5 h on stream. These results evidenced the applicability of bimetallic catalysts prepared purely by colloidal techniques, not only in the semi-hydrogenation of substrates in liquid phase but also with substrates of industrial relevance in the gas phase.

Important insights were also obtained by DFT calculations and for instance, Lopez et al. studied the effect of the presence of a second metal on the properties of Pd in the semi-hydrogenation of acetylene in excess ethylene [9]. As expected, the adsorption of molecules was preferred on Pd sites, and that of acetylene was stronger than for ethylene. Interestingly, alloying the Pd phase with metals such as Au or Cu stabilized the adsorption of both substrate and product while others such as Bi enhanced the thermodynamic selectivity by only favouring the adsorption of acetylene. In addition, the formation of subsurface hydride (β-PdH phase) was reduced on all the studied promoters, which consequently could limit the over-hydrogenation reaction.

10.4 Effect of the Stabilizer

For, the stabilization of M-NPs is applied in the semi-hydrogenation of alkynes, a wide variety of organic compounds such as surfactants, ligands, polymers, ionic liquids among other compounds have been utilized. According to the intrinsic properties of the stabilizer, prevention of NPs agglomeration can be obtained by steric hindrance or electrostatic repulsion [31]. Moreover, the stabilizer can provide steric or electronic modifications of the NPs surface, thus influencing the performance in catalysis [30, 97–99]. At this point, it is important to mention that the effect of the stabilizer depends on the reaction media (gas or liquid phase). In gas phase, the catalysts are dried, which results in the collapse of the stabilizer on the metal surface and possibly the blockage of active sites. Differently, in liquid phase, the stabilizers are flexible enough to permit the access of substrates to the metal surface. For instance, Witte et al. reported that the CO uptake observed in chemisorption of PdNPs stabilized by hexadecyl(2-hydroxyethyl)dimethyl ammonium dihydrogenphosphate (HHDMA) was suppressed when the NPs were isolated and dried [100].

Polymers have been classically employed for the steric stabilization of M-NPs due to their capacity to provide a protective layer that wraps the NPs. Hirai et al. reported that for PVP-stabilized PdNPs, the thickness of such a layer was proportional to the molecular weight of the polymer (thickness of 4–16 nm, for M$_w$ from 6000 to 574,000); however, the catalytic activity for the hydrogenation of cyclooctadiene was not sensitive to the thickness of the polymer layer [98]. Several authors have addressed the application of PVP-stabilized Pd NPs in the semi-hydrogenation of alkynes evidencing moderate to good alkene selectivities independently of the PVP:Pd ratio [7, 97]. Niu et al. reported the evaluation of polyethylene glycol stabilized Ru NPs in the semi-hydrogenation of methyl propiolate [101]. For this system, the alkene selectivity resulted highly sensitive to the temperature and pressure. Using the same polymer, Zharmagambetova et al. prepared ZnO-supported Ni [102] and Pd

[103] NPs and their evaluation in the semi-hydrogenation of alkynols. Interestingly, catalysts prepared in the presence of PEG were ca. 4 times more active compared to those prepared in absence of the polymer.

If amines have been the classical additives for the moderation of the reactivity of palladium catalysts, the incorporation of amine groups in the structure of polymer stabilizer could play the double role of stabilizing the NPs and enhancing the alkene selectivity. Following this approach, Long et al. reported the preparation of PdNPs supported on a mesoporous silica material functionalized with branched poly(ethyleneimine) polymers (PEI) composite and its application in the selective hydrogenation of alkynes. Interestingly, the over-hydrogenation rate was significantly reduced by increasing the support porosity and the molecular weight and branching structure of the polymer [104]. More recently, Yamashita et al. reported the preparation of a yolk-shell nanostructured composite composed of Pd NPs stabilized by PEI, confined in hollow silica spheres and their application in the semi-hydrogenation of alkynes [105]. Using a one-pot method, the yolk-shell nanostructured Pd-PEI-silica composite (Pd + PEI@HSS) was fabricated as a Pd core of ca. 5–9 nm in diameter and a porous silica shell of ca. 30–50 nm thickness. This composite material provided a high alkene selectivity in the semi-hydrogenation of phenylacetylene to styrene due to the strong poisoning effect of amine groups of the PEI. Using a novel approach, Studer et al. reported a facile light-mediated preparation of small polymer-coated Pd NPs and their application as catalysts for alkyne semi-hydrogenation [106, 107]. The photoactive polymers acted as reagents for the photochemical reduction of Pd ions and as stabilizers for the Pd NPs generated in situ. These materials revealed efficient hydrogenation catalysts with high activity and Z-selectivity in the semi-hydrogenation of alkynes.

Very recently, Xinwu et al. reported the preparation of sulphur-containing polymer-supported palladium nanoparticles (Pd/SPMB) using cross-linked poly(N,N-methylene bis(acrylamide)) (SPMB) as platform and stabilizer of the NPs [76]. This material exhibited excellent activity, stability, and selectivity in the semi-hydrogenation of both internal and terminal alkynes. Notably, the selectivity enhancement towards alkene was attributed to the coverage of sulphide and thiolate on the Pd surface with the appropriate degree of poisoning.

Micelles are another family of widely applied stabilizers of M-NPs. Depending on the solvent´s polarity, the structure of the amphiphilic compound and the metallic precursor, it is possible to tune the hydrophobic or hydrophilic environment of the miscelle´s core that embed the M-NPs. For instance, the semi-hydrogenation of 2-butyne-1,4-diol catalysed by PdNPs stabilized in the micelle core of poly(ethylene-oxide)-block-poly-2-vinylpiridine (PEO-b-P2VP) was reported by Kiwi-Minsker et al. [108]. The structural characteristics of the amphiphilic co-polymer permitted the stabilization of PdNPs by the pyridine units inside the core of the micelle, while the PEO chains constituting the shell were accountable of its dispersion in aqueous media (Fig. 10.9). The high selectivity was ascribed to a modification of the Pd surface with the 2-vinylpyridine units in the micelle´s core. Using a simple ultra-filtration procedure, the authors demonstrated the recyclability of the micelles.

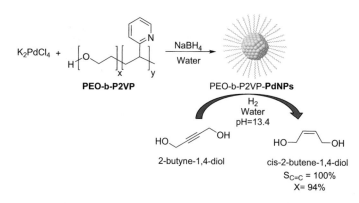

Fig. 10.9 Semi-hydrogenation of 2-butyne-1,4-diol catalysed by PdNPs stabilized in the micelle core of poly(ethylene-oxide)-block-poly-2-vinylpiridine (PEO-b-P2VP)

Following a related strategy, Mizugaki et al. reported the preparation of Pd NPs stabilized inside the core of spherical micelle-like structures constituted by poly(amidoamine) (PAMAM) dendrons with a pyridine core and alkyl end groups [109]. These NPs demonstrated activity in the semi-hydrogenation of alkenes, providing for instance 98% selectivity to the Z-alkene at 96% conversion for the case of 1-phenyl-1-propyne. More recently, Karakhanov et al. reported the semi-hydrogenation of phenylacetylene by Pd nanocatalysts encapsulated into dendrimer (PPI) networks [110]. These catalysts demonstrated superior selectivity towards styrene compared to traditional Lindlar catalyst, which were maintained under prolonged reaction times.

Thanks to their amphiphilic nature, ionic liquids (IL) are ideal candidates as dispersing solvents and/or stabilizers of M-NPs. Moreover, catalytic systems based on ILs can often be recycled by solvent extractions [111]. For the semi-hydrogenation of alkynes, Dupont et al. reported the preparation of monodisperse PdNPs of ca. 7.3 nm using 1-butyronitrile-3-methylimidazolium-N-bis(trifluoromethanesulfonyl) imide [(BCN)MI.NTf$_2$] as solvent and stabilizer [112]. These NPs provided good performance in the semi-hydrogenation of phenylacetylene (95% selectivity at 97% conversion), and could be recycled up to 4 times. The authors ascribed the good selectivity of these NPs to electronic modifications at the metal surface induced by the coordination of the nitrile groups of the ionic liquids. More recently, the same author reported the semi-hydrogenation of 2-pentyne catalysed by PdNPs stabilized by ILs and imidazolium salts containing pyridine or phosphine functionalities (Fig. 10.10) [113]. The hydrogenation of 2-pentyne in the presence of any of these P- and N-containing ILs displayed higher alkene selectivity in comparison with reference NPs prepared in the absence of ligand (up to 87% vs. 20% both at ca. 95% of conversion).

Recently, Van Leeuwen et al. reported the use of NiNPs stabilized by several imidazolium-amidinate ligands for selective hydrogenation of alkynes [114]. An influence on the activity and selectivity was observed depending on the ligand used as a stabilizer. The NPs stabilized with the strongest donor ligand ($R = -OMe$) was

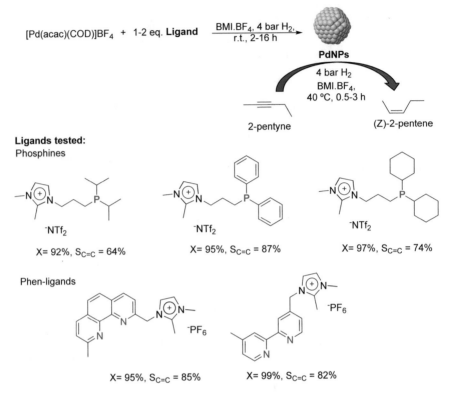

Fig. 10.10 Semi-hydrogenation of 2-pentyne catalysed by PdNPs stabilized by imidazolium salts provided with P or N groups

the most active catalyst while the one containing an electron-withdrawing group $(R = -Cl)$ was the slowest but at the same time exhibited the best selectivity towards the (Z)-3-hexene formation at full conversion (90% under 1 bar H_2 and r.t.). The previous enhancement of the selectivity could be attributed to the higher coverage of the metal surface by the ligand. Fe NPs stabilized by acetonitrile or nitrile-functionalized IL were also reported as nanocatalysts for the semi-hydrogenation of substituted diphenylacetylenes [84]. Interestingly, low alkene selectivities were obtained when the substrate incorporated electron-withdrawing groups. Due to the biphasic nature of the catalytic system, easy separation and recycling of the catalyst were possible.

An example of a highly structured catalytic system was reported by Peng et al. The hybrid Pd/IL/MOF was prepared by initial formation of a copper metal-organic framework in the presence of the IL (1,1,3,3-tetramethylguanidinium trifluoroac-etate, TMGT), followed by the reduction of the palladium precursor [115]. The IL was expected to stabilize and anchor the Pd NPs to the MOF. This material exhibited excellent alkene selectivity in the semi-hydrogenation of phenylacetylene (>99% at full conversion), which was ascribed to the interactions between the metal surface and the nitrogen atoms of the IL.

Classical ligands employed in coordination chemistry have also been used for the stabilization of M-NPs [116, 117].

Among the first reports addressing the application of colloidal catalysts for hydrogenation reactions, Caubere et al. described what they called "Complex Reducing Agents (CRA)" [118]. Such reagents consisted in metal colloids prepared by reduction of metal salts using strong reducing agents such as $NaBH_4$, $LiAlH_4$ and NaH. For instance, a nickel catalyst named "Nic" by the authors was synthesized via the reduction of $Ni(OAc)_2$ using NaH in t-AmOH (under argon atmosphere) and catalysed the hydrogenation of 1-octyne and phenylacetylene providing alkene selectivities up to 90% at full conversion [119]. Interestingly, differences in reactivity of the NPs were observed as a function of the reaction solvent [120]. In subsequent reports, the authors addressed the effect of the alkyne structure, the effect of quinoline as additive [120], the reactivity of internal and terminal alkynols [121] and the evaluation of Pd-based catalysts [122]. In the latter, PdNPs were prepared by reduction of $Pd(OAc)_2$ using NaH in the presence of t-AmOH (Fig. 10.11). Although the resulted NPs required the presence of nitrogenated additives in order to moderate the alkene selectivity, it can be highlighted that PdNPs prepared in the absence of the t-AmOH (and tested under the same conditions) evidenced extended over-hydrogenation issues. This observation suggested an effect of the in situ formed alkoxide on the reactivity of the resulting NPs.

Godard et al. reported a method for the preparation of reusable Pd nanocatalysts immobilized on paper for the semi-hydrogenation of alkynes [123]. The catalyst was prepared using a one-pot approach, by decomposition of $Pd_2(dba)_3$ (dba = dibenzylideneacetone) under H_2 in the presence of tricyclohexylphosphine (PCy_3) and multi-walled carbon nanotubes (MWCNTs). Although the paper composite required the presence of quinoline in order to moderate the selectivity in the semi-hydrogenation of alkynes, the catalyst could be easily recovered and recycled. Koptyug et al. also reported the modification of a supported Cu/SiO catalyst with tricyclohexylphosphine and its evaluation in the semi-hydrogenation of 1-butyne [124]. Although the surface

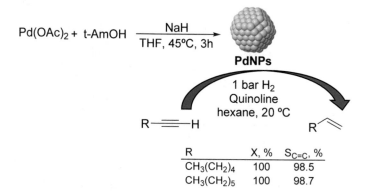

R	X, %	$S_{C=C}$, %
$CH_3(CH_2)_4$	100	98.5
$CH_3(CH_2)_5$	100	98.7

Fig. 10.11 Semi-hydrogenation of terminal alkynes catalysed by the Pd NPs reported by Caubere et al.

modification reduced the catalytic activity compared to the unmodified catalyst, it remarkably increased the alkene selectivity.

Gomez et al. reported the preparation of small zero-valent nickel nanoparticles (1.2 nm) stabilized by cinchona-based alkaloids and TPPTS (tris(3-sulfophenyl)phosphine trisodium salt) and their application in the selective hydrogenation of alkynes [125]. The NiNPs were synthesized from the organometallic precursor [Ni(cod)$_2$] in neat glycerol under hydrogen pressure. The colloidal NiNPs dispersed in glycerol demonstrated remarkable activity and selectivity in the hydrogenation of internal alkynes under 3 bar H$_2$ and 100 °C. The catalytic phase was recycled at least ten times without loss of activity, affording in each case metal-free organic products. Other functional groups such as nitro, nitrile and formyl groups were efficiently hydrogenated to the corresponding anilines, benzylamines and benzylalcohols, respectively, (77–95% yields).

When interested in the use of water as the media for either the preparation of the M-NPs or the catalytic reaction, surfactants are frequently the choice of stabilizing agent [30]. Their amphiphilic nature, conditioned with polar and lipophilic moieties, provides steric stabilization to the M-NPs. One of the most relevant examples of the use of surfactants as stabilizers of MPs for semi-hydrogenation of alkynes is the commercially available Pd NanoSelect catalyst [100, 126]. The catalyst, which comprises Pd NPs of ca. 6 nm immobilized on titanium silicate or activated carbon, is prepared via a two-step methodology. First, the NPs colloid is prepared in water by reduction of a Pd salt using HHDMA (Hexadecyl(2-hydroxyethyl)dimethylammonium dihydrogen phosphate) as stabilizer and reducing agent and subsequently, the suspension is impregnated on the desired support (Fig. 10.12a) [126]. The resulting catalyst exhibited outstanding activity and selectivity in the hydrogenation of alkenes and alkynes. The authors suggested a double layer of HHDMA surrounding the PdNPs with polar groups pointing inside and outside the micelle. Such a distribution is typical of the formation of micelles by

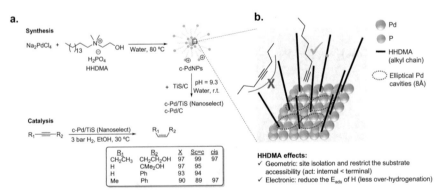

Fig. 10.12 a Nanoselect Pd catalyst for the semi-hydrogenation of alkynes. **b** Electronic and geometric effects of HHDMA over the Pd surface. Reproduced with permission from Ref. [15]. Copyright 2017 Royal Society of Chemistry

surfactants and is the responsible of the dispersion of the NPs in aqueous media [100]. In a subsequent study, the same authors reported the application of PdNPs using chiral tetraalkyl ammonium phosphate stabilizers for the semi-hydrogenation of 3-hexyn-1-ol [127]. No relevant differences in catalytic performance where observed in comparison with HHDMA.

To understand the origin of the high alkene selectivity displayed by the Pd Nanoselect catalyst, Pérez-Ramírez et al. investigated the selectivity patterns and accessibility constraints by means of DFT calculations [128]. According to the computed structure, the hydrogen phosphate groups of the HHDMA appear firmly adsorbed on the Pd atoms. Such a configuration resulted in the isolation of Pd ensembles (which can be considered a geometric effect of the stabilizer), and a reduction of the hydride coverage (electronic effect, Fig. 10.12b). Additionally, the authors compared the catalytic activity of terminal and internal alkynes, and proposed that the adsorbed HHDMA and its distribution on the Pd surface create accessibility constrains which condition the catalytic activity (Fig. 10.12b). These observations reflect the essential role of the stabilizer in the catalytic performance of M-NPs [128]. In a subsequent report, the same authors discussed the effect of the amount of stabilizer on the performance of catalysts analogous to Pd Naloselect prepared with distinct HHDMA contents (0.3–16.8 wt%) [129]. Higher activities were obtained at higher HHDMA contents during the semi-hydrogenation of 1-hexyne, and with inputs from DFT, the authors proposed that the configuration of the adsorbed HHDMA on the Pd NPs highly depended on the stabilizer concentration: when the HHDMA concentration is low, the surfactant lies flat on the metal surface thus blocking the accessibility of the substrate to the active sites, while at higher surfactant concentrations, its configuration is the one described previously (with cavities open to receive reagents).

More recently, Scarso et al. reported a method for the preparation of Pd nanoparticles in aqueous medium stabilized by anionic sulfonated surfactants [130]. The nanoparticles were prepared from $Pd(OAc)_2$ solutions in the presence of anionic surfactant and using hydrogen gas as the reducing agent. The aqueous PdNP suspensions were tested in the dechlorination of aromatic substrates, hydrogenation and dihydroxylation of carbonyl groups and semi-hydrogenation of alkynes. In all cases, the micellar medium was crucial for stabilizing the metal nanoparticles, dissolving substrates, steering product selectivity and enabling the recycling.

Regarding the semi-hydrogenation of acetylene catalysed by colloidal NPs or supported catalysts prepared through colloidal approaches, scarce reports are identified in the literature. Ji et al. reported the preparation of water-soluble Pd NPs stabilized by three polymers, hydroxyethyl cellulose (HEC), polyquaternium-10 (quaternized hydroxyethyl cellulose) and sodium carboxymethyl cellulose (CMC) and its application in the acetylene semi-hydrogenation [131]. From the tested polymers, the NPs stabilized by CMC evidenced the highest performance of the series (50% selectivity at full conversion). Kiwi-Minsker et al. investigated the effect of the stabilizer on the performance of carbon nanofibres supported PdNPs during the semi-hydrogenation of acetylene [39]. In this study, polyvinylalcohol (PVA), sodium di-2-ethylhexylsulfosuccinate (AOT) and polyvinyl pyrrolidone (PVP) were evaluated as stabilizers. Differences in activity were justified by geometric and electronic effects induced by the stabilisers (AOT < PVA ≈ PVP). Regarding the alkene selectivity, a certain pattern was correlated to the particles size but not to the stabilizer. Interestingly, in view of the detrimental effect of treating the catalysts with ozone/H_2 (for the removal of the surface stabilizer), the authors suggested a positive role of the stabilizer. However, other studies recommend the removal of the stabilizer prior to the reaction to maximize the available active sites. For instance, a UV–ozone (UVO) treatment was employed for the surface cleaning of PVP-stabilized Pd nanocubes supported on carbon [132]. This time, the stabilizer removal resulted in a four-fold increase of the activity during the semi-hydrogenation of acetylene.

Therefore, although the stabilizer can provide improvements of the catalyst performance, its selection is a critical part of the catalyst design and should consider the target alkyne. For instance, in liquid phase, it is generally assumed that a certain mobility of the capping agent is beneficial for the reagents to access the metal surface. In this case, a common strategy to improve the alkene selectivity has been the use of nitrogen-containing stabilizers. In contrast, gas–solid reactions appear highly sensitive to the presence of organic residues at the metal surface, thus involving the use surface cleaning procedures prior to catalysis (sometimes just calcination or thermal treatments). For gas phase applications, stabilizers provided only with easy decomposable moieties (e.g. based on C, H, O, and N) are recommended, thus avoiding the possible blockage of active sites.

10.5   Other Parameters Affecting the Selectivity

10.5.1   Subsurface H and C

The occurrence of over-hydrogenation issues during the semi-hydrogenation of alkynes has been commonly linked with the existence of surface or subsurface hydrides [133]. The low activation barrier for H_2 dissociation combined with the small size of the hydrogen atoms makes them ready to easily populate either the surface or the subsurface of palladium catalysts [134].

Analogous versions of carbon (surface and subsurface) have been also identified as a modifier of the performance of catalysts and particularly relevant for those used in the semi-hydrogenation of acetylene. According to DFT calculations, subsurface carbon displays an improved thermodynamic factor on Pd surface, which justifies the generally observed positive impact on catalysis [134, 135]. Regarding its localization within the NPs, Kiwi-Minsker et al. reported the favoured formation of carbides on step sites of small particles [37]. Depending on the reaction conditions, the formation of oligomeric hydrocarbons (C_{8+}, coke) are commonly observed during the semi-hydrogenation of acetylene. Some authors have observed an improvement of the alkene selectivity with the formation of such a carbon deposits on specific sites of the metal surface [23].

10.5.2   Structure of the Substrate

The structure of the substrate has proved to also define to some extent the alkene selectivity. Geometric hindrance at the triple bond (internal vs. terminal) or electronic properties (determined by the presence of donating or withdrawing groups) could condition the adsorption strength of the substrate to the metal surface, thus influencing its propensity to be or not selectively hydrogenated. Furthermore, the interplay between the substrate structure and the stabilizer properties (e.g. steric hindrance, site isolation, etc.) determine the substrate-metal interaction, thus increasing the complexity of the phenomena and preventing the identification of general patterns.

For instance, during the semi-hydrogenation of terminal alkynes, the hydrogenation of the triple bond is commonly slower than that of the alkene product (for Pt, Rh, Ni, Pd) [136, 137]. Moreover, the hydrogenation of terminal alkenes proceeds faster than that of internal ones due to accessibility constraints to the metal surface [91, 138]. Conversely, in a recent report, the semi-hydrogenation of 4-octyne catalysed by a Pd-based catalyst proceeded faster than that of 1-octyne (full hydrogenation after 10 and 30 min, respectively, at 3 bar H_2, 30 °C) [62]. Comparison of the reactivity with an iso-structural internal alkyne (dimethyl acetylenedicarboxylate) suggested that the presence of electron-withdrawing groups might result in an important decrease in the hydrogenation rate (47% conversion after 3 h at 3 bar H_2, 30 °C). This argument could also explain the larger reaction rate observed for 4-octyne, which contains a more electron rich triple bond than that of 1-octyne. A similar behaviour was also reported by Corma et al., who evaluated the semi-hydrogenation of a series of phenylacetylenes substituted in *para* positions catalysed by Ni_2P NPs [3, 4]. The lowest activity and selectivity was observed for the substrate provided with the most electron-withdrawing NO_2 group ($X = 3\%$, and $S_{C=C} = 43\%$). Another common reactivity pattern is the observation of retarded hydrogenation rate when hydroxyl groups are present in the alkyne structure [79]. For instance, based on experimental evidence, Kelsen et al. proposed that the hydrogenation rate of Fe NPs is decreased by the presence of hydroxyl groups in the alkynol substrate, a phenomena that favoured the alkene selectivity [85].

10.5.3   Effect of the Use of Additives

The use of additives has been one of the traditional strategies to improve the alkene selectivity in the semi-hydrogenation of alkynes. As a general view, a successful additive should advantageously compete with the alkene product for active sites and thus prevent the over-hydrogenation reaction. Its efficiency is normally a function of their concentration in the reaction media and their adsorption properties at the NPs surface versus those of reagents and products. The use of nitrogenated compounds was early employed for this purpose. For instance, Brown et al. studied the effect of a series of amine additives in the semi-hydrogenation of alkynes catalysed by P_2Ni

Fig. 10.13 Relationship between of C=C and NH$_2$ adsorption energies at the surface of a Pt nanoclusters and the selectivity in catalysis. Reproduced with permission from Ref. [15]. Copyright 2017 Royal Society of Chemistry

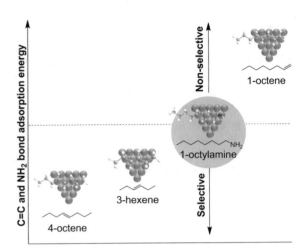

[139]. From the tested amines, ethylenediamine evidenced the best improvement in alkene selectivity.

In a comprehensive report, Shevchenko et al. studied with experimental and DFT insights, the impact of amine additives on the semi-hydrogenation of 4-octyne catalysed by Pt and CoPt$_3$ nanoclusters [140, 141]. By comparing the adsorption energies of reagents and products with those of the amine additives, the authors observed that their relative positioning would determine the catalytic performance (Fig. 10.13). Accordingly, the effective additive requires an adsorption energy larger than the alkene product (therefore could displace it from the metal surface), but lower than that of the alkyne substrate otherwise it would hinder the catalyst reactivity.

Employing an unconventional metal for this transformation, Rossi et al. reported the semi-hydrogenation of alkynes catalysed by AuNPs promoted with a series of nitrogenated bases [142]. Outstandingly, an important increase of the activity (while maintaining full selectivity at full conversion) was observed with the addition of stoichiometric amounts of the base, being piperazine the one with the largest positive effect. Combining insights from DFT and experimental results, a hydrogenation mechanism was proposed and consisted in the formation of frustrated Lewis pairs at the additive-Au interface, activating heterolytically the H$_2$ molecule, and therefore permitting the highly selective hydrogenation of the alkyne substrate. Another example addressing the effect of inorganic bases in the semi-hydrogenation of 2-methyl-3-butyn-2-ol catalysed by supported PdNPs was reported by Kiwi-Minsker et al. [143]. Interestingly, the addition of potassium or sodium hydroxides to the reaction media boosted both the activity and selectivity, a phenomenon attributed to the base induced site separation and favoured reactant adsorption.

In the palladium-catalysed semi-hydrogenation of acetylene, CO has been by excellence the main additive or process modifier employed in the industry [144]. Based on experimental and DFT insights, Lopez et al. reported that under the presence of CO, the Pd surface is covered by a blanket of this molecule which in turn limits the availability of hydrides at both the surface and the subsurface, and restricts the formation of Pd ensembles at the surface [134, 145]. Other reports addressed the poisoning effect of CO in the reduction of active sites and therefore the activity as a function of the reaction conditions (T and P) [9]. Additional care must be taken, when considering the use of CO in bimetallic formulations due to deep metal reorganization. For instance, Lopez et al. investigated the effect of CO on PdM alloys (M = Cu, Ag, Au, Zn, Ga, Sn, Pb and Bi) by DFT calculations and concluded that none of the studied cases evidenced satisfactory resistance against metal segregation under high CO pressures [9]. Other reports have addressed also the evaluation of alkali, sulphur and amine compounds as modifiers in this reaction [144].

10.5.4   Effect of the Support

Crespo-Quesada et al. thoroughly discussed the role of the support at meso/microlevel in the selective hydrogenation of alkynes [10]. Although the main function of the carrier is to assure the metal dispersion and prevent the metal sintering during catalysis, depending on the synthetic strategy, it might highly influence properties such as the metal distribution, particle size, which in turn would affect the catalytic performance [146].

Furthermore, textural or electronic properties of the support have an impact on important phenomena such as the diffusion of reaction molecules from the media to the active sites or charge transfer processes which concerns the adsorption modes at the metal surface. Under special conditions, the support could even modify the structure of the active phase. For instance, depending on the reducibility of the support, thermal treatments might induce the migration of reduced species towards the metal surface, giving rise to what is known as strong metal-support interactions (SMSI). Such a phenomenon might result in site blocking or the generation of new bimetallic phases [144, 147].

The presence of specific surface functionalities within the support can be induced by means of pre-treatment steps. For example, treating carbon-based supports with strong acids (e.g. HNO_3, HCl), bases (e.g. NaOH) or oxidizing agents (e.g. H_2O_2, ozone), can cause the formation of oxygenated functionalities, which in turn might influence the dispersion or the reactivity of the metallic phase [148]. The following paragraphs address the effect of the support on the reactivity of nanocatalysts prepared by colloidal techniques in the semi-hydrogenation of alkynes.

In relation with the SMSI, Semagina et al. studied the effect of performing reductive thermal treatments on a Pd/ZnO/SMFs (sintered metal fibres) catalyst prior the semi-hydrogenation of 2-methyl-3-buten-2-ol (MBE) [149]. Superior alkene selectivity was observed for the reduced catalyst under hydrogen at 773 K (95% at full conversion) when compared to a Pd/Al_2O_3 catalyst. This observation was attributed to the formation of a bimetallic PdZn phase during the high-temperature reductive treatment with enhanced thermodynamic selectivity. More recently, Wang et al. reported the preparation of a Pd-In/In_2O_3 catalyst by an impregnation method [150].

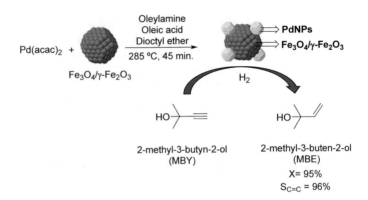

Fig. 10.14 Synthesis of Pd/FeO$_x$ NPs and their application in the semi-hydrogenation of 2-methyl-3-butyn-2-ol (MBY)

The reduction of the parent material (Pt/In$_2$O$_3$) under defined temperatures permitted the selective poisoning of edge and corner sites of Pd nanoparticles by reduced In atoms which resulted in a remarkable enhancement of the alkene selectivity in the semi-hydrogenation of relevant molecules such as dehydrolinalool when compared with a non-poisoned catalyst (95 vs. 89% at 99% of conversion).

As a strategy for recycling colloidal catalyst based on Pd, hybrids with iron oxide moieties have been reported as easily recoverable catalysts by simply using magnetic fields. For instance, Bronstein et al. reported the deposition of Pd NPs on preformed iron oxide NPs and their application in the semi-hydrogenation of 2-methyl-3-butyn-2-ol [151]. Interestingly, the hybrid Pd/FeOx catalyst displayed a higher activity compared to the monometallic Pd NPs (95 vs. 6% conversion), without affecting the alkene selectivity (96%, Fig. 10.14). The authors attributed this result to a possible electron transfer from the iron oxide NPs towards the Pd NPs.

Following an analogous approach, Uberman et al. studied the application of a recoverable catalyst made of PdNPs deposited on a colloidal magnetic support (silica-coated magnetite functionalized with amido groups, FFSiNH$_2$) in the semi-hydrogenation of 3-butyn-1-ol [152]. The hybrid material evidenced moderate to good selectivity under mild conditions (82% at 98% conversion). Corma et al. demonstrated that semi-hydrogenation of alkynes is possible by using merely small iron oxide NPs (either Fe$_2$O$_3$ or FeO) deposited on slightly acidic inorganic oxides (e.g. nanosized ZnO, TiO$_2$, ZrO$_2$,) [153]. In spite of the simplicity of concept, this material provided good performance in the semi-hydrogenation of alkynes in liquid phase and acetylene in ethylene-rich mixtures.

10.5.5   Interplay Metal/Oxide and Encapsulation

The interface between an oxide and a metal revealed of paramount relevance in several modern technologies, including microelectronics, electrodes for electrochemistry, corrosion protection and catalysis of supported or hybrid metal particles [154]. For several catalytic transformations, the metal/oxide interface was described as the active site for catalysis. Phenomena such as charge transfer (from the oxide to the metal or vice versa) as well as surface strain (due to differences in their crystalline lattices) give rise to highly active interface sites with distinctive catalytic properties [155].

A remarkable example evidencing the highly specific reactivities that might arise by interactions between metal particles with oxides (e.g. support carrier) was reported by Yang et al. [156]. A multi-walled carbon nanotubes (MWCNT) supported Fe_3O_4–Cu_2O–Pd nanocomposite, prepared by the deposition of CuNPs and Fe_3O_4 on the MWCNTs followed by in situ galvanic substitution between Cu atoms and $PdCl_4^{2+}$ resulted in active and highly selective catalysts in the semi-hydrogenation of terminal alkynes, but unreactive when using internal alkynes. Control experiments demonstrated that Cu_2O was essential for the prevention of the over-hydrogenation reaction although no insights were given regarding the nature of the active sites. A few years later, a step forward in the understanding of the nature of the Pd/Cu_2O catalysts was made by Zheng et al., who reported the preparation of atomically dispersed Pd atoms supported on Cu_2O and their evaluation in the semi-hydrogenation of terminal alkynes [157]. Characterization by EXAFS revealed the formation of a two-coordinated Pd(I) species formed during the deposition of Pd by galvanic displacement of Cu(I) (Fig. 10.15). Mechanistic studies suggested that the formed Pd–(O–Cu)$_2$ species was critical for the activation and heterolytic splitting of H_2 into Pd–$H^{\delta-}$ and O–$H^{\delta+}$ species. Moreover, the adsorption of alkenes on H_2-preadsorbed Pd(I) is meant to be relatively weak, preventing the over-hydrogenation thus favouring the selectivity. The same feature prevented the coordination of internal alkynes, thus justifying their inactivity for hydrogenation.

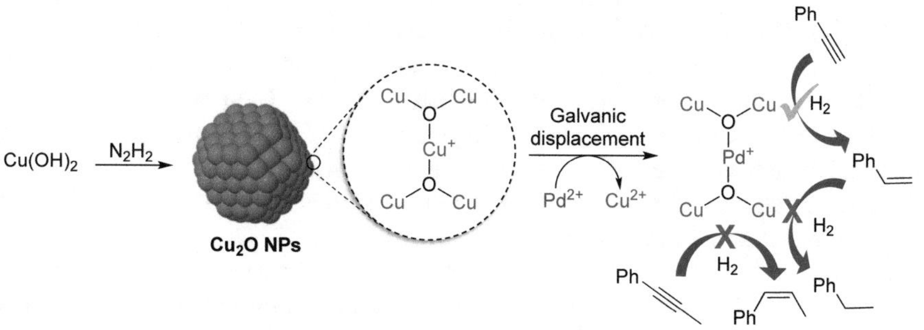

Fig. 10.15 Cu_2O-supported atomically dispersed Pd atoms for semi-hydrogenation of terminal alkynes

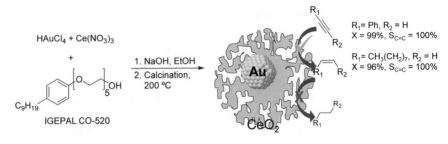

Fig. 10.16 Au@CeO₂ NPs for selective hydrogenation of alkynes. Reproduced with permission from Ref. [15]. Copyright 2017 Royal Society of Chemistry

Recent trends of alternative catalysts for hydrogenation reactions include the use of CeO₂ [158, 159] or Au [142, 160] as active phases. For the case of Ceria, although high alkene selectivities were reported, utilization of these systems is still restricted by the severe reaction conditions required in terms of temperature and pressure [161, 162]. The combination of both Au and CeO₂ phases has demonstrated remarkable synergistic reactivities in catalysis. For instance, Mitsudome et al. reported the preparation of Au@CeO₂ NPs using a one-step redox-coprecipitation method, and their application in the semi-hydrogenation of alkynes (Fig. 10.16) [158]. When tested in the semi-hydrogenation of phenylacetylene (at r.t. and 30 bar H₂), outstanding alkene selectivity was obtained (99% at full conversion) even after extended reaction times. The maximization of interfacial sites due to the core@shell nanostructure was proposed by the authors as responsible for the observed reactivity.

Using a similar approach, Datye et al. reported the effect of the preparation methodology of nickel-doped ceria catalysts on the selective hydrogenation of acetylene in the presence of ethylene [163]. Coprecipitation and sol–gel techniques allowed the homogeneous incorporation of nickel into the ceria lattice. Using these catalysts, ethane formation was prevented even at full acetylene conversion while coke formation was significantly reduced. In an attempt to combine the possible promoting effect of CeO₂ and the dilution by a second metal phase, very recently, Patarroyo et al. reported the preparation of hollow PdAg-CeO₂ heterodimer nanocrystals [164]. Colloidal solutions of highly monodisperse and well-defined hollow PdAg-CeO₂ heterodimer NCs were produced via galvanic replacement reaction (GRR) between Ag-CeO₂ NCs and K₂PdCl₆ precursor. The promotion of Ag on the moderation of the reactivity of the Pd phase, in combination with the creation of interfacial sites with the CeO₂ moiety in the same nanostructure, was related with the good catalytic performance of these NPs, in particular in the hydrogenation of terminal and internal aliphatic alkynes.

Another rational but still uncommon approach to improve the alkene selectivity was reported by Gulyaeva et al. which consisted in the encapsulation of small Pd NPs (1–2 nm) below the surface of fibreglass [165]. This material afforded remarkable ethylene selectivity in the semi-hydrogenation of acetylene in ethylene-rich mixtures, which was ascribed to the selective adsorption of polarizable molecules (e.g. only acetylene) within the glass. Moreover, low activity towards oligomerization was detected for this material thus suggesting high resistance against deactivation via this pathway. This report highlights that encapsulation of the active phase could be a promising strategy for the selective transformations of small molecules such as acetylene. Extending this concept, Qin et al. proposed the utilization of a porous TiO_2/Pt/TiO_2 sandwich catalyst for the semi-hydrogenation of alkynes [166]. This catalyst displayed enhanced alkene selectivity when compared with the uncovered Pd catalyst, indicating once more that encapsulation is an efficient strategy for the moderation of the reactivity of highly reactive (normally unselective) metal phases. Similarly, Beller et al. reported the deposition of nitrogen-doped graphitic layers onto the surface of CoNPs immobilized on silica (Co/phen@SiO_2-800) and the evaluation of the resulting catalyst in the semi-hydrogenation of alkynes [167]. Good to excellent alkene selectivities were observed for both, internal and terminal alkynes.

More recently, Gong et al. reported the activation and spill-over of hydrogen on sub-1 nm palladium nanoclusters confined within sodalite zeolite for the semi-hydrogenation of acetylene [168]. According to the authors, the small pore of the sodalite is crucial as it only allows H_2 diffusion into the channels to reach the encapsulated Pd nanoclusters and thus avoids over-hydrogenation to form ethane.

In an attempt to rationalize the effects of carbon supports in the semi-hydrogenation of acetylene catalysed by single atoms and small metal clusters, Richards et al. reported the deposition of Pd on carbon nanotubes or carbon nanofibres [169]. Stabilization of the most finely dispersed palladium (at the atom scale) was observed on carbon nanofibres with a stack structure in comparison with other carbon supports. This catalyst demonstrated the highest selectivity in the semi-hydrogenation of acetylene to ethylene. In addition, the doping of carbon nanofibres by nitrogen atoms resulted in the relative increase of the ethylene selectivity but with a concomitant decrease of the activity.

10.5.6   External Parameters

In the semi-hydrogenation of alkynes, the selection of the adequate reaction conditions (P and T) play a fundamental role in the reaction kinetics, at the time of defining the desired conversion level which is indeed directly linked with the alkene selectivity. Generation of excessive subsurface hydrides under relatively high hydrogen pressures with concomitant over-hydrogenation consequences have been frequently documented for Pd catalysts [133]. Conversely, catalysts based on Ag, Au and CeO_2, with intrinsic lower hydrogen activation activity, normally require higher pressures [158, 162].

Frequently underestimated, reaction parameters such as the mixing rate (relevant for liquid phase) might importantly impact the reaction output, and a careful consideration should be taken at the beginning of experimentation. For instance, Bruehwiler et al. investigated the reaction kinetics and their interplay with the mass transfer for the semi-hydrogenation of 2-methyl-3-butyn-2-ol catalysed by Pd/$CaCO_3$ [170].

When stirring rates higher than 1500 rpm were tested, no gas–liquid (G–L) mass-transfer limitations were observed while the alkenol selectivity increased with the stirring rate. In contrast, minor variations of the hydrogen liquid–solid (L–S) were registered, while the substrate L–S mass transfer turned relevant under high hydrogen concentration and high conversions.

Less intuitive parameters such as the heating method have also revealed to affect the catalysts reactivity in this reaction. For instance, Wu et al. reported the effect of using microwave irradiation as heating method in the aqueous phase semi-hydrogenation of 2-butyne-1,4-diol catalysed by a Pd/Bohemite catalyst [171]. The presented catalytic system heated by microwave irradiation evidenced enhanced conversion when compared to normal heating while the selectivity was maintained. The same authors also investigated the impact of ultrasound in the semi-semi-hydrogenation of phenylacetylenes employing the mentioned catalyst [172]. Interestingly, higher alkene selectivity was observed by the application of ultrasound, a phenomena ascribed to the accelerated desorption of the alkene product from the metal surface thus preventing the over-reduction reaction.

Very recently, Fukazawa et al. reported an unconventional approach for the electrocatalytic hydrogenation of alkynes in a proton-exchange membrane reactor (PEM) [173]. The alkene selectivity resulted sensitive to the catalyst material and the cathode potential. From the tested catalysts (Pd, Pt and Rh based), Pd/carbon was the more selective one, with the highest current efficiencies at relatively low electrolysis potentials (too negative potentials favoured the over-hydrogenation reaction). Although quite demanding in terms of the technical infrastructure required for the electrocatalytic reaction, this approach enables the hydrogenation employing only water as hydrogen source, and might represent an alternative when pressurized hydrogen or reactors are not available.

10.6 Conclusions

The preparation of well-defined nanostructured materials is nowadays accessible thanks to the cumulative efforts in the area of materials and catalysis. The availability of such tools, combined with the understanding of the structural and molecular parameters that rule the reactivity in the semi-hydrogenation of alkynes, permits the rationalization of highly structured nanostructures tailor-made for the specific reaction of application.

From this, not exhaustive but comprehensive review of the relevant literature, it can be summarized that most of the strategies evaluated to enhance the alkene selectivity can be classified in one of the following three categories (Fig. 10.17).

First, we can distinguish those assigned as "surrounding strategy" that includes the use of unreactive materials that wrap, surround or simply accompany the active phase (e.g. the solid support). Reaction conditions are included in this category as global parameters that can condition energy states of the whole catalyst, availability of hydrogen (and reactants for gas phase reaction) or diffusional phenomena.

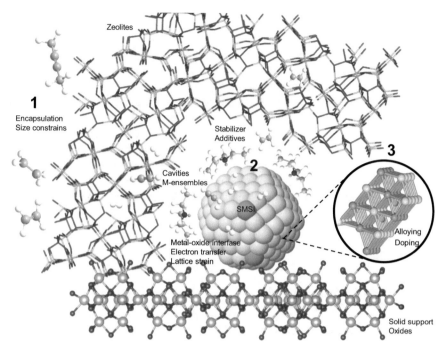

Fig. 10.17 Schematic representation of several strategies employed for the selectivity enhancement in the semi-hydrogenation of alkynes

A step closer to the active phase are the "surface strategy" that comprises the utilization of all types of modifiers that can be located on top of the active phase like stabilizers, additives, a second metal, etc. For instance, depending on the affinity for the metal surface of specific functional groups present in the modifiers (typically containing N), strong adsorption might result in the formation of geometric ensembles with restricted accessibility of reactants, as well as exert electronic disruptions in the d-band properties of the active metal, potentially affecting the energetic balance of the whole mechanistic path or the formation of subsurface species (e.g. H or C), all of them, resulting in effects on the alkene selectivity. Poisoning of unselective sites (typically low coordination atoms) has been extensively exploited as a strategy framed in this category, either by total or partial coverage with an unreactive metal (e.g. core@shell structures), dosing nitrogenated compounds (in liquid phase) or carbon monoxide (in gas phase), among others.

Finally, we can distinguish the strategy based on the modifications of structural properties of the active phase. This includes the modification of the internal part of the active phase for instance by doping the subsurface with small atoms (e.g. B, N, C), or the formation of alloyed systems. Any of the mentioned modifications would impact directly the d-band structure of the resulting metallic structure, with direct impact on the adsorption properties towards reagents and products. These effects are more dramatic for highly diluted active phases by a second metal, where large site isolation additionally occur.

While the semi-hydrogenation of alkynes has generally evidenced structure sensitivity, the utilization of sub-nanometer entities such as small metal clusters and single atoms accessible by current synthetic methodologies have uncovered a new type of potential catalysts of special value for demanding reactions such as the semi-hydrogenation of acetylene. In this line, modern unconventional catalytic systems such as cationic nickel or palladium single sites (with unusual oxidation states such as Pd(I)), as well as hybrid systems comprised by metal-oxide entities, (with potential to display synergistic reactivities at the metal interface) have demonstrated a huge potential in selective hydrogenation reactions, but are still at early stages of exploitation and therefore there is plenty of room for future developments.

References

1. Marteel-Parrish AE, Abraham MA (2013) Understanding the issues. In: Green chemistry and engineering. Wiley, pp 1–19. https://doi.org/10.1002/9781118720011.ch1
2. Descorme C, Gallezot P, Geantet C, George C (2012) Heterogeneous catalysis: a key tool toward sustainability. ChemCatChem 4(12):1897–1906. https://doi.org/10.1002/cctc.201 200483
3. Carenco S, Leyva-Pérez A, Concepción P, Boissière C, Mézailles N, Sanchez C, Corma A (2012) Nickel phosphide nanocatalysts for the chemoselective hydrogenation of alkynes. Nano Today 7(1):21–28. https://doi.org/10.1016/j.nantod.2011.12.003
4. Carenco S, Le Goff XF, Shi J, Roiban L, Ersen O, Boissière C, Sanchez C, Mézailles N (2011) Magnetic core–shell nanoparticles from nanoscale-induced phase segregation. Chem Mater 23(8):2270–2277. https://doi.org/10.1021/cm200575g
5. Planken KL, Kuipers BWM, Philipse AP (2008) Anal Chem 80:8871
6. Hori J, Murata K, Sugai T, Shinohara H, Noyori R, Arai N, Kurono N, Ohkuma T (2009) Highly active and selective semihydrogenation of alkynes with the palladium nanoparticles-tetrabutylammonium borohydride catalyst system. Adv Synth Catal 351(18):3143–3149. https://doi.org/10.1002/adsc.200900721
7. Yarulin A, Yuranov I, Cárdenas-Lizana F, Abdulkin P, Kiwi-Minsker L (2013) Size-effect of Pd-(Poly(N-vinyl-2-pyrrolidone)) nanocatalysts on selective hydrogenation of alkynols with different alkyl chains. J Phys Chem C 117(26):13424–13434. https://doi.org/10.1021/jp402258s
8. Vilé G, Baudouin D, Remediakis IN, Copéret C, López N, Pérez-Ramírez J (2013) Silver nanoparticles for olefin production: new insights into the mechanistic description of propyne hydrogenation. ChemCatChem 5(12):3750–3759. https://doi.org/10.1002/cctc.201300569
9. Lopez N, Vargas-Fuentes C (2012) Promoters in the hydrogenation of alkynes in mixtures: insights from density functional theory. Chem Commun 48(10):1379–1391. https://doi.org/10.1039/c1cc14922a
10. Crespo-Quesada M, Cárdenas-Lizana F, Dessimoz A-L, Kiwi-Minsker L (2012) Modern trends in catalyst and process design for alkyne hydrogenations. ACS Catal 2(8):1773–1786. https://doi.org/10.1021/cs300284r
11. Odom TW, Pileni M-P (2008) Nanoscience. Acc Chem Res 41(12):1565–1565. https://doi.org/10.1021/ar800253n
12. Beaumont SK (2014) Recent developments in the application of nanomaterials to understanding molecular level processes in cobalt catalysed Fischer-Tropsch synthesis. Phys Chem Chem Phys 16(11):5034–5043. https://doi.org/10.1039/c3cp55030c

13. Liu L, Corma A (2018) Metal catalysts for heterogeneous catalysis: from single atoms to nanoclusters and nanoparticles. Chem Rev 118(10):4981–5079. https://doi.org/10.1021/acs.chemrev.7b00776

14. Huang X, Xia Y, Cao Y, Zheng X, Pan H, Zhu J, Ma C, Wang H, Li J, You R, Wei S, Huang W, Lu J (2017) Enhancing both selectivity and coking-resistance of a single-atom Pd1/C3N4 catalyst for acetylene hydrogenation. Nano Res 10(4):1302–1312. https://doi.org/10.1007/s12274-016-1416-z

15. Delgado JA, Benkirane O, Claver C, Curulla-Ferre D, Godard C (2017) Advances in the preparation of highly selective nanocatalysts for the semi-hydrogenation of alkynes using colloidal approaches. Dalton Trans 46(37):12381–12403. https://doi.org/10.1039/C7DT01607G

16. Liu Y, Hu L, Chen H, Du H (2015) An alkene-promoted borane-catalyzed highly stereoselective hydrogenation of alkynes to Give Z- and E-alkenes. Chem Eur J 21(8):3495–3501. https://doi.org/10.1002/chem.201405388

17. Alonso F, Osante I, Yus M (2007) Highly selective hydrogenation of multiple carbon–carbon bonds promoted by nickel(0) nanoparticles. Tetrahedron 63(1):93–102. https://doi.org/10.1016/j.tet.2006.10.043

18. Huang B, Wang T, Yang Z, Qian W, Long J, Zeng G, Lei C (2017) Iron-based bimetallic nanocatalysts for highly selective hydrogenation of acetylene in N,N-dimethylformamide at room temperature. ACS Sustain Chem Eng 5(2):1668–1674. https://doi.org/10.1021/acssuschemeng.6b02413

19. Liang S, Hammond GB, Xu B (2016) Supported gold nanoparticles catalyzed cis-selective semihydrogenation of alkynes using ammonium formate as the reductant. Chem Commun 52(35):6013–6016. https://doi.org/10.1039/C6CC01318J

20. Slack ED, Gabriel CM, Lipshutz BH (2014) A palladium nanoparticle-nanomicelle combination for the stereoselective semihydrogenation of alkynes in water at room temperature. Angew Chem Int Ed 53(51):14051–14054. https://doi.org/10.1002/anie.201407723

21. Zhong J-J, Liu Q, Wu C-J, Meng Q-Y, Gao X-W, Li Z-J, Chen B, Tung C-H, Wu L-Z (2016) Combining visible light catalysis and transfer hydrogenation for in situ efficient and selective semihydrogenation of alkynes under ambient conditions. Chem Commun 52(9):1800–1803. https://doi.org/10.1039/C5CC08697C

22. Lian J, Chai Y, Qi Y, Guo X, Guan N, Li L, Zhang F (2020) Unexpectedly selective hydrogenation of phenylacetylene to styrene on titania supported platinum photocatalyst under 385 nm monochromatic light irradiation. Chin J Catal 41(4):598–603. https://doi.org/10.1016/S1872-2067(19)63453-4

23. Duca D, Arena F, Parmaliana A, Deganello G (1998) Hydrogenation of acetylene in ethylene rich feedstocks: comparison between palladium catalysts supported on pumice and alumina. App Catal A Gen 172(2):207–216. https://doi.org/10.1016/S0926-860X(98)00123-9

24. Osswald J, Giedigkeit R, Jentoft RE, Armbrüster M, Girgsdies F, Kovnir K, Ressler T, Grin Y, Schlögl R (2008) Palladium–gallium intermetallic compounds for the selective hydrogenation of acetylene: part I: preparation and structural investigation under reaction conditions. J Catal 258(1):210–218. https://doi.org/10.1016/j.jcat.2008.06.013

25. Yarulin AE, Crespo-Quesada RM, Egorova EV, Kiwi-Minsker LL (2012) Structure sensitivity of selective acetylene hydrogenation over the catalysts with shape-controlled palladium nanoparticles. Kinet Catal 53(2):253–261. https://doi.org/10.1134/s0023158412020152

26. Zhang Y, Diao W, Monnier JR, Williams CT (2015) Pd-Ag/SiO$_2$ bimetallic catalysts prepared by galvanic displacement for selective hydrogenation of acetylene in excess ethylene. Catal Sci Technol 5(8):4123–4132. https://doi.org/10.1039/c5cy00353a

27. Börger L, Cölfen H (1999) Prog Colloid Polym Sci 113:23

28. Molnár Á, Sárkány A, Varga M (2001) Hydrogenation of carbon–carbon multiple bonds: chemo-, regio- and stereo-selectivity. J Mol Catal A Chem 173(1–2):185–221. https://doi.org/10.1016/S1381-1169(01)00150-9

29. Bond GC (1962) Catalysis by metals. Academic Press, New York

30. Roucoux A, Schulz J, Patin H (2002) Reduced transition metal colloids: a novel family of reusable catalysts? Chem Rev 102(10):3757–3778. https://doi.org/10.1021/cr010350j

31. Ott LS, Finke RG (2007) Transition-metal nanocluster stabilization for catalysis: a critical review of ranking methods and putative stabilizers. Coord Chem Rev 251(9–10):1075–1100. https://doi.org/10.1016/j.ccr.2006.08.016

32. Astruc D, Lu F, Aranzaes JR (2005) Nanoparticles as recyclable catalysts: the frontier between homogeneous and heterogeneous catalysis. Angew Chem Int Ed 44(48):7852–7872. https://doi.org/10.1002/anie.200500766

33. Bond GC (1991) Supported metal catalysts: some unsolved problems. Chem Soc Rev 20(4):441–475. https://doi.org/10.1039/CS9912000441

34. Semagina N, Renken A, Kiwi-Minsker L (2007) Palladium nanoparticle size effect in 1-hexyne selective hydrogenation. J Phys Chem C 111(37):13933–13937. https://doi.org/10.1021/jp073944k

35. Semagina N, Renken A, Laub D, Kiwi-Minsker L (2007) Synthesis of monodispersed palladium nanoparticles to study structure sensitivity of solvent-free selective hydrogenation of 2-methyl-3-butyn-2-ol. J Catal 246(2):308–314. https://doi.org/10.1016/j.jcat.2006.12.011

36. Hu J, Zhou Z, Zhang R, Li L, Cheng Z (2014) Selective hydrogenation of phenylacetylene over a nano-Pd/α-Al2O3 catalyst. J Mol Catal A Chem 381:61–69. https://doi.org/10.1016/j.molcata.2013.10.008

37. Ruta M, Semagina N, Kiwi-Minsker L (2008) Monodispersed Pd nanoparticles for acetylene selective hydrogenation: particle size and support effects. J Phys Chem C 112(35):13635–13641. https://doi.org/10.1021/jp803800w

38. Teschner D, Vass E, Hävecker M, Zafeiratos S, Schnörch P, Sauer H, Knop-Gericke A, Schlögl R, Chamam M, Wootsch A, Canning AS, Gamman JJ, Jackson SD, McGregor J, Gladden LF (2006) Alkyne hydrogenation over Pd catalysts: a new paradigm. J Catal 242(1):26–37. https://doi.org/10.1016/j.jcat.2006.05.030

39. Lamey D, Prokopyeva I, Cárdenas-Lizana F, Kiwi-Minsker L (2014) Impact of organic-ligand shell on catalytic performance of colloidal Pd nanoparticles for alkyne gas-phase hydrogenation. Catal Today 235:79–89. https://doi.org/10.1016/j.cattod.2014.03.006

40. Crespo-Quesada M, Yarulin A, Jin M, Xia Y, Kiwi-Minsker L (2011) Structure sensitivity of alkynol hydrogenation on shape- and size-controlled palladium nanocrystals: which sites are most active and selective? J Am Chem Soc 133(32):12787–12794. https://doi.org/10.1021/ja204557m

41. Zhang L, Zhou M, Wang A, Zhang T (2020) Selective hydrogenation over supported metal catalysts: from nanoparticles to single atoms. Chem Rev 120(2):683–733. https://doi.org/10.1021/acs.chemrev.9b00230

42. Pei GX, Liu XY, Yang X, Zhang L, Wang A, Li L, Wang H, Wang X, Zhang T (2017) Performance of Cu-alloyed Pd single-atom catalyst for semihydrogenation of acetylene under simulated front-end conditions. ACS Catal 7(2):1491–1500. https://doi.org/10.1021/acscatal.6b03293

43. Feng Q, Zhao S, Wang Y, Dong J, Chen W, He D, Wang D, Yang J, Zhu Y, Zhu H, Gu L, Li Z, Liu Y, Yu R, Li J, Li Y (2017) Isolated single-atom Pd sites in intermetallic nanostructures: high catalytic selectivity for semihydrogenation of alkynes. J Am Chem Soc 139(21):7294–7301. https://doi.org/10.1021/jacs.7b01471

44. Huang F, Deng Y, Chen Y, Cai X, Peng M, Jia Z, Ren P, Xiao D, Wen X, Wang N, Liu H, Ma D (2018) Atomically dispersed Pd on nanodiamond/graphene hybrid for selective hydrogenation of acetylene. J Am Chem Soc 140(41):13142–13146. https://doi.org/10.1021/jacs.8b07476

45. Dai X, Chen Z, Yao T, Zheng L, Lin Y, Liu W, Ju H, Zhu J, Hong X, Wei S, Wu Y, Li Y (2017) Single Ni sites distributed on N-doped carbon for selective hydrogenation of acetylene. Chem Commun 53(84):11568–11571. https://doi.org/10.1039/C7CC04820C

46. Lin R, Albani D, Fako E, Kaiser SK, Safonova OV, López N, Pérez-Ramírez J (2019) Design of single gold atoms on nitrogen-doped carbon for molecular recognition in alkyne semi-hydrogenation. Angew Chem Int Ed 58(2):504–509. https://doi.org/10.1002/anie.201805820

47. Huang F, Deng Y, Chen Y, Cai X, Peng M, Jia Z, Xie J, Xiao D, Wen X, Wang N, Jiang Z, Liu H, Ma D (2019) Anchoring Cu1 species over nanodiamond-graphene for semi-hydrogenation of acetylene. Nat Commun 10(1):4431. https://doi.org/10.1038/s41467-019-12460-7

48. Ji S, Chen Y, Zhao S, Chen W, Shi L, Wang Y, Dong J, Li Z, Li F, Chen C, Peng Q, Li J, Wang D, Li Y (2019) Atomically dispersed ruthenium species inside metal-organic frameworks: combining the high activity of atomic sites and the molecular sieving effect of MOFs. Angew Chem Int Ed 58(13):4271–4275. https://doi.org/10.1002/anie.201814182

49. Chai Y, Wu G, Liu X, Ren Y, Dai W, Wang C, Xie Z, Guan N, Li L (2019) Acetylene-selective hydrogenation catalyzed by cationic nickel confined in zeolite. J Am Chem Soc 141(25):9920–9927. https://doi.org/10.1021/jacs.9b03361

50. Hamilton JF, Baetzold RC (1979) Catalysis by small metal clusters. Science 205(4412):1213–1220. https://doi.org/10.1126/science.205.4412.1213

51. Abdollahi T, Farmanzadeh D (2018) Selective hydrogenation of acetylene in the presence of ethylene on palladium nanocluster surfaces: a DFT study. Appl Surf Sci 433:513–529. https://doi.org/10.1016/j.apsusc.2017.10.085

52. Abdollahi T, Farmanzadeh D (2018) Graphene-supported Cu11 nanocluster as a candidate catalyst for the selective hydrogenation of acetylene: a density functional study. J Alloys Compd 735:117–130. https://doi.org/10.1016/j.jallcom.2017.11.051

53. Abdollahi T, Farmanzadeh D (2018) Activity and selectivity of Ni nanoclusters in the selective hydrogenation of acetylene: a computational investigation. C R Chim 21(5):484–493. https://doi.org/10.1016/j.crci.2018.01.002

54. Naoki T, Hu Y, Yiukihide S (2008) Recent progress in bimetallic nanoparticles: their preparation, structures and functions. In: Metal nanoclusters in catalysis and materials science. Elsevier, Amsterdam, pp 49–75. http://dx.doi.org/10.1016/B978-044453057-8.50005-2

55. Bronstein LM, Chernyshov DM, Volkov IO, Ezernitskaya MG, Valetsky PM, Matveeva VG, Sulman EM (2000) Structure and properties of bimetallic colloids formed in polystyrene-block-poly-4-vinylpyridine micelles: catalytic behavior in selective hydrogenation of dehydrolinalool. J Catal 196(2):302–314. https://doi.org/10.1006/jcat.2000.3039

56. Rieker T, Hanprasopwattana A, Datye A, Hubbard P (1999) Langmuir 15:638

57. Yarulin A, Yuranov I, Cárdenas-Lizana F, Alexander DTL, Kiwi-Minsker L (2014) How to increase the selectivity of Pd-based catalyst in alkynol hydrogenation: effect of second metal. App Catal A Gen 478:186–193. https://doi.org/10.1016/j.apcata.2014.04.003

58. Calver CF, Dash P, Scott RWJ (2011) Selective hydrogenations with Ag–Pd catalysts prepared by galvanic exchange reactions. ChemCatChem 3(4):695–697. https://doi.org/10.1002/cctc.201000346

59. Mitsudome T, Urayama T, Yamazaki K, Maehara Y, Yamasaki J, Gohara K, Maeno Z, Mizugaki T, Jitsukawa K, Kaneda K (2016) Design of core-Pd/shell-Ag nanocomposite catalyst for selective semihydrogenation of alkynes. ACS Catal 6(2):666–670. https://doi.org/10.1021/acscatal.5b02518

60. Sulman E, Matveeva V, Usanov A, Kosivtsov Y, Demidenko G, Bronstein L, Chernyshov D, Valetsky P (1999) Hydrogenation of acetylene alcohols with novel Pd colloidal catalysts prepared in block copolymers micelles. J Mol Catal A Chem 146(1–2):265–269. https://doi.org/10.1016/S1381-1169(99)00100-4

61. Diaz de los Bernardos M, Perez-Rodriguez S, Gual A, Claver C, Godard C (2017) Facile synthesis of NHC-stabilized Ni nanoparticles and their catalytic application in the Z-selective hydrogenation of alkynes. Chem Commun 53(56):7894–7897. http://dx.doi.org/10.1039/c7cc01779k

62. Lomelí-Rosales DA, Delgado JA, Díaz de los Bernardos M, Pérez-Rodríguez S, Gual A, Claver C, Godard C (2019) A general one-pot methodology for the preparation of mono- and bimetallic nanoparticles supported on carbon nanotubes: application in the semi-hydrogenation of alkynes and acetylene. Chem Eur J 25(35):8321–8331. http://dx.doi.org/10.1002/chem.201901041

63. Buxaderas E, Volpe MA, Radivoy G (2019) Selective semi-hydrogenation of terminal alkynes promoted by bimetallic Cu–Pd nanoparticles. Synthesis 51(6):1466–1472. https://doi.org/10.1055/s-0037-1610318

64. Ren M, Li C, Chen J, Wei M, Shi S (2014) Preparation of a ternary Pd-Rh-P amorphous alloy and its catalytic performance in selective hydrogenation of alkynes. Catal Sci Technol 4(7):1920–1924. https://doi.org/10.1039/c4cy00338a

65. Miyazaki M, Furukawa S, Takayama T, Yamazoe S, Komatsu T (2019) Surface modification of PdZn nanoparticles via galvanic replacement for the selective hydrogenation of terminal alkynes. ACS Appl Nano Mater 2(5):3307–3314. https://doi.org/10.1021/acsanm.9b00761

66. Hu M, Zhao S, Chen C, Chen W, Zhu W, Cheong W-C, Wang Y, Peng Q, Li J, Li Y, Hu M, Zhou K, Zhao S, Liu S, Liang C, Yu Y (2018) MOF-confined Sub-2 nm atomically ordered intermetallic PdZn nanoparticles as high-performance catalysts for selective hydrogenation of acetylene. Adv Mater e1801878

67. Mashkovsky IS, Markov PV, Bragina GO, Rassolov AV, Baeva GN, Stakheev AY (2017) Intermetallic Pd1-Zn1 nanoparticles in the selective liquid-phase hydrogenation of substituted alkynes. Kinet Catal 58(4):480–491. https://doi.org/10.1134/S0023158417040139

68. Zimmermann RR, Hahn T, Reschetilowski W, Armbruester M (2017) Kinetic parameters for the selective hydrogenation of acetylene on GaPd2 and GaPd. ChemPhysChem 18(18):2517–2525. https://doi.org/10.1002/cphc.201700535

69. Johnston SK, Bryant TA, Strong J, Lazzarini L, Ibhadon AO, Francesconi MG (2019) Stabilization of $Pd_{3-x}In_{1+x}$ polymorphs with Pd-like crystal structure and their superior performance as catalysts for semi-hydrogenation of alkynes. ChemCatChem (Ahead of Print). https://doi.org/10.1002/cctc.201900391

70. Rao D-M, Zhang S-T, Li C-M, Chen Y-D, Pu M, Yan H, Wei M (2018) The reaction mechanism and selectivity of acetylene hydrogenation over Ni–Ga intermetallic compound catalysts: a density functional theory study. Dalton Trans 47(12):4198–4208. https://doi.org/10.1039/C7DT04726F

71. Aviziotis IG, Duguet T, Soussi K, Heggen M, Lafont M-C, Morfin F, Mishra S, Daniele S, Boudouvis AG, Vahlas C (2018) Chemical vapor deposition of $Al_{13}Fe_4$ highly selective catalytic films for the semi-hydrogenation of acetylene. Phys Status Solidi A 215(2). https://doi.org/10.1002/pssa.201700692

72. Kojima T, Kojima T, Kameoka S, Tsai A-P, Fujii S, Ueda S, Ueda S (2018) Catalysis-tunable Heusler alloys in selective hydrogenation of alkynes: a new potential for old materials. Sci Adv 4(10):eaat6063

73. Xiu J-h, Chen X, Yang Z-h, Cao H, Bai Z-x, Liang C-h (2017) Chemical synthesis of Al-Co intermetallic compound catalysts for selective hydrogenation of alkyne. Fenzi Cuihua 31(4):325–333

74. Chan CWA, Mahadi AH, Li MM-J, Corbos EC, Tang C, Jones G, Kuo WCH, Cookson J, Brown CM, Bishop PT, Tsang SCE (2014) Interstitial modification of palladium nanoparticles with boron atoms as a green catalyst for selective hydrogenation. Nat Commun 5:5787. https://doi.org/10.1038/ncomms6787. http://www.nature.com/articles/ncomms6787#supplementary-information

75. Burch R, Lewis FA (1970) Absorption of hydrogen by palladium + boron and palladium + silver + boron alloys. Trans Faraday Soc 66:727–735. https://doi.org/10.1039/TF9706600727

76. Zhang Y, Wen X, Shi Y, Yue R, Bai L, Liu Q, Ba X (2019) Sulfur-containing polymer as a platform for synthesis of size-controlled pd nanoparticles for selective semihydrogenation of alkynes. Ind Eng Chem Res 58(3):1142–1149. https://doi.org/10.1021/acs.iecr.8b04913

77. Albani D, Chen Z, Mitchell S, Perez-Ramirez J, Shahrokhi M, Lopez N, Hauert R (2018) Selective ensembles in supported palladium sulfide nanoparticles for alkyne semi-hydrogenation. Nat Commun 9(1):2634

78. McCue AJ, Guerrero-Ruiz A, Ramirez-Barria C, Rodriguez-Ramos I, Anderson JA (2017) Selective hydrogenation of mixed alkyne/alkene streams at elevated pressure over a palladium sulfide catalyst. J Catal 355:40–52. https://doi.org/10.1016/j.jcat.2017.09.004

79. Phua P-H, Lefort L, Boogers JAF, Tristany M, de Vries JG (2009) Soluble iron nanoparticles as cheap and environmentally benign alkene and alkyne hydrogenation catalysts. Chem Commun 25:3747–3749. https://doi.org/10.1039/b820048c

80. Rangheard C, de Julian Fernandez C, Phua P-H, Hoorn J, Lefort L, de Vries JG (2010) At the frontier between heterogeneous and homogeneous catalysis: hydrogenation of olefins and alkynes with soluble iron nanoparticles. Dalton Trans 39(36):8464–8471. https://doi.org/10. 1039/c0dt00177e

81. Gregori BJ, Schwarzhuber F, Pöllath S, Zweck J, Fritsch L, Schoch R, Bauer M, Jacobi von Wangelin A (2019) Stereoselective alkyne hydrogenation by using a simple iron catalyst. ChemSusChem 12(16):3864–3870. https://doi.org/10.1002/cssc.201900926

82. Hudson R, Riviere A, Cirtiu CM, Luska KL, Moores A (2012) Iron-iron oxide core-shell nanoparticles are active and magnetically recyclable olefin and alkyne hydrogenation catalysts in protic and aqueous media. Chem Commun 48(27):3360–3362. https://doi.org/10.1039/c2c c16438h

83. Cirtiu CM, Raychoudhury T, Ghoshal S, Moores A (2011) Systematic comparison of the size, surface characteristics and colloidal stability of zero valent iron nanoparticles pre- and post-grafted with common polymers. Colloids Surf A 390(1–3):95–104. https://doi.org/10. 1016/j.colsurfa.2011.09.011

84. Gieshoff TN, Welther A, Kessler MT, Prechtl MHG, Jacobi von Wangelin A (2014) Stereoselective iron-catalyzed alkyne hydrogenation in ionic liquids. Chem Commun 50(18):2261–2264. https://doi.org/10.1039/c3cc49679a

85. Kelsen V, Wendt B, Werkmeister S, Junge K, Beller M, Chaudret B (2013) The use of ultrasmall iron(0) nanoparticles as catalysts for the selective hydrogenation of unsaturated C–C bonds. Chem Commun 49(33):3416–3418. https://doi.org/10.1039/c3cc00152k

86. Liu Y, Zhao J, Feng J, He Y, Du Y, Li D (2018) Layered double hydroxide-derived Ni–Cu nanoalloy catalysts for semi-hydrogenation of alkynes: Improvement of selectivity and anti-coking ability via alloying of Ni and Cu. J Catal 359:251–260. https://doi.org/10.1016/j.jcat. 2018.01.009

87. Bruno JE, Dwarica NS, Whittaker TN, Hand ER, Guzman CS, Dasgupta A, Chen Z, Rioux RM, Chandler BD (2020) Supported Ni–Au colloid precursors for active, selective, and stable alkyne partial hydrogenation catalysts. ACS Catal 10(4):2565–2580. https://doi.org/10.1021/ acscatal.9b05402

88. Brown CA, Brown HC (1963) The reaction of sodium borohydride with nickel acetate in aqueous solution—a convenient synthesis of an active nickel hydrogenation catalyst of low isomerizing tendency. J Am Chem Soc 85(7):1003–1005. https://doi.org/10.1021/ja0089 0a040

89. Brown HC, Brown CA (1963) The reaction of sodium borohydride with nickel acetate in ethanol solution—a highly selective nickel hydrogenation catalyst. J Am Chem Soc 85(7):1005–1006. https://doi.org/10.1021/ja00890a041

90. Brown HC, Brown CA (1966) Catalytic hydrogenation—I: the reaction of platinum metal salts with sodium borohydride—new active platinum metal catalysts for hydrogenation. Tetrahedron 22, Supplement 8(0):149–164. http://dx.doi.org/10.1016/S0040-4020(01)821 80-3

91. Brown CA (1970) The hydrogenation of hex-1-yne over borohydride-reduced catalysts. Evidence concerning the mechanism of reduction. J Chem Soc D Chem Commun 3:139–140. https://doi.org/10.1039/c29700000139

92. Albani D, Karajovic K, Tata B, Li Q, Mitchell S, Lopez N, Perez-Ramirez J (2019) Ensemble design in nickel phosphide catalysts for alkyne semi-hydrogenation. ChemCatChem 11(1):457–464. https://doi.org/10.1002/cctc.201801430

93. Zhao M (2016) Fabrication of ultrafine palladium phosphide nanoparticles as highly active catalyst for chemoselective hydrogenation of alkynes. Chem Asian J 11(4):461–464. https:// doi.org/10.1002/asia.201500939

94. Zhao M, Ji Y, Wang M, Zhong N, Kang Z, Asao N, Jiang W-J, Chen Q (2017) Composition-dependent morphology of bi- and trimetallic phosphides: construction of amorphous Pd–Cu–Ni–P nanoparticles as a selective and versatile catalyst. ACS Appl Mater Interfaces 9(40):34804–34811. https://doi.org/10.1021/acsami.7b08082

95. Shi X, Wen X, Nie S, Dong J, Li J, Shi Y, Zhang H, Bai G (2020) Fabrication of Ni₃N nanorods anchored on N-doped carbon for selective semi-hydrogenation of alkynes. J Catal 382:22–30. https://doi.org/10.1016/j.jcat.2019.12.005

96. Zhang Y, Diao W, Williams CT, Monnier JR (2014) Selective hydrogenation of acetylene in excess ethylene using Ag- and Au–Pd/SiO₂ bimetallic catalysts prepared by electroless deposition. App Catal A Gen 469:419–426. https://doi.org/10.1016/j.apcata.2013.10.024

97. Evangelisti C, Panziera N, D'Alessio A, Bertinetti L, Botavina M, Vitulli G (2010) New monodispersed palladium nanoparticles stabilized by poly-(N-vinyl-2-pyrrolidone): preparation, structural study and catalytic properties. J Catal 272(2):246–252. https://doi.org/10.1016/j.jcat.2010.04.006

98. Hirai H, Yakura N (2001) Protecting polymers in suspension of metal nanoparticles. Polym Adv Technol 12(11–12):724–733. https://doi.org/10.1002/pat.95

99. Telkar MM, Rode CV, Chaudhari RV, Joshi SS, Nalawade AM (2004) Shape-controlled preparation and catalytic activity of metal nanoparticles for hydrogenation of 2-butyne-1,4-diol and styrene oxide. App Catal A Gen 273(1–2):11–19. https://doi.org/10.1016/j.apcata.2004.05.056

100. Witte PT, Boland S, Kirby F, van Maanen R, Bleeker BF, de Winter DAM, Post JA, Geus JW, Berben PH (2013) NanoSelect Pd catalysts: what causes the high selectivity of these supported colloidal catalysts in alkyne semi-hydrogenation? ChemCatChem 5(2):582–587. http://dx.doi.org/10.1002/cctc.201200460

101. Niu M, Wang Y, Li W, Jiang J, Jin Z (2013) Highly efficient and recyclable ruthenium nanoparticle catalyst for semihydrogenation of alkynes. Catal Commun 38:77–81. https://doi.org/10.1016/j.catcom.2013.04.015

102. Zharmagambetova A, Zamanbekova A, Jumekeyeva A, Tumabayev N (2017) Study of nickel catalysts in hydrogenation of acetylene alcohols at low-temperature. Izv Nats Akad Nauk Resp Kaz, Ser Khim Tekhnol 4:65–72

103. Zharmagambetova AK, Zamanbekova AT, Darmenbayeva AS, Auyezkhanova AS, Jumekeyeva AI, Talgatov ET (2017) Effect of polymers on the formation of nanosized palladium catalysts and their activity and selectivity in the hydrogenation of acetylenic alcohols. Theor Exp Chem 53(4):265–269. https://doi.org/10.1007/s11237-017-9524-8

104. Long W, Brunelli NA, Didas SA, Ping EW, Jones CW (2013) Aminopolymer-silica composite-supported Pd catalysts for selective hydrogenation of alkynes. ACS Catal 3(8):1700–1708. https://doi.org/10.1021/cs3007395

105. Kuwahara Y, Kango H, Yamashita H (2019) Pd nanoparticles and aminopolymers confined in hollow silica spheres as efficient and reusable heterogeneous catalysts for semihydrogenation of alkynes. ACS Catal 9(3):1993–2006. https://doi.org/10.1021/acscatal.8b04653

106. Maesing F, Wang X, Nuesse H, Klingauf J, Studer A (2017) Facile light-mediated preparation of small polymer-coated palladium-nanoparticles and their application as catalysts for alkyne semi-hydrogenation. Chem Eur J 23(25):6014–6018. https://doi.org/10.1002/chem.201603297

107. Maesing F, Nuesse H, Klingauf J, Studer A (2017) Light Mediated preparation of palladium nanoparticles as catalysts for alkyne cis-semihydrogenation. Org Lett 19(10):2658–2661. https://doi.org/10.1021/acs.orglett.7b00999

108. Semagina N, Joannet E, Parra S, Sulman E, Renken A, Kiwi-Minsker L (2005) Palladium nanoparticles stabilized in block-copolymer micelles for highly selective 2-butyne-1,4-diol partial hydrogenation. App Catal A Gen 280(2):141–147. https://doi.org/10.1016/j.apcata.2004.10.049

109. Mizugaki T, Murata M, Fukubayashi S, Mitsudome T, Jitsukawa K, Kaneda K (2008) PAMAM dendron-stabilised palladium nanoparticles: effect of generation and peripheral groups on particle size and hydrogenation activity. Chem Commun 2:241–243. https://doi.org/10.1039/b710860e

110. Karakhanov EA, Maximov AL, Zolotukhina AV (2019) Selective semi-hydrogenation of phenyl acetylene by Pd nanocatalysts encapsulated into dendrimer networks. Mol Catal 469:98–110. https://doi.org/10.1016/j.mcat.2019.03.005

111. Olivier-Bourbigou H, Magna L, Morvan D (2010) Ionic liquids and catalysis: recent progress from knowledge to applications. App Catal A Gen 373(1–2):1–56. https://doi.org/10.1016/j.apcata.2009.10.008
112. Venkatesan R, Prechtl MHG, Scholten JD, Pezzi RP, Machado G, Dupont J (2011) Palladium nanoparticle catalysts in ionic liquids: synthesis, characterisation and selective partial hydrogenation of alkynes to Z-alkenes. J Mater Chem 21(9):3030–3036. https://doi.org/10.1039/c0jm03557b
113. Leal BC, Consorti CS, Machado G, Dupont J (2015) Palladium metal nanoparticles stabilized by ionophilic ligands in ionic liquids: synthesis and application in hydrogenation reactions. Catal Sci Technol 5(2):903–909. https://doi.org/10.1039/C4CY01116C
114. López-Vinasco AM, Martínez-Prieto LM, Asensio JM, Lecante P, Chaudret B, Cámpora J, van Leeuwen PWNM (2020) Novel nickel nanoparticles stabilized by imidazolium-amidinate ligands for selective hydrogenation of alkynes. Catal Sci Technol 10(2):342–350. https://doi.org/10.1039/C9CY02172H
115. Peng L, Zhang J, Yang S, Han B, Sang X, Liu C, Yang G (2015) The ionic liquid microphase enhances the catalytic activity of Pd nanoparticles supported by a metal-organic framework. Green Chem 17(8):4178–4182. https://doi.org/10.1039/C5GC01333J
116. Philippot K, Chaudret B (2003) Organometallic approach to the synthesis and surface reactivity of noble metal nanoparticles. C R Chim 6(8–10):1019–1034. https://doi.org/10.1016/j.crci.2003.07.010
117. Chaudret B (2005) Organometallic approach to nanoparticles synthesis and self-organization. C R Phys 6(1):117–131. https://doi.org/10.1016/j.crhy.2004.11.008
118. Caubère P (1983) Complex reducing agents (CRA's)—versatile, novel ways of using sodium hydride in organic synthesis. Angew Chem Int Ed Engl 22(8):599–613. https://doi.org/10.1002/anie.198305991
119. Brunet JJ, Gallois P, Caubere P (1977) Activation of reducing agents. Sodium hydride containing complex reducing agents. VII NiC, a new heterogeneous Ni hydrogenation catalyst. Tetrahedron Lett 18(45):3955–3958. http://dx.doi.org/10.1016/S0040-4039(01)83401-8
120. Brunet JJ, Gallois P, Caubere P (1980) Activation of reducing agents. Sodium hydride containing complex reducing agents. 12. New convenient, highly active, and selective nickel hydrogenation catalysts. J Organ Chem 45(10):1937–1945. https://doi.org/10.1021/jo01298a036
121. Gallois P, Brunet JJ, Caubere P (1980) Activation of reducing agents. Sodium hydride containing complex reducing agents. 13. Selective heterogeneous hydrogenation of polyfunctional substrates over Nic. J Organ Chem 45(10):1946–1950. https://doi.org/10.1021/jo01298a037
122. Brunet JJ, Caubere P (1984) Activation of reducing agents. Sodium hydride containing complex reducing agents. 20. Pdc, a new, very selective heterogeneous hydrogenation catalyst. J Org Chem 49(21):4058–4060. https://doi.org/10.1021/jo00195a037
123. Montiel L, Delgado JA, Novell M, Andrade FJ, Claver C, Blondeau P, Godard C (2016) A simple and versatile approach for the fabrication of paper-based nanocatalysts: low cost, easy handling, and catalyst recovery. ChemCatChem 8(19):3041–3044. https://doi.org/10.1002/cctc.201600666
124. Salnikov OG, Liu H-J, Fedorov A, Burueva DB, Kovtunov KV, Coperet C, Koptyug IV (2017) Pairwise hydrogen addition in the selective semihydrogenation of alkynes on silica-supported Cu catalysts. Chem Sci 8(3):2426–2430. https://doi.org/10.1039/C6SC05276B
125. Reina A, Favier I, Pradel C, Gómez M (2018) Stable zero-valent nickel nanoparticles in glycerol: synthesis and applications in selective hydrogenations. Adv Synth Catal 360(18):3544–3552. https://doi.org/10.1002/adsc.201800786
126. Witte P, Berben P, Boland S, Boymans E, Vogt D, Geus J, Donkervoort J (2012) BASF nanoSelect™ technology: innovative supported Pd- and Pt-based catalysts for selective hydrogenation reactions. Top Catal 55(7–10):505–511. https://doi.org/10.1007/s11244-012-9818-y
127. Kirby F, Moreno-Marrodan C, Baán Z, Bleeker BF, Barbaro P, Berben PH, Witte PT (2014) NanoSelect precious metal catalysts and their use in asymmetric heterogeneous catalysis. ChemCatChem 6(10):2904–2909. https://doi.org/10.1002/cctc.201402310

128. Vilé G, Almora-Barrios N, Mitchell S, López N, Pérez-Ramírez J (2014) From the Lindlar catalyst to supported ligand-modified palladium nanoparticles: selectivity patterns and accessibility constraints in the continuous-flow three-phase hydrogenation of acetylenic compounds. Chem Eur J 20(20):5926–5937. https://doi.org/10.1002/chem.201304795

129. Albani D, Vile G, Mitchell S, Witte PT, Almora-Barrios N, Verel R, Lopez N, Perez-Ramirez J (2016) Ligand ordering determines the catalytic response of hybrid palladium nanoparticles in hydrogenation. Catal Sci Technol 6(6):1621–1631. https://doi.org/10.1039/C5CY01921D

130. La Sorella G, Sperni L, Canton P, Coletti L, Fabris F, Strukul G, Scarso A (2018) Selective hydrogenations and dechlorinations in water mediated by anionic surfactant-stabilized Pd nanoparticles. J Org Chem 83(14):7438–7446. https://doi.org/10.1021/acs.joc.8b00314

131. Zhang H, Yang Y, Dai W, Yang D, Lu S, Ji Y (2012) An aqueous-phase catalytic process for the selective hydrogenation of acetylene with monodisperse water soluble palladium nanoparticles as catalyst. Catal Sci Technol 2(7):1319–1323. https://doi.org/10.1039/c2cy20179h

132. Crespo-Quesada M, Andanson J-M, Yarulin A, Lim B, Xia Y, Kiwi-Minsker L (2011) UV–ozone cleaning of supported poly(vinylpyrrolidone)-stabilized palladium nanocubes: effect of stabilizer removal on morphology and catalytic behavior. Langmuir 27(12):7909–7916. https://doi.org/10.1021/la201007m

133. Borodziński A, Bond GC (2006) Selective hydrogenation of ethyne in ethene-rich streams on palladium catalysts. Part 1. Effect of changes to the catalyst during reaction. Catal Rev 48(2):91–144. https://doi.org/10.1080/01614940500364909

134. García-Mota M, Bridier B, Pérez-Ramírez J, López N (2010) Interplay between carbon monoxide, hydrides, and carbides in selective alkyne hydrogenation on palladium. J Catal 273(2):92–102. https://doi.org/10.1016/j.jcat.2010.04.018

135. Studt F, Abild-Pedersen F, Bligaard T, Sørensen RZ, Christensen CH, Nørskov JK (2008) On the role of surface modifications of palladium catalysts in the selective hydrogenation of acetylene. Angew Chem Int Ed 47(48):9299–9302. https://doi.org/10.1002/anie.200802844

136. Delannay F (1984) Characterization of heterogeneous catalysts

137. Bae W, Mehra RK (1998) J Inorg Biochem 70:125

138. Brown CA, Ahuja VK (1973) Catalytic hydrogenation. VI. Reaction of sodium borohydride with nickel salts in ethanol solution. P-2 Nickel, a highly convenient, new, selective hydrogenation catalyst with great sensitivity to substrate structure. J Organ Chem 38(12):2226–2230. https://doi.org/10.1021/jo00952a024

139. Brown CA, Ahuja VK (1973) "P-2 nickel" catalyst with ethylenediamine, a novel system for highly stereospecific reduction of alkynes to cis-olefins. J Chem Soc Chem Commun 15:553–554. https://doi.org/10.1039/c39730000553

140. Kwon SG, Krylova G, Sumer A, Schwartz MM, Bunel EE, Marshall CL, Chattopadhyay S, Lee B, Jellinek J, Shevchenko EV (2012) Capping ligands as selectivity switchers in hydrogenation reactions. Nano Lett 12(10):5382–5388. https://doi.org/10.1021/nl3027636

141. Shevchenko EV, Talapin DV, Rogach AL, Kornowski A, Haase M, Weller H (2002) Colloidal synthesis and self-assembly of CoPt$_3$ nanocrystals. J Am Chem Soc 124(38):11480–11485. https://doi.org/10.1021/ja025976l

142. Fiorio JL, Lopez N, Rossi LM (2017) Gold-ligand catalyzed selective hydrogenation of alkynes into cis-alkenes via H$_2$ heterolytic activation by frustrated Lewis pairs. ACS Catal. https://doi.org/10.1021/acscatal.6b03441

143. Nikoshvili LZ, Bykov AV, Khudyakova TE, LaGrange T, Heroguel F, Luterbacher JS, Matveeva VG, Sulman EM, Dyson PJ, Kiwi-Minsker L (2017) Promotion effect of alkali metal hydroxides on polymer-stabilized pd nanoparticles for selective hydrogenation of C–C triple bonds in alkynols. Ind Eng Chem Res 56(45):13219–13227. https://doi.org/10.1021/acs.iecr.7b01612

144. Borodziński A, Bond GC (2008) Selective Hydrogenation of ethyne in ethene-rich streams on palladium catalysts, part 2: steady-state kinetics and effects of palladium particle size, carbon monoxide, and promoters. Catal Rev 50(3):379–469. https://doi.org/10.1080/01614940802142102

145. López N, Bridier B, Pérez-Ramírez J (2008) Discriminating reasons for selectivity enhancement of CO in alkyne hydrogenation on palladium. J Phys Chem C 112(25):9346–9350. https://doi.org/10.1021/jp711258q

146. Uzio D, Berhault G (2010) Factors governing the catalytic reactivity of metallic nanoparticles. Catal Rev 52(1):106–131. https://doi.org/10.1080/01614940903510496

147. Corma A, Serna P, Concepción P, Calvino JJ (2008) Transforming nonselective into chemoselective metal catalysts for the hydrogenation of substituted nitroaromatics. J Am Chem Soc 130(27):8748–8753. https://doi.org/10.1021/ja800959g

148. Moreno-Castilla C, Ferro-Garcia MA, Joly JP, Bautista-Toledo I, Carrasco-Marin F, Rivera-Utrilla J (1995) Activated carbon surface modifications by nitric acid, hydrogen peroxide, and ammonium peroxydisulfate treatments. Langmuir 11(11):4386–4392. https://doi.org/10.1021/la00011a035

149. Semagina N, Grasemann M, Xanthopoulos N, Renken A, Kiwi-Minsker L (2007) Structured catalyst of Pd/ZnO on sintered metal fibers for 2-methyl-3-butyn-2-ol selective hydrogenation. J Catal 251(1):213–222. https://doi.org/10.1016/j.jcat.2007.06.028

150. Mao S, Zhao B, Wang Z, Gong Y, Lü G, Ma X, Yu L, Wang Y (2019) Tuning the catalytic performance for the semi-hydrogenation of alkynols by selectively poisoning the active sites of Pd catalysts. Green Chem 21(15):4143–4151. https://doi.org/10.1039/C9GC01356C

151. Easterday R, Leonard C, Sanchez-Felix O, Losovyj Y, Pink M, Stein BD, Morgan DG, Lyubimova NA, Nikoshvili LZ, Sulman EM, Mahmoud WE, Al-Ghamdi AA, Bronstein LM (2014) Fabrication of magnetically recoverable catalysts based on mixtures of Pd and iron oxide nanoparticles for hydrogenation of alkyne alcohols. ACS Appl Mater Interfaces 6(23):21652–21660. https://doi.org/10.1021/am5067223

152. Uberman PM, Costa NJS, Philippot K, C. Carmona R, Dos Santos AA, Rossi LM (2014) A recoverable Pd nanocatalyst for selective semi-hydrogenation of alkynes: hydrogenation of benzyl-propargylamines as a challenging model. Green Chem 16(10):4566–4574. http://dx.doi.org/10.1039/c4gc00669k

153. Tejeda-Serrano M, Cabrero-Antonino JR, Mainar-Ruiz V, Lopez-Haro M, Hernandez-Garrido JC, Calvino JJ, Leyva-Perez A, Corma A (2017) Synthesis of supported planar iron oxide nanoparticles and their chemo- and stereoselectivity for hydrogenation of alkynes. ACS Catal 7(5):3721–3729. https://doi.org/10.1021/acscatal.7b00037

154. Ruiz Puigdollers A, Schlexer P, Tosoni S, Pacchioni G (2017) Increasing oxide reducibility: the role of metal/oxide interfaces in the formation of oxygen vacancies. ACS Catal 7(10):6493–6513. https://doi.org/10.1021/acscatal.7b01913

155. Schlexer P, Ruiz Puigdollers A, Pacchioni G (2019) Role of metal/oxide interfaces in enhancing the local oxide reducibility. Top Catal 62(17):1192–1201. https://doi.org/10.1007/s11244-018-1056-5

156. Yang S, Cao C, Peng L, Zhang J, Han B, Song W (2016) A Pd-Cu$_2$O nanocomposite as an effective synergistic catalyst for selective semi-hydrogenation of the terminal alkynes only. Chem Commun 52(18):3627–3630. https://doi.org/10.1039/c6cc00143b

157. Liu K, Qin R, Zhou L, Liu P, Zhang Q, Jing W, Ruan P, Gu L, Fu G, Zheng N (2019) Cu$_2$O-supported atomically dispersed Pd catalysts for semihydrogenation of terminal alkynes: critical role of oxide supports. CCS Chemistry 1(2):207–214. https://doi.org/10.31635/ccschem.019.20190008

158. Mitsudome T, Yamamoto M, Maeno Z, Mizugaki T, Jitsukawa K, Kaneda K (2015) One-step synthesis of core-gold/shell-ceria nanomaterial and its catalysis for highly selective semihydrogenation of alkynes. J Am Chem Soc 137(42):13452–13455. https://doi.org/10.1021/jacs.5b07521

159. Wang M-M, He L, Liu Y-M, Cao Y, He H-Y, Fan K-N (2011) Gold supported on mesostructured ceria as an efficient catalyst for the chemoselective hydrogenation of carbonyl compounds in neat water. Green Chem 13(3):602–607. https://doi.org/10.1039/C0GC00937G

160. Shao L, Huang X, Teschner D, Zhang W (2014) Gold supported on graphene oxide: an active and selective catalyst for phenylacetylene hydrogenations at low temperatures. ACS Catal 4(7):2369–2373. https://doi.org/10.1021/cs5002724

344 J. A. Delgado and C. Godard

161. Vilé G, Dähler P, Vecchietti J, Baltanás M, Collins S, Calatayud M, Bonivardi A, Pérez-Ramírez J (2015) Promoted ceria catalysts for alkyne semi-hydrogenation. J Catal 324:69–78. https://doi.org/10.1016/j.jcat.2015.01.020

162. Vile G, Perez-Ramirez J (2014) Beyond the use of modifiers in selective alkyne hydrogenation: silver and gold nanocatalysts in flow mode for sustainable alkene production. Nanoscale 6(22):13476–13482. https://doi.org/10.1039/c4nr02777a

163. Riley C, De La Riva A, Zhou S, Wan Q, Peterson E, Artyushkova K, Farahani MD, Friedrich HB, Burkemper L, Atudorei N-V, Lin S, Guo H, Datye A (2019) Synthesis of nickel-doped ceria catalysts for selective acetylene hydrogenation. ChemCatChem 11(5):1526–1533. https://doi.org/10.1002/cctc.201801976

164. Patarroyo J, Delgado JA, Merkoçi F, Genç A, Sauthier G, Llorca J, Arbiol J, Bastus NG, Godard C, Claver C, Puntes V (2019) Hollow PdAg-CeO$_2$ heterodimer nanocrystals as highly structured heterogeneous catalysts. Sci Rep 9(1):18776. https://doi.org/10.1038/s41598-019-55105-x

165. Gulyaeva YK, Kaichev VV, Zaikovskii VI, Suknev AP, Bal'zhinimaev BS (2015) Selective hydrogenation of acetylene over Pd/Fiberglass catalysts: kinetic and isotopic studies. App Catal A Gen 506:197–205. http://dx.doi.org/10.1016/j.apcata.2015.09.016

166. Liang H, Zhang B, Ge H, Gu X, Zhang S, Qin Y (2017) Porous TiO$_2$/Pt/TiO$_2$ sandwich catalyst for highly selective semihydrogenation of alkyne to olefin. ACS Catal 7(10):6567–6572. https://doi.org/10.1021/acscatal.7b02032

167. Chen F, Kreyenschulte C, Radnik J, Lund H, Surkus A-E, Junge K, Beller M (2017) Selective semihdrogenation of alkynes with N-graphitic-modified cobalt nanoparticles supported on silica. ACS Catal 7(3):1526–1532. https://doi.org/10.1021/acscatal.6b03140

168. Wang S, Zhao Z-J, Chang X, Zhao J, Tian H, Yang C, Li M, Fu Q, Mu R, Gong J (2019) Activation and spillover of hydrogen on sub-1 nm palladium nanoclusters confined within sodalite zeolite for the semi-hydrogenation of alkynes. Angew Chem Int Ed 58(23):7668–7672. https://doi.org/10.1002/anie.201903827

169. Chesnokov VV, Podyacheva OY, Richards RM (2017) Influence of carbon nanomaterials on the properties of Pd/C catalysts in selective hydrogenation of acetylene. Mater Res Bull 88:78–84. https://doi.org/10.1016/j.materresbull.2016.12.013

170. Bruehwiler A, Semagina N, Grasemann M, Renken A, Kiwi-Minsker L, Saaler A, Lehmann H, Bonrath W, Roessler F (2008) Three-phase catalytic hydrogenation of a functionalized alkyne: mass transfer and kinetic studies with in situ hydrogen monitoring. Ind Eng Chem Res 47(18):6862–6869. https://doi.org/10.1021/ie800070w

171. Wu Z, Calcio Gaudino E, Rotolo L, Medlock J, Bonrath W, Cravotto G (2016) Efficient partial hydrogenation of 2-butyne-1,4-diol and other alkynes under microwave irradiation. Chem Eng Process 110:220–224. https://doi.org/10.1016/j.cep.2016.10.016

172. Wu Z, Cravotto G, Gaudino EC, Giacomino A, Medlock J, Bonrath W (2017) Ultrasonically improved semi-hydrogenation of alkynes to (Z-)alkenes over novel lead-free Pd/Boehmite catalysts. Ultrason Sonochem Part B 35:664–672. http://dx.doi.org/10.1016/j.ultsonch.2016.05.019

173. Fukazawa A, Minoshima J, Tanaka K, Hashimoto Y, Kobori Y, Sato Y, Atobe M (2019) A new approach to stereoselective electrocatalytic semihydrogenation of alkynes to Z-alkenes using a proton-exchange membrane reactor. ACS Sustain Chem Eng 7(13):11050–11055. https://doi.org/10.1021/acssuschemeng.9b01882

Chapter 11
Selective Hydrogenation of Aldehydes and Ketones

Israel Cano and Piet W. N. M. van Leeuwen

Abstract The selective hydrogenation of α,β-unsaturated aldehydes and ketones catalyzed by metal nanoparticles is reviewed with an emphasis on the recent advances in this area. Key issue is the selectivity for allyl alcohol formation. General trends in the use of conditions, ligands and co-catalysts were observed. In spite of the progress achieved in the last two decades, there is still a quest for cost-effective, selective and fast new catalysts.

Keywords Chemoselective hydrogenation · Metal nanoparticles · α,β-unsaturated carbonyl compounds · Unsaturated aldehydes · Unsaturated ketones · Unsaturated alcohol · Au NPs · Lewis acids · Oxide supports · Heterolytic cleavage of H_2

11.1 Introduction

Selective hydrogenation of α,β-unsaturated aldehydes and ketones has been a long-standing theme in catalysis as it is scientifically interesting which function will be hydrogenated, but also because partial hydrogenation is of industrial importance. The target molecule is usually the allylic alcohol, as the other half-product, the saturated aldehyde or ketone, is thermodynamically more stable than the former and via isomerization one can always arrive at full selectivity to the saturated aldehyde or ketone (Scheme 11.1). In terms of selectivity, therefore, the product of most interest is the unsaturated alcohol. High selectivity to the allyl alcohol at high conversion requires that all rates of reaction in Scheme 11.1 are very low except the one that leads to allyl alcohol. All five steps are very common and straightforward.

I. Cano (✉)
Departamento de Física Aplicada, Facultad de Ciencias, Universidad de Cantabria, 39005 Santander, Spain
e-mail: israel.canorico@unican.es

P. W. N. M. van Leeuwen
LPCNO, Institut National des Sciences Appliquées-Toulouse, 135 Avenue de Rangueil, Toulouse 31077, France
e-mail: vanleeuw@insa-toulouse.fr

© Springer Nature Switzerland AG 2020
P. W. N. M. van Leeuwen and C. Claver (eds.), *Recent Advances in Nanoparticle Catalysis*, Molecular Catalysis 1, https://doi.org/10.1007/978-3-030-45823-2_11

Scheme 11.1 General scheme for the hydrogenation of α,β-unsaturated aldehydes

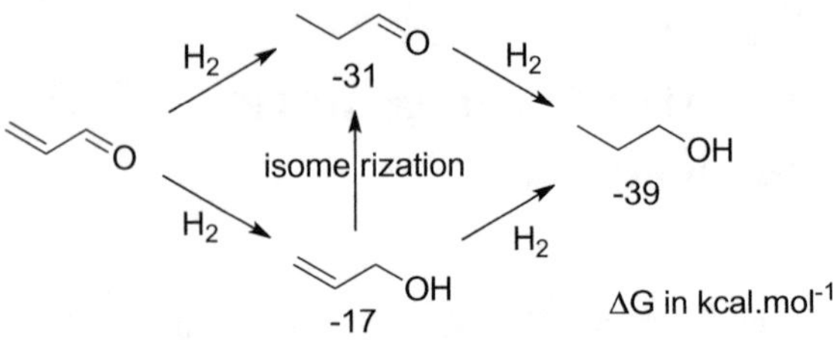

By contrast, Cao et al. proposed an alternative mechanism to aldehyde formation based on DFT calculations for the reaction of crotonaldehyde on Pt (111). This study indicated that the formation of butanal does not involve the hydrogenation of the C=C bond, but instead proceeds via 1,4 hydrogen addition yielding the vinyl alcohol, which quickly isomerizes to butanal [1]. Energy differences are small, in accord with the modest selectivity often encountered (Scheme 11.2).

Both the unsaturated alcohol and the saturated carbonyl product may have industrial interest. An example in which the two products can potentially be used via different routes is the BASF process to menthol (Scheme 11.3). The molecule desired is actually an aldehyde, R-citronellal and the chirality complicates the synthesis. It can be made via two routes: the BASF route, selective hydrogenation of neral or

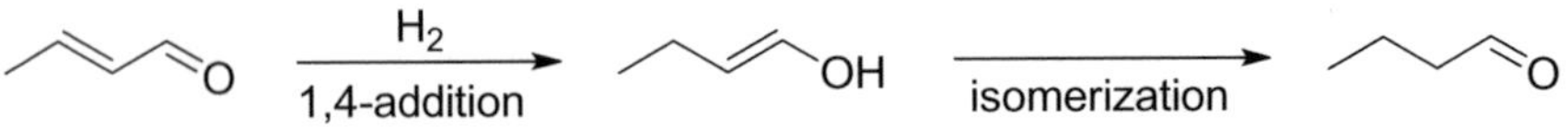

Scheme 11.2 Enol route to aldehyde

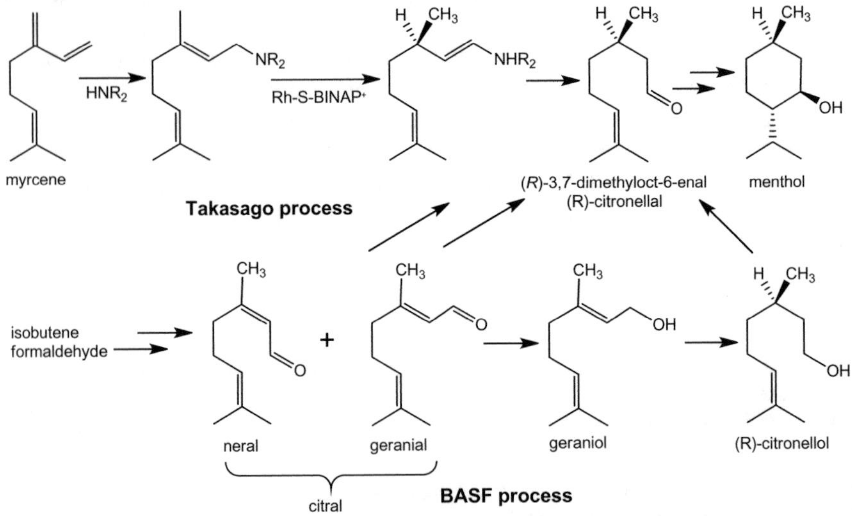

Scheme 11.3 Selective hydrogenations in the synthesis of menthol

geranial to R-citronellal with a chiral homogeneous Rh catalyst, or by asymmetric hydrogenation of geraniol with a chiral Ru complex, followed by dehydrogenation of the alcohol. The former route is the commercial one, which requires separation of neral and geranial as these give opposite enantiomers in the chiral hydrogenation reaction. In the Takasago route for menthol, R-citronellal is also the key intermediate, but this is prepared via asymmetric isomerization with Rh/BINAP from the allylic amine and not via an asymmetric hydrogenation of the C=C bond [2].

A widely studied substrate and test case for many catalytic systems is cinnamaldehyde (Scheme 11.4). The hydrogenation products are cinnamyl alcohol, 3-phenylpropanal and 3-phenylpropanol. Throughout the abbreviations shown in the scheme will be used, in imitation of most authors in this area. All compounds of the scheme are used in the fragrance industry. CAL is made from benzaldehyde and acetaldehyde. Classical conversions from CAL to COL involve Meerwein–Ponndorf–Verley reductions and NaBH$_4$ reductions.

One might wonder if certain metals have a preference for giving either HCAL or COL. In their review on this topic, Gallezot and Richard, in 1998, reported that the selectivity for COL production from CAL hydrogenation follows, for the unmodified metals, the sequence Os > Ir > Pt > Ru > Rh > Pd, a trend perhaps true until the late 1970s. It was thought that this order could be correlated with the width of the d-band of the metal [3]. The best metal catalyst, Os on carbon, gave as much as 95% selectivity to COL (isopropanol, 100 °C, 70 bar H$_2$) [4]. Gallezot and Richard stated, however, in their overview that the situation was far more complicated and that solvents, metal additives, acidity, supports, temperature, etc., were capable of changing completely the selectivity for the whole range of COL to HCAL. In this context, herein we report on the metal nanoparticle catalyzed partial hydrogenation of unsaturated aldehydes and ketones making a selection of the most typical examples, in search of general features of metals and systems.

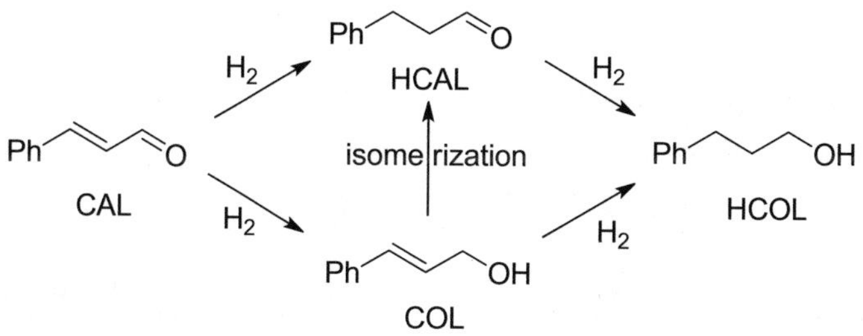

Scheme 11.4 Hydrogenation of cinnamaldehyde

11.2 Hydrogenation of Unsaturated Aldehydes and Ketones Catalyzed by Gold Nanoparticles

In comparison with traditional metals employed in hydrogenation chemistry (group 8–10 metals), gold is not a typical catalyst for hydrogenation reactions. This is not surprising, given that Au has shown a low ability to dissociate hydrogen due to its high resistance to oxidation. However, on the nanoscale this is no longer the case, and Au has exhibited hydrogenation activity with a strong preference for the reduction of C=O over C=C bonds [5]. As a consequence, in the last decade, gold has attracted a resurgent interest by researchers working in selective hydrogenations.

11.2.1 Homogeneous Gold Nanoparticles

There are only a few reports on hydrogenation by homogeneous gold nanoparticles (AuNPs) and most of them concern the reduction of a nitro group, most often with $NaBH_4$. In 2007, De Vos and co-workers employed polyvinylpyrrolidone stabilized AuNPs dispersed in amide solvents for the chemoselective hydrogenation of α,β-unsaturated aldehydes and ketones [6]. The hydrogenation of crotonaldehyde with the use of AuNPs of 7 nm in size and synthesized in N,N–dimethyl formamide led to 92% conversion and 73% selectivity to crotyl alcohol (Table 11.1), whereas 64% selectivity to the allylic alcohol product was obtained in the hydrogenation of mesityl oxide. Interestingly, the use of amide solvents provides improved colloidal stability and thus preservation of the nanodispersion, which allows efficient recycling of the AuNPs by ultrafiltration. The versatility of this nanodispersion was demonstrated in later work, in which a wide range of unsaturated aldehydes and ketones were hydrogenated in moderate to good selectivities [7]. In a further step, the addition of

Table 11.1 Influence of Lewis-acid cations on the hydrogenation of crotonaldehyde catalyzed by Au^0 NPs

Catalyst	Time (h)	Conversion (%)	Selectivity (%)
Au^0	5	29	78
$Au^0 + Fe^{3+}$	5	35	81
$Au^0 + Zn^{2+}$	5	37	86
Au^0	40	92	73
$Au^0 + Fe^{3+}$	28	90	79
$Au^0 + Zn^{2+}$	28	92	83

Reaction conditions: PVP/Au = 6, crotonaldehyde/Au = 200. Fe^{3+}/Au = 0.6, Zn^{2+}/Au = 0.4

Lewis-acid metal cations to this catalytic system enhances the activity and chemoselectivity towards the allylic alcohol product (Table 11.1), since the activation of the C=O group by Zn^{2+} or Fe^{3+} cations leads to an increase in the hydrogenation rate of the carbonyl group [8].

Following this pioneering work, kinetic studies were developed for the selective hydrogenation of cinnamaldehyde (CAL) catalyzed by AuNPs stabilized with a water-soluble imidazolium-based ionic polymer [9]. A first-order reaction regarding substrate consumption was observed, while a moderate selectivity of 73% for the cinnamyl alcohol (COL) product was obtained. Based on previous reports, the authors suggest that the AuNPs are negatively charged by electron donation from the polymer and, consequently, display a preferential interaction with the carbonyl group over the olefinic bond.

Unsupported thiolate-stabilized gold $Au_{25}(SR)_{18}$, $Au_{38}(SR)_{24}$ and $Au_{25}(SR)_{18}$ clusters showed moderate selectivity towards the unsaturated alcohol in the hydrogenation of benzalacetone at 60 °C [10]. However, decreasing the temperature to 0 °C, the ultrasmall and atomically precise $Au_{25}(SR)_{18}$ clusters (with 0.97 nm as metal-core diameter) were found to be a highly selective catalyst for a range of α,β-unsaturated carbonyl compounds, providing the corresponding allylic alcohols with moderate yields and complete selectivities (except acrolein, with 91% selectivity, Table 11.2) [11]. The $Au_{25}(SR)_{18}$ clusters exhibit a core–shell structure in which 12 Au atoms are in the shell and 13 metal atoms form an electron-rich icosahedral core. Only 12 of the 20 triangular faces of the icosahedron are capped by the external gold atoms, and thus there are 8 'hole' sites that can operate as active sites. These electron-rich 'hole' sites may activate the C=O functionality, and the low coordination-character of surface gold atoms could favour the adsorption and dissociation of H_2, leading to a cooperative effect that promotes the selective reduction of the carbonyl group (Scheme 11.5).

Table 11.2 Chemoselective hydrogenation of α,β-unsaturated carbonyl compounds by $Au_{25}(SR)_{18}$ clusters

Substrate	Conversion (%)	Selectivity (%)
R_1: Me; R_2, R_3: H; R_4: Ph	22	100
R_1, R_2, R_3: H; R_4: Ph	38	100
R_1, R_4: H; R_2, R_3: Me	43	100
R_1, R_2: H; R_3, R_4: Me	44	100
R_1, R_3, R_4: H; R_2: Me	39	100
R_1, R_2, R_3: Me; R_4: H	29	100
R_1, R_2, R_3, R_4: H	46	91

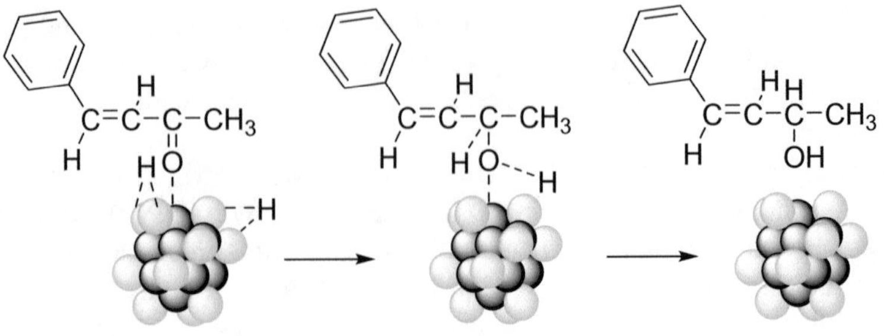

Scheme 11.5 Proposed mechanism of the chemoselective hydrogenation of α,β-unsaturated ketones and aldehydes to unsaturated alcohols catalyzed by $Au_{25}(SR)_{18}$ clusters

The activation of dihydrogen by $Au_{25}(SR)_{18}$ clusters may indeed occur by a ligand–metal cooperation (involving protonation of the bound thiolate) rather than a purely Au-mediated process. Following this approach, van Leeuwen and co-workers introduced an innovative strategy in which AuNPs ligated by secondary phosphine oxides (SPOs) were employed as catalysts for the chemoselective hydrogenation of substituted aldehydes [12]. The SPO ligand plays a key role and is directly related to the catalytic process, acting as heterolytic activator for dihydrogen cooperatively with a neighbouring metal atom (Scheme 11.6). A series of control experiments showed that the SPO ligand is a prerequisite for the observed catalytic activity, thus asserting a heterolytic hydrogenation mechanism. This ligand–metal cooperative effect allowed to obtain high conversions and complete selectivities for a wide range of substrates, including α,β-unsaturated aldehydes (Table 11.3).

In a subsequent study, a series of AuNPs stabilized with different SPOs were prepared in order to investigate the ligand effect on the morphology and catalytic properties [13]. An extensive characterization showed significant differences in the size, morphology and catalytic behaviour depending on the characteristics of the substituents in the ligand. In particular, CP-MAS NMR spectroscopy provided insight into the polarization of the P=O or POH bond in the different types of AuNPs. Nanoparticles (NPs) ligated by aromatic phosphine oxides, which contain only Au(I) atoms and SPO anions in the surface, show a strong polarity of the P=O bond and are highly selective in the hydrogenation of several α,β-unsaturated aldehydes,

Scheme 11.6 a Heterolytic cleavage of hydrogen on SPO-stabilized AuNPs, **b** hydrogenation of aldehydes

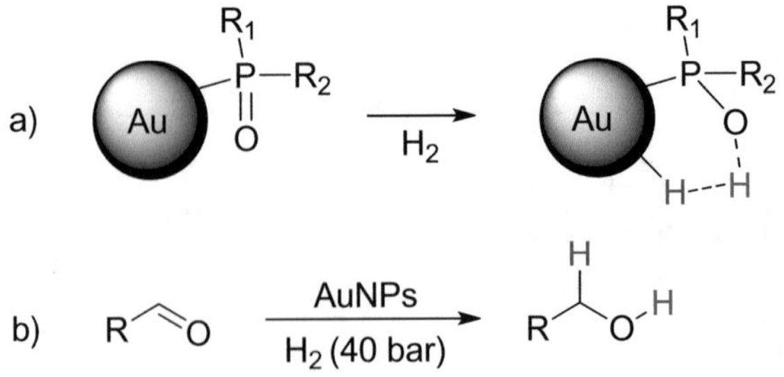

Table 11.3 Chemoselective hydrogenation of α,β-unsaturated aldehydes by SPO-stabilized AuNPs

Substrate	Product	Conversion (%)	Selectivity (%)
Ph⌇⌇O (cinnamaldehyde)	Ph⌇⌇OH	>99	>99
⌇⌇⌇O	⌇⌇⌇OH	>95	>99
(branched enal)	(branched enol)	>95	99
⌇⌇⌇O	⌇⌇⌇OH	>95	>99
(citral)	(citral alcohol)	>95	>99
⌇⌇O (acrolein)	⌇⌇OH	61	>99
⌇⌇⌇≡O	⌇⌇⌇≡OH	>95	>99
(aryl aldehyde)	(aryl alcohol)	42	>99

Reaction conditions: AuNPs (0.01 mmol Au), substrate (0.5 mmol), THF/Hexane (5 mL), 18 h, 40–60 °C, 40 bar H_2

such as CAL (cinnamaldehyde), citral and *trans*-2-hexen-1-al. Conversely, nanoparticles stabilized with alkyl-substituted SPOs also present POH bonds and display a lower polarity of the P=O bond, hindering the heterolytic cleavage of dihydrogen. These AuNPs contain both Au (I) and Au(0) atoms at the surface and thus exhibit different active sites. As a consequence, they are less selective and active for aldehyde hydrogenation.

Of particular chemical interest was the chemoselective hydrogenation of acrolein to the corresponding unsaturated alcohol by SPO-stabilized AuNPs [12]. This transformation can be considered as a challenging one since both C=C and C=O bonds are essentially isosteric. In fact, DFT-PW91 studies [14] of acrolein hydrogenation on a Au_{20} cluster model and Au (110) surface showed that the variations in selectivity are related to the different adsorption modes of acrolein: through both C=C and C=O bonds on Au(110), and via C=C to low-coordinated metal atoms on edges and apexes for Au_{20}. Thus, some preference for the hydrogenation of the C=C bond could be expected in small AuNPs such as those stabilized by *tert*-butyl(naphthalen-1-yl)phosphine oxide (1.2 nm size). However, the allyl alcohol product was obtained with complete selectivity (Table 11.3), which reveals that the hydrogenation of unsaturated aldehydes by SPO-stabilized AuNPs occurs by a different mechanism and thus supports a ligand to metal cooperation in which the SPO undergoes a crucial role.

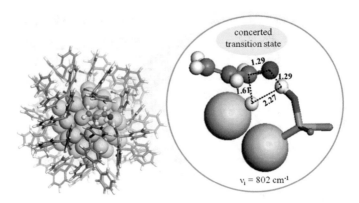

Fig. 11.1 Left: DFT adsorption configuration of 27 SPO ligands (Ph$_2$PO) on Au$_{55}$. Right: structure, distances (Å), and imaginary frequency (cm^{-1}) for the transition state associated with the concerted mechanism in the reduction of C=O group of acrolein on SPO–stabilized AuNP. The inset shows the four-membered ring formation and distances. Yellow spheres: Au atoms; stick models: adsorbed SPO ligands. Colour codes: P, pink; O, red; C, grey; H, white. Reprinted with permission from Ref. [15]. Copyright 2017 American Chemical Society

This was demonstrated through DFT studies performed by López and van Leeuwen [15]. The authors identified a cooperative effect at the AuNP-SPO interface (as a Frustrated Lewis Pair) that promotes the heterolytic cleavage of dihydrogen and its kinetically favourable addition to the C=O group of acrolein via a concerted mechanism to generate the unsaturated alcohol (Fig. 11.1). The activity is determined by the difference in basicity between SPO ligand and substrate and the active sites available on the NP.

11.2.2   Supported Gold Nanoparticles

As supported metallic nanoparticles (MNPs) have traditionally provided saturated aldehydes, the development of new approaches to achieve the formation of unsaturated alcohols with high selectivities is very desirable. In this way, different strategies have been described in the last years for the selective hydrogenation of unsaturated aldehydes by supported AuNPs, such as the addition of Lewis acid or ligand additives or the use of a second metal.

TiO$_2$-supported AuNPs. Catalysts based on metal oxide-supported AuNPs have attracted much interest in recent decades. Among different metal oxides employed to immobilize AuNPs, titania (TiO$_2$) is probably the most extensively investigated material, since catalytic systems composed of this support display special physical and chemical features, such as chemical stability, acid–base properties and a strong metal–support interaction.

Yoshitake and co-workers supported gold on mesoporous titania (MT) [16]. The resulting AuNPs of 3 nm in size are highly dispersed and smaller than those supported on commercial anatase (5 nm). This Au/MT system showed 82% selectivity towards the allyl alcohol product in the vapour-phase hydrogenation of crotonaldehyde, whereas the catalyst containing commercial anatase only reached 49% selectivity (Table 11.4). In addition, there seems to be a correlation between catalytic performance and ratio of positively charged Au species (Table 11.4), which increases with the heating temperature employed in the synthesis of mesoporous titania.

An enhancement in the activity was also observed for the hydrogenation of CAL catalyzed by TiO_2-supported AuNPs of 8–9 nm in size and partially covered by Ir [17]. This Au-Ir/TiO_2 system is fivefold more active than the analogous monometallic Au/TiO_2 catalyst, although the selectivity did not improve and stayed constant at 83–84%. XPS studies revealed a transfer of electron density from Ir to Au, in such a way that these electron-enriched Au sites could interact with the electron-deficient carbonyl carbon, while electropositive Ir atoms would activate the electron-rich carbonyl oxygen. Nevertheless, an increase in the selectivity should also be expected for this hypothesis.

Conversely, AuNPs supported on TiO_2 can also be employed for the preferential reduction of the conjugated C=C bond of α,β-unsaturated carbonyl compounds such as carvone [18], although the reported selectivity values were not high (<63%) and additional strategies are required to improve the results. In this context, higher selectivities towards the hydrogenation of the conjugated C=C bond were observed by the use of bimetallic Pd and Au nanoparticles (Pd/Au/TiO_2) supported on TiO_2 and physical mixtures of the corresponding monometallic Au/TiO_2 and Pd/TiO_2 catalysts [19, 20]. These systems were applied for the hydrogenation of citral in supercritical CO_2 (scCO_2), leading to the selective formation of citronellal. An increase in both activity and selectivity was achieved by the use of Pd/Au/TiO_2 catalysts, as compared to those shown by monometallic Au/TiO_2 and Pd/TiO_2 (Table 11.5). Similarly, the physical mixture of Au/TiO_2 and Pd/TiO_2 also provided an enhancement in catalytic behaviour. A hydrogen spillover is believed to be responsible for the improvement of hydrogenation rate noted in the physical mixture. In addition, the authors suggest that an electron transfer between the two metals produces the enhancement in the catalytic performance of bimetallic Pd/Au/TiO_2 catalysts. Interestingly, the use of hexane as a solvent gives worse results, which highlights the advantages of scCO_2.

Table 11.4 Vapour-phase hydrogenation of crotonaldehyde catalyzed by AuNPs supported on MT and commercial anatase

Catalyst	Heating temperature	Conversion (%)	Selectivity (%)
Au/MT200	200	1.1	68
Au/MT300	300	2.4	68
Au/MT400	400	6.9	82
Au/anatase	–	1.6	49

Conditions: Au/MT (0.1 g), Au/anatase (0.3 g), H_2/crotonaldehyde = 24, H_2 flow (4 L/h)

Table 11.5 Hydrogenation of citral by TiO_2-supported Au and Pd catalysts in $scCO_2$

Catalyst	Time (h)	Conversion (%)	Selectivity (%)	TOF (h^{-1})
Au(0)/TiO$_2$	4	8	67	26
Pd(II)/TiO$_2$	4	24	62	956
Pd(0)/TiO$_2$	4	27	72	1075
Pd(II)/TiO$_2$ + Au(0)/TiO$_2$	0.5	85	80	2059
Pd(0)/TiO$_2$ + Au(0)/TiO$_2$	1.5	85	89	686
Pd(II)/Au(0)/TiO$_2$	0.5	43	61	948
Pd (0)/Au(0)/TiO$_2$	4	55	88	152

Reaction conditions: Catalyst (16.7 mg for single component; 16.7 + 16.7 mg for multicomponent samples), citral (2 mmol), 40 bar H_2, 80 bar CO_2, 80 °C

In the same way, Piccolo and co-workers reported the selective formation of the saturated aldehyde in the hydrogenation of CAL catalyzed by Au NPs (Au/TiO$_2$), Rh NPs (Rh/TiO$_2$) and bimetallic Rh–Au NPs (RhAu/TiO$_2$) supported on rutile TiO$_2$ nanorods [21]. A moderate selectivity of 61% was obtained with the monometallic Rh/TiO$_2$ catalyst, being slightly improved through the use of bimetallic RhAu/TiO$_2$ (Table 11.6). CO-FTIR studies showed the presence of Lewis acid sites on Rh/TiO$_2$ that were attributed not only to the support but also to electropositive Rh species. Indeed, the ratio of these $Rh^{\delta+}$ sites decreases in RhAu/TiO$_2$ catalysts due to the stabilization of Rh^0 induced by Au, which can be associated with the enhancement in the catalytic behaviour observed for bimetallic systems.

SiO$_2$-supported AuNPs. Because SiO$_2$ is 'inert' support and interacts weakly with gold, the most important recent advances to improve the catalytic performance of Au/SiO$_2$ catalysts in the hydrogenation of unsaturated carbonyl compounds have been achieved by the addition of a second metal. In this line, the amino groups of aminopropyltrimethoxysilane (APTMS) were employed as linker to support Au and Au-In nanoparticles on mesoporous SiO$_2$ SBA-15 [22]. The Au and Au-In NPs exhibit a diameter of *ca.* 2 nm, whereas in the absence of APTMS the obtained

Table 11.6 Hydrogenation of cinnamaldehyde catalyzed by various TiO_2-supported Rh, Au and RhAu catalysts

Catalyst	Au:Rh atomic ratio	NP size (nm)	Conversion (%)	Selectivity (%)[a]
Rh/TiO$_2$	0:100	2.4 ± 0.6	25	61
Au$_{13}$Rh$_{87}$/TiO$_2$	13:87	2.4 ± 0.9	27	69
Au$_{37}$Rh$_{63}$/TiO$_2$	37:63	3.2 ± 1.2	37	64
Au$_{57}$Rh$_{43}$/TiO$_2$	57:43	4.0 ± 1.3	18	52
Au/TiO$_2$	100:0	2.8 ± 0.8	11	48

Reaction conditions: Catalyst (1 mg), CAL (2 mmol), H_2 (10 bar), 2-propanol (100 mL), 50 °C, 3 h
[a]Selectivity at 20% conversion

AuNPs present a diameter of about 31 nm. The AuNPs immobilized on APTMS-SBA-15 (Au/APTMS-SBA-15) are more active and selective than Au/SBA-15 for the hydrogenation of crotonaldehyde (Table 11.7). A transfer of electron density from APTMS to gold is proposed as a cause of this improved selectivity, in such a way that these electron-enriched AuNPs favour the activation of the C=O group. Other authors invoke Lewis acidic Au^{3+} sites for higher selectivity to the alcohol, *vide infra*. In addition, the alloyed Au-In NPs/APTMS-SBA-15 display higher selectivity towards the crotyl alcohol product than Au/APTMS-SBA-15. The authors suggest that the C=O group is preferentially activated through a synergistic effect of electron-deficient indium and electron-rich gold sites.

A similar effect due to different phenomena was observed for the vapour-phase hydrogenation of crotonaldehyde catalyzed by Ag-doped AuNPs supported on SBA-15 [23]. The Ag-decorated Au/SBA-15 catalysts (8AuxAg/SBA-15, with 8% Au and x% Ag) provided higher TOF and selectivity values than those shown by monometallic 8Au/SBA-15 systems (Table 11.8), whereas Ag/SBA-15 is only slightly active. An enhancement in the chemisorption of crotonaldehyde was noted by in situ FTIR analyses, which is associated with a stronger binding of the C=O group on Ag sites. Additional experiments suggested that dihydrogen and C=C bonds are preferentially activated on Au sites. It was proposed that the active hydrogen species are generated on Au sites and then react with crotonaldehyde through a spillover mechanism (Fig. 11.2).

Table 11.7 Liquid-phase hydrogenation of crotonaldehyde catalyzed by Au and Au-In APTMS-SBA-15 systems

Catalyst	In/Au wt%	Time (h)	Conversion (%)	Selectivity (%)[a]
$Au_{5.0}$/SBA-15	0	4	8	3
$Au_{5.0}$/APTMS-SBA-15	0	2	97	55
$Au_{5.0}In_{0.75}$APTMS-SBA-15	0.15	5	94	75

Reaction conditions: Catalyst (1 mg, 4.5–5% Au), crotonaldehyde (1 mL), H_2 (20 bar), hexane (49 mL), 120 °C
[a]Crotyl alcohol selectivity

Table 11.8 Vapour-phase continuous hydrogenation of crotonaldehyde catalyzed by Au/SBA-15 and AuAg/SBA-15 systems

Catalyst	Au wt%	Ag wt%	Conversion (%)	Selectivity (%)[a]
8Au/SBA-15	8	0	13.1	55.3
8Au0.5Ag/SBA-15	8	0.5	68.8	72.3
8Au0.75Ag/SBA-15	8	0.75	71.7	70.4

Reaction conditions: Catalyst (200 mg), H_2/crotonaldehyde = 50, reaction pressure: 20 bar, 304 min on stream, 120 °C
[a]Crotyl alcohol selectivity

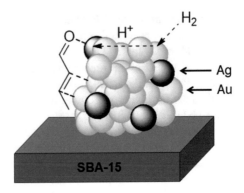

Fig. 11.2 Possible surface adsorption and following activation of H_2 and crotonaldehyde on AuAg/SBA-15

On the other hand, Rojas and co-workers [24] employed a monometallic system based on SiO_2-supported AuNPs for the selective hydrogenation of benzalacetone and CAL (Table 11.9). 79% selectivity was achieved for the cinnamyl alcohol product, although the reported conversions are very low. DRIFT analyses suggested the presence of $Au^{\delta-}$ sites that could bind to the electron-deficient carbonyl carbon of CAL and facilitate the selective hydrogenation of the C=O group, but firm evidence is lacking.

AuNPs immobilized on carbon materials. Currently, carbon supports are one of the most used materials for heterogeneous catalysis, since their properties can be adapted to specific requirements. Activated carbons, nanofibers, fullerenes, xerogels, carbon nanotubes (CNTs) and multiwalled carbon nanotubes (MWCNTs) are some examples of carbon materials employed to immobilize metals and develop catalytic systems for chemoselective hydrogenation of unsaturated carbonyl compounds. In this line, monometallic Au and Pd and bimetallic Cu–Au and Pd–Au nanoparticles were immobilized on reduced graphene oxide (rGO) [25]. The resulting rGO-supported catalysts were applied for the hydrogenation of benzalacetone with sodium borohydride ($NaBH_4$) as reducing agent. In all cases, complete selectivity values towards the allyl alcohol product were obtained. The activity increases in the order Au/rGO < Pd/rGO < Cu–Au/rGO < Cu–Pd/rGO, which can be associated with the decrease of NP size, better dispersion and higher ratio of defects with better homogeneity. In addition, the authors suggest that an improved charge distribution in the density of d-band electrons on bimetallic NPs could be involved.

Table 11.9 Hydrogenation of unsaturated carbonyl compounds catalyzed by AuNPs/SiO_2

Substrate	Conversion (%)	Selectivity C=O (%)	Selectivity C=C (%)
Cinnamaldehyde	7.5	46	54
Benzalacetone	3.6	79	19

Conditions: 0.2 g catalyst, 50 mL substrate (0.05 M), 6.2 bar H_2, 363 K, ethanol, 3 h

On the contrary, the C=C bond of CAL was selectively hydrogenated by the use of AuNPs (*ca.* 3.2 nm) confined in the cavity of CNTs [26]. 91% selectivity was reached at 95% conversion in a process for which a homolytic cleavage of H_2 was proposed. A better result was obtained in the hydrogenation of CAL catalyzed by separate Au and Pd nanoparticles immobilized on ordered mesoporous carbon (OMC) [27]. The AuPd-OMC catalyst displayed higher activity and selectivity values than those shown by the monometallic Pd-OMC system, while the analogous Au-OMC was completely inactive (Table 11.10). This synergistic effect is proposed to occur through a hydrogen spillover mechanism in which PdNPs interact with dihydrogen and generate activate hydrogen species, and AuNPs act as active hydrogen acceptors. In addition, the larger amount of saturated alcohol observed for Pd-OMC suggests that the presence of Au leads to a dilution of active Pd species and thus avoids further substrate hydrogenation.

A comparison of the results obtained using AuNPs supported on different carbon materials indicates that the type of carbon support can significantly affect the selectivity (Table 11.11), which highlights the versatility of carbon materials to be tailored to specific needs. However, an effect of the nature of the reducing agent should not be discarded.

Hydrotalcite-supported AuNPs. Hydrotalcite (HT) is a naturally occurring mineral of chemical composition: $Mg_6Al_2(OH)_{16}CO_3 \cdot 4H_2O$. Hydrotalcite-like compounds are mixed magnesium and aluminium hydroxycarbonates composed of positively

Table 11.10 Hydrogenation of cinnamaldehyde by OMC-supported Au, Pd and AuPd catalysts

Catalyst	Metal wt%		Size (nm)		TOF (h^{-1})	Conversion (%)	Selectivity (%)[a]	
	Au	Pd	Au	Pd			C=C	C=O
Pd-OMC	0	0.2	–	8.2	45	4.5	60.2	24.2
AuPd-OMC	0.25	0.2	2.5	6.5	120	19.9	96.2	2.1
Au-OMC	3.5	0	2.3	–	0	0	–	–

Reaction conditions: Catalyst (100 mg), CAL (0.5 g), isopropanol (20 mL), H_2 (10 mL/min), 40 °C, 2 h
[a]Selectivity towards saturated alcohol = 100 − %C=C − %C=O

Table 11.11 Hydrogenation of benzalacetone and cinnamaldehyde catalyzed by AuNPs supported on different carbon materials

Support	NP size	Reducing agent	Conv. (%)	Sel. C=O (%)	Sel. C=C (%)	References
rGO	10	$NaBH_4$	35[a]	>99	–	[25]
CNT	3.2	H_2	95[b]	–	91	[26]
OMC	2.3	H_2	0[c]	–	–	[27]

Reaction conditions:
[a]Catalyst (2 mg), benzalacetone (1.5 mM), $NaBH_4$ (1 M), toluene (5 mL), ethanol (5 mL), 0 °C, 3 h
[b]Catalyst (5 mg, 1.6–1.7% Au), CAL (0.2 mmol), toluene (1 mL), 8 bar H_2, 90 °C, 2 h
[c]Same reaction conditions as Table 11.10

charged two-dimensional hydroxide layers and interlayer regions containing charge compensating anions. HTs are promising materials as basic supports which, when calcined, generate mixed Mg–Al–O oxides with strong Lewis basic character.

AuNPs immobilized on hydrotalcite (2%Au/Mg$_2$AlO) were employed as catalyst in the chemoselective hydrogenation of several α,β-unsaturated aldehydes, such as CAL [28] and crotonaldehyde [29] (Table 11.12). A decrease in activity and selectivity was observed at higher calcination temperatures of both Mg$_2$AlO support and Au/Mg$_2$AlO catalyst, which can be correlated with the Au^{3+}/Au^0 ratio. At lower calcination temperatures, the Au^{3+}/Au^0 ratio increases, and thus the Lewis acidity is higher. These Lewis acid sites may favour the adsorption and subsequent activation of the polar carbonyl group, leading to increased activity and selectivity.

Complete selectivities towards the unsaturated alcohol product were obtained by Kaneda and co-workers in the hydrogenation of carbonyl compounds catalyzed by CeO$_2$-covered AuNPs supported on hydrotalcite (Au@CeO$_2$/HT) [30]. A multi-gram synthesis based on a redox-coprecipitation method allows to obtain core–shell Au@CeO$_2$ nanoparticles, which show spherical AuNPs of 5 nm in the core and a shell of CeO$_2$ with *ca.* 4 nm in thickness (Fig. 11.3).

The HT-supported Au@CeO$_2$ catalyst is highly selective for the hydrogenation of unsaturated aldehydes and ketones (Table 11.13). IR and deuteration studies demonstrated that a core–shell cooperative effect leads to the formation of Au–H$^{\delta-}$ and CeO$_2$–H$^{\delta+}$ hydrogen species through a heterolytic activation of H$_2$, thus favouring the selective reduction of the C=O group.

Table 11.12 Liquid phase hydrogenation of α,β-unsaturated aldehydes by 2%Au/Mg$_2$AlO

Substrate	P (bar)	Conversion (%)	Unsaturated alcohol selectivity (%)
Cinnamaldehyde	12.0	93.2	86.4
Citral	9.3	56.4	97.5
2-Hexen-1-al	9.3	24.2	72.4
Crotonaldehyde	9.3	19.6	66.3

Reaction conditions: 120 °C, 2 h, catalyst (0.5 g), substrate (2 mL), ethanol (78 mL)

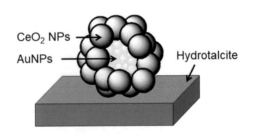

Fig. 11.3 Representative scheme of AuNPs@CeO$_2$ supported on hydrotalcite

Table 11.13 Hydrogenation of unsaturated carbonyl compounds by Au@CeO$_2$/HT

Substrate	Product	Time	Yield (%)	Selectivity (%)
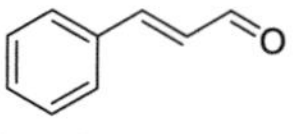	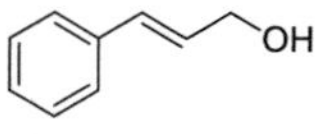	25	>99	>99
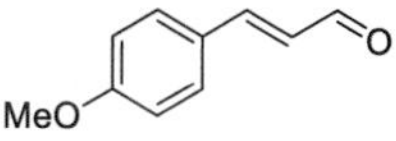	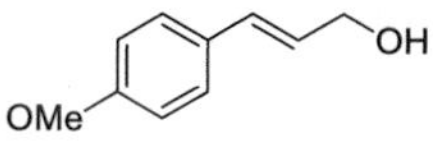	40	>99	>99
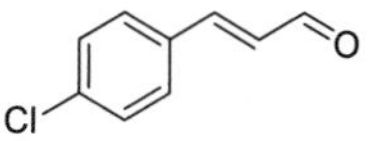	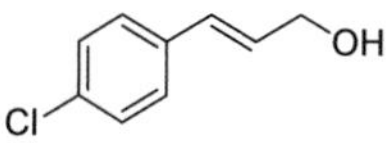	35	>99	>99
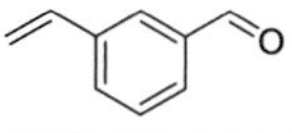	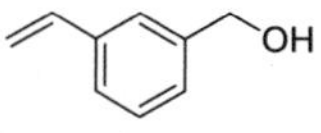	30	>99	>99
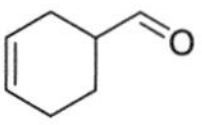	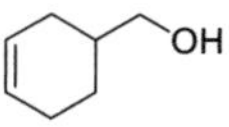	30	97	>99
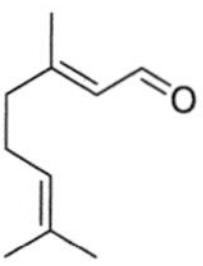	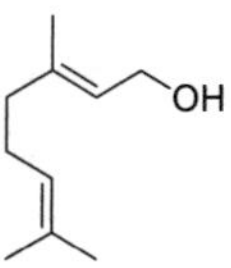	40	>99	>99

Reagents and conditions: Au@CeO$_2$/HT (50 mg, 0.0042 mmol Au), substrate (0.3 mmol), toluene (4 mL), 120 °C, 30 bar H$_2$

Other supports. In addition to typical TiO$_2$, SiO$_2$, hydrotalcite and carbon-based supports, other materials have been described in the last years for the selective hydrogenation of α,β-unsaturated carbonyl compounds by immobilized AuNPs. Some of these supports not only serve to disperse the catalyst but also interact with the metal and thus have an influence on the catalysis. For example, the use of magnetite (Fe$_3$O$_4$) is an interesting option to prepare Au-supported catalysts with additional benefits [31]. Gold was deposited on hematite materials with a particular flowerlike morphology. The following integrative reduction allowed to obtain the AuNPs/Fe$_3$O$_4$ systems. The as-prepared flowerlike AuNPs/Fe$_3$O$_4$ catalysts display higher activity and selectivity in the hydrogenation of crotonaldehyde than AuNPs supported on particulate Fe$_3$O$_4$ (Table 11.14). This increased activity can be attributed to the heteroepitaxial growth of AuNPs on flowerlike magnetite along the (111) planes. These AuNPs are smaller (2.6–2.7 nm) and more uniform than those grown on particulate magnetite (3.5–3.8 nm).

A perimeter interface mechanism involving the hydroxyl groups on magnetite and electropositive Au$^{\delta+}$ species on Au–O–Fe linkage is believed to be responsible for the improved selectivity. A CH$_3$–CH=CH–CHOH intermediate is generated through the attack of the H$^+$ from hydroxyl group on carbonyl oxygen. This proton is replaced by one H$^+$ species resulting from the heterolytic cleavage of H$_2$ on interfacial Au–O–Fe linkage, while the corresponding hydride is added to the carbonyl carbon thus

Table 11.14 Liquid phase hydrogenation of crotonaldehyde by AuNPs/Fe₃O₄ catalysts

Fe source	Catalyst	Conversion (%)	Time (h)	Crotyl alcohol selectivity (%)
$FeCl_3$	Au/Fe_3O_4–Cl-f	>99	7.5	78
$Fe(NO_3)_3$	Au/Fe_3O_4–NO_3-f	>99	6.5	76
$FeCl_3$	Au/Fe_3O_4–Cl-p	>99	17.5	72
$Fe(NO_3)_3$	Au/Fe_3O_4–NO_3-p	>99	20.5	68

Reaction conditions: 120 °C, 2 h, catalyst (0.4 g), crotonaldehyde (2.42 mmol), hexane (20 mL), H_2 (20 bar). Au/Fe_3O_4-f: flowerlike Au/Fe_3O_4; Au/Fe_3O_4-p: particulate Au/Fe_3O_4

generating the crotyl alcohol product. Importantly, the system can be magnetically separated and reused up to six cycles without any significant loss in the activity.

On the other hand, Larese and co-workers claim that the concentration of positively charged Au atoms is not a key parameter in the hydrogenation of α,β-unsaturated aldehydes catalyzed by AuNPs supported on zinc oxide [32]. Through an exhaustive study of three different types of Au/ZnO systems (Au/rod-tetrapod ZnO, Au/porous ZnO and Au/ZnO–CP prepared by coprecipitation), it is concluded that small and mound-shaped AuNPs, a nanosized ZnO support with surface defects, and Au–ZnO interaction are the crucial parameters to reach an efficient catalytic behaviour. The best results (94.9% conversion and 100% selectivity towards the allylic alcohol in CAL to COL hydrogenation) were provided by the Au/ZnO–CP catalyst, which exhibits the smallest size and highest dispersion of AuNPs. In addition, this support displays the smallest ZnO NPs, which boosts the Au–ZnO interaction. However, no influence of cationic gold concentration is observed by the use of several Au/ZnO-CP systems with different $Au^{\delta+}/Au^0$ ratios.

A different approach to achieve high selectivity in the hydrogenation of α,β-unsaturated aldehydes was proposed by Hupp and co-workers [33]. AuNPs with a size of 2.2 nm were encapsulated in a zeolitic imidazolate framework (ZIF-8). Such an Au@ZIF-8 system showed 95% selectivity towards the unsaturated alcohol in the hydrogenation of crotonaldehyde, although the reported activities are not very high (TOF = 12.5 h^{-1}; 15% conversion under 5 bar H_2 at 80 °C). This selectivity is associated with the small aperture exhibited by ZIF-8 (3.4 Å width), which only allows access to the terminal groups of crotonaldehyde (methyl and carbonyl groups), thus preventing the hydrogenation of the C=C bond. In addition, the encapsulation avoids NP aggregation and allows catalyst recycling.

Conversely, the C=C bond of benzalacetone was selectively hydrogenated by bimetallic Au–Pt nanoparticles of 3–4 nm in size supported on CeO_2 microspheres of ca. 150 nm ($AuPt/CeO_2$) [34]. The presence of Au favours the formation of Au–Pt alloyed NPs with better shape and size distribution, and boosts the catalytic performance of Pt. Indeed, the $AuPt/CeO_2$ catalysts are more active and selective than the analogous monometallic Pt/CeO_2 system (Table 11.15), reaching a range of 90–95% selectivity. As the monometallic Au/CeO_2 system does not present catalytic

Table 11.15 Hydrogenation of benzalacetone catalyzed by Au/CeO$_2$, Pt/CeO$_2$ and AuPt/CeO$_2$ catalysts

Catalyst	Metal wt%	Yield (%)	Selectivity C=C (%)	TON$_{Pt}$ (%)[a]
Au/CeO$_2$	1.80	0.9	0	–
Pt/CeO$_2$	1.20	71.9	87	1171
Au$_{0.31}$Pt$_{0.69}$/CeO$_2$	1.74	78.5	93	1342
Au$_{0.48}$Pt$_{0.52}$/CeO$_2$	2.30	87.7	90	1428
Au$_{0.59}$Pt$_{0.41}$/CeO$_2$	2.89	96.7	90	1574
Au$_{0.76}$Pt$_{0.24}$/CeO$_2$	4.95	85.8	93	1397
Au$_{0.87}$Pt$_{0.13}$/CeO$_2$	9.38	61.8	95	1007

Reaction conditions: Catalyst (10 mg), benzalacetone (1 mmol), ethanol (3 mL), 25 °C, 4 h, 1 bar H$_2$
[a] Assuming that Pt is the only responsible for the catalytic activity

activity, this study shows how the catalytic performance of an active metal can be improved with an inactive metal.

In contrast, Dupont and co-workers described the selective formation of saturated aldehydes and ketones in the hydrogenation of α, β-unsaturated carbonyl compounds catalyzed by AuNPs immobilized on ionic liquid-hybrid γ-Al$_2$O$_3$ supports [35]. AuNPs of ca. 6.6 nm in size were deposited on this material by sputtering-deposition employing gold foils. This monometallic gold catalyst showed a strong preference for the reduction of the conjugated C=C bond of several unsaturated aldehydes and ketones (Table 11.16). The ionic liquid acts as a cage that surrounds the AuNPs and avoids the AuNP–Al$_2$O$_3$ interaction, leading to a change in the reaction kinetics in comparison with that shown by AuNPs supported on γ-Al$_2$O$_3$ (Au/Al$_2$O$_3$). Indeed, AuNPs immobilized on ionic liquid-hybrid γ-Al$_2$O$_3$ exhibit the reactivity expected

Table 11.16 Hydrogenation of unsaturated carbonyl compounds by AuNPs supported on ionic liquid-hybrid γ-Al$_2$O$_3$

Substrate	Product	Conversion (%)	Selectivity (%)	TOF (h^{-1})
(CH=CH-CHO)	(CH$_2$-CH$_2$-CHO)	>99[a]	95	108
Ph–CH=CH–CHO	Ph–CH$_2$–CH$_2$–CHO	94[b]	>99	60
(cyclohexenone)	(cyclohexanone)	>99[a]	96	126
(methyl cyclohexenone)	(methyl cyclohexanone)	21[b]	>99	36

Reaction conditions: Au (0.5 μmol), substrate/Au = 250, m-xylene (5 mL), 100 °C, 25 bar H$_2$
[a] 24 h
[b] 30 h

for 'naked' AuNPs (NPs that are surrounded by weakly coordinating species and thus present accessible active sites), where a homolytic H_2 activation mechanism was demonstrated through kinetic and deuterium labelling investigations. On the other hand, these studies revealed a heterolytic cleavage of dihydrogen for the Au/Al_2O_3 system.

11.3 Hydrogenation of Unsaturated Aldehydes and Ketones Catalyzed by Silver Nanoparticles

Like gold, silver displays a low capacity for hydrogen activation, but silver nanoparticles (AgNPs) have been shown to exhibit high selectivity to the allylic alcohol product in the hydrogenation of α, β-unsaturated aldehydes and ketones. However, in comparison with Au, there is very little precedent for hydrogenation of unsaturated carbonyl compounds, in particular with Ag nanoparticles dispersed in solution.

Claus and co-workers investigated the catalytic performance of AgNPs supported on SiO_2 and TiO_2 for the gas-phase hydrogenation of crotonaldehyde [36] and acrolein [37] (Scheme 11.7). 59% selectivity towards the crotyl alcohol was obtained in the hydrogenation of crotonaldehyde by $AgNPs/SiO_2$, for which no dependence on the catalyst structure was observed. On the other hand, a strong influence of the structure over the catalytic behaviour was detected for AgNPs immobilized on TiO_2, since nanoparticles with sizes of ~3 and ~1.5 nm lead to different selectivities (53 vs. 28% towards crotyl alcohol, respectively) and TOF values. A similar trend was observed in the hydrogenation of acrolein, but the selectivities towards the unsaturated alcohol were even lower (42 vs. 27% selectivity for AgNPs with ~3 and ~1.5 nm size, respectively). The origin of this difference could be attributed to the combined effect of the increased particle coverage by TiO_x in smaller NPs and the higher ratio of Ag (111) planes for larger particles (assuming that the hydrogenation of the C=O bond takes place at surface atoms located in faces). Although firm evidences were lacking, these results were consistent with those reported by Meyer and co-workers for the hydrogenation of acrolein catalyzed by a series of AgNPs supported on SiO_2 [38]. An increase in activity and selectivity towards the allyl alcohol (up to 37%) with a growth in the nanoparticle size was observed. The authors suggested that acrolein is adsorbed through the C=O bond on flat Ag (111) surfaces, which are present in

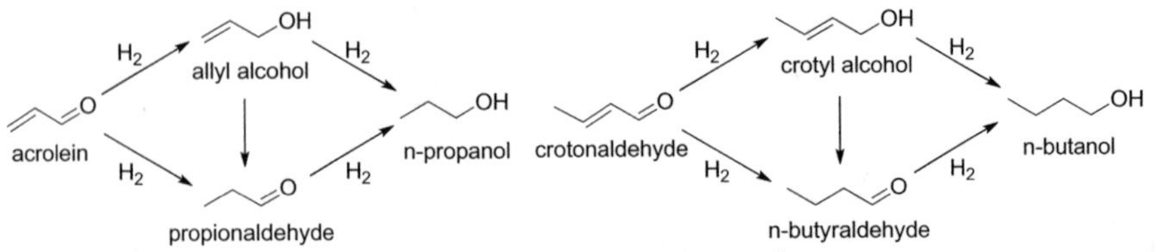

Scheme 11.7 Reaction pathways for the hydrogenation of acrolein and crotonaldehyde

higher ratio in larger NPs, thus leading to the activation of the carbonyl group. In addition, higher partial pressure of acrolein (increased acrolein coverage) produces a rise in the selectivity. This effect could be attributed to a change in the adsorption geometry of the substrate from parallel to the surface to expose only the C=O group.

In a subsequent work of Claus [39], an in-depth study of the different factors that determine the activity and selectivity in the gas-phase hydrogenation of acrolein by AgNPs supported on SiO_2, Al_2O_3 and ZnO was performed. Influence of the nature of the support on the selectivity was observed, which can be ascribed to the specific acidity and/or defect sites of each support at the silver-support interface (Table 11.17). The selectivity is also affected by the degree of acrolein coverage (reaction pressure) for the same reason commented above. In addition, it was concluded that low coordination Ag atoms (edges, kinks), the presence of subsurface oxygen due to oxygen pretreatment, and the interaction with the oxide support result in electropositive $Ag^{\delta+}$ surface sites that favour the adsorption of the polar C=O bond, and thus the selective formation of unsaturated alcohol. Similarly, the activity is influenced by all these factors that determine the amount of electron-deficient silver sites available for the activation of dihydrogen. Finally, the best selectivity towards the allyl alcohol product (61%, Table 11.17) was obtained by the addition of indium. The presence of oxidized indium generates more positively charged $Ag^{\delta+}$ surface sites or produces indium ions that may behave as Lewis acid sites and assist the selective hydrogenation of the C=O group.

The implication of low-coordinated and electropositive Ag atoms (edges, apexes) in the adsorption and activation of the carbonyl group, and thus in the selective formation of the allyl alcohol, was supported by a combined experimental and theoretical study on the liquid-phase hydrogenation of crotonaldehyde by $AgNPs/SiO_2$ [40]. DFT calculations of crotonaldehyde chemisorption on Ag_{19} cluster model and Ag (111) surface indicate that the most favourable adsorption mode is the σ-binding of the oxygen atom of the C=O bond to low coordination Ag atoms, which leads to its activation. DFT studies also demonstrated that smaller AgNPs favour the chemoselectivity towards the carbonyl functionality since a decrease in the size of the nanoparticle involves an increase in the number of low coordination sites. This was confirmed by catalytic experiments (Table 11.18). However, this is in contrast with the work of Meyer mentioned above in which an opposite effect of the size on the activity and selectivity was reported [38].

Table 11.17 Selective formation of allyl alcohol in gas-phase acrolein hydrogenation catalyzed by supported AgNPs

Catalyst	Particle size (nm)	P (bar)	Selectivity (%)
7.5% Ag/SiO_2	2.5	10	39
5% Ag/Al_2O_3	11	10	42
5% Ag/ZnO	Broad distribution	10	50
9% Ag 0.75% In/SiO_2	5	20	61

Table 11.18 Influence of particle size in the chemoselective hydrogenation of crotonaldehyde and cinnamaldehyde to allyl alcohol catalyzed by AgNPs/SiO$_2$

Size	Crotonaldehyde[a]		Cinnamaldehyde[b]	
	Conversion (%)	Selectivity (%)	Conversion (%)	Selectivity (%)
4.8 ± 1.1	38	59	19	70
11.3 ± 1.8	30	49	13	60
n.d.[c]	27	44	9	55

Reaction conditions: catalyst (0.6 g), isopropanol (80 mL), 30 bar H$_2$
[a]Crotonaldehyde (5 mL), 140 °C, 7 h
[b]CAL (1 mL), 140 °C, 7 h
[c]Not determined due to NP agglomeration

A less frequently used oxide support, CeO$_2$, was employed to immobilize a core–shell AgNPs@CeO$_2$ nanocomposite containing AgNPs (10 nm size) in the core and spherical CeO$_2$ nanoparticles (3 nm diameter) in the shell (Fig. 11.4) [41]. The gaps between CeO$_2$ NPs in the shell allow the access of substrate molecules to the Ag sites in the core, and a cooperative effect between AgNPs and basic sites of CeO$_2$ promotes the heterolytic cleavage of H$_2$ and therefore the selective reduction of the C=O group. This system provided the highest chemoselectivities reported to date for the hydrogenation of unsaturated aldehydes catalyzed by AgNPs (THF, 150 °C, 15 bar H$_2$). In addition, the catalyst can be easily recovered by filtration without loss of efficiency.

Metal–organic frameworks (MOFs) can also be employed as an interesting alternative to typical oxide supports. AgNPs immobilized on a zirconium-based MOF, UIO-66, showed complete conversion and 66% selectivity to the unsaturated alcohol COL in the hydrogenation of CAL [42]. AgNPs/SiO$_2$ used as reference displays an increased selectivity (71%), but a characterization of both systems after catalytic reactions by powder XRD indicates that the UIO-66 MOF provide a better stabilization. Indeed, the catalytic system AgNPs/UIO-66 was reused up to 5 runs with only a slight decrease in the selectivity, showing good recycling behaviour.

The use of Pd–Ag alloys is another innovative strategy to improve the catalytic performance of AgNPs in the selective hydrogenation of unsaturated aldehydes [43]. AgNPs/SiO$_2$ (8% Ag) doped with 0.01% Pd displayed higher activity and selectivity

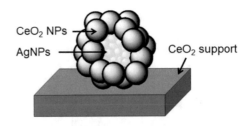

Fig. 11.4 Representative scheme of AgNPs@CeO$_2$ nanocomposite

in the hydrogenation of acrolein than those shown by the analogous monometallic system with the same particle size. The Pd atoms were dispersed and isolated as single atoms in the AgNPs, providing active sites that reduce the energy barrier for hydrogen dissociation, as was supported by DFT studies. Such calculations also suggest that the selectivity increases because the single Pd sites promote the adsorption of acrolein in a more favourable configuration for the hydrogenation of the C=O group (Scheme 11.8).

Finally, the work of De Vos and co-workers is worth noting [7], in which good selectivities towards the allyl alcohol product in the hydrogenation of carbonyl compounds catalyzed by AgNPs were reported (Table 11.19). A wide range of α, β-unsaturated aldehydes and ketones were hydrogenated by the use of non-supported AgNPs (4 nm size) stabilized with polyvinylpyrrolidone. These NPs were efficiently recycled by ultrafiltration, as they were dispersed in amide solvents that improve the colloidal stability and preserve the nanodispersion. In addition, Ag–Au bimetallic nanocolloids (Au/Ag ratio: 1.5; 6 nm size) were also prepared in order to combine the high activity observed for the AgNPs with the high chemoselectivity displayed by AuNPs described in the same research work. The alloyed Au–Ag NPs showed an intermediate activity between Ag and Au nanocolloids, but higher selectivity than those presented by both nanocatalysts (Table 11.19), which proves the synergistic effect of the two metals.

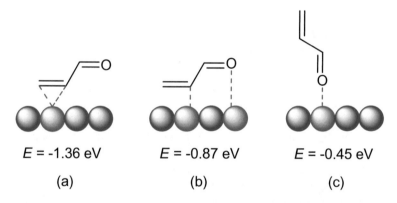

| E = -1.36 eV | E = -0.87 eV | E = -0.45 eV |
| (a) | (b) | (c) |

Scheme 11.8 Adsorption of acrolein on isolated Pd atoms: **a** through the C=C bond with horizontal orientation; **b** through the O atom and C=C bond with horizontal orientation; **c** through the C=O bond with vertical orientation. Ag (blue), Pd (red)

Table 11.19 Hydrogenation of carbonyl compounds catalyzed by Ag/DMA, Au/DMF and Ag–Au/DMF colloidal nanoparticles

Substrate	Ag/DMA		Au/DMA		Ag–Au/DMF	
	Conv. (%)	Sel. (%)	Conv. (%)	Sel. (%)	Conv. (%)	Sel. (%)
Crotonaldehyde	94 (20 h)	66	93 (20 h)	71	91 (12 h)	79
Methacrolein	92 (24 h)	69	94 (30 h)	75	96 (24 h)	82
trans-2-Hexen-1-al	93 (24 h)	77	92 (24 h)	80	93 (18 h)	88
Citral	93 (28 h)	72	94 (30 h)	75	95 (24 h)	79
Cinnamaldehyde	94 (48 h)	93	94 (56 h)	94	91 (36 h)	96
Perillaldehyde	97 (48 h)	80	93 (48 h)	87	94 (36 h)	90
Mesityl oxide	92 (24 h)	56	90 (20 h)	66	91 (18 h)	67
3-Methyl-3-penten-2-one	96 (30 h)	66	91 (28 h)	70	93 (24 h)	74
Benzalacetone	92 (18 h)	57	91 (20 h)	66	92 (16 h)	71
β-Ionone	92 (32 h)	61	90 (32 h)	71	91 (24 h)	75
α-Ionone	93 (40 h)	62	92 (40 h)	74	91 (30 h)	77
Isophorone	93 (96 h)	45	92 (96 h)	52	93 (72 h)	57
Carvone	94 (56 h)	47	90 (56 h)	57	91 (42 h)	60

Reaction conditions: 40 bar H_2, 333 K. Aldehydes: substrate/Ag = 200, substrate/Au = 100, substrate/Ag–Au = 100. Ketones: substrate/Ag = 100, substrate/Au = 500, substrate/Ag–Au = 50

11.4 Pt Nanoparticles in the Hydrogenation of α,β-Unsaturated Aldehydes and Ketones

More than a century ago, Pt colloids were already used as hydrogenation catalysts in solution. Indeed, they were an important tool in organic chemistry for the hydrogenation of nitro-groups, ketones, alkenes, alkynes, nitriles, etc. The colloids were prepared in situ by reduction of, e.g. H_2PtCl_6, but the observed reproducibility was low and complete separation of the catalyst after the reaction was cumbersome. Adams initiated a search for a better alternative and he proposed the use of finely divided $PtO_2 \cdot H_2O$, which is reduced to the actual catalyst 'Pt black' under H_2 or upon heating [44]. This catalyst is employed until today as the Adams catalyst, mainly in solution for organic synthesis. By far the most practical solution associated with catalysis of colloids is the immobilization of metal particles on solid supports. The renewed interest in colloids or nanoparticles of the last two decades is based on the controlled synthesis of nanoparticles, with control over size, shape, exposed faces, the effect of ligands on the catalytic properties, etc. Still, for practical purposes, the nanoparticles are immobilized in many ways to aid separation. This adds a new variable to the catalyst system, namely the interaction of the metal NP/ligand with the support. Many studies have been devoted to supported Pt catalysts because of their high activity and moderate selectivity. Among them, reducible oxides seem to lead to higher selectivity for allylic alcohols.

11.4.1 Unsupported Pt NPs

Large octahedral and cubic, water-soluble Pt NPs and bimetallic $PtNi_2$ NPs (12 nm size) were synthesized in a closed vessel from their acetylacetonates in benzylalcohol at 150 °C with PVP or oleylamine as the stabilizer [45]. They were used for the hydrogenation of various substrates, among them benzalacetone (4-phenyl-3-buten-2-one). Interestingly, only the alkene bond of the enone was hydrogenated under mild conditions (THF, 1 bar H_2, room temperature). The PVP capped particles were most active and $PtNi_2$ NPs were six times more active than monometallic ones. In addition, the particles with (1,1,1) facets exposed were the most active.

Liu and co-workers prepared small Pt NPs (1 nm) stabilized also by PVP by microwave or conventional heating, which showed low selectivity in the hydrogenation of unsaturated aldehydes, (ethanol/water, 60 °C, 40 bar H_2). Interestingly, addition of Fe^{3+} ions to the reaction mixture raised the selectivities to remarkably high values [46] (Crotonaldehyde 0 → 84%, Citronellal 31 → 100%, Cinnamaldehyde 24 → 89%). The authors did not comment on this effect; it seems that an oxidized surface switches the selectivity from alkene to aldehyde hydrogenation. Indeed, the non-modified Pt NPs were very active catalysts for 1-heptene and cyclohexene hydrogenation.

Pt NPs surrounded by oleylamine were generated from $Pt(acac)_2$ via reduction with tetrabutylammonium borohydride. In addition, PtSn NPs incorporating Sn(0) and partly covered by SnO_2 were prepared the same way in the presence of $SnCl_2 \cdot 2H_2O$ [47]. In the CAL to COL reaction the selectivity to unsaturated alcohol increases substantially (92%) at high tin content (25 °C, 1 bar H_2, butanol). The authors proposed that Pt serves for the generation of hydrides, while Sn ions activate the aldehyde.

Large Pt NPs (6 nm size) protected by 3,3′-thiodipropionic acid [bis(2-carboxyethyl) sulphide] showed low activity and good selectivity (83%) in the CAL to COL hydrogenation [48]. In this instance, the ligand reduces catalytic activity.

Soluble Pt NPs (2 nm size) ligated by imidazolium-amidinates (Fig. 11.5) were prepared by Martinez-Prieto et al. [49]. These ligands coordinate as N,N'-κ^2 chelates to a densely covered surface as was established by ^{15}N NMR on smaller particles (1.2 nm, to avoid Knight shifts) and DFT calculations. The nanoparticles were employed as catalysts for the hydrogenation of several substrates, of which 4-phenyl-3-buten-2-one and 3-methyl-2-cyclohexenone are of relevance here. They are very active alkene hydrogenation catalysts (styrene, THF, 60 °C, 5 bar H_2, initial TOF

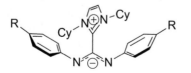

Fig. 11.5 Imidazolium-amidinates employed to stabilize Pt NPs

20 s^{-1}). However, the Pt NPs show moderate activity for aldehyde and ketone hydrogenation, which results in the reduction of the C=C bond in both substrates with 81% and 68% selectivity, respectively, at total conversion, the remainder being the doubly hydrogenated product, the saturated alcohol. It is thought that the electron-rich surface coordinates only weakly to the ketone function, while back-donation enhances alkene coordination.

Rodionov and co-workers prepared bimetallic PtFe MNPs stabilized by carboxylic acids and perfluorinated carboxylic acids from $Pt(acac)_2$ and $Fe(CO)_5$ with 1,2-hexadecanediol as reducing agent [50]. Particle sizes varied from 3.2 to 3.6 nm, independent of the metal ratio. It was proposed that the carboxylic acids coordinate to Fe, leaving Pt free. The selectivity of CAL to COL depends strongly on Fe content and the type of acid used (perfluorodecalin, n-hexane, 1 bar H_2, 50 °C). High Fe contents (0.33) and long-chain (C10) perfluoroacids gave the highest selectivity, up to 94%, much higher than that shown by the Fe (0.13) catalyst prepared this way, 64%.

Zheng, Fu, et al. studied the effect of the nature of the capping amine ligand on Pt_3Co NPs in the hydrogenation of CAL [51]. Thus, for oleylamine a very homogeneous monodispersion of NPs with 8.2 nm size was obtained by CO reduction of acetylacetonate precursors. The truncated octahedrally shaped NPs gave a highly ordered TEM as shown in Fig. 11.6. An 8 nm NP contains about 3000 surface atoms and analysis shows slightly less than 500 oleylamine molecules on the surface which, according to DFT, can form ordered arrays on the surface. >90% COL was obtained in the hydrogenation of CAL (n-butanol, 25 °C, 1 bar H_2), which the authors explained by the coordination of the carbonyl group to the surface interchelating in the stacks, while coordination of the alkene is prohibited by the stacked amines. For shorter chain amines the selectivity gradually diminished (octylamine 80%, butylamine 20% COL in addition to HCAL and HCOL) while the rate increased, in support of the mechanism proposed. This catalytic system also provided >90% selectivity towards the unsaturated alcohol in the hydrogenation of citral.

11.4.2 Organic Supports

Polymerization of 4-tritylaniline and formaldehyde with $FeCl_3$ as a catalyst yields hollow polymeric tubes with a diameter of 80–100 nm [52]. They were used as support for Pt NPs (3–4 nm) generated the usual way in water from chloroplatinate and sodium borohydride. This catalytic system (2 wt% metal) was explored for the hydrogenation of unsaturated aldehydes (isopropanol, 60 °C, 20 bar). Modest selectivities for cinnamaldehyde (COL < 72%) and 3-methyl-2-butenal (3-methyl-2-butenol < 50%) were observed, whereas furfural was converted selectively (100%) to furfuryl alcohol.

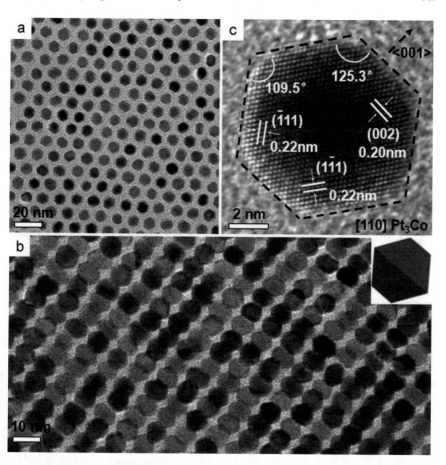

Fig. 11.6 TEM images of 8.2 nm Pt3Co nanoparticles. **a** A monolayer assembly. **b** A multilayer assembly. **c** HRTEM image of a single Pt3Co nanoparticle along the [110] zone axis. Inset in **b** is a model of truncated octahedron along the [110] zone axis with two vertices in the h001i direction cutting off from the octahedron. Reproduced with permission from reference [51]. Copyright 2012 Wiley-VCH Verlag GmbH & Co. KGaA, Weinheim

11.4.3   Pt on Metal

Pt NPs (3.0–3.8 nm) deposited on Au@SiO$_2$ were employed as catalysts for CAL hydrogenation, leading to the preferential formation of the saturated aldehyde HCAL (150 °C, 10 bar H$_2$) [53]. A similar result was reported by the same authors in which they stressed the rate enhancement by the Pt@Au@SiO$_2$ catalyst compared to other catalysts in which one component was lacking [54].

11.4.4 Metal–Organic Frameworks as Supports

MOFs have become important supports for MNPs and nanoclusters (NCs), as they confine the metal clusters inside their pores and also influence catalyst and substrates through their close contact. Making uniform particles inside MOFs is not trivial, but many researchers have succeeded in doing so. Huang et al. used UiO-66 [55] and UiO-66-NH$_2$, which are built up from [Zr$_6$O$_4$(OH)$_4$] clusters linked with 1,4-benzenedicarboxylic acid and the amino-substituted diacid, respectively, forming a cubic porous structure. The material presents channels and two types of cavities of 0.7 and 1.1 nm [56]. Pt was introduced as a water solution of K$_2$PtCl$_4$ and then reduced at high temperatures to yield 1–10% wt Pt@UiO-66-NH$_2$ with particle size 1.16 nm. It was shown that the NPs were indeed inside the porous MOF. The catalysts were used for the hydrogenation of CAL and, as Table 11.20 shows, the MOF host has a favourable effect on rate and selectivity (methanol, 10% Et$_3$N, 40 bar H$_2$, 25 °C, various times).

Higher loadings of Pt gave higher COL selectivity. The authors suggested that access of CAL in the pores was restricted, such that the alkene function cannot coordinate onto the Pt surface. On the other hand, Pt prepared outside, in PVP, and precipitated on UiO-66 was the most active catalyst, but the least selective for COL. In addition, the conditions applied by Huang show a good result for Pt/C (COL 71.9%), higher than usual for C supports as we will see later (Sect. 4.8).

Li et al. reported on the catalysis of Pt NPs in MOFs for the hydrogenation of unsaturated aldehydes [57]. Thus, H$_2$PtCl$_6$ was reduced with NaBH$_4$ in the presence of PVP to give Pt NPs of 2.3 nm, upon which immediately MIL-101, containing Cr^{3+} Lewis acid sites, was added. 1%Pt NPs showed an even distribution on the MIL surface. This catalytic system was used for the hydrogenation of CAL (ethanol or isopropanol, 1 bar H$_2$, 25 °C) to give >99% HCAL, a result deviating considerably from the common hydrogenation results and it contrasts especially with the effect of Lewis acids in other cases. The authors ascribe the formation of HCAL to the interaction of the ketone with the Lewis acid sites, but in view of other results, we propose that there is no close contact between the Lewis acids and Pt/PVP sitting on

Table 11.20 Catalytic performance of Pt@UiO-66-NH$_2$ catalysts in cinnamaldehyde hydrogenation

Sample	Activity (mol mol Pt^{-1} h^{-1})	Conv. (%)	Selectivity (%)			
			HCAL	HCOL	COL	Other
10.7%Pt/UiO-66-NH$_2$	26.3	98.7	2.5	4.0	91.7	1.8
3.3%Pt/UiO-66-NH$_2$	77.4	34.2	8.7	2.5	88.8	0.0
0.97%Pt/UiO-66-NH$_2$	170	49.4	18.1	1.8	80.0	0.0
5%Pt/C	108	41.3	21.4	4.7	71.9	2.0
4.2%Pt/UiO-66-NH$_2^a$	297	71.6	35.5	8.5	52.2	3.8

^{a}Outside PVP

top of the MIL-101. In a subsequent publication, Li and co-workers employed MIL-100 with Fe as the cation and placed 3 nm Pt NPs inside followed by an outer layer of MIL-100, Pt/MIL-100@MIL-100 [58]. The latter were used as catalysts in the hydrogenation of CAL under 1 bar H_2 and room temperature (RT). The selectivity to the desired COL could be enhanced up to 96%, which shows a significant increase compared to that of the uncoated Pt/MOFs (55%).

Zhao, Yuan and co-workers also used MIL-101 (Fe, Cr) as supports to immobilize Pt NPs, which were applied for CAL to COL hydrogenation [59]. These sandwich MIL-101@Pt@MIL-101 nanostructures were obtained by precipitating ex-situ prepared 2.8 nm Pt/PVP NPs on top of MIL crystalline particles (300 nm, Fe or Cr), that were subsequently covered by a thin layer of MIL-101 (2–20 nm, Fe or Cr), thus sandwiching the Pt NPs between two MOFs. The new catalysts show excellent selectivity towards the COL product (30 bar H_2, 25 °C, ethanol/water, MIL-101(Fe)@Pt@MIL-101(Fe) (9.2 nm shell) conversion 94.3%, selectivity to COL 97%, TOF 13.3). The authors assigned the effect to Lewis acid interaction with the substrate.

Hu and co-workers have put Pt@MIL-101 inside a hydrophobic shell of polymerized Fe-porphyrins (FeP-CMP = conjugated micro- and mesoporous polymer) in order to increase the wettability of the catalyst with hydrophobic substrates such as CAL [60]. Pt NPs were prepared ex-situ in ethanol-water with PVP as stabilizer and precipitated onto MIL-101. The outer shell was synthesized via a Suzuki coupling of tetra(p-bromophenyl)chloro-ironporphyrin and 1,4-benzenediboronic acid (Scheme 11.9). Adsorption of CAL was more than twice as high in the new material as in Pt@MIL-101. Typical data for just two materials are shown in Table 11.21.

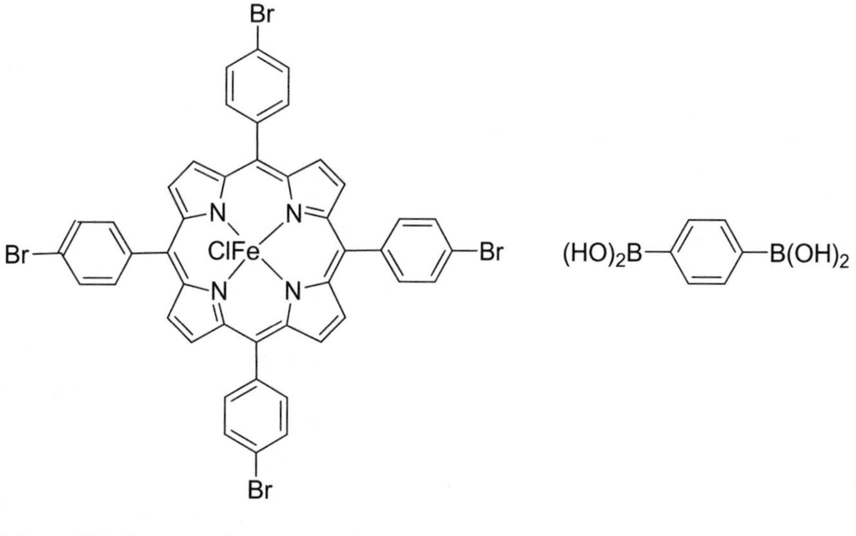

Scheme 11.9 Precursors for organic polymeric layer

Table 11.21 MIL-101@Pt catalysts in CAL hydrogenation

Catalyst	Conversion (%)	COL%	HCAL%	TOF $(Pt^{-1}\ h^{-1})^a$
MIL-101@Pt	15.0	26.3	76.7	203.4
MIL-101@Pt@FeP-CMP	97.6	97.3	2.6	1516.1

Conditions: Methanol, triethylamine, 25 °C, 30 bar H_2
^{a}Regarding CAL conversion

Thus, the rate increased considerably thanks to the porphyrin shell, but also the selectivity is excellent. The latter was ascribed to the interaction of the aldehyde oxygen atom with the iron Lewis sites of FeP.

11.4.5   Pt on Oxide Supports

The MOF material NH_2-MIL-125(Ti) (500 nm) was converted to a titanium oxide/hydroxide porous support at high temperature in alcohols in the presence of various amino acids [61]. The new material consisted of layered titanate $H_2Ti_8O_{17}$ and anatase TiO_2, which is much more robust than the MOF precursor. Pt_3Co was deposited as NPs (<10 nm size) on this support from ethanol solution at 176 °C. At 74.8% conversion of CAL it was converted to COL with 97% selectivity (ethanol, 2 bar, 80 °C). At full conversion, the selectivity dropped to 89%.

Zhu and Zaera studied the structure sensitivity of cinnamaldehyde hydrogenation catalyzed by Pt NPs supported on SiO_2 (Pt/SiO_2) with various metal loadings, between 0.5 and 5.0 Pt wt% [62]. To avoid the influence of stabilizing agents or ligands on the catalytic properties of the NPs the authors chose conventional synthetic methods, impregnation of silica by H_2PtCl_6 followed by reduction and calcination. The average particle size varied from 1.3 to 2.5 nm. Below 80% conversion, the ratio of the primary products remained constant. The low Miller index surfaces were found the most active ones. The observed selectivities were independent of particle size and amounted to 20% for COL, 55% to phenylpropanal (HCAL), and the remainder being phenylpropane and phenylpropene (10 bar, 27 °C, isopropanol). Thus, bare Pt NP surfaces show a low selectivity for COL and the molecules on the surface play an important role. In a subsequent study, Weng and Zaera investigated $Pt@Al_2O_3$ covered with layers of SiO_2 by atomic layer deposition (ALD) [63]. After six ALD cycles, complexation of CO to Pt was reduced only by one third. Activity dropped by this procedure but after 3 ALD cycles COL selectivity was as high as 85% (vs. 26% for zero ALD). This means that half a monolayer of silica was deposited, and the authors proposed that the gain in COL selectivity was due to the Bronsted sites created this way. Han and co-workers studied Pt on alumina as catalysts for CAL hydrogenation (10 bar H_2, 30 °C, methanol/water) [64] and they reported 46% CAL to COL at 50% conversion for the conventional catalysts. In addition, modified Pt NPs were prepared by $NaBH_4$ reduction in water in the presence of aspartic acid after which alumina was added. As one can see from Table 11.22 this has an enormous effect on the selectivity of the catalyst on n-Al$_2$O$_3$ but not for mesoporous Al$_2$O$_3$. Other unsaturated aldehydes showed similar improvements. The authors proposed steric interactions as likely cause for this selectivity change.

Medlin and co-workers modified a conventional Pt/Al$_2$O$_3$ catalyst with various thiols in order to improve the COL selectivity of the catalyst [65]. Reactions were run at 50 °C, 40 bar with hydrogen gas, in ethanol as the solvent. For the unmodified catalyst conversion to HCAL was twice as fast as that to COL (always below 23%), while at high conversion COL was converted to HCOL. Application of thiols led to considerable changes. The nature of the thiol used turned out to be of major importance; COL selectivity at 50% conversion decreases in the range 3-phenyl-1-propanethiol > 4-phenyl-1-butanethiol, 2-phenyl-1-ethanethiol > benzyl mercaptan, dodecanethiol > uncoated, thiophenol > an 'aged' catalyst of 3-phenyl-1-propanethiol, thiophene (Scheme 11.10). The authors explained the effect with non-bonding interactions between the thiol substituents on the surface and the substrate, 3-phenyl-1-propanethiol giving the best 'host' for steering the oxygen atom to the Pt surface.

Pt@TiO$_2$-SBA-15 catalysts with a loading of TiO$_2$ in the range 25–60% wt were synthesized by grafting a titanium precursor on preformed mesoporous silica followed by Pt deposition (4.6 nm size) [66]. These materials were used for the partial hydrogenation of citral (70 °C, 70 bar H$_2$, isopropanol). The presence of TiO$_2$ results in lower conversions but higher selectivity to unsaturated alcohols not exceeding 25%

Table 11.22 Catalytic performance of different Pt/Al$_2$O$_3$ catalysts for the hydrogenation of cinnamaldehyde

Catalyst	Time (h)	Conversion (%)	Selectivity (%)		
			COL	HCAL	HCOL
Pt/n-Al$_2$O$_3$	2	50	46	40	14
Pt-aa/n-Al$_2$O$_3$[a]	2	76	94	3	3
Pt/mesoporous-Al$_2$O$_3$	1	53	38	49	13
Pt-aa/mesoporous-Al$_2$O$_3$[a]	2	98	19	49	32

[a]aa: aspartic acid

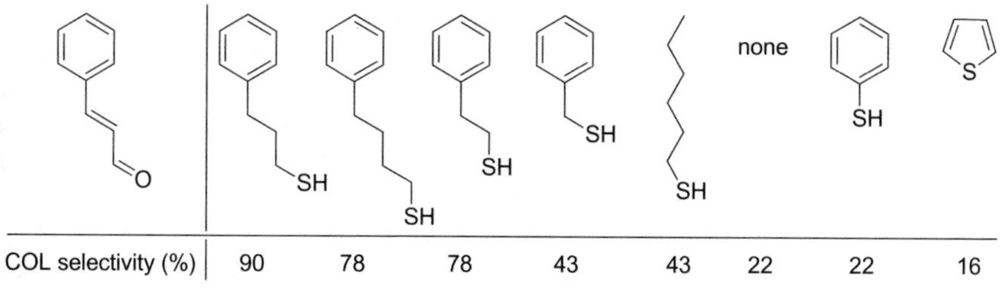

COL selectivity (%)	90	78	78	43	43	22	22	16

Scheme 11.10 CAL to COL selectivity at 50% conversion of thiol-modified Pt/Al$_2$O$_3$ catalysts

(vs. only 5% for Pt@SBA). Decomposition of citral leads to CO, which reduces the rate of hydrogenation.

FeO$_x$-doped Pt catalysts supported on 15 wt% Al$_2$O$_3$@SBA-15 composites were studied by Li and co-workers in the hydrogenation of CAL [67]. The added iron oxides raised the positive charge on Pt and on nearby Fe sites, which was held responsible for the increase in the selectivity for COL, from 40% for Pt@SBA-15 to 77%, at almost full conversion (isopropanol/water, 20 bar H$_2$, 90 °C). In a related publication, Al$_2$O$_3$ was replaced by 15% TiO$_2$ that was deposited onto mesoporous silica, SBA-15 [68]. The best result obtained in this instance with the three-component catalyst under the same conditions was 86% selectivity to COL at 90% conversion; at higher conversion somewhat more HCOL was produced and the selectivity for the desired COL product dropped.

Somorjai et al. studied the influence of thin films (30 nm) of Co$_3$O$_4$ support on the hydrogenation of crotonaldehyde catalyzed by Pt NPs and compared these with SiO$_2$ and TiO$_2$ [69]. SiO$_2$ was considered the support with least interaction with Pt NPs (4.6 nm size), and this catalyst showed the lowest activity and zero selectivity for crotyl alcohol. Pt@Co$_3$O$_4$ showed the highest activity, TiO$_2$ being in between, but the selectivity for the unsaturated alcohol was modest at 20%, butanol 20%, the remainder 60% is butyraldehyde (gas-phase, 1 Torr of crotonaldehyde, 100 Torr of hydrogen, 120 °C). Under the reaction conditions, the Pt particles are fully reduced to the metallic state and the cobalt oxide surface is partially reduced, likely to Co^{2+}. It was thought that atomic hydrogen spills over onto the oxide to carry out the low-temperature reduction. This reduced surface, it was stated, contains sites for the adsorption and selective reaction of crotonaldehyde to the alcohol products. Thus, more in general, reducible oxide supports enhance enol selectivity and velocity.

Li and co-workers also studied Pt NPs in the order of 5 nm size that interact strongly with Co oxides [70]. Although the support used is amorphous carbon, the topic fits better in the discussion of Pt/Co$_x$O than under the heading carbon supports (4.8). First-layered double hydroxides were made (Mg, Al and variable Co content) containing intercalated chloroplatinic acid which were calcined in the presence of glucose. Cobalt oxides were observed at the surface of the Pt NPs, which according to XPS contained 30–50% divalent Pt. Four catalysts were prepared with a Co content relative to Pt (4.8% wt) of 0, 0.3, 0.6 and 0.9 (Table 11.23). The amount of oxides in the catalysts was not reported. A remarkable difference between the catalysts was found in the hydrogenation of CAL (Table 11.23).

The presence of CoO improves the catalyst enormously, both in terms of selectivity and activity. Catalysts Co$_{0.3}$ and Co$_{0.9}$ perform about the same, but Co$_{0.6}$ really stands out as it is 10 times more active than Pt only and 99% selectivity to the desired allylic alcohol product. Aliphatic enals reacted equally well when Co$_{0.6}$Pt/C was used as the hydrogenation catalyst. The authors assign this result to the higher Pt(0) content of catalyst Co$_{0.6}$ (68% vs. 54% for Co$_0$Pt).

Related to the last publications is the work by Xiang, Qin and co-workers [71]. They decorated Pt nanoparticles with Fe oxide by atomic layer deposition in a precise way. First, Pt was deposited (2.7 nm) from an organometallic precursor and ozone on Al$_2$O$_3$ (60–100 nm), followed by ferrocene and ozone, both in several cycles.

Table 11.23 Catalytic performance for the CAL hydrogenation over Pt-Co$_x$O catalysts

Catalyst	TOF (s^{-1})[a]	CAL conversion (%)	Selectivity (%)		
			COL	HCAL	HCOL
Co$_0$Pt/C	0.44	32	37	61	2
Co$_{0.3}$Pt/C	2.69	69	76	21	3
Co$_{0.6}$Pt/C	4.19	100	99	0	1
Co$_{0.9}$Pt/C	1.75	71	82	14	4

Reaction conditions: Ethanol, 80 °C, 20 bar H$_2$
[a]TOF value for CAL conversion determined based on the number of surface Pt sites within the initial 30 min (surface atoms amount to approximately 20% of the total Pt for this size)

These materials were employed as catalysts for the hydrogenation of CAL (ethanol, 60 °C, 20 bar H$_2$). The authors found by DFT calculations and DRIFTS analyses that low-coordinated Pt sites were selectively blocked by ALD of Fe oxide. The result was an increase in selectivity to COL from 45% for bare Pt to 84% for the best iron oxide covered Pt NPs. It was proposed that the same mechanism may be operative in other catalysts. One could also imagine that the blocking function could be exerted by organic ligands, cf. supported TiCl$_3$ catalysts and their interaction with esters and siloxanes [72].

Huang and co-workers used Co(OH)$_2$ as a support for Pt and Pt$_3$Co NPs [73]. Co(OH)$_2$ is a layered double hydroxide that attracted attention because of its large surface area. The nanosheet compounds were synthesized in one step from metal salts in methanol, water and octylamine by heating at 220 °C. These nanosheets have a thickness of 3.5 nm and a diameter of 400 nm, and the Pt NPs are 4 nm in size. The catalysts were found to be highly selective for the hydrogenation of CAL (ethanol, 70 °C, 5 bar H$_2$). At 99.6% conversion, the highest observed selectivity was 91.3% to COL.

The effect of acidic supports on NP catalysis is well established, as is the effect of oxides on the hydrogenation of unsaturated aldehydes. Thus, Bhogeswararao and Srinivas studied the effect of ceria and zirconia on the hydrogenation of CAL by Pt NPs [74]. Pt was deposited on ceria, zirconia and mixed supports by classic methods, yielding 5–12 nm particles. XPS showed Pt^0 and partially oxidized Pt. Hydrogenation of CAL was conducted in isopropanol, RT, 20 bar H$_2$. The as-prepared catalyst Pt@Zr/CeO$_2$ gave a selectivity of CAL to COL in the range 60–70%. Surprisingly, the treatment of the material with a water solution of NaOH increased the initial selectivity to 97% COL (93% at 96% conversion). Nevertheless, the authors concluded that 'the acidity of the support (due to the presence of ZrO$_2$ component) and higher electron density at Pt (due to CeO$_2$ component) are responsible for the higher catalytic activity and selectivity of Pt'.

Liang and co-workers used atomic layer deposition (ALD) of cobalt oxide on Pt as a modification of Pt@MWCNT [75]. Small Pt NPs on MWCNT were obtained (1.2–1.7 nm size) by ALD with [(MeCp)PtMe$_3$] as the Pt precursor and oxygen (O$_2$) as the other reactant at 300 °C in a fluidized bed reactor. Co was deposited in 1:1

Table 11.24 Hydrogenation of CAL by PtCo/MWCNT

Entry	Catalyst	Conversion (%)	Selectivity to COL (%)
1	Pt/MWCNT	46.0	62.2
2	Co/Pt/MWCNT	93.3	93.4
3	Co/MWCNT	11.2	76.6
4	Co/Pt/Al$_2$O$_3$/MWCNT	33.1	87.3
5	1 + 3	43.4	53.1

Reaction conditions: 1.87 mg of metal (Pt and Co), CAL (0.5 g), 2-propanol (30 mL), 80 °C, 12 h, 10 bar H$_2$

ratio to Pt likewise from Cp$_2$Co in H$_2$. A small part (<10%) of Co occurred as Co^{2+}. Three typical results are collected in Table 11.24. Clearly, a cooperative effect of Co and MWCNT can be seen on the selectivity and rate of CAL to COL conversion by Pt NPs. The authors proposed a CNT assisted electron transfer from CO to Pt, based on the results of entries 4 and 5, as supported by DFT calculations. In entry 4 a layer of Al$_2$O$_3$, also deposited by ALD, isolates the NPs from the conductive CNT. Hydrogenation of other aldehydes was affected the same way.

11.4.6 Light-Induced Hydrogen Transfer

Pt and Au NPs were deposited on TiO$_2$ both yielding 5 nm sized particles [76]. These materials were used as photocatalyst for the transfer of hydrogen from alcohols to CAL with visible light and UV light 365 nm at RT. The TiO$_2$ support without noble metal showed some activity for the reaction and low selectivity to COL. The Au catalyst showed good activity under visible light irradiation. The selectivity depended strongly on the alcohol donor used, isopropanol being the most selective one for COL (isopropanol 100%, n-propanol 69%, ethanol 58%, benzyl alcohol 47%). This effect of the donor suggests that at least one of the hydrogen atoms is directly transferred from the donor alcohol to CAL, as proposed for Meerwein–Ponndorf–Verley reductions. Pt@TiO$_2$ only shows high activity when UV irradiation is applied and selectivity can also reach 100% at full conversion.

11.4.7 Carbon Supports

Carbon materials are attractive supports for MNPs as they can be modified in several ways and these modifications can influence the properties of the Pt NP catalysts. Deposition of Pt salts via reduction usually requires the partial oxidation of the carbon material in order to facilitate the wetting of the support by the aqueous solution of, e.g. H$_2$PtCl$_6$, a common starting material. Pt on graphite prepared this way gave a low selectivity to crotyl alcohol in gas-phase hydrogenation of acrolein [77]. The use of chloride and tin as promoters led to an improvement in the selectivity to crotyl alcohol up to 25%. In the last decades, single-walled carbon nanotubes (CNT or SWCNT), multiwalled CNTs (MWCNTs), and graphite nanofibers (GNFs) have been a material of choice as support and comparisons were made with commercial activated carbon (AC). Thus, Serp and co-workers deposited several metals on these new materials and studied their properties as catalysts for cinnamaldehyde to cinnamyl alcohol hydrogenation (CAL to COL, Table 11.25). They employed a surface organometallic approach to deposit platinum and ruthenium on various carbon supports by means of $[Pt(CH_3)_2(\eta^4\text{-}C_8H_{12})]$ and $[Ru(\eta^4\text{-}C_8H_{12})(\eta^6\text{-}C_8H_{10})]$. Even for these precursors oxidation of carbon with nitric acid was used prior to deposition [78].

The untreated new materials are clearly more active but less selective to COL than the classic AC support, but heat treatment has a favourable effect on Pt@MWCNT (2.3 nm) and the selectivity is now 66% at almost full conversion. Ru@C (1.7 nm) was worse in all respects. A bimetallic Pt/Ru catalyst (1–7 nm size) gave the best performance, Pt/Ru@MWCNT 79/93%, for the heat-treated material. The authors ascribed the better performance of heat-treated MWCNTs to the increase in electronic density around metal particles which promotes the adsorption and activation of CAL. Similar bimetallic PtRu@MWCNT prepared in THF or scCO$_2$ by Serp, Gomez et al. did not lead to major improvements [79]. In a subsequent publication, Pt NPs were deposited on MWCNTs which had been activated by three different procedures, nitric acid treatment, ball-milling and air oxidation to modify their surface chemistry and morphology [80]. IR studies showed that the supports contained different amounts of oxides and carboxylic acids and this influenced the properties of the catalyst, prepared as above from the same organometallic precursor. Pt NP sizes varied from 2 to 20 nm and the largest gave the highest selectivity to COL. The selectivity to COL at 50% CAL conversion varied from 4 to 69% in heptane as the solvent, both extremes for air treated carbon. Carboxylic acids are efficient for anchoring of the precursor, but the presence of these oxygenated groups in the final catalyst is detrimental to selectivity. A post-reduction at high temperature removes most of these functionalities, increases the Pt NP size to 15 nm, and gives a higher selectivity to COL.

Table 11.25 CAL conversions and selectivity to COL for different carbon-supported Pt, Ru and Pt/Ru catalysts

Support	2% Pt	2% Ru	Pt/Ru
	Conv./Selec. (%)	Conv./Selec. (%)	Conv./Selec. (%)
SWCNT	85/28	65/31	
MWCNT	95/32	66/35	44/55
MWNT heated	97/66		79/93
GNF	96/14	78/18	
AC	20/62	22/51	15/64

Reaction conditions: Isopropanol, 100 °C, 2 h, 20 bar H$_2$

Hou and co-workers reported on the use of small Pt NPs (2.5 nm size) deposited on the surface of few-layered reduced graphene oxide (RGO). The Pt NPs were obtained via direct ethylene glycol reduction of H_2PtCl_6 in aqueous solution [81]. In the series of alcohol solvents, the rate decreased drastically in methanol to 2-pentanol, but the selectivity increased from 74 to 94% (20 bar H_2, 70 °C). Ethanol, with 74% conversion in 2 h and selectivity 85% at 0.04% Pt (TOF = 27 min^{-1}), showed the best compromise. In addition, the selectivity is retained at full conversion.

Han et al. compared similar catalysts on CNT (Pt NPs of 3.5 nm size) and RGO@SiO_2 (3 nm) obtained from H_2PtCl_6 in aqueous solution by $NaBH_4$ reduction [82]. CAL hydrogenation was carried out under the same conditions as the next report. The Pt/CNT catalyst showed 62% selectivity to COL, whereas the Pt/RGO@SiO_2 system only 48%, which seems in line with the explanations given below and above.

Li, Zaera, Zhu and co-workers used Pt NPs supported on graphene as catalyst for CAL hydrogenation [83]. Pt NPs on several homo-made graphenes were generated by sodium borohydride reduction of chloroplatinic acid in water. Mean particle sizes of 3–4 nm were obtained. Catalysis was carried out at 10 bar H_2, in isopropanol as solvent at 60 °C. The best result was 88% selectivity to COL at 92% conversion. One factor of importance to be mentioned is the high amount of Pt^0 for the best catalyst. DFT studies indicated the bonding of C=O functions to Pt^0 is stronger than that of C=C bonds. Higher crystalline graphene gave higher selectivity to COL, which might be due to stronger phenyl coordination to the support, dragging the C=C bond away from Pt on the Pt-graphene edges.

A highly active Pt catalyst on porous support of SiC/C was reported by Li et al. [84]. The catalyst was prepared from chloroplatinic acid in water and reduction by sodium formate giving Pt NPs of 2.2–2.6 nm. Heat treatment under hydrogen was applied at 400 °C. From a range of supports tested this one gave the most active catalyst for CAL to COL hydrogenation at 25 °C, in isopropanol, 20 bar H_2, TOF 2400 h^{-1}. At 40 °C, the TOF was twice as high. However, the selectivity was lower than those described in the previous report (80%). In contrast to the previous study, the high preference for COL was assigned to the cationic Pt ions at the surface as found by XPS.

Xiao, Ye and co-workers studied the hydrogenation of 3-methyl-2-butenal (3-methylcrotonaldehyde, 3-MeCal) to the corresponding allylic alcohol 3-MeCol (Scheme 11.11) catalyzed by Pt NPs immobilized on MWCNTs (Pt/MWCNT) under conditions similar to those used above for CAL (15 bar H_2, 80 °C, ethanol, 10%

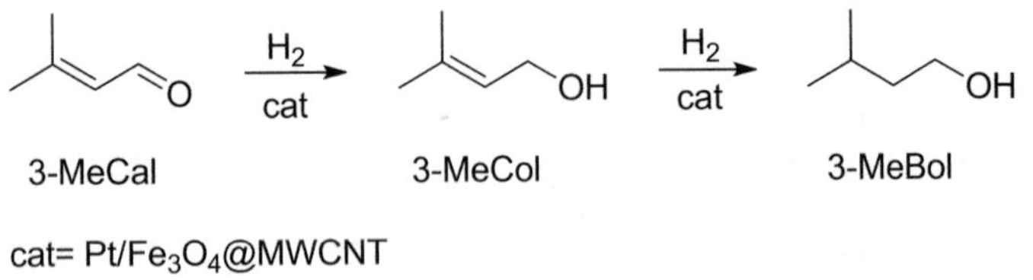

3-MeCal   3-MeCol   3-MeBol

cat= Pt/Fe_3O_4@MWCNT

Scheme 11.11 Hydrogenation of 3-methylcrotonaldehyde by Pt/Fe_3O_4@MWCNT

water) [85]. Pt was deposited as 2.6 nm NPs on MWCNTs by reduction in two-phase water/toluene in the presence of oleylamine as stabilizer. Next magnetic Fe_3O_4 was also deposited on the material, which enabled them to separate catalyst and product mixture magnetically. At conversions of 60–80%, selectivities to 3-MeCol of 99–97% were obtained, the remainder being mostly 3-MeBol. Thus, the triply substituted alkene of 3-MeCal gives better results than the doubly substituted one in CAL, containing also the phenyl group in conjugation with the alkene.

Dai and co-workers reported on the influence of water on the CAL hydrogenation by Pt–Fe catalysts on CNTs [86]. The MNPs were prepared from the corresponding acetylacetonates by heating at 180 °C under CO in benzyl alcohol in the presence of CNTs. XPS showed that Pt was mainly Pt(0) (6.7 nm in Pt_3Fe) and Fe consisted of mixtures of di- and trivalent metal. In the absence of Fe smaller Pt NPs were formed (2.7 nm size). The latter gave poor selectivity for CAL to COL (50%) at a low rate in isopropanol, but for Pt_3Fe this increased to over 90% with a ten times higher rate (isopropanol, 60 °C, 20 bar H_2). When the reaction was carried out in water, the selectivity increased for both catalysts (97% for Pt_3Fe), while for Pt_3Fe in addition the rate tripled. When the reaction was carried out in D_2O, deuterium was found to be incorporated in the hydrogenated alkene, but this was mainly the case for Fe containing catalysts. The water molecule was proposed to act as a bridge to facilitate hydrogen exchange between the CAL species and the Pt sites.

Hydrogenation of citral (Scheme 11.3) on Pt or Pd on activated carbon gave complicated mixtures of products and this will not be discussed [87].

SnO_2 NPs (4.8–5.8 nm size) were first coated onto the surface of reduced graphene oxide (rGO), and then very small Pt_3Sn NPs (0.6–1.2 nm size) were deposited on top of the SnO_2 NPs by the microwave-assisted decomposition of H_2PtCl_6 in ethylene glycol [88]. This catalyst exhibits a high selectivity for the CAL to COL hydrogenation, 92.5% (ethanol, 70 °C, 20 bar H_2). Crotonaldehyde (90%), citral (82%), furfural (>99%) and perillyl aldehyde (83%) also gave excellent results. According to XPS, the PtSn catalyst contains less Pt(II) than the tin-free catalyst.

11.5 Palladium Nanoparticles

11.5.1 Pd NPs in Solution

Pd NPs have been much less studied than their Pt counterparts, perhaps because the 'natural' preference for Pd catalysts seems to be the hydrogenation of the C=C bond in enones and enals. The remarkable success of many polar modifications of the Pt environment that led to very high selectivity to unsaturated alcohols might stimulate more work on Pd as well.

Pd NPs were made water-soluble by sticking cyclodextrins (CD) to their surface [89]. β-CDs can act as hosts for the targeted substrate molecules and thus the Pd surface would be accessible for the substrates. Alcohol groups in CDs were replaced

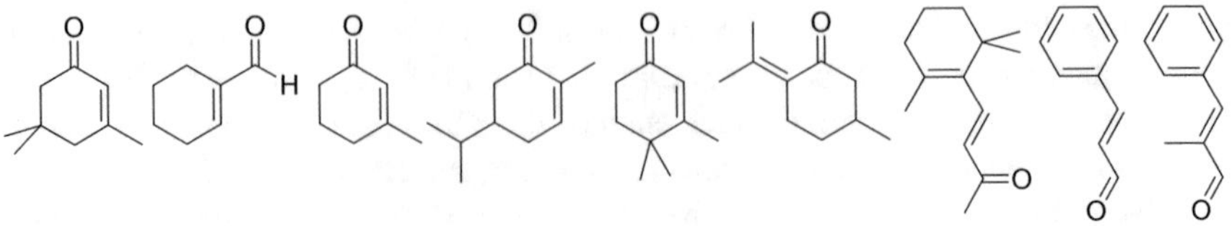

Scheme 11.12 Unsaturated ketones and aldehydes all yielding C–C saturated under Pd hydrogenation

by thiols, which coordinate strongly to Pd NPs. All substrates shown in Scheme 11.12 gave full hydrogenation to the saturated ketone or aldehyde in a two-phase reaction, in which the catalyst remains fully in the water phase (water, 20 bar H_2, 25 °C, 1–5 h). While the product may not be the most desired one, it presents an interesting system.

11.5.2 Organic Supports

Pd NPs immobilized on polystyrene modified with diphenylphosphino groups and imidazolium salts, or the same further modified with PEG chains, were used as hydrogenation catalysts [90]. Pd NPs were 2.3 and 1.9 nm in size, respectively. CAL was hydrogenated in various solvents and, except toluene, all gave active catalysts with selectivities to HCAL from 58 to 85% (5 bar H_2, 25 °C). When inorganic bases were added to the catalysts in water the selectivity to HCAL increased to 100%. Complete conversion was reached in 1 h, i.e. the TOF is higher than 200 h^{-1}.

11.5.3 Pd in Ionic Liquids (ILs)

Pd NPs in an imidazolium-based ionic liquid, [bmim][BF_4] and immobilized on silica were used for the hydrogenation of the cis/trans mixture of citral (hexane as bulk solvent, 100 °C, 10 bar H_2) by Mikkola et al. in a detailed kinetic study on the diffusion phenomena in this complicated multi-phase system [91]. The primary product is citronellal, which is directly converted to dihydrocitronellal with 90% selectivity. The selectivity to the primary product did not exceed 45% in the course of the reaction. Instead of silica also carbon (active carbon cloth, ~1500 m^2 g^{-1}) was used as a support and the results were basically the same for various ILs under the same conditions [92]. In addition, hydrogenation of CAL under the same conditions gave 82–94% HCAL, showing again that Pd NPs (5–10 nm size, estimated from photograph) have a strong preference for C=C bond hydrogenation.

11.5.4   Pd on Oxides

Pd tetrahedrons (13 nm size) were deposited on MoO_{3-x} nanosheets (50–100 nm) in a one-pot solvothermal method [93]. These materials were used for the selective hydrogenation of α,β-unsaturated aldehydes with high conversion (97%) and selectivity (96%) to their saturated aldehydes. Thus 3-methylcrotonaldehyde was converted to 3-methylbutanal and a few % 3-methylbutanol (ethanol, 20 bar H_2, 30 °C).

Five Pd catalysts on alumina were prepared from different precursors which led to NP sizes from 1.2 to 8.5 nm [94]. In the gas-phase hydrogenation of crotonaldehyde the primary product was in all cases butanal, which was converted into butanol at a much lower in four cases (150 °C, 1.2 bar H_2). Butanal was obtained with the NPs (2.5 nm) produced from $PdCl_2$; the second stage depends more severely on the structure of the catalyst and it was proposed that this catalyst contained chlorides on the edges.

Pd, Pd–Au and Au NPs (3.9 nm size) stabilized by tetraalkylammonium salts were embedded in silica via a sol-gel method and used as catalysts in various hydrogenations [95]. In the case of CAL, the selectivity of pure Pd for COL was 18, HCAL 28, HCOL 56%; for Au the numbers were 3, 33 and 58%, respectively. Interestingly, for the alloyed catalysts the selectivities changed to 50, 20 and 30%, thus much higher for COL. It was suggested that oxygen coordination to Au was the cause of this effect.

Spherical Fe_3O_4@C core–shell composites with core diameter about 350 nm and shell thickness about 10 nm were prepared by carbonization of glucose around Fe_3O_4 microspheres [96]. By loading Pd onto the spherical Fe_3O_4@C core–shell composites via the ethylene glycol reduction method, Pd/Fe_3O_4@C catalysts with Pd NP diameters ranging from 7.9 to 9.1 nm were prepared. The particles are magnetically responsive and can be separated from the reaction medium this way. These catalysts were used in the CAL hydrogenation, where they gave only HCAL and HCOL (resp. 72%, 28%) (80 °C, ethanol, water, 14 bar H_2).

Zhu, Fu and co-workers deposited Pd NPs and tungsten nitride (WN) on various supports [97]. SBA-15 was found to ensure the best contact between the two types of nanoparticles (sizes 13.6 nm for WN, and 16.8 nm for Pd). In the presence of WN high selectivity to HCAL was reached (90%) at 70% conversion of CAL, whereas conversion to HCOL was almost completely suppressed. In the absence of WN mostly HCOL was obtained. Both catalysts show high TOF (400–700 h^{-1}, isopropanol, 40 °C, 10 bar H_2).

11.5.5   Pd on Carbon

Xu, Pham-Huu, et al. prepared a structured catalyst of graphene felt and oxidized graphene felt (OGF) on which Pd NPs were deposited (~3.8 nm size). They compared the materials with commercial Pd/AC used as a powder in the hydrogenation of

CAL [98]. The structured catalyst was used as a stirrer allowing easy separation of the catalyst and products after the batch reaction. The reaction was carried out in dioxane, 80 °C and 1 bar H_2. All three materials show selectivity to HCAL of around 90%, but the OGF catalyst was far more stable than the commercial Pd/AC catalyst, the non-oxidized GF catalyst being in between. Pd/OGF does not sinter during the reaction while the particles size of Pd/GF increased considerably, from 12 to 42 nm. XPS showed that Pd/OGF contained more oxidized Pd; while this adds to the stability it did not increase the selectivity to COL.

Drelinkiewicz and co-workers studied the effect of alloying Pd/C with Au with the aim to reduce alkene hydrogenation in favour of aldehyde hydrogenation of CAL [99]. Large MNPs (6–8 nm size) were prepared from the chloride salts by reduction with hydrazine by an inverse micelle method in the presence of Vulcan carbon. All catalysts exhibit high CAL conversions, but at higher Au content a decrease in the rate of CAL conversion was observed and the Pd/Au 1:2 catalyst with the highest Au/Pd ratio is 5-times less active than monometallic Pd/C. Hydrogenation yielded only HCAL and HCOL under the conditions applied (toluene, room temperature, 1 bar H_2). It was established that hydrogenation of COL to HCOL was relatively fast.

11.5.6 Pd as Hydrogen Transfer Catalyst on Silica

In view of the propensity of Pd/H_2 to hydrogenate the C=C bond in unsaturated aldehydes and ketones, the use of alcohols as hydrogen donors seems a good option; the two hydrogen atoms are abstracted by the catalyst in a heterolytic fashion and chances are that they will be transferred to the substrate in the same way. Alternatively, a concerted transfer of the hydrogen atom from carbon to carbon can take place if alcohols are the donor, as mentioned above for the Meerwein–Ponndorf–Verley reaction. Ying and co-workers designed Pd catalysts for various Pd catalyzed reactions by immobilizing Pd NPs via urea ligands attached covalently to silicious mesocellular foams (MCF) [100]. MCF possesses a three-dimensional, interconnected pore structure with ultra-large pores (24–42 nm) connected by windows of 9–22 nm and a very large surface area. Pd NPs (2–3 and 4–6 nm size) were obtained from Pd (OAc)$_2$ by simply heating. Inter alia these catalysts were used for the hydrogen transfer from formic acid to 4-phenyl-3-buten-2-one to give 4-phenyl-3-buten-2-ol in 99% yield (fivefold excess of formic acid/triethylamine v/v = 1:1, ethyl acetate, 25 °C). No hydrogenation of the alkene was observed. On the contrary, when 6 bar H_2 was applied, 99.9% of 4-phenylbutan-2-one was obtained (methanol/ethanol, 25 °C), i.e. the 'normal' behaviour of a Pd catalyst.

11.6 Hydrogenation of Unsaturated Aldehydes and Ketones Catalyzed by Iridium Nanoparticles

Among the transition-metal nanoparticles usually employed in hydrogenation catalysis, iridium has attracted large interest due to its high activity and low tendency to oxidation under atmospheric conditions. However, iridium nanoparticles (IrNPs) have shown limited efficacy in the selective hydrogenation of unsaturated carbonyl compounds, although great efforts have been made in recent years to develop highly chemoselective IrNPs-based systems. The majority of these systems consist of IrNPs immobilized on oxide supports or oxygenated surfaces, for which the hydrogenation process occasionally occurs through a metal–support interaction.

Faria and co-workers reported moderate selectivity to the COL product (68%) in the hydrogenation of CAL catalyzed by IrNPs supported on MWCNTs [101]. Interestingly, removal of oxygenated surface groups used to anchor the IrNPs gave an increase in the selectivity (from 56 to 68%). The authors suggest that this process may produce a transfer of electron density from the graphite sheets to the metal and therefore lead to a stronger metal–support interaction. This increase in the electron density on the metal could hinder the hydrogenation of the C=C group due to repulsion effects, thus favouring the formation of the unsaturated alcohol.

Lower selectivities towards the unsaturated alcohol were observed in the hydrogenation of CAL and benzalacetone catalyzed by IrNPs immobilized on SiO_2 (Table 11.26) [24]. DRIFT analyses suggested the presence of Ir^0 and $Ir^{\delta+}$ sites, of which the latter could facilitate the adsorption of CAL in a favourable geometry for the hydrogenation of the C=O group. However, an increase in the polarization of the C=O bond due to the presence of these positively charged $Ir^{\delta+}$ species is also possible. On the other hand, the steric impediment generated by the methyl substituent in benzalacetone could avoid the reduction of the C=O group, giving, as a result, the predominant reduction of the C=C bond. Along this line, $IrNPs/SiO_2$ doped with gold were employed as catalyst in citral hydrogenation in order to study the effect of metal composition on the catalytic behaviour [102]. Different characterization techniques pointed to a transfer of electron density from iridium to gold and an increase in the ratio of $Ir^{\delta+}$ species as a function of the gold content. The best activity was observed for the catalytic system with the highest $Ir^{\delta+}/Ir^0$ ratio. In addition, very high selectivities to the unsaturated alcohol were achieved (98–99%).

The use of additives that promote the heterolytic cleavage of H_2 through a cooperative mechanism with the metal is an interesting strategy to selectively reduce

Table 11.26 Hydrogenation of carbonyl compounds catalyzed by $IrNPs/SiO_2$

Substrate	Conversion (%)	Selectivity C=O (%)	Selectivity C=C (%)
Cinnamaldehyde	21.6	57	40
Benzalacetone	76.0	6	86

Conditions: 0.2 g catalyst, 50 mL substrate (0.05 M), 6.2 bar H_2, 363 K, ethanol, 3 h

polarized double bonds such as carbonyl groups. In this context, Tomishige and co-workers employed oxidized Re clusters (ReO_x) to partially cover IrNPs supported on SiO_2 [103]. The Ir atoms and the oxide anions of ReO_x operate in tandem to heterolytically cleave the H_2 molecule into H^+ and H^-, thus favouring the transfer of these species to the C=O group that coordinates to the Re cation. This Ir–ReO_x/SiO_2 system is very active (initial TOF = 2016 h^{-1} in crotonaldehyde hydrogenation, H_2O, 70 °C, 80 bar H_2) and shows high selectivities in the reduction of several α,β-unsaturated aldehydes (Table 11.27). In addition, the system can be easily recycled by filtration.

In subsequent work, several α,β-unsaturated aldehydes and ketones were hydrogenated with high activities (initial TOF = 720 h^{-1}) and selectivities by a Fe cation modified Ir/MgO system [104]. Kinetic and control experiments, together with an exhaustive catalyst characterization, indicated that the real active sites are the interface among Ir^0, Ir^{4+} and Fe^{2+} ions on the MgO support (Scheme 11.13). A heterolytic activation of H_2 was proposed to occur on Ir^0 metal in the vicinity of Fe^{2+}–O^{2-} pair site, leading to the formation of H^- and H^+ species. Similarly, the authors suggest that dipole–dipole interactions facilitate the adsorption of the substrate through the C=O group on Ir^{4+} sites near these hydride and proton species. As a consequence,

Table 11.27 Chemoselective hydrogenation of unsaturated aldehydes by Ir–ReO_x/SiO_2

Substrate	Product	Conversion (%)	Yield (%)	Selectivity (%)
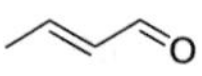	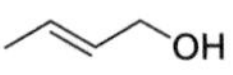	99	90	91
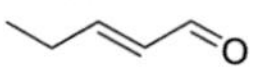	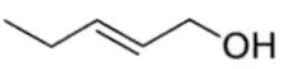	>99	90	90
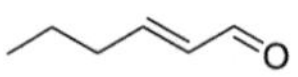	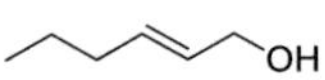	95	87	92
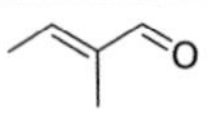	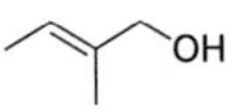	96	89	93
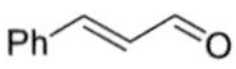	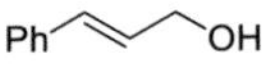	99	91	96
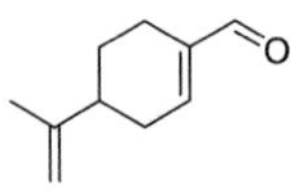	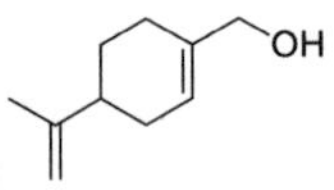	85	70	82
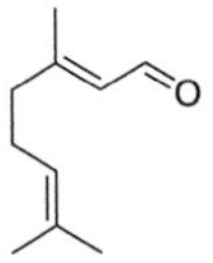	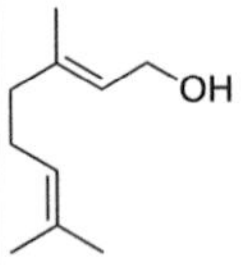	91	88	97
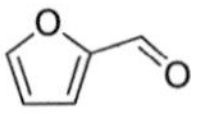	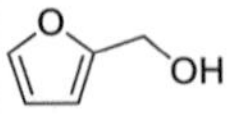	>99	97	>99

Conditions: Ir–ReO_x/SiO_2 (50 mg), substrate (3 mmol), H_2O (3 mL), 5–8 h, 30 °C, 8 bar H_2

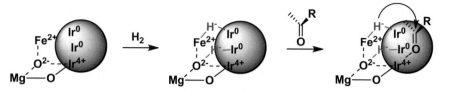

Scheme 11.13 Proposed mechanism for the hydrogenation of carbonyl compounds over Ir/MgO + Fe(NO₃)₃

such H$^-$ and H$^+$ species generate a six-membered intermediate with the C=O bond of the substrate, which results in the selective reduction of the carbonyl functionality.

A heterolytic cleavage of dihydrogen was also proposed in the only system reported to date about hydrogenation of unsaturated aldehydes by non-supported iridium nanoparticles [105, 106]. SPO–stabilized IrNPs showed high activity and very high selectivity in the hydrogenation of cinnamaldehyde (TOF = 41 h^{-1}; 99% selectivity) and 2-octynal (TOF = 18 h^{-1}; 96% selectivity). It was suggested that the SPO ligand acts as a heterolytic activator for H$_2$ through a metal–ligand cooperative mechanism, the same way as mentioned above for Au [15]. Interestingly, no reaction was observed for the alkyne substrate when an analogous Ir–SPO organometallic complex was employed as catalyst, which highlights the robustness of IrNPs.

Finally, it is worth mentioning the work of Luo on vapour-phase (80 °C) hydrogenation of crotonaldehyde catalyzed by IrNPs immobilized on different types of support. In a first study [107], a series of Ir/TiO$_2$ systems were obtained at different reduction temperatures of H$_2$IrCl$_6$ in H$_2$ (100–500 °C). A strong influence of this temperature on activity and selectivity was found for the different catalysts, and a model for adsorption of crotonaldehyde was proposed (Fig. 11.7). The carbonyl oxygen atom interacts with Lewis acid sites (σ_2), and Ir^0 NPs are the adsorption sites for the C=C group (π) and carbonyl carbon atom (σ_1). Catalysts reduced at low temperatures show a large number of Lewis acid sites due to a large content of Cl$^-$ species and high Ir$^{\delta+}$/Ir^0 ratio. These Lewis acid sites can interact with the carbonyl oxygen through a strong σ_2 bond, inhibiting or slowing the desorption of adsorbed crotonaldehyde molecules and thus suppressing the activity. This could result in a strong interaction of Ir^0 sites with C=C bonds (π bonds) and the formation of chlorinated by-products, leading to a reduction of selectivity. On the other hand,

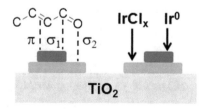

Fig. 11.7 Proposed adsorption mechanism of crotonaldehyde on Ir/TiO$_2$ systems

the authors proposed, high reduction temperatures (e.g. 500 °C) lead to a weak Lewis acid character of the active sites due to a high $Ir^0/Ir^{\delta+}$ ratio, which results in a strong σ_1 interaction with the carbonyl carbon that hinders the desorption of crotonaldehyde. Furthermore, the electron density on Ir^0 species and the IrNP size increase with the reduction temperature. These electron-enriched Ir^0 sites can promote the interaction with the electron-deficient carbonyl carbon and be beneficial for the selectivity. However, large IrNPs can interact simultaneously with both C=C (π) and C=O (σ_1) groups and thus lead to a decreased selectivity.

In conclusion, a suitable reduction temperature is required in terms of activity and selectivity. Indeed, the best results were observed for the Ir/TiO_2 catalyst obtained at 300 °C (Table 11.28), which shows appropriate IrNP size and electron density on Ir^0 species, as well as moderate surface acidity and interaction between Ir sites and C=O groups.

This influence of surface Lewis acidity on the selectivity was also observed for SiO_2-supported iridium systems [108], in which different metal loadings were examined. Ir/SiO_2 catalysts with low iridium content exhibit small NPs and high ratio of $Ir^{\delta+}$ species (strong Lewis acidity), while systems with high iridium loading show larger NPs and thus an increased $Ir^0/Ir^{\delta+}$ ratio (weak Lewis acidity). In line with the model proposed above, the selectivity towards the crotyl alcohol product increases with the iridium loading, reaching a maximum of 77.6% with 3% Ir, after which it starts to decrease.

In a further step, this Ir/SiO_2 system was modified with FeO_x clusters, which effectively promoted both activity and selectivity towards the crotyl alcohol product in the vapour-phase hydrogenation of crotonaldehyde [109]. An improvement in the catalytic behaviour of the system containing 3%Ir and 0.1%Fe ($3Ir/0.1Fe/SiO_2$) was observed compared with that of bare Ir/SiO_2 (Table 11.29).

The FeO_x-promoted Ir/SiO_2 catalyst is obtained by a sequential impregnation procedure in such a way that new active sites are generated at the Ir–FeO_x interface (Fig. 11.8). The C=O group is adsorbed on Ir^0–FeO_x sites through interactions of Ir^0 with the electron-deficient carbonyl carbon (σ_1) and FeO with the electron-rich

Table 11.28 Hydrogenation of crotonaldehyde catalyzed by Ir/TiO_2 systems reduced at different temperatures

Reduction temperature (°C)	Cl:Ir[a]	TOF (h^{-1})	Crotyl alcohol selectivity (%)
100	4.2	1.08	28.9
200	2.4	7.92	69.7
300	1.8	13.68	74.6
400	0.9	5.76	70.4
500	0.44	4.68	66.9

Reaction conditions: Catalyst (100 mg), H_2 (26 mL/min), crotonaldehyde (0.0106 bar), 80 °C
[a]Molar ratio Cl:Ir

Table 11.29 Hydrogenation of crotonaldehyde catalyzed by $3Ir/SiO_2$ and $3Ir/0.1Fe/SiO_2$ catalysts

Catalyst	Ir:Fe content (%)	TOF (h^{-1})	Crotyl alcohol selectivity (%)
$3Ir/SiO_2$	3:0	13.3	78.3
$3Ir/0.1Fe/SiO_2$	3:0.1	64.8	90.8

Reaction conditions: Catalyst (200 mg), H_2 (26 mL/min), crotonaldehyde (0.0106 bar), 80 °C. The Ir:Fe content refers to the weight percentage in the catalyst

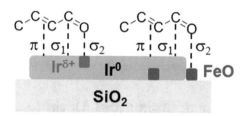

Fig. 11.8 Proposed adsorption mechanism of crotonaldehyde catalyzed by $Ir–FeO_x/SiO_2$ systems

carbonyl oxygen (σ_2). As a result, the activation of the C=O functionality on Ir^0–FeO_x sites by this di-σ_{CO} adsorption mode is favoured over that produced on Ir^0–$Ir^{\delta+}$ sites, leading to enhanced activity and selectivity.

In the same way, the catalytic behaviour of this Ir/SiO_2 system is also improved by the appropriate promotion with CrO_x and combined CrO_x–FeO_x oxides [110], although the reported activities and selectivities are lower than those observed for the $Ir–FeO_x/SiO_2$ catalysts previously described. The catalytic performance is very dependent on CrO_x and FeO_x content, since high ratios of additives strengthen the σ_1 and σ_2 bonds (Fig. 11.8), which hinders the desorption of products and block the active sites, thus deactivating the catalyst.

The use of FeO_x as additive may also have a beneficial effect on the catalytic activity of IrNPs immobilized on other types of support, such as boron nitride (BN) [111]. In fact, 91.2% selectivity towards allylic alcohol product was achieved in the hydrogenation of crotonaldehyde mediated by FeO_x-promoted Ir/BN catalysts, as compared to 59.5% selectivity obtained with bare Ir/BN systems (Table 11.30). Kinetic experiments showed higher intrinsic rate and crotonaldehyde adsorption equilibrium constants for Ir^0–FeO_x sites than those for Ir^0–$Ir^{\delta+}$, which proves that as in $Ir–FeO_x/SiO_2$ systems [109], the activation of the C=O group is favoured in the former. In addition, these studies suggest a heterolytic cleavage of H_2 on Ir^0 sites. However, high FeO_x ratios partially cover the iridium surface and, as commented above, hinder the desorption of products, which thus leads to a decreased activity (Table 11.30).

Table 11.30 Vapour-phase hydrogenation of crotonaldehyde catalyzed by Ir/BN catalysts

Catalyst	Ir:Fe content (%)	TOF (h^{-1})	Crotyl alcohol selectivity (%)
1Ir/BN	1:0	5.4	59.5
3Ir/BN	3:0	4.0	55.5
3Ir/0.01Fe/BN	3:0.01	9.0	61.9
3Ir/0.03Fe/BN	3:0.03	11.9	82.7
3Ir/0.05Fe/BN	3:0.05	24.1	84.4
3Ir/0.1Fe/BN	3:0.1	16.2	89.9
3Ir/0.30Fe/BN	3:0.3	7.9	91.2

Conditions: Catalyst (200 mg), H$_2$ flow (26 mL/min), crotonaldehyde (1%, 0.0106 bar), 80 °C. Data recorded at 3 h of reaction. The Ir:Fe content refers to the wt% in the catalyst

11.7 Hydrogenation of Unsaturated Aldehydes and Ketones Catalyzed by Rhodium and Ruthenium Nanoparticles

Because rhodium and ruthenium are very active metals as hydrogenation catalysts, the chemoselective reduction of unsaturated carbonyl compounds mediated by rhodium or ruthenium nanoparticles (RhNPs/RuNPs) is difficult to achieve and only a few examples have been reported in which the C=C or C=O group was selectively hydrogenated. In this context, pyridine- and phosphine-stabilized Pt–Ru NPs were confined in MWCNTs (PtRuNP@MWCNT) [79]. These bimetallic systems were applied for CAL hydrogenation, leading to the formation of COL with moderate selectivities (Table 11.31). Interestingly, the most selective catalyst corresponds to that with the highest particle size, which was attributed to the adsorption geometry of CAL on the NP surface. It was proposed that a steric repulsion due to the aromatic ring hinders the interaction between NP surface and C=C bond, thus enabling the selective reduction of the C=O functionality (Fig. 11.9). However, this effect would not be as marked in smaller NPs, for which both C=C and C=O groups could approach the NP surface.

Table 11.31 Hydrogenation of cinnalmaldehyde catalyzed by various PtRuNP@MWCNT catalysts

Pt:Ru ratio in non-supported NP (wt%)	NP size	TOF (h^{-1})[a]	Selectivity (%)[a]
20:9	3.0	10.4	71
1:1	3.0	8.8	71
47:17	2.4	8.5	67
16:7	5.3	13.5	76
19:12	2.9	10.5	68
41:17	4.3	8.5	72

Conditions: Catalyst (35 mg), CAL (7.6 mmol), 2-propanol (50 mL), 20 bar H$_2$, 70 °C .[a] Data at 50% conversion

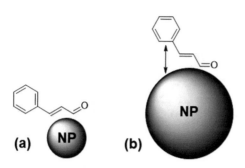

Fig. 11.9 Adsorption of cinnamaldehyde on **a** a small nanoparticle without steric repulsion, and **b** a large nanoparticle with steric repulsion

This was also suggested by Hermans and co-workers for monometallic RuNPs, bimetallic Ru–Pt and Ru–Au NPs and multimetallic Ru–Pt–Au NPs immobilized on CNTs and graphene [112]. A dependence of selectivity on the NP size was observed in the hydrogenation of CAL, in such a way that larger NPs provided higher selectivities to COL. In addition, bimetallic Ru–Pt nanoparticles displayed an increase in the selectivity towards the unsaturated alcohol in comparison with that shown by the monometallic Ru catalyst, whereas the presence of Au in Ru–Au NPs shifts the selectivity towards hydrocinnamaldehyde. In this way, a maximum selectivity of 75% was obtained for Ru–Pt nanoparticles immobilized on MWNTs. Surprisingly, a strong influence of solvent nature on the selectivity was seen. For instance, Ru–Pt nanoparticles supported on MWNTs exhibit 51% selectivity towards COL in a mixture isopropanol/water, while the use of toluene as solvent provides the saturated aldehyde product with 71% selectivity.

In subsequent work, RuNPs and Ru–Pt NPs of 2.2–3.4 nm supported onto nanotubes (CNTs) and nanofibers (CNFs) were employed as catalysts in CAL hydrogenation (isopropanol/water, 120 °C, 50 bar H_2) [113]. As previously described [112], an increase in the selectivity towards the allylic alcohol product was obtained as a result of platinum addition. A transfer of electron density from Ru to Pt is proposed to occur, in such a way that these electropositive Ru sites may activate the C=O functionality. In addition, CNFs-supported NPs exhibit better selectivities than those shown by NPs dispersed on CNTs. Most importantly, a marked increase in the activity was observed for Ru–Pt NPs immobilized on CNFs, which is attributed to the confinement of bimetallic nanoparticles inside the fibres.

Similar selectivity towards the unsaturated alcohol product (73%) was observed in the hydrogenation of CAL catalyzed by Au-doped RuNPs supported on mesoporous silica MCM-41 (Ru–Au/MCM-41) [114]. An increase in both activity and selectivity was achieved by the use of this bimetallic catalyst, as compared to those shown by the monometallic Ru/MCM-41 system (Table 11.32). Other α,β-unsaturated carbonyl compounds such as citral, crotonaldehyde and benzalacetone were also reduced with good selectivities (69–95%).

Table 11.32 Hydrogenation of cinnamaldehyde catalyzed by Ru/MCM-41 and Ru–Au/MCM-41 systems

Catalyst	Ru wt%	Au wt%	Conversion (%)	Selectivity (%)
Ru/MCM-41	2.0	0	14	30
Ru–Au/MCM-41	2.0	0.24	61	73

Reaction conditions: Catalyst (50 mg), CAL (0.5 g), ethanol (20 mL), 30 bar H_2, 4 h, 70 °C

Different analyses suggested an increase in the ratio of Ru atoms present as close-packed planes and showed that the presence of Au leads to a better dispersion of RuNPs, which resulted in improved activity. Furthermore, XPS and CO-FTIR studies demonstrated a transfer of electron density from Ru to Au. These electropositive $Ru^{\delta+}$ species can act as Lewis acid sites and favour the adsorption and subsequent activation of the polar C=O bond, and thus the selective formation of unsaturated alcohol.

In contrast with these modest selectivities, Kitano and Hosono reported high selectivities for the solvent-free hydrogenation of α,β-unsaturated aldehydes catalyzed by electride-supported alloyed Ru–Fe NPs [115]. Bimetallic Ru–Fe NPs of 15 nm in size were deposited on $[Ca_{24}Al_{28}O_{64}]^{4+} \cdot (e^-)_4$ (C12A7:e$^-$), an inorganic electride with anionic electrons in the positively charged framework, by·Chemical Vapour Deposition. This Ru–Fe/C12A7:e$^-$ catalyst was applied in aldehyde hydrogenation, giving high selectivities for a wide range of substrates (Table 11.33). Mechanistic studies revealed an electron donation from the electride to the Ru–Fe NPs which intensifies the repulsive interaction between C=C bond of the substrate and metal-d orbitals, thus promoting a vertical adsorption geometry of the aldehyde through the C=O group. In addition, FTIR experiments showed the formation of electropositive $Fe^{\delta+}$ sites that may activate the C=O functionality and favour the selective reduction of this group. However, the authors cannot conclude whether dihydrogen is cleaved via a homolytic or heterolytic mechanism.

On the other hand, Ru and Rh NPs can also be employed for the preferential reduction of the conjugated C=C bond of α,β-unsaturated carbonyl compounds. For instance, products of importance in fine chemistry such as 4-(6,-methoxy-2,-naphthyl)-butan-2-one (Nabumetone™, anti-inflammatory drug) and 2-acetyl-5,8-dimethoxy-1,2,3,4-tetrahydronaphthalene (precursor of antitumor anthracyclinic products) were synthesized through selective reduction of the corresponding α,β-unsaturated ketone catalyzed by RhNPs immobilized on γ-Al₂O₃ (Scheme 11.14) [116]. The Rh nanoparticles with 1.1 nm size were obtained by Metal Vapour Synthesis and are additionally stabilized with trioctylamine (Rh(TOA)/γ-Al₂O₃). This recyclable Rh(TOA)/γ-Al₂O₃ system showed complete selectivity in the synthesis of Nabumetone™ under very mild conditions (25 °C, 1 bar H_2), while 2-acetyl-5,8-dimethoxy-1,2,3,4-tetrahydronaphthalene was obtained with 85% selectivity.

Along this line, Süss-Fink and co-workers described the selective formation of several saturated ketones by the use of RuNPs (*ca.* 7 nm) intercalated in hectorite [117]. This catalytic system showed high activity and very high selectivity in the

Table 11.33 Hydrogenation of α,β-unsaturated aldehydes catalyzed by Ru–Fe/C12A7:e⁻

Substrate	Product	Yield (%)	Selectivity (%)
(cinnamaldehyde structure)	*(cinnamyl alcohol structure, OH)*	96.2	96.7
(α-methyl cinnamaldehyde structure)	*(structure, OH)*	88.9	95.2
(para-X substituted cinnamaldehyde structure)	*(para-X substituted cinnamyl alcohol, OH)*	X = OMe: 89.2[a] X = F: 92.2[a] X = Cl: 91.3[a] X = Br: 85.5[a] X = NO₂: 73.1[b]	93.6 95.4 94.2 91.4 91.7
(structure)	*(structure, OH)*	83[c]	92.9
(structure)	*(structure, OH)*	93.4[d]	96.8
(structure)	*(structure, OH)*	97.8[d]	95.5
(structure)	*(structure, OH)*	90.1[e]	75.0
(structure)	*(structure, OH)*	96.7[d]	93.1
(structure)	*(structure, OH)*	44.1[e]	87.3
(structure)	*(structure, OH)*	39.6[f]	96.4

Reaction conditions: Catalyst (100 mg, 1 wt% Ru, 1 wt% Fe), substrate (8 mmol), 20 bar H₂, 12 h, 130 °C
[a] 48 h
[b] Substrate (4 mmol), THF (5 mL), 48 h
[c] 110 °C, 12 h
[d] Substrate (4 mmol), 90 °C, 36 h
[e] 40 bar H₂, 48 h
[f] 36 h

hydrogenation of different aliphatic α,β-unsaturated ketones (Table 11.34). The authors suggest that the support modifies the electron density on the metal, which favours the adsorption and subsequent activation of the C=C bond at the surface of the nanoparticles.

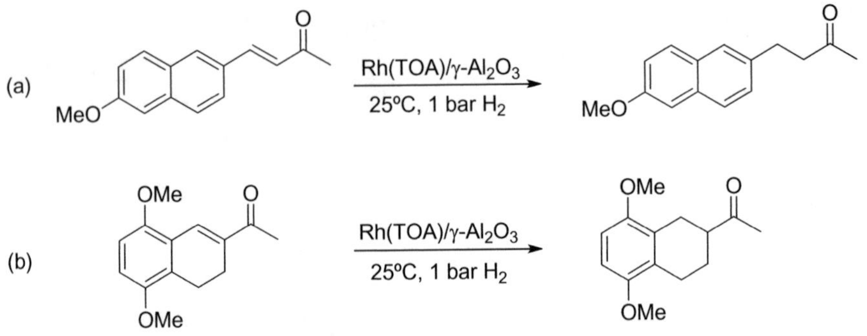

(a)

(b)

Scheme 11.14 Synthesis of **a** Nabumetone™ and **b** 2-acetyl-5,8-dimethoxy-1,2,3,4-tetrahydronaphthalene. Conditions: toluene (20 mL), 25 °C, 1 bar H$_2$; **a** catalyst (0.00442 mg atom Rh), substrate (4.42 mmol); **b** catalyst (0.0043 mg atom Rh), substrate (4.3 mmol)

Table 11.34 Hydrogenation of α,β-unsaturated ketones by RuNPs intercalated in hectorite

Substrate	Product	P (bar)	Conversion (%)	Selectivity (%)	TOF (h^{-1})	TON
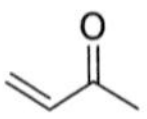	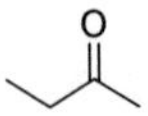	1	100[a]	>99	822	765
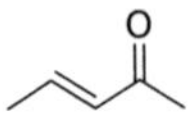	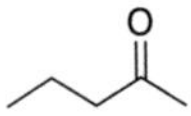	7	100[b]	>99	1254	3825
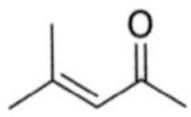	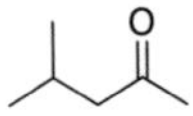	10	100[b]	>99	1212	91,800

Reaction conditions: RuNPs/hectorite (0.0159 mmol Ru in 5 mL ethanol), ketone (12.2 mmol), ethanol (15 mL), 35 °C
[a]2 h
[b]1 h

11.8 Co, Ni and Cu Nanoparticles

The more abundant and cheaper metals Co, Ni and Cu have been used much less than their noble counterparts in the partial hydrogenation of unsaturated aldehydes and ketones. The nanoparticles of these metals are prone to oxidation and one must use vacuum lines and Schlenk techniques to work with these materials as catalysts. Sometimes the oxides are also candidate catalysts or supports. As for the other metals, a broad variation of activities and selectivities is observed.

11.8.1 Cobalt NPs

In 2005 Qiao, Fan et al. reported on the use of amorphous Co/Fe/B species (approximately equal molar ratios) as catalysts for the hydrogenation of crotonaldehyde [118]. The catalyst was made by reduction of the corresponding chlorides with NaBH$_4$, giving NPs of 9 nm size. Having established that Co/B was not very selective for the reaction (20% crotyl alcohol at 89% conversion), the new catalyst was examined (ethanol, 100 °C, 10 bar H$_2$). The catalyst contains both zerovalent and cationic Fe and Co. At 95% conversion, the selectivity for crotyl alcohol amounted to 67%, slightly lower than the initial selectivity as slowly also the fully hydrogenated alcohol (butanol) was formed.

An amorphous alloy Co–B (50 nm size) was obtained by reduction of Co salts with NaBH$_4$ under sonication [119]. Sonication gave a much more active catalyst than the one prepared in the absence of ultrasound. Optimal CAL to COL was 92%, entry 5 in Table 11.35, but prolonged hydrogenation gave conversion to HCOL. Interestingly, in this case, the hydrogenation of COL did not start before nearly all CAL had been consumed.

Mesoporous Co–B amorphous alloy catalysts were prepared by reduction of Co salts with NaBH$_4$ in the presence of CTAB, which led to a much higher surface area [120]. Optimal result for COL was 84% (Co$_{67}$B$_{33}$), entry 6 in Table 11.35.

High loadings of Co as large NPs (50 nm) on SiO$_2$ were also used for the hydrogenation of crotonaldehyde by Touroude and co-workers [121]. Catalytic experiments were performed in a glass fixed-bed reactor, operating at atmospheric pressure under a continuous flow of H$_2$ and 10% crotonaldehyde, at 80 °C. The catalyst contained both Co metal and Co oxides, the ratio of which depended on the reduction temperature. Mildly treated catalysts yielded a maximum selectivity for crotyl alcohol of 88%, while for strongly reduced catalysts this went down to below 40%. The non-reduced oxide was not active in crotonaldehyde hydrogenation and thus it

Table 11.35 Hydrogenation of CAL by different Co catalysts in ethanol

Catalyst	T (°C)	P (bar)	Time (h)	Conversion (%)	Selectivity (%)		
					COL	HCAL	HCOL
Co@C–N	90	10	4	17.5	10.0	73.1	16.9
Co$_2$Si@C–N	90	10	4	82.5	75.1	5.7	19.2
CoSi@C–N	90	10	4	66.2	72.2	4.7	22.2
Co/SiO$_2$[a]	120	10	1	20.0	38.0	26.2	35.8
Co–B sonic.	110	10	1	99.0	92.1		
Co–B mesop.	100	10	5.5	97	84		14
CoGa$_3^b$ DLH	100	20	6	95	96		

[a]In methanol. Classic catalyst for comparison [124]
[b]In isopropanol

was speculated that Co^{++} would coordinate the aldehyde and Co (0) was needed for the H_2 dissociation which then could spillover to the oxide for the reduction.

Shi, Wei and co-workers took a different approach in their effort to reduce alkene hydrogenation and favour aldehyde hydrogenation with citral (the mixture of geranial and neral) as the substrate [122]. They calculated by DFT that both the alkene and the aldehyde group coordinate to a Co surface in a $\eta^2-\mu_2$ fashion. They argued that separating Co atoms by Sn should decrease alkene coordination and increase C=O coordination. Thus four samples Co, $Co_{2.9}Sn_2$, CoSn and $CoSn_2$ (18–20 nm size) were obtained on support prepared from double-layered hydroxides of Co, Zn and Al by urea decomposition that after heating maintained the original platelet structure (20 μm). The four catalysts were studied in the hydrogenation of citral (isopropanol, 160 °C, 40 bar H_2). All Sn-containing catalysts are more than an order of magnitude slower than Co. The latter gives under these conditions up to 90% of citronellol, which diminishes by further exposure as the last double bond is also hydrogenated. For the Sn catalysts, the maximum selectivities to geraniol/nerol were 60% $Co_{2.9}Sn_2$, 68% CoSn and 62% $CoSn_2$, proving the authors' hypothesis qualitatively.

Related to Co NPs, we turn to Co silicides on N-doped carbon ($Co_xSi@N-C$) made and used by Peng, Liang et al. for the hydrogenation of CAL [123]. A high-surface area MOF containing Co (ZIF-67, based on 2-methylimidazolate) was carbonized to give Co NPs (6–10 nm) inside the porous N-carbon. This was treated with an organosilane to obtain Co_2Si and CoSi NPs of 12–15 nm in size. The Co content is around 9%. Three catalysts were considered, Co, a Co_2Si rich one, and one rich in CoSi. Typical results in ethanol are shown in Table 11.35. The silicide improves both the activity of Co and the selectivity to COL, but the selectivity to HCOL would further increase unfortunately if higher conversions are desired.

The favourable effect of intermetallic compounds of Co immobilized on layered double hydroxides in the conversion of α,β-unsaturated aldehydes was further exploited recently by Zhang, Wei and co-workers [125]. Here they made $CoIn_3$ and $CoGa_3$ nanoparticles with a size of 20 nm on top of the discs stemming from the hydroxides (4 μm). The $CoGa_3$ catalyst showed a selectivity of 96% (entry 7, Table 11.35) in the hydrogenation from cinnamaldehyde (CAL) to cinnamyl alcohol (COL), significantly higher than $CoIn_3$ (80%) and the monometallic Co catalyst (42%). Measurements and calculations substantiate once more that the electropositive element serves as an active site and facilitates the adsorption of the polarized C=O group, while C=C bond adsorption on the Co site is depressed.

Yang and co-workers developed a support of biomass-derived carbon from pyrolyzed bamboo shoots which was derivatized with cobalt oxide NPs and used as catalysts for the hydrogenation α,β-unsaturated aromatic and aliphatic ketones [126]. The Co NPs (20 nm) have a core of Co and a shell of Co oxides. The catalysts, denoted as $CoO_x@NC$, were used in water for the hydrogenation of a broad spectrum of unsaturated ketones, of which several are important pharmaceutical hydrogenated chalcone intermediates. Reactions were carried out in the presence of tetrabutylammonium iodide as a surfactant as otherwise the educt would react very sluggishly (water, 110 °C, 20 bar H_2). The products were the saturated ketones with >99% selectivity, a little surprising in the context of the results in this chapter as we are dealing with ideal conditions for 'heterolytic' hydrogenations. Isomerization of potential allyl alcohol product intermediates was not studied.

We consider one example of hydrogen transfer from isopropanol to CAL using Co_3O_4 NPs as catalyst reported by Schüth and co-workers [127]. With the procedure followed, small cobalt oxide NPs were obtained (3 nm size) in microporous carbon. Indeed, hydrogen retains in this process its heterolytic character and unsaturated alcohols (furfuryl alcohol, COL, geraniol, nerol) were obtained with high selectivity. CAL to COL selectivity was over 95% at 99% conversion (isopropanol, 120 °C). The catalyst is relatively slow at a TOF of 1 h^{-1}, our estimation, but orders of magnitude cheaper than a Pt catalyst, and hydrogen transfer avoids high-pressure reactors.

11.8.2 Nickel NPs

Ni NPs (6 nm size) inside and outside Al_2O_3 nanotubes (100 nm diameter) were prepared by ALD on carbon nanocoils used as a sacrificial template [128]. The catalysts containing Ni inside were several times more active in the hydrogenation of CAL (isopropanol, 80 °C, 20 bar H_2). This was ascribed to a stronger interaction between Ni and the support inside the tubes. Indeed, the catalysts with Ni outside the tubes showed considerably more leaching in the recycling experiments. The selectivity is >97% to HCAL for both catalysts at all conversions.

Chiang and co-workers studied NiB/PVP NPs in the hydrogenation of α,β-unsaturated aldehydes [129]. The nanoparticles were made as described above for CoB from a nickel salt and $NaBH_4$ reduction in water or ethanol in the presence of PVP with several MWs. The Ni_2B nanoparticles obtained were amorphous and by TEM their size was estimated 3–5 nm. The composition of the surface indicated a Ni/B ratio of 3–4, with B mostly trivalent. Hydrogenation reactions were carried out in ethanol, at various temperatures, 9 bar H_2, and furfural, crotonaldehyde and citral served as substrates. Crotonaldehyde gave >90% butanal, also at low conversions and 60 °C. Citral (80 °C) provided up to 87% citronellal at 80% conversion, but prolonged reaction times led to further hydrogenation. As usual, furfural gave high selectivity to furfuryl alcohol, even in cyclohexane as a solvent. In a subsequent paper, NiB NPs stabilized by CTBA were used, but again the main products of the hydrogenation of citral were citronellal and citronellol [130].

Another attempt to improve the performance of Ni in the present hydrogenation reactions was reported by Shimazu et al. [131], who alloyed Ni and Sn for this purpose, supported on $Al(OH)_3$. The NiSn alloy catalysts at various Ni/Sn ratios were synthesized by the hydrothermal treatment of a mixture of Raney Ni supported on aluminium hydroxide and $SnCl_2–H_2O$ in $EtOH/H_2O$. Ni_3Sn, Ni_3Sn_2 and Ni_3Sn_4 alloy phases were formed and the size of NPs was 4–6 nm. As for the NiB catalysts, hydrogenation of furfural to furfuryl alcohol was straightforward for NiSn catalysts (isopropanol, 30 bar H_2, 180 °C). CAL was hydrogenated to COL also at a relatively high temperature (160 °C) in a yield of 92% with catalyst Ni_3Sn. This favourable effect of Sn had been noted before by Delbecq and Sautet for Pt_3Sn. Their DFT

Table 11.36 Hydrogenation of CAL using NiIr4, NiOs4, Ni, Ir and Os NPs

Catalyst[a]	COL yield (%)	Product selectivity (%)				TOF (h^{-1})
		COL	HCAL	HCOL	Other[b]	
NiIr$_4$(9)	19	23	2	3	72	419
NiOs$_4$(7)	11	28	1	5	66	158
Ni(9)	2	3	19	19	59	150
Ir(4)	14	22	3	2	73	675
Os(4)	48	52	22	12	14	952

Conditions: Methanol, 3 h, 80 °C, 10 bar H$_2$
[a]Mean diameter in parentheses (nm)
[b]Acetals

calculations showed that in the presence of the electropositive Sn, the more electron-rich Pt coordinated preferably to the aldehyde function in η^2 or on top bonding mode, especially for C-substituted substrates as in prenal, leading to hydrogenation of the C=O bond [132].

In view of the good results obtained with Os and Ir in some instances, Feldmann and co-workers modified these MNPs with Ni and compared the alloys and the pure metals [133]. The nanoparticles were prepared at 300 °C in oleylamine, which served as the solvent, as the stabilizing agent to control particle nucleation and growth, and as the reducing agent for the reduction of the starting salts to the MNPs. As one can see in Table 11.36, Os NPs gave the highest rate and selectivity to COL. Unfortunately, the reactions were carried out in methanol, which leads to substantial acetal formation that does not react further.

Li and Zeng prepared nickel silicate hollow spheres of submicrometer size which were covered by mesoporous silica with the aid of CTAC [134]. Subsequently, the nickel salt was reduced in hydrogen and Ni NPs (4 nm size) formed sandwiched between the two mesoporous spheres (mSiO$_2$@Ni/SiO$_2$@mSiO$_2$), pore size 2.3 nm, specific surface area 200–700 m^2 g^{-1}. This robust catalyst was used in the hydrogenation of CAL (ethanol, 80 °C, 35 bar H$_2$) and found to be highly selective, unfortunately towards HCAL (>99%).

11.8.3 Copper NPs

Copper becomes an active hydrogenation catalyst only when the particle size is below 10 nm [135]. Cu NPs on oxide supports are prone to sintering and thus measures have to be taken to prevent this. Ungureanu and Mehdi achieved this by synthesizing Cu NPs (2 nm size, 10% wt) on mesoporous supports by the use of polyether-functionalized ordered mesoporous silica as organic–inorganic hybrid supports [136]. The SBA-15 like hybrid materials (pores 9 nm, specific surface area 900 m^2 g^{-1}) were used as catalysts for the hydrogenation of CAL, showing excellent activity. For the untreated material, at 50% conversion the HCAL selectivity was 90% and that for COL only 10%. The selectivity towards COL could be improved to 60 at 50% conversion by heat treatment of the material at the cost of activity (ten-fold).

In the next publication, Ungureanu, Royer, et al. described Cu/SBA-15 which contained residual polyethers from the hybrid materials in their pores [137]. Cu contents were 5, 10 and 20% wt treated at various calcination temperatures, or the organic material was extracted. A mean crystallite size of 25 nm for CuO was observed, larger than the pores of SBA-15 and thus it was thought that considerable amounts of CuO were sitting on the outer surface. Extraction of the organic material gave Cu NPs after reduction of which the size varied strongly with the loading (1.7, 9.1, 12.3 nm). The sintered, calcined material had by far the lowest activity in CAL hydrogenation (isopropanol, 10 bar H_2, 130 °C). The other well-dispersed catalysts showed good activity, but the highest selectivity towards COL was 28%, the main products being HCAL and HCOL. Earlier the authors had reported on the favourable effect of adding Cr_2O_3 to Cu/SBA-15 systems, which enhanced the selectivity for COL to 52% at low conversions (propylene carbonate, 1 bar H_2 flow, 150 °C) [138]. The improved chemoselectivity was explained by the existence of dual Cu^0–Cr^{3+} sites to which C=O can coordinate.

The last example concerns a hydrogen transfer reaction with 12 nm Cu NPs supported on MgO particles of 30–50 nm size obtained by reduction of CuO with H_2 at 350 °C [139]. Cyclohexanol was a convenient donor for CAL hydrogenation to >99% COL at 180 °C. Aliphatic substrates such as crotonaldehyde and prenal showed lower selectivities towards the allyl alcohols (78 and 85%, respectively) at full conversion. The nanostructure is important for the activity as conventional Cu catalysts give no reaction. TOFs reported were 10–20 h^{-1}. It was suggested that both Cu and Mg^{2+} participate in a cyclic Meerwein–Ponndorf–Verley like mechanism.

11.9   Concluding Remarks

In the last decade, the enormous progress in metallic nanoparticle-based catalysts has opened new possibilities into the chemoselective hydrogenation of α, β-unsaturated carbonyl compounds. However, the selective formation of allyl alcohol, the product with the most industrial interest, is still fraught with difficulties. Some metals show higher intrinsic selectivity towards the carbonyl functionality than others, but improvements are necessary to obtain the target product with acceptable yields and rates. In this context, several modifications have shown to be successful to increase the selectivity. Among them, the most common strategy consists in the use of Lewis acid species (addition of metal cations or electropositive metal species generated in situ in the catalytic system) that favour the adsorption and subsequent activation of the polar C=O group. Along this line, acidic and/or oxide supports can produce an analogous effect. Similarly, the use of bimetallic nanoparticles has been revealed as an efficient strategy to improve the selectivity. In most cases, this enhancement is achieved by an electron transfer between the two metals which makes one of them more electron-deficient and thus activates the carbonyl functionality. However, the mechanism of action of alloyed MNPs may be different, such as a cooperation between an electron-deficient and an electron-rich metal, or a synergistic effect in which one metal serves for the generation of hydrogen species (for example, through a spillover mechanism) and the other metal activates the C=O group or acts as active hydrogen acceptor. As regards solvents, many authors have reported increased allyl alcohol selectivity for protic solvents, but there are notable exceptions; when the catalyst has a strong intrinsic preference for heterolytic transfer, the solvent is less important.

In addition, a steric impediment can avoid the reduction of the internal C=C bond, thus favouring the selective hydrogenation of the C=O functionality. This effect can be achieved in different ways: bulky groups in the vicinity of the C=C bond, flat metal surfaces, steric repulsion between large nanoparticles and aromatic substituents on the substrate, etc. Finally, it is worth noting the use of additives and ligands that promote the heterolytic cleavage of H_2 through a cooperative effect with the metal, and thus lead to the selective reduction of the polar C=O group. This successful approach has allowed to obtain very high selectivities towards the unsaturated alcohol product, either with supported and unsupported MNPs.

In summary, important developments in the partial hydrogenation of unsaturated aldehydes and ketones mediated by metallic nanoparticles have been recently described. However, new approaches and strategies are required to make this process worthwhile from an industrial point of view. Due to the increasing number of researchers working in the area of MNP catalysis, there is no doubt that new and exciting advances in this transformation will appear in the near future.

References

1. Cao XM, Burch R, Hardacre C, Hu P (2011) Reaction mechanisms of crotonaldehyde hydrogenation on Pt(111): density functional theory and microkinetic modeling. J Phys Chem C 115:19819–19827
2. Leffingwell JC, Leffingwell D (2011) Chiral chemistry in flavours & fragrances. Speciality Chem Mag March 2011:30–33. http://www.leffingwell.com/menthol13/menthol_basf.htm. Accessed 3 May 2019
3. Gallezot P, Richard D (1998) Selective hydrogenation of α,β-unsaturated aldehydes. Catal Rev Sci Engin 40:81–126
4. Rylander PN, Steele DR (1968) Hydrogenation of unsaturated aldehydes to unsaturated alcohols. US Patent 3,655,777 to Engelhard Minerals and Chemicals Corp. April 11, 1972
5. Mitsumode T, Kaneda K (2013) Gold nanoparticle catalysts for selective hydrogenations. Green Chem 15:2636–2654
6. Mertens PGN, Poelman H, Ye X, Vankelecom IFJ, Jacobs PA, De Vos DE (2007) Au0 nanocolloids as recyclable quasihomogeneous metal catalysts in the chemoselective hydrogenation of α,β-unsaturated aldehydes and ketones to allylic alcohols. Catal Today 122:352–360
7. Mertens PGN, Vandezande P, Ye X, Poelman H, Vankelecom IFJ, De Vos DE (2009) Recyclable Au0, Ag0 and Au0–Ag0 nanocolloids for the chemoselective hydrogenation of α,β-unsaturated aldehydes and ketones to allylic alcohols. Appl Catal A 355:176–183

8. Mertens PGN, Wahlen J, Ye X, Poelman H, De Vos DE (2007) Chemoselective C=O hydrogenation of α,β-unsaturated carbonyl compounds over quasihomogeneous and heterogeneous nano-Au0 catalysts promoted by Lewis acidity. Catal Lett 118:15–21

9. Biondi I, Laurenczy G, Dyson PJ (2011) Synthesis of gold nanoparticle catalysts based on a new water-soluble ionic polymer. Inorg Chem 50:8038–8045

10. Zhu Y, Qian H, Zhu M, Jin R (2010) Thiolate-protected Au_n nanoclusters as catalysts for selective oxidation and hydrogenation processes. Adv Mater 22:1915–1920

11. Zhu Y, Qian H, Drake BA, Jin R (2010) Atomically precise Au25(SR)18 nanoparticles as catalysts for the selective hydrogenation of α,β-unsaturated ketones and aldehydes. Angew Chem Int Ed 49:1295–1298

12. Cano I, Chapman AM, Urakawa A, van Leeuwen PWNM (2014) Air-Stable gold nanoparticles ligated by secondary phosphine oxides for the chemoselective hydrogenation of aldehydes: crucial role of the ligand. J Am Chem Soc 136:2520–2528

13. Cano I, Huertos MA, Chapman AM, Buntkowsky G, Gutmann T, Groszewicz PB, van Leeuwen PWNM (2015) Air-stable gold nanoparticles ligated by secondary phosphine oxides as catalysts for the chemoselective hydrogenation of substituted aldehydes: a remarkable ligand effect. J Am Chem Soc 137:7718–7727

14. Li Z, Chen Z-X, He X, Kang G-J (2010) Theoretical studies of acrolein hydrogenation on Au_{20} nanoparticle. J Phys Chem 132:184702

15. Almora-Barrios N, Cano I, van Leeuwen PWNM, López N (2017) Concerted chemoselective hydrogenation of acrolein on secondary phosphine oxide decorated gold nanoparticles. ACS Catal 7:3949–3954

16. Yoshitake H, Saito N (2013) Selective hydrogenation of crotonaldehyde on Au supported on mesoporous titania. Microporous Mesoporous Mater 168:51–56

17. Zhao J, NI J, Xu J, Xu J, Cen J, Li X (2014) Ir promotion of TiO_2 supported Au catalysts for selective hydrogenation of cinnamaldehyde. Catal Commun 54:72–76

18. Demidova YS, Suslov EV, Simakova OA, Volcho KP, Salakhutdinov NF, Murzin DY (2015) Selective carvone hydrogenation to dihydrocarvone over titania supported gold catalyst. Catal Today 241:189–194

19. Liu R, Yu Y, Yoshida K, Li G, Jiang H, Zhang M, Zhao F, Fujita S-i, Arai M (2010) Physically and chemically mixed TiO_2-supported Pd and Au catalysts: unexpected synergistic effects on selective hydrogenation of citral in supercritical CO_2. J Catal 269:191–200

20. Liu R, Zhao F (2010) Selective hydrogenation of citral over Au-based bimetallic catalysts in supercritical carbon dioxide. Sci China Chem 53:1571–1577

21. Konuspayeva Z, Berhault G, Afanasiev P, Nguyen T-S, Giorgio S, Piccolo L (2017) Monitoring in situ the colloidal synthesis of AuRh/TiO_2 selective-hydrogenation nanocatalysts. J Mater Chem A 33:17360–17367

22. Yan Q-Y, Zhu Y, Tian L, Xie S-H, Pei Y, Li H, Li H-X, Qiao M-H, Fan K-N (2009) Preparation and characterization of Au-In/APTMS-SBA-15 catalysts for chemoselective hydrogenation of crotonaldehyde to crotyl alcohol. Appl Catal A 369:67–76

23. Lin H, Zheng J, Zheng X, Gu Z, Yuan Y, Yang Y (2015) Improved chemoselective hydrogenation of crotonaldehyde over bimetallic AuAg/SBA-15 catalyst. J Catal 330:135–144

24. Rojas H, Diaz G, Martinez JJ, Castañeda C, Gomez-Cortes A, Arenas-Alatorre J (2012) Hydrogenation of α,β-unsaturated carbonyl compounds over Au and Ir supported on SiO_2. J Mol Cat A-Chem 363–364:122–128

25. Sivaranjan K, Padmaraj O, Santhanalakshmi J (2017) Synthesis of highly active rGO-supported mono and bi-metallic nanocomposites as catalysts for chemoselective hydrogenation of α,β-unsaturated ketone to alcohol. New J Chem 42:1725–1735

26. Zhang X, Guo YC, Zhang ZC, Gao JS, Xu CM (2012) High performance of carbon nanotubes confining gold nanoparticles for selective hydrogenation of 1,3-butadiene and cinnamaldehyde. J Catal 292:213–226

27. Gu H, Xu X, Chen A-a, Ao P, Yan X (2013) Separate deposition of gold and palladium nanoparticles on ordered mesoporous carbon and evaluation of their catalytic activity for cinnamaldehyde hydrogenation under atmospheric condition. Catal Commun 41:65–69

28. You K-J, Chang C-T, Liaw B-J, Huang C-T, Chen Y-Z (2009) Selective hydrogenation of α,β-unsaturated aldehydes over Au/MgxAlO hydrotalcite catalysts. Appl Catal A 361:65–71
29. Chen H-Y, Chang C-T, Chiang S-J, Liaw B-J, Chen Y-Z (2010) Selective hydrogenation of crotonaldehyde in liquid-phase over Au/Mg$_2$AlO hydrotalcite catalysts. Appl Catal A 381:209–215
30. Urayama T, Mitsudome T, Maeno Z, Mizugaki T, Jitsukawa K, Kaneda K (2016) Green, multi-gram one-step synthesis of core–shell nanocomposites in water and their catalytic application to chemoselective hydrogenations. Chem Eur J 22:17962–17966
31. Zhu Y, Tian L, Jiang Z, Pei Y, Xie S, Qiao M, Fan K (2011) Heteroepitaxial growth of gold on flowerlike magnetite: an efficacious and magnetically recyclable catalyst for chemoselective hydrogenation of crotonaldehyde to crotyl alcohol. J Catal 281:106–118
32. Chen H, Cullen DA, Larese JZ (2015) Highly efficient selective hydrogenation of cinnamalde-hyde to cinnamyl alcohol over gold supported on zinc oxide materials. J Phys Chem C 119:2885–28894
33. Stephenson CJ, Whitford CL, Stair PC, Farha OK, Hupp JT (2016) Chemoselective hydrogenation of crotonaldehyde catalyzed by an Au@ZIF-8 composite. ChemCatChem 8:855–860
34. Xu Y, Liu L, Chong H, Yang S, Xiang J, Meng X, Zhu M (2016) The key gold: enhanced platinum catalysis for the selective hydrogenation of α,β-unsaturated ketone. J Phys Chem C 120:12446–12451
35. Luza L, Rambor CP, Gual A, Fernandes JA, Eberhardt D, Dupont J (2017) Revealing hydrogenation reaction pathways on naked gold nanoparticles. ACS Catal 7:2791–2799
36. Claus P, Hofmeister H (1999) Electron microscopy and catalytic study of silver catalysts: structure sensitivity of the hydrogenation of crotonaldehyde. J Phys Chem B 103:2766–2775
37. Grünert W, Brückner A, Hofmeister H, Claus P (2004) Structural properties of Ag/TiO$_2$ catalysts for acrolein hydrogenation. J Phys Chem B 108:5709–5717
38. Wei H, Gomez C, Liu J, Guo N, Wu T, Lobo-Lapidus R, Marshall CL, Miller JT, Meyer RJ (2013) Selective hydrogenation of acrolein on supported silver catalysts: a kinetics study of particle size effects. J Catal 298:18–26
39. Bron M, Teschner D, Knop-Gericke A, Jentoft FC, Kröhnert J, Hohmeyer J, Volckmar C, Steinhauer B, Schlögl R, Claus P (2007) Silver as acrolein hydrogenation catalyst: intricate effects of catalyst nature and reactant partial pressures. Phys Chem Chem Phys 9:3559–3569
40. Yan X, Wang A, Wang X, Zhang T, Han K, Li J (2009) Combined experimental and theo-retical investigation on the selectivities of Ag, Au, and Pt catalysts for hydrogenation of crotonaldehyde. J Phys Chem C 113:20918–20926
41. Mistudome T, Matoba M, Mizugaki T, Jitsukawa K, Kaneda K (2013) Core–shell AgNP@CeO$_2$ nanocomposite catalyst for highly chemoselective reductions of unsaturated aldehydes. Chem Eur J 19:5255–5258
42. Plessers E, De Vos DE, Roeffaers MBJ (2016) Chemoselective reduction of α,β-unsaturated carbonyl compounds with UiO-66 materials. J Catal 340:136–143
43. Aich P, Wei H, Basan B, Kropf AJ, Schweitzer NM, Marshall CL, Miller JT. Meyer R (2015) Single-atom alloy Pd–Ag catalyst for selective hydrogenation of acrolein. J Phys Chem C 119:18140–18148
44. Voorhees V, Adams R (1922) The use of the oxides of platinum for the catalytic reduction of organic compounds I. J Am Chem Soc 44:1397–1405
45. Wu Y, Cai S, Wang D, He W, Li Y (2012) Syntheses of water-soluble octahedral, trun-cated cctahedral, and cubic Pt–Ni nanocrystals and their structure–activity Study in model hydrogenation reactions. J Am Chem Soc 134:8975–8981
46. Tu W, He B, Liu H, Luo X, Liang X (2005) Catalytic properties of polymer-stabilized colloidal metal nanoparticles synthesized by microwave irradiation. Chin J Polym Sci 23:211–217
47. Rong H, Niu Z, Zhao Y (2015) Structure evolution and associated catalytic properties of Pt–Sn bimetallic nanoparticles. Chem Eur J 21:12034–12041
48. Pournara A, Kovala-Demertzi D (2014) Platinum/3,3′-thiodipropionic acid nanoparticles as recyclable catalysts for the selective hydrogenation of trans-cinnamaldehyde. Catal Commun 43:57–60

49. Martinez-Prieto LM, Cano I et al (2017) Zwitterionic amidinates as effective ligands for platinum nanoparticle hydrogenation catalysts. Chem Sci 8:2931–2941
50. Vu KB, Bukhryakov KV, Anjum DH, Rodionov VO (2015) Surface-bound ligands modulate chemoselectivity and activity of a bimetallic nanoparticle catalyst. ACS Catal 5:2529–2533
51. Wu B, Huang H, Yang J, Zheng N, Fu G (2012) Selective hydrogenation of α,β-unsaturated aldehydes catalyzed by amine-capped platinum-cobalt nanocrystals. Angew Chem Int Ed 51:3440–3443
52. Bhanja P, Liu X, Modak A (2017) Pt and Pd nanoparticles immobilized on amine-functionalized hypercrosslinked porous Polymer nanotubes as selective hydrogenation catalyst for α,β-unsaturated aldehydes. Chem Select 2:7535–7543
53. Sun KQ, Hong YC, Zhang GR, Xu BQ (2011) Synergy between Pt and Au in Pt-on-Au nanostructures for chemoselective hydrogenation catalysis. ACS Catal 1:1336–1346
54. Hong YC, Sun KQ, Zhang GR et al (2011) Fully dispersed Pt entities on nano-Au dramatically enhance the activity of gold for chemoselective hydrogenation catalysis. Chem Commun 47:1300–1302
55. http://www.chemtube3d.com/solidstate/MOF-UiO66.html
56. Guo Z, Xiao C, Maligal-Ganesh RV et al (2014) Pt nanoclusters confined within metal–organic framework cavities for chemoselective cinnamaldehyde hydrogenation. ACS Catal 4:1340–1348
57. Liu H, Li Z, Li Y (2015) Chemoselective hydrogenation of cinnamaldehyde over a Pt-Lewis acid collaborative catalyst under ambient conditions. Ind Eng Chem Res 54:1487–1497
58. Liu H, Chang L, Chen L, Li Y (2016) Nanocomposites of platinum/metal–organic frameworks coated with metal–organic frameworks withrRemarkably enhanced chemoselectivity for cinnamaldehyde hydrogenation. ChemCatChem 8:946–951
59. Zhao M, Yuan K et al (2016) Metal–organic frameworks as selectivity regulators for hydrogenation reactions. Nature 539:76–80
60. Yuan K, Song T et al (2018) Effective and selective catalysts for cinnamaldehyde hydrogenation: hydrophobic hybrids of metal–organic frameworks, metal nanoparticles, and micro- and mesoporous polymers. Angew Chem Int Ed 57:5708–5713
61. Gu Z, Chen L, Li X, Chen L, Zhang Y, Duan C (2019) NH2-MIL-125(Ti)-derived porous cages of titanium oxides to support Pt–Co alloys for chemoselective hydrogenation reactions. Chem Sci 10:2111–2117
62. Zhu Y, Zaera F (2014) Selectivity in the catalytic hydrogenation of cinnamaldehyde promoted by Pt/SiO2 as a function of metal nanoparticle size. Catal Sci Technol 4:955–962
63. Weng Z, Zaera F (2108) Sub-monolayer control of mixed-oxide support composition in catalysts via atomic layer deposition: selective hydrogenation of cinnamaldehyde promoted by (SiO$_2$-ALD)-Pt/Al$_2$O$_3$. ACS Catal 8:8513–8524
64. Liu H, Mei Q, Li S et al (2018) Selective hydrogenation of unsaturated aldehydes over Pt nanoparticles promoted by the cooperation of steric and electronic effects. Chem Commun 54:908–911
65. Kahsar KR, Schwartz DK, Medlin JW (2014) Control of metal catalyst selectivity through specific noncovalent molecular interactions. J Am Chem Soc 136:520–526
66. Ekou T, Especel C, Royer S (2011) Catalytic performances of large pore Ti-SBA15 supported Pt nanocomposites for the citral hydrogenation reaction. Catal Today 173:44–52
67. Pan H, Li J, Lu J et al (2017) Selective hydrogenation of cinnamaldehyde with PtFex/Al$_2$O$_3$@SBA-15 catalyst. J Catal 354:24–36
68. Xue Y, Yao R, Li J et al (2017) Efficient Pt–FeO$_x$/TiO$_2$@SBA-15 catalysts for selective hydrogenation of cinnamaldehyde to cinnamyl alcohol. Catal Sci Technol 7:6112–6123
69. Kennedy G, Melaet G, Han HL et al (2016) In situ spectroscopic investigation into the active sites for crotonaldehyde hydrogenation at the Pt nanoparticle−Co$_3$O$_4$ interface. ACS Catal 6:7140–7147
70. Li C, Ke C, Han R, Fan G, Yang L, Li F (2018) The remarkable promotion of in situ formed Pt-cobalt oxide interfacial sites on the carbonyl reduction to allylic alcohols. Mol Catal 455:78–87

71. Hu Q, Wang S, Gao Z, Li Y et al (2017) The precise decoration of Pt nanoparticles with Fe oxide by atomic layer deposition for the selective hydrogenation of cinnamaldehyde. App Catal B Environ 218:591–599
72. Chadwick JC (1995) In: Fink G, Mülhaupt R, Brintzinger HH (eds) Ziegler catalysts. Recent scientific innovations and technological improvements. Springer, Berlin, pp 427–440
73. Wang H, Bai S, Pi Y, Shao Q, Tan Y, Huang X (2019) A strongly coupled ultrasmall Pt3Co nanoparticle-ultrathin Co(OH)2 nanosheet architecture enhances selective hydrogenation of α,β-unsaturated aldehydes. ACS Catal 9:154–159
74. Bhogeswararao S, Srinivas D (2012) Intramolecular selective hydrogenation of cinnamaldehyde over CeO2–ZrO2-supported Pt catalysts. J Catal 285:31–40
75. Wang X, He Y, Liu Y, Park Y, Liang X (2018) Atomic layer deposited Pt-Co bimetallic catalysts for selective hydrogenation of α,β-unsaturated aldehydes to unsaturated alcohols. J Catal 366:61–69
76. Ma Y, Li Z (2018) Coupling plasmonic noble metal with TiO2 for efficient photocatalytic transfer hydrogenation. Appl Surf Sci 452:279–285
77. Bachiller-Baeza B, Guerrero-Ruiz A, Rodríguez-Ramos I (2000) Role of the residual chlorides in platinum and ruthenium catalysts for the hydrogenation of α,β-unsaturated aldehydes. Appl Catal A 192:289–297
78. Vu H, Gonçalves F, Philippe R et al (2006) Bimetallic catalysis on carbon nanotubes for the selective hydrogenation of cinnamaldehyde. J Catal 240:18–22
79. Castillejos E, Jahjah M, Favier I, Orejón A, Pradel C, Teuma E, Masdeu-Bultó AM, Serp P, Gómez M (2012) Synthesis of platinum–ruthenium nanoparticles under supercritical CO2 and their confinement in carbon nanotubes: hydrogenation applications. ChemCatChem 4:118–122
80. Solhy A, Machado BF, Beausoleil J et al (2008) MWCNT activation and its influence on the catalytic performance of Pt/MWCNT catalysts for selective hydrogenation. Carbon 46:1194–1207
81. Shi J, Nie R, Chen P, Hou Z (2013) Selective hydrogenation of cinnamaldehyde over reduced graphene oxide supported Pt catalyst. Catal Commun 41:101–105
82. Han Q, Liu Y, Wang D, Yuan F, Niu X, Zhu Y (2016) Effect of carbon supported Pt catalysts on selective hydrogenation of cinnamaldehyde. Hindawi J Chem. ID 4563832
83. Ji X, Niu X, Li B et al (2014) Selective hydrogenation of cinnamaldehyde to cinnamal alcohol over platinum/graphene catalysts. ChemCatChem 6:3246–3253
84. Yao R, Li J, Wu P, Li X (2016) The superior performance of a Pt catalyst supported on nanoporous SiC–C composites for liquid-phase selective hydrogenation of cinnamaldehyde. RSC Adv 6:81211–81218
85. Song S, Yu J, Xiao Q, Ye X, Zhong Y, Zhu W (2014) A stepwise loading method to magnetically responsive Pt-Fe3O4/MCNT catalysts for selective hydrogenation of 3-methylcrotonaldehyde. Nanoscale Res Lett 9:677
86. Dai Y, Gao X, Chu X et al (2018) On the role of water in selective hydrogenation of cinnamaldehyde to cinnamyl alcohol on PtFe catalysts. J Catal 364:192–203
87. Lederhos CR, Badano JM, Quiroga ME et al (2012) Supported metal nanoparticles on activated carbon for α,β-unsaturated aldehyde hydrogenation. Curr Org Chem 16:2782–2790
88. Shi J, Zhang M, Du W, Ning W, Hou Z (2015) SnO2-isolated Pt3Sn alloy on reduced graphene oxide: an efficient catalyst for selective hydrogenation of C=O in unsaturated aldehydes. Catal Sci Technol 5:3108–3112
89. Mhadgut SC, Palaniappan K, Thimmaiah M et al (2005) A metal nanoparticle-based supramolecular approach for aqueous biphasic reactions. Chem Commun 2005:3207–3209
90. Doherty S, Knight JG, Backhouse T et al (2017) Highly efficient aqueous phase chemoselective hydrogenation of α,β-unsaturated aldehydes catalysed by phosphine-decorated polymer immobilized IL-stabilized PdNPs. Green Chem 19:1635–1641
91. Mikkola JP, Warnå J, Virtanen P, Salmi T (2007) Effect of internal diffusion in supported ionic liquid catalysts: interaction with kinetics. Ind Eng Chem Res 46:3932–3940

92. Virtanen P, Karhu H, Kordas K, Mikkola JP (2007) The effect of ionic liquid in supported ionic liquid catalysts (SILCA) in the hydrogenation of α,β-unsaturated aldehydes. Chem Engin Sci 62:3660–3671
93. Zhou X, Zhou HY, Cheang TY et al (2017) Monodisperse Pd nanotetrahedrons on ultrathin MoO_{3-x} nanosheets as excellent heterogeneous catalyst for chemoselective hydrogenation reactions. J Phys Chem C 121:27528–27534
94. McInroy AR, Uhl A, Lear T (2011) Morphological and chemical influences on alumina-supported palladium catalysts active for the gas phase hydrogenation of crotonaldehyde. J Chem Phys 134:214704–214715
95. Parvulescu VI, Parvulescu V, Endruschat U et al (2006) Characterization and catalytic-hydrogenation behavior of SiO_2-embedded nanoscopic Pd, Au, and Pd–Au alloy colloids. Chem Eur J 12:2343–2357
96. Yu J, Yan L, Tu G (2014) Magnetically responsive core–shell $Pd/Fe_3O_4@C$ composite catalysts for the hydrogenation of cinnamaldehyde. Catal Lett 144:2065–2070
97. Wang D, Zhu Y, Fu H et al (2016) Synergistic effect of tungsten nitride and palladium for the selective hydrogenation of cinnamaldehyde at the C=C bond. ChemCatChem 8:1718–1726
98. Xu Z, Duong-Viet C, Pham-Huu C et al (2019) Macroscopic graphite felt containing palladium catalyst for liquid-phase hydrogenation of cinnamaldehyde. App Catal B Environ 244:128–139
99. Szumełda T, Drelinkiewicz A, Kosydar R, Gurgul J (2014) Hydrogenation of cinnamaldehyde in the presence of PdAu/C catalysts prepared by the reverse "water-in-oil" microemulsion method. Appl Catal A 487:1–15
100. Erathodiyil N, Ooi S, Seayad AM et al (2008) Palladium nanoclusters supported on propylurea-modified siliceous mesocellular foam for Coupling and hydrogenation reactions. Chem Eur J 14:3118–3125
101. Machado BF, Gomes HT, Serp P, Kalck P, Faria JL (2010) Liquid-phase hydrogenation of unsaturated aldehydes: enhancing selectivity of multiwalled carbon nanotube-supported catalysts by thermal activation. ChemCatChem 2:190–197
102. Rojas HA, Martinez JJ, Diaz G, Gomez-Cortes A (2015) The effect of metal composition on the performance of Ir–Au/TiO$_2$ catalysts for citral hydrogenation. Appl Catal A 503:196–202
103. Tokonami K, Nakagawa Y, Tomishige K (2013) Rapid synthesis of unsaturated alcohols under mil conditions by highly selective hydrogenation. Chem Commun 49:7034–7036
104. Tumura M, Yonezawa D, Oshino T, Nakagawa Y, Tomishige K (2017) In situ formed Fe cation modified Ir/MgO catalyst for selective hydrogenation of unsaturated carbonyl compounds. ACS Catal 7:5103–5111
105. Cano I, Martínez-Prieto LM, Fazzini PF, Coppel Y, Chaudret B, van Leeuwen PWNM (2017) Characterization of secondary phosphine oxide ligands on the surface of iridium nanoparticles. Phys Chem Chem Phys 19:21655–21662
106. Cano I, Martínez-Prieto LM, Chaudret B, van Leeuwen PWNM (2017) Iridium versus iridium: nanocluster and monometallic catalysts carrying the same ligand behave differently. Chem Eur J 23:1444–1450
107. Chen P, Lu J-Q, Xie G-Q, Hu G-S, Zhu L, Luo L-F, Huan W-X, Luo M-F (2012) Effect of reduction temperature on selective hydrogenation of crotonaldehyde over Ir/TiO$_2$ catalysts. Appl Catal A 433–434:236–242
108. Hong X, Li B, Wang Y, Lu J, Hu G, Luo M (2013) Stable Ir/SiO$_2$ catalyst for selective hydrogenation of crotonaldehyde. Appl Surf Sci 270:394–399
109. Yu Q, Bando KK, Yuan J-F, Luo C-Q, Jia A-P, Hu G-S, Lu J-Q, Luo M-F (2016) Selective hydrogenation of crotonaldehyde over Ir–FeO$_x$/SiO$_2$ catalysts: enhancement of reactivity and stability by Ir–FeO$_x$ interaction. J Phys Chem C 120:8663–8673
110. Yuan J-F, Luo C-Q, Yu Q, Jia A-P, Hu G-S, Lu J-Q, Luo M-F (2016) Great improvement on the selective hydrogenation of crotonaldehyde over CrO$_x$- and FeO$_x$-promoted Ir/SiO$_2$ catalysts. Catal Sci Technol 6:4294–4305
111. Ning X, Xu Y-M, Wu A-Q, Tang C, Jia A-P, Luo M-F, Lu J-Q (2019) Kinetic study of selective hydrogenation of crotonaldehyde over Fe-promoted Ir/BN catalysts. Appl Surf Sci 463:463–473

112. Vriamont C, Haynes T, McCague-Murphy E, Pennetreau F, Riant O, Hermans S (2015) Covalently and non-covalently immobilized clusters onto nanocarbons as catalysts precursors for cinnamaldehyde selective hydrogenation. J Catal 329:389–400

113. Mager N, Libioulle P, Carlier S, Hermans S (2018) Water-soluble single source precursors for homo- and hetero-metallic nanoparticle catalysts supported on nanocarbons. Catal Today 301:153–163

114. Li R, Zhao J, Gan Z, Jia W, Wu C, Han D (2018) Gold Promotion of MCM-41 supported ruthenium catalysts for selective hydrogenation of α,β-unsaturated aldehydes and ketones. Catal Lett 148:267–276

115. Ye T-N, Li J, Kitano M, Sasase M, Hosono H (2016) Electronic interactions between a stable electride and a nano-alloy control the chemoselective reduction reaction. Chem Sci 7:5969–5975

116. Evangelisti C, Panziera N, Vitulli M, Pertici P, Balzano F, Uccello-Barretta G, Salvadori P (2008) Supported rhodium nanoparticles obtained by metal vapour synthesis as catalysts in the preparation of valuable organic compounds. Appl Catal A 339:84–92

117. Khan F-A, Vallat A, Süss-Fink G (2012) Highly selective C=C bond hydrogenation in α,β-unsaturated ketones catalyzed by hectorite-supported ruthenium nanoparticles. J Mol Cat A-Chem 355:168–173

118. Pei Y, Wang J, Fu Q, Guo P, Qiao M, Yan S, Fan K (2005) A non-noble amorphous Co–Fe–B catalyst highly selective in liquid phase hydrogenation of crotonaldehyde to crotyl alcohol. New J Chem 29:992–994

119. Li H, Li H, Zhang J, Dai W, Qiao M (2007) Ultrasound-assisted preparation of a highly active and selective Co-B amorphous alloy catalyst in uniform spherical nanoparticles. J Catal 246:301–307

120. Li H, Yang H, Li H (2007) Highly active mesoporous Co–B amorphous alloy catalyst for cinnamaldehyde hydrogenation to cinnamyl alcohol. J Catal 251:233–238

121. Djerboua F, Benachoura D, Touroude R (2005) On the performance of a highly loaded Co/SiO$_2$ catalyst in the gas phase hydrogenation of crotonaldehyde. Thermal treatments-catalyst structure–selectivity relationship. Appl Catal A 282:123–133

122. Zhou J, Shi S, Wei M et al (2016) Synthesis of Co–Sn intermetallic nanocatalysts toward selective hydrogenation of citral. J Mater Chem A 4:12825–12832

123. Zhang L, Chen X, Peng Z, Liang C et al (2018) Cobalt silicides nanoparticles embedded in N-doped carbon as highly efficient catalyst in selective hydrogenation of cinnamaldehyde. Chem Select 3:1658–1666

124. Raj KJA, Prakash MG, Elangovan T, Viswanathan B (2012) Selective hydrogenation of cinnamaldehyde over cobalt supported on alumina, Silica and Titania. Catal Lett 142:87–94

125. Yang Y, Zhang X, Wei M et al (2018) Selective hydrogenation of cinnamaldehyde over Co-Based intermetallic compounds derived from layered double hydroxides. ACS Catal 8:11749–11760

126. Song T, Ma Z, Yang Y (2019) Chemoselective hydrogenation of α,β-unsaturated carbonyls catalyzed by biomass-derived cobalt nanoparticles in water. ChemCatChem 11:1313–1319

127. Wang G-H, Deng X, Schüth F et al (2016) Co3O4 nanoparticles supported on mesoporous carbon for selective transfer hydrogenation of α,β-unsaturated aldehydes. Angew Chem Int Ed 55:11101–11105

128. Gao Z, Dong M, Qin Y et al (2015) Multiply confined nickel nanocatalysts produced by atomic layer deposition for hydrogenation reactions. Angew Chem Int Ed 127:9134–9138

129. Liaw BJ, Chiang SJ, Tsai CH, Chen YZ (2005) Preparation and catalysis of polymer-stabilized NiB catalysts on hydrogenation of carbonyl and olefinic groups. Appl Catal A 284:239–246

130. Chiang SJ, Liaw BJ, Chen YZ (2007) Preparation of NiB nanoparticles in water-in-oil microemulsions and their catalysis during hydrogenation of carbonyl and olefinic groups. Appl Catal A 319:144–152

131. Rodiansono R, Hara T, Ichikuni N, Shimazu S (2012) A novel preparation method of NiSn alloy catalysts supported on aluminium hydroxide: application to chemoselective hydrogenation of unsaturated carbonyl compounds. Chem Lett 41:769–771

132. Delbecq F, Sautet P (2003) Influence of Sn additives on the selectivity of hydrogenationof α-β-unsaturated aldehydes with Pt catalysts: a density functional study of molecular adsorption. J Catal 220:115–126

133. Egeberg A, Dietrich C, Feldmann C et al (2017) Bimetallic nickel-iridium and nickel-osmium alloy nanoparticles and their catalytic performance in hydrogenation reactions. ChemCatChem 9:3534–3543

134. Li B, Zeng HC (2018) Formation-cum-intercalation of Ni and its alloy nanoparticles within mesoporous silica for robust catalytic reactions. ACS Appl Mater Interfaces 10:29435–29447

135. Yoshida K, González-Arellano C, Luque R, Gai PL (2010) Efficient hydrogenation of carbonyl compounds using low-loaded supported copper nanoparticles under microwave irradiation. Appl Catal A 379:38–44

136. Rudolf C, Ungureanu A, Mehdi A et al (2015) An efficient route to prepare highly dispersed metallic copper nanoparticles on ordered mesoporous silica with outstanding activity for hydrogenation reactions. Catal Sci Technol 5:3735–3745

137. Dragoi B, Ungureanu A, Royer S et al (2017) Highly dispersed copper (oxide) nanoparticles prepared on SBA-15 partially occluded with the P123 surfactant: toward the design of active hydrogenation catalysts. Catal Sci Technol 7:5376–5385

138. Dragoi B, Ungureanu A, Chirieac A et al (2013) Enhancing the performance of SBA-15-supported copper catalysts by chromium addition for the chemoselective hydrogenation of trans-cinnamaldehyde. Catal Sci Technol 3:2319–2329

139. Siddqui N, Sarkar B, Bal R et al (2017) Highly selective transfer hydrogenation of α,β-unsaturated carbonyl compounds using Cu-based nanocatalysts. Catal Sci Technol 7:2828–2837

Chapter 12
Ligand Effects in Ruthenium Nanoparticle Catalysis

Luis M. Martínez-Prieto and Piet W. N. M. van Leeuwen

Abstract This chapter comprehends the recent advances on the influence of nitrogen-, oxygen-, phosphorus-, and carbon-ligands in ruthenium nanoparticle catalysis with some discussion toward other metals when required. It focuses on the influence of stabilizing ligands that able to modify the nanoparticle catalytic properties and it discusses how the catalytic properties of supported nanoparticles can be modified through their functionalization with ligands.

Keywords Ruthenium · Metal nanoparticles · Ligand effects · Surface chemistry · Ligand-stabilized nanoparticles · Supported ligand-functionalized nanoparticles · Nanocatalysis

12.1 Introduction

The application of metal nanoparticles (MNPs) in catalysis has undergone a real revolution during the last decades [1]. This boom resides in the many advantages of MNPs, which combine the benefits of both heterogeneous and homogeneous catalysts. As well as heterogeneous catalysts, MNPs present high stability (stable at harsh reaction conditions) and can be easily recovered from the reaction medium (recyclables), but they have a higher surface area than conventional metallic surface materials. Therefore, MNPs display a high surface/volume ratio, which means a large number of surface sites. Although MNPs are sometimes more selective than common heterogeneous catalysts, they still present a large variety of surface atoms with different nature and activity, which is reflected in a great diversity of active

L. M. Martínez-Prieto (✉)
Instituto de Tecnología Química (ITQ), Universitat Politècnica de València-Consejo Superior de Investigaciones Científicas (UPV-CSIC), Av. de los Naranjos S/N, Puerta L, Edificio 6C, 46022 Valencia, Spain
e-mail: luismiguel.martinez@csic.es

P. W. N. M. van Leeuwen
LPCNO, Institut National des Sciences Appliquées-Toulouse, 135 Avenue de Rangueil, 31077 Toulouse, France
e-mail: vanleeuw@insa-toulouse.fr

© Springer Nature Switzerland AG 2020
P. W. N. M. van Leeuwen and C. Claver (eds.), *Recent Advances in Nanoparticle Catalysis*, Molecular Catalysis 1, https://doi.org/10.1007/978-3-030-45823-2_12

407

sites [2]. As these different catalytically active sites can transform the reactants in different ways, the manner to control MNP selectivity is always a challenge.

An efficient strategy to control the activity and selectivity of an MNP is the use of ancillary ligands that can transform the NP surface. As in organometallic complexes, surface ligands are able to modify the electronic and steric properties of MNPs, and therefore modify their catalytic properties [3]. In fact, there are many similarities between a ligand coordinated to an MNP and in a molecular complex [4]. For example, the coordination modes of the ancillary ligands at the NP surface are very similar to the ones found in organometallic chemistry. Indeed, the chemical adsorptions between the ligand and the MNP surface are normally due to the donation of a lone pair of electrons of the ancillary ligand to the metal, in the same way as in coordination chemistry. In general, efficient ligands for molecular complex stabilization are also good as MNP stabilizer. In addition to all of the above, the interaction between the reactants and the surface ligands can also influence the catalytic activity of the catalysts; depending on the ligand coordinated at the MNP surface, it can boost or inhibit the reactivity of a certain functional group [5]. Therefore, the use of ancillary ligands in MNP catalysis opens the door of a new way to improve the selectivity and activity of a catalyst, until now mainly exploited by homogeneous catalysis.

Ruthenium nanoparticles (Ru NPs) are nanostructures of interest in catalysis as they are active in a great number of catalytic processes, such as hydrogenations, oxidations, C-H activation reactions, and Fischer–Tropsch synthesis [6]. Ru NPs are also very convenient for studying ligand effects in MNP catalysis since they easily accommodate ligands on their surface [7]. Moreover, surface studies by FT-IR or solid-state NMR (ruthenium presents a negligible Knight shift) can be easily performed on Ru NPs by using CO as a probe molecule. For all of this, this chapter will focus on Ru NP catalysis but with some discussion toward other metals when required.

The key to control the potential of ancillary ligands on Ru NP catalysis is to comprehend the synergy between the ligands and the ruthenium and the way that both interact with the substrate, which ultimately enables the enhancement of the catalyst activity/selectivity. Therefore, the aim of this chapter is to give an overview of the recent advances on the influence of nitrogen-, oxygen-, phosphorus-, and carbon-ligands in Ru NP catalysis. In principle, the Ru NP catalysts will be divided into: (i) Ligand-stabilized Ru NPs, where the ligands do not only stabilize Ru NPs but also are able to modify their catalytic properties and (ii) supported ligand-functionalized Ru NPs, where the catalytic properties of supported Ru NPs are modified through the coordination of the ligands on the metal surface.

12.2   Ligand-Stabilized Ru NPs

An interesting approach to tune the reactivity of colloidal MNPs is the use of organic ligands as stabilizer, which can also transform the NP surface [8]. Different synthetic approaches to stabilize colloidal MNPs with ancillary ligands have been developed during the last years. In short, ligand-stabilized MNPs can mainly be obtained by three different strategies: (i) by direct reduction of an organometallic complex stabilized by the ancillary ligand [9], (ii) by ligand exchange [10] of preformed MNPs, and (iii) by reduction of a metallic precursor in the presence of the corresponding ligand. This metal precursor can be a metallic salt [11], which needs the use of a reducing agent ($NaBH_4$, hydrazines or amineboranes), a carbonyl complex (by thermal decomposition) [12], or a high-energy organometallic precursor (Scheme 12.1) [13]. The latter has the advantage of its easy decomposition under mild conditions, which permits a good control on MNP formation and produces clean surface nanoparticles (free from possible contaminants from the metal salt or the reducing agent) [14]. By this route, a large variety of RuNPs ligated by different organic ligands (amines, phosphines, thiolates, N-heterocyclic carbenes, carboxylic acids, etc.) can be easily obtained [7].

12.2.1   Amines and Nitrogen Donor-Stabilized Ru NPs

12.2.1.1   Amine Ligands

One of the most frequently used ligands to stabilize MNPs are amines, and in particular oleylamine, which can also act as a reducing agent [15]. Metin et al. prepared monodisperse Ru NPs by thermal decomposition of ruthenium(III) acetylacetonate ($Ru(acac)_3$) in the presence of oleylamine. These oleylamine-capped Ru NPs supported on aluminum oxide (Al_2O_3) demonstrated to be a highly active and robust

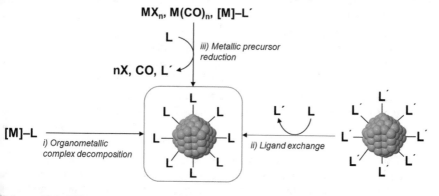

Scheme 12.1   Different approaches to synthesize ligand-stabilized MNPs

catalyst in the hydrolytic dehydrogenation of ammonia borane in water [16]. However, the interactions between amine ligands and Ru NPs are labile. In solution, the amines coordinate reversibly to the ruthenium surface, leading to a continuous exchange between coordinated and free ligands. For example, Lee et al. observed that the interaction between ethylenediamine and the surface of Ru NPs is weak, losing their solubility in water with time due to the gradual decoordination of the amine ligand from the metal surface [17]. This fast ligand exchange was experimentally confirmed by Chaudret and co-workers [18]. A dynamic equilibrium between the free ligand and the coordinated one was observed by ^{13}C NMR studies on worm-like Ru NPs stabilized with high concentrations of long alkyl amines (C_8, C_{12}, and C_{16}).

Depending on the amount and type of amine used as stabilizer, it is also possible to control the morphology of Ru NPs. Tilley et al. obtained hourglass-shaped Ru NPs by using dodecylamine as stabilizer (Fig. 12.1a–d, left) [19]. Moreover, by using a mixture of hexadecylamine (HDA) and hexadecanoic acid (PA), Lignier et al. obtained Ru nanostars with large surface area and good dispersion (Fig. 12.1e–f , right), which were presented as promising non-supported Ru NPs for catalytic applications, since they were able to activate C–C and C–O bonds [20]. As a whole, we can say that amines are not the most appropriate ligands to obtain small and spherical Ru NPs, due to their weak coordination to the Ru surface. However, novel nano-objects can be reached with these amine ligands and may present peculiar reactivities, which demonstrates the high potential of amines in Ru NP catalysis.

Philippot in collaboration with Zahmakıran stabilized Ru NPs with 3-aminopropyltriethoxysilane (APTS), which demonstrated to be highly active, isolable, and reusable catalysts in the dehydrogenation of dimethylamine borane, which is considered an efficient hydrogen storage molecule [21, 22]. The coordination of this ligand through the amine group was evidenced by solution NMR studies, as was previously observed on Ru NPs capped with long alkyl amines [17]. It was the first example of a reusable heterogeneous catalyst for the dehydrogenation of dimethylamine borane, the higher the APTS/Ru ratio is, the lower the catalytic activity is; this is a matter of active metal surface coverage. Ru NPs capped with the same bifunctional amine (APTS) were also used as building blocks to synthesize composite ruthenium-containing silica nanomaterials (RuO_2@SiO2; Fig. 12.2), which showed a promising activity in the aerobic oxidation of benzyl alcohol [23].

In 2011, Salas et al. presented a new catalytic system based on Ru NPs stabilized by ionic liquids (ILs: $[C_1C_nH_{2n+1}Im][NTf_2]$ (where $n = 2$; 4; 6; 8; 10) and long alkyl chain amines (1-octylamine, 1-hexadecylamine) [24]. ILs help the formation of very small NPs inside their non-polar domains (nanoreactors), while amines increase their stability. Indeed, one could evidence the influence of the amine ligands in the control of the size and the stability of these Ru NPs. These NPs showed a high degree of recyclability in the hydrogenation of toluene (up to 10 cycles), without the formation of agglomerations or any loss of activity. The presence of amine ligands increases the NP stability without reducing the catalytic activity, due to their weak interaction with the NP surface. By using a similar catalytic system (RuNPs stabilized by in $[C_1C_4Im][NTf_2]$) but functionalized with different compounds (1-octylamine,

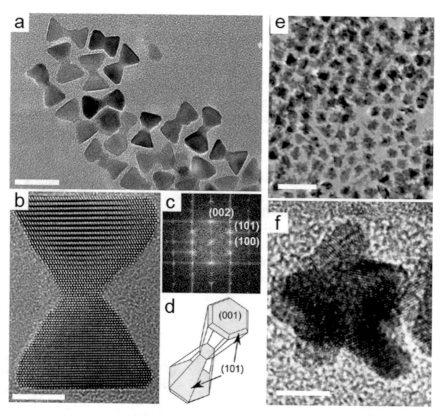

Fig. 12.1 Left: Ruthenium hourglass nanocrystals synthesized after 70 h. **a** TEM image of ruthenium hourglass nanocrystals, scale bar = 20 nm. **b** HRTEM image of a single, highly crystalline, ruthenium hourglass nanocrystal, scale bar = 5 nm. **c** FFT of the ruthenium hourglass nanocrystal shown in (**b**). **d** Schematic of a single ruthenium hourglass nanocrystal showing termination by {001} and {101} facets. Adapted from Ref. [19] with permission from ACS. Right: **a** TEM images of Ru nanostar particles prepared from Ru3(CO)12 with a 1:3 ratio of PA/HDA at 160 °C for 6 h, scale bar = 50 nm. **b** HRTEM image of a nanostar, scale bar = 5 nm. Adapted from Ref. [20] with permission from ACS

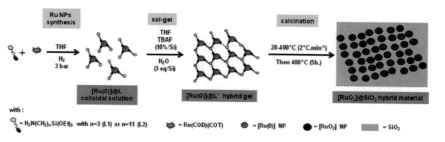

Fig. 12.2 A schematic representation of the synthesis of RuO2@SiO2 nanocomposites. Adapted from Ref. [23] with permission from RSC

412   L. M. Martínez-Prieto and P. W. N. M. van Leeuwen

PPhH$_2$, PPh$_2$H and H$_2$O), Salas et al. observed how these ligands influence the
catalytic activity and selectivity of the catalyst [25]. In particular, they showed that
σ-donor ligands (1-octylamine and H$_2$O) increase the activity of the IL-stabilized Ru
NPs in the hydrogenation of unsaturated compounds (1,3-cyclohexadiene, limonene
and styrene), whereas π-acceptor ones (PPhH$_2$ and PPh$_2$H) decrease it. This work
highlights how the activity of small Ru NPs can be controlled by the σ–π donor
character of the coordinating ligands, in the same way as in homogeneous catalysis.

Water-soluble Ru NPs stabilized by cyclodextrins (RaMeCDs) grafted with chiral amino acids [l-alanine (Ala) and l-leucine (Leu); Fig. 12.3] were prepared by
Roucoux and co-workers by two methodologies: (i) one-pot method, where ruthenium trichloride was reduced by hydrogen in the presence of the chiral cyclodextrins
and (ii) cascade method, where a ruthenium salt was first reduced with NaBH$_4$, and
the resulting Ru NPs were then stabilized by RaMeCDs [26]. Here, the nature of the
ligand and the synthetic method influence the size, dispersion, and surface properties
of the resulting Ru NPs. The catalytic properties of these systems were studied in the
hydrogenation of pro-chiral compounds such as methyl 2-acetamidoacrylate, ethyl
pyruvate, acetophenone, and m-methylanisole. Although no significant enantiomeric

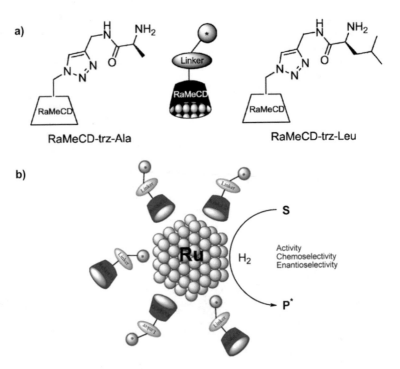

Fig. 12.3 **a** Structure of grafted-cyclodextrins with chiral amino acid moieties like l-alanine (Ala)
and l-leucine (Leu). **b** Stabilization of nanoparticles by randomly methylated β-cyclodextrins grafted
with chiral amino acid moieties investigated in biphasic hydrogenation of pro-chiral compounds.
Adapted from Ref. [26] with permission from Elsevier

excess was observed in any case (less than 5%), interesting differences in terms of activity were found; Ru NPs containing RaMeCD-Leu were more active than the ones ligated by RaMeCD-Ala.

12.2.1.2 Nitrogen Donor Ligands

In contrast with amines, nitrogen donor compounds such as pyridines derivatives can coordinate strongly to the ruthenium surface, as Philippot and co-workers reported during the last years. It was observed that 4-(3-phenylpropyl)-pyridine-stabilized small Ru NPs through a dual coordination mode. Specifically, this ligand coordinates to the Ru surface through the pyridinic N atom (σ-bond) and the phenyl group (π-interaction) [27]. This coordination mode was later confirmed by DFT calculations in a similar system: a $Ru_{55}H_{53}$ model NP capped by 4-phenylpyridine (Fig. 12.4). These 4-(3-phenylpropyl)-pyridine-stabilized Ru NPs deposited on glassy carbon were demonstrated to be a highly active catalyst for the hydrogen evolution reaction (HER) in both acidic and basic media. Here, the presence of the ancillary ligand allowed preserving the nanostructured character of the material during the catalysis, which led to one of the best HER electrocatalysts reported to date [28]. The role of the capping ligands on this Ru NP catalysis is not only to increase the stability of the nanoparticles, they can also modify their electronic properties as was corroborated by DFT investigations.

In 2006, Philippot and co-workers reported a series of Ru NPs stabilized with different chiral N-donor ligands (Fig. 12.5) for the reduction of organic pro-chiral unsaturated substrates [29]. Although the enantiomeric excess was very low, it is one of the first examples about the induction of enantioselectivity through stabilizing chiral ligands on MNPs, usually the domain of molecular catalysts, but earlier used on metal catalysts, vide infra. By comparison with Ru complexes bearing the same chiral N-donor ligands, remarkable differences in the asymmetric induction were

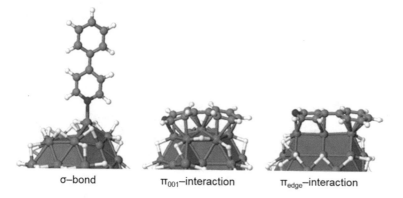

σ–bond π_{001}–interaction π_{edge}–interaction

Fig. 12.4 σ and π coordination modes of 4-phenylpyridine on a $Ru_{55}H_{53}$ model NP. Adapted from Ref. [28] with permission from ACS

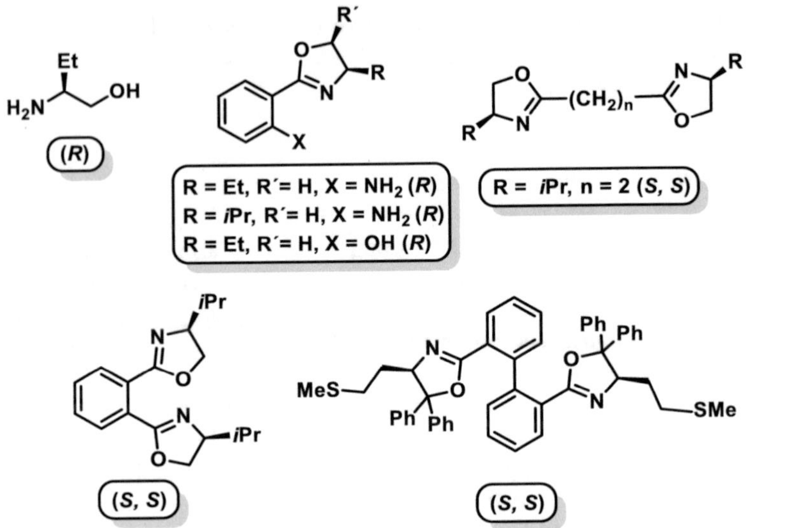

Fig. 12.5 Chiral N-donor ligands: β-amino alcohol, mono(oxazolines), and bis(oxazolines). Adapted from Ref. [29] with permission from ACS

observed. The poor enantioselectivity shown by Ru NPs points to a fluxional behavior of the ligands at the Ru surface. In any case, this study shows the importance of using the appropriate ligand in each case, depending primarily on the goal of the catalysis.

A novel N-donor ligand to stabilize ultra-small Ru NPs has been recently reported by Martínez-Prieto et al. It is an adduct of an N-heterocyclic carbene (NHCs) and a carbodiimide (NCN), i.e., 1,3-dicyclohexylimidazolium-2-di-p-tolylcarbodiimide (ICy.(p−tol)NCN), which due to its zwitterionic character coordinates strongly to metal surfaces (Fig. 12.6, left). A clear correlation was observed between the quantity of ligand used as stabilizer and the size of the NPs, producing very small Ru NPs (ca. 1.0 nm) with 0.2 eq. of the amidinate ligand per Ru [30]. The latter presented less available faces than the larger ones (ca. 1.3 nm), as was confirmed by studying the

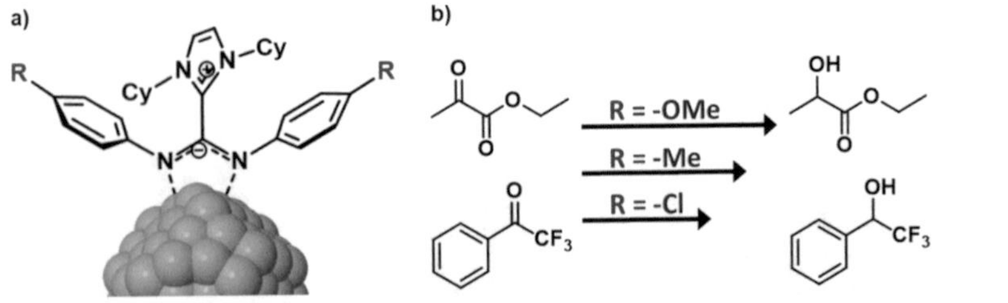

Fig. 12.6 Left: Imidazolium amidinate ligand coordinate at metal surface. Right: Ligand effect in the hydrogenation of activated ketones with Pt NPs stabilized by the zwitterionic adduct of N N,N-dicyclohexylimidazolidene and diarylcarbodiimide (ICy.(Ar)NCN; Ar = p-tol, p-anisyl, p-ClC6H4)

coordination of CO, [31] and higher selectivity in the hydrogenation of styrene to ethylbenzene. Therefore, by playing with the amount and the donor strength of ligand, it is possible to obtain ultra-small Ru NPs, which present a reactivity at the frontier between homogeneous and heterogeneous catalysts. Recently Martínez-Prieto et al. also reported that it is possible to control the surface properties of the nanoparticles by modulating the electronic character of the N-substituents of these amidinate ligands [32]. It was evidenced how the catalytic behavior of Pt NPs on the hydrogenation of activated ketones was influenced by the electron donor/acceptor ($-$Me, $-$OMe, $-$Cl) groups in the N-aryl substituents (Fig. 12.6, right). The nanoparticles stabilized with the strongest donor ligand (R $=$ $-$OMe) demonstrated to be the most active catalyst, while the slowest catalyst was the system containing the electron-withdrawing group (R $=$ $-$Cl). A comparable catalytic trend has been recently observed for Ni NPs stabilized with the same donor/acceptor amidinate ligands in semi-hydrogenation reactions of alkynes [33]. The latter two publications are manifest examples of how small changes on the stabilizing ligands are capable of modifying the MNP catalytic properties.

Although the next example is not strictly an ancillary ligand, the N-donor capacity of the stabilizing polymer, poly(4-vinylpyridine) (PVPy), strongly influences the catalytic behavior of Ru NPs. In particular, Ru NPs immobilized on PVPy reported by Sanchez-Delgado et al. display a peculiar reactivity in the hydrogenation of arenes and N-heteroaromatic compounds [34]. Experimental evidence points to a dual mechanism where two hydrogenation pathways operate at the same time. In path a, which takes place on sites A, the basic pyridine groups assist the heterolytic H_2 cleavage and the N-heteroaromatic moieties are preferable hydrogenated. On the other hand, in path b (sites B), where simple aromatics are hydrogenated, the hydrogen is homolytically dissociated, such as on conventional metallic surfaces (Fig. 12.7). A similar dual homolytic/heterolytic mechanism was observed on Ru NPs supported on basic supports such as MgO (Ru/MgO) [35] or reduced graphene oxide functionalized with NH_2 groups (Ru/NH_2-rGO) [36]. Here, as well as in the above-mentioned case,

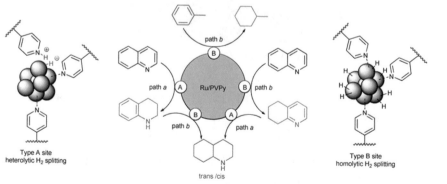

Fig. 12.7 Model of the dual-site structure of Ru/PVPy catalyst and the corresponding hydrogenation pathways. Reproduced from Ref. [34] with permission from RSC

two mechanisms operate in parallel in the hydrogenation of quinoline (Ru/MgO), palmitic acid (Ru/NH$_2$-rGO), or acetophenone (Ru/NH$_2$-rGO).

A short while ago, Axet et al. investigated the influence of the ligands on the hydrogenation of nitrobenzene using Ru NPs as catalyst [37]. In particular, the ruthenium nanoparticles used in this study were stabilized by two donor ligands, which are directly coordinated to the metal surface (HDA = hexadecylamine and IPr = 1,3-bis(2,6- diisopropylphenyl)imidazol-2-ylidene), a polymer that presents almost no interaction with the ruthenium surface (PVP = poly(vinylpyrrolidone)), and a fullerene (C60), which is an electron-acceptor ligand (Fig. 12.8). The basicity of the metal centers was studied by FT-IR using the CO as probe molecule, confirming that Ru-IPr and Ru-HDA are the more electron-rich Ru NPs and Ru-C$_{60}$ is the less basic one, being Ru-PVP in the middle, as was expected. Also, a clear ligand effect was evidenced in the selective hydrogenation of nitrobenzene to cyclo-hexylamine. Ru NPs hydrogenate stepwise nitrobenzene first to aniline and then to cyclohexylamine. The nanocatalyst stabilized with the electron-attracting ligand, Ru-C$_{60}$, promotes the hydrogenation of the N-phenylhydroxylamine intermediate, which is the rate-determining step, and shows the best activity and selectivity (TOF

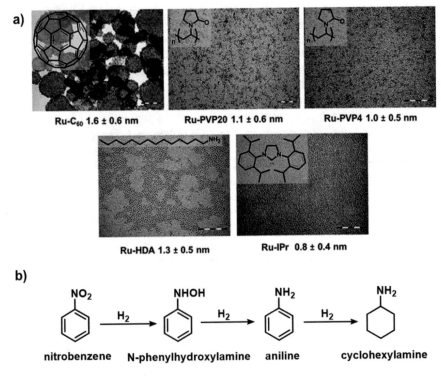

Fig. 12.8 a TEM images of Ru-C$_{60}$, Ru-PVP$_{20}$, Ru-PVP4, Ru-HDA, and Ru-IPr (scale bar = 50 nm). **b** Simplified reaction pathway for hydrogenation of nitrobenzene to cyclohexylamine. Reproduced from Ref. [37] with permission from ACS

= 136.9 h⁻¹; 95% selectivity toward aniline). On the contrary, Ru-IPr is the less
active and selective catalyst (TOF = 64.8 h⁻¹; 68% selectivity toward aniline). Ru-
IPr together with Ru-HDA showed the highest selectivity in the hydrogenation of
aniline to cyclohexylamine (97 and 95%, respectively).

12.2.2   Alcohol and Oxygen Donor-Stabilized Ru NPs

12.2.2.1   Alcohols

Alcohols, and in particular polyols such as ethylene glycol, have been extensively
used in the synthesis of metal nanoparticles [38, 39]. Chaudret et al. obtained Ru
NPs by decomposition of Ru(COD)(COT) in neat alcohol or THF/alcohol mixtures,
without the necessity of any other stabilizer. Furthermore [40, 41], depending on
the type of alcohol (polarity and alkyl chain length) it was possible to control the
growth of the nanoparticles, obtaining Ru NPs with different sizes, shapes, and
reactivities. For example, Ru NPs prepared in a MeOH/THF (10/90) solvent mixture
showed catalytic activity in the hydrogenation of benzene to cyclohexane under mild
conditions.

Alcohols, as well as amines, are labile ligands that coordinate weakly at the metal
surface. For this reason, they are ideal stabilizers for synthetic procedures based on
ligand exchange. For example, Chaudret and Salmeron reported the synthesis of 3 nm
Ru NPs capped with [bis(diphenylphosphino)butane (dppb)] following a two-step
synthetic route (Fig. 12.9) [42]. First, Ru(COD)(COT) was decomposed under H₂
in neat heptanol, obtaining weakly heptanol-stabilized 3 nm Ru NPs [43]. Then, by
adding 0.1 equivalents of dppb in a heptanol/THF mixture, a ligand exchange was
performed and dppb-stabilized Ru NPs of 3 nm were obtained. In this way, it was
possible to synthesize larger Ru NPs stabilized by the strong donor ligand dppb,
since the direct decomposition of Ru(COD)(COT) in the presence of dppb produces
2 nm Ru NPs.

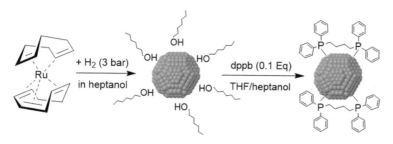

Fig. 12.9 Two-step route followed for the synthesis of Ru-hept-dppb NPs (~3 nm). Inspired from
Ref. [42] with permission from ACS

12.2.2.2 Oxygen Donor Ligands

Roucoux and Monflier described the preparation of RuNPs in the presence of various methylated cyclodextrins (Me-CDs) by reduction of $RuCl_3$ with $NaBH_4$ [44]. The resulting NPs (between 1 and 2.5 nm) were employed as catalysts in the chemoselective hydrogenation of arene derivatives. Here, the Me-CDs have a multiple role: (i) as stabilizer, (ii) as phase-transfer agent, because catalysis is performed in biphasic medium, and iii) as reactant host in the selective hydrogenation reactions. The nature of the Me-CD [type (α, β, γ) and substitution degree] (Fig. 12.10) controls the

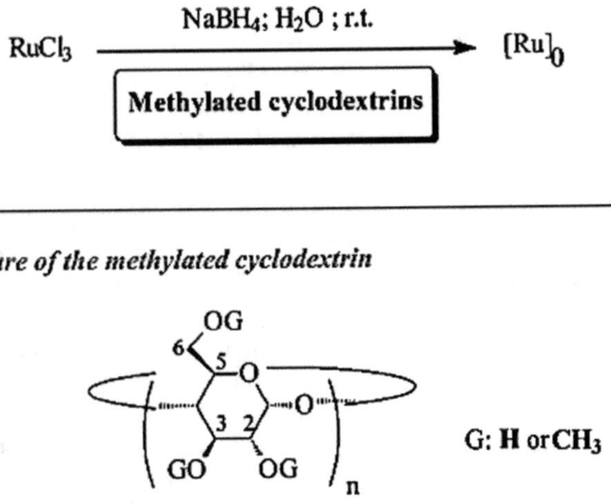

$$RuCl_3 \xrightarrow[\text{Methylated cyclodextrins}]{NaBH_4;\ H_2O\ ;\ r.t.} [Ru]_0$$

Structure of the methylated cyclodextrin

G: **H** or **CH₃**

Abbreviations	n	Carbon bearing the OCH₃ group	Average number of OH group substituted per glucopyranose unit
Me-α-CD	6	2, 3, 6	1.8
Me-β-CD (1.8)	7	2, 3, 6	1.8
Me-β-CD (0.7)	7	2	0.7
Me-γ-CD	8	2, 3, 6	1.8

Fig. 12.10 Preparation of RuNPs in the presence of various methylated cyclodextrins. Reproduced from Ref. [44] with permission from RSC

catalytic properties of the nanoparticles, and in particular their selectivity. For example, using Ru/Me-α-CD, the aromatic rings were not hydrogenated at all, obtaining selectively ethylbenzene in styrene hydrogenation. However, Ru/Me-γ-CD showed a total hydrogenation of the aromatic rings, ethylcyclohexene being the unique reaction product. In the case of Ru/Me-β-CD, the substitution degree governs the selectivity; while a low degree of substitution leads to total hydrogenation of aromatic rings, a higher degree of substitution blocks arene hydrogenation in benzene derivatives with an alkyl chain of more than two carbon atoms. This was explained by differences in the hydrophobic host cavities of Me-β-CDs. The hydrogenation of other substrates (olefins, ketones, and disubstituted arenes) was also accomplished by using the same catalytic system [45]. Recyclability studies revealed a high stability of the catalyst in the hydrogenation of ethyl pyruvate.

Philippot and Poteau combined experimental and computational studies to investigate the surface of Ru NPs capped with a new stabilizing ligand, ethanoic acid [46]. DFT calculations, supported by experimental observations, showed that these capping ligands coordinate to the ruthenium surface as ethanoate anions (Fig. 12.11). CH_3COOH-functionalized Ru NPs were presented as promising catalyst for the hydrogen evolution reaction (HER). The co-existence of carboxylates and hydrides on the NP surface decreases the theoretical H_2 dissociative adsorption Gibbs free energy to values close to zero (-2.0 to -3.0 kcal mol^{-1}) and makes these carboxylate-stabilized NPs potentially effective HER catalysts.

Fig. 12.11 Representation of a Ru nanoparticle (55 atoms) ligated by 15 ethanoates and 32 hydrides $[Ru_{55}(CH_3COO)_{15}H_{32}]$. Reproduced from Ref. [46] with permission from RSC

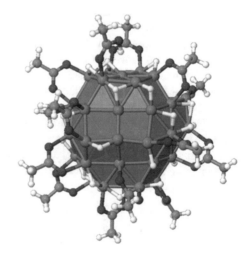

12.2.3   Phosphorous Donor-Stabilized Ru NPs

12.2.3.1   Phosphine Ligands

Phosphines have been broadly used in organometallic chemistry to stabilize metal complexes, and during the last decade, they have also been employed to stabilize metal nanoparticles. One of the first example of phosphine-stabilized Ru NPs was reported by Chen et al. who used TPPTS [$P(m$-$C_6H_4SO_3Na)_3$, 3,3′,3‴-phosphinetriyltribenzenesulfonate trisodium salt; Fig. 12.12a] [47]. TPPTS-stabilized Ru NPs were prepared by reduction of $RuCl_3 \cdot 3H_2O$ under H_2 pressure in ethanol. These 5 nm Ru NPs were employed as catalyst in the asymmetric hydrogenation of aromatic ketones with the use of a chiral modifier, (1R, 2R)-DPENDS [disodium salt of sulfonated (1R, 2R)-1,2-diphenyl-1,2-ethylene-diamine; Fig. 12.12b] and an ionic liquid/water as reaction medium (Ru/DPENDS 1:9, large excess KOH). Full conversion and 85% of enantiomeric excess (*ee*) were obtained in the asymmetric hydrogenation of acetophenone. Leaching was very low, but experiments with a homogeneous counterpart gave the same *ee*, albeit at a lower rate.

Shortly after, Chaudret et al. described the use of diphosphines [dppb: 1,4-bis(diphenylphosphino)butane and dppd: 1,10-bis(diphenylphosphino)decane] to stabilize small Ru NPs (1.5–1.9 nm) by decomposition of Ru(COD)(COT) under H_2 [48]. The coordination and partial hydrogenation of phosphine ligands were proven by solid-state NMR. The surface hydrides, which originate from the synthetic process, were accurately quantified by titration of Ru NPs with olefins (2-norbornene, 1-octene); Ru/dppb contains 1.1 hydrides per surface ruthenium atom. By NMR and desorption methods, the ability of these NPs to split C-C bonds was also demonstrated, since methyl groups were detected at their surface after reaction with ^{13}C-ethylene. Interesting surface studies using CO as a probe molecule (FT-IR and solid-state NMR) were carried out with Ru-dppb, demonstrating the importance of this technique to understand the NP surface through the location of the active sites [29]. Furthermore, the selective poisoning of Ru-dppb with bridging CO groups (Ru/CO$_b$/dppb) produces nanocatalysts unable to hydrogenate aromatic rings, and they hydrogenated styrene selectively into ethylbenzene (Table 12.1, entry 2) [49]. Another effective way to control the selectivity of Ru-dppb in arene hydrogenation is

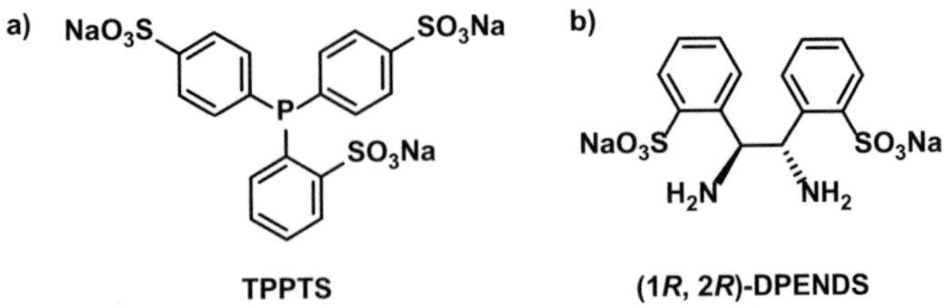

Fig. 12.12 a TPPTS: $P(m$-$C_6H_4SO_3Na)_3$; 3,3′,3‴-phosphinetriyltribenzenesulfonate trisodium salt and **b** DPENDS: (1R, 2R)-1,2-diphenyl-1,2-ethylene-diamine

Table 12.1 Hydrogenation of styrene with Ru/dppb, Ru/CO$_b$/dppb, Ru/CO/dppb, and Ru/dppb/Sn NPs as catalysts

Entry	Catalyst	Product ratio A:B:C (%)
1.	Ru/dppb	0:0:100
2.	Ru/CO$_b$/dppb	0:99:1
3.	Ru/CO/dppb	100:0:0
4.	Ru/dppb/Sn$_{0.2}$	0:99:1

through the selective poisoning of some specific actives sites with tributyltin hydride [(n-C$_4$H$_9$)$_3$SnH] [50]. These tin-decorated ruthenium nanoparticles (Ru/dppb/Sn) showed a high selectivity to ethylbenzene in the hydrogentation of styrene, due to the formation of μ^3-bridging "SnR" groups on the ruthenium surface that block the NP faces necessary for the hydrogenation of arenes (Table 12.1, entry 4).

In 2014, Chaudret in collaboration with Salmeron investigated the influence of the dppb ligand in a model CO hydrogenation reaction by using a series of Ru NPs with different stabilizers (dppb and PVP) and sizes (1.3, 1.9, and 3.1 nm) [42]. It was demonstrated that the coordination of the dppb at the Ru surface modifies both the selectivity and activity, whereas the size increase did not affect significantly the catalytic properties of these Ru NPs. The catalysis was carried out under mild conditions [3 bar of syngas (1:1 molar mixture of H$_2$:CO), 150–180 °C] and monitored by gas phase NMR. Ambient-pressure XPS experiments showed that diphosphine ligands were partially oxidized during catalysis, but they increased the activity of the nanocatalysts in comparison with polymer-stabilized Ru NPs (Ru-PVP). This study reveals the high importance of the stabilizing ligands to modulate both the catalytic activity and selectivity of Ru NPs.

Chaudret, Limbach, and Buntkowsky carried out a complete NMR study (gas phase and solid-state NMR) on the reaction of Ru-dppb, RuPt-dppb, and Pt-dppb with D$_2$ [51]. By gas phase NMR, they mostly observed the formation of HD, due to the isotopic exchange between D$_2$ and surface hydrides that appear during the NP synthesis. In addition, partially deuterated butane and cyclohexane were detected coming from the decomposition of the dppb at the Ru surface, which indicates the capability of these NPs to split C-P bonds. To sum up, this study identifies the great variety of side reactions that can be found on the metal surface of these dppb-stabilized MNPs under D$_2$ pressure (Fig. 12.13), such as C–P bond cleavage (A1 and A2), benzene hydrogenation (B), deuteration of cyclohexene and butane by C–H bond activation (C1 and C2), butane hydrogenation, D$_2$ adsorption (D), isotopic exchange with surface hydrides (E), and dppb hydrogenation. The C–H activation/deuteration of different alkanes with this Ru/dppb was further investigated by the same groups. It was

Fig. 12.13 Reactions taking place on dppb-stabilized nanoparticles when exposed to D_2. Adapted from Ref. [51] with permission from Wiley

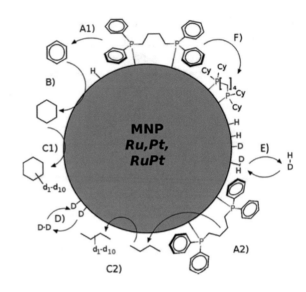

observed that the deuteration of cyclopentane is much more favored than deuteration of cyclohexane or other alkanes [52]. After experimental evidences obtained from the deuteration experiments, they proposed that the higher reactivity of cyclopentane is due to surface/substrate interactions. However, the details could not be definitely clarified by DFT calculations (using a $Ru_{13}H_{17}$ model cluster) as both cyclopentane and cyclohexane presented similar binding and activation enthalpies.

Castillón et al. compared the activity and selectivity of Ru and Rh NPs stabilized with two different phosphines, dppb and PPh_3, in the hydrogenation of a series of aromatic ketones [53]. Remarkable effects in terms of stabilizing ligands and nature of the metal were described. While for Ru NPs, the interaction substrate/metal and the consequent hydrogenation occurs in both functional groups (ketones and aromatic rings), for Rh NPs the interaction with the arene is favored. Rh NPs hydrogenate selectively the aromatic group when the carbonyl group is far enough from the ring, whereas Ru NPs hydrogenate both of them in a competitive way, independently of the substrate structure. The stabilizing ligands also influence the selectivity of the catalysts, i.e., hydrogenolysis products were only observed when Rh-PPh_3 was used as catalyst and for Ru NPs, the higher steric hindrance of the ligand increases the hydrogenation of ketone groups. Thus, it is another clear example about the ability of ancillary ligands to modulate the surface chemistry and catalytic properties of metal nanoparticles. The same group, in collaboration with Chaudret and co-workers, investigated the coordination mode of phosphines, triphenylphosphine oxide, and triphenyl phosphite at the Ru surface through their selective catalytic deuteration [54]. Specifically, using Ru-PVP NPs as catalysts, phenyl- and phenyl-alkyl phosphines compounds were selectively deuterated in ortho position. This is due to the strong phosphorus-Ru interaction, which orientates the ortho position close to the Ru surface and facilitates the H/D exchange (Fig. 12.14, left). On the other hand,

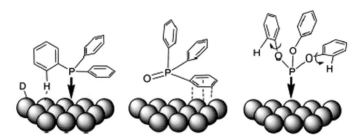

Fig. 12.14 Proposed coordination mode of the ligands studied to the Ru NP surface. Reproduced from Ref. [54] with permission from ACS

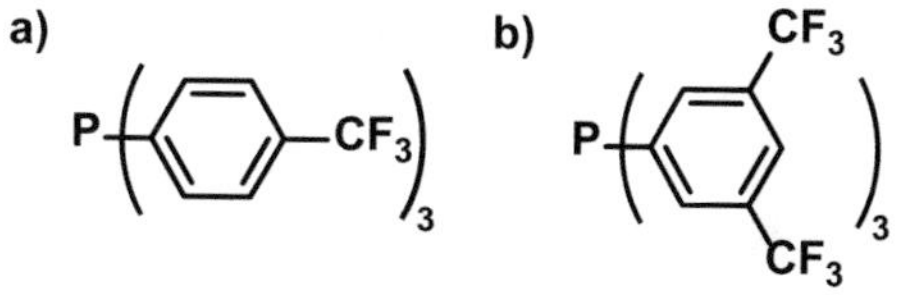

Fig. 12.15 Fluorinated phosphines used to stabilize Ru NPs in Ref. [55]

triphenylphosphine oxide was totally reduced, because of intense π-interactions of the aromatic rings and metal surface (Fig. 12.14, center). Lastly, arylphosphites did not react at all, neither labeling nor reduction occurred, since the oxygen atoms keep the aromatic rings away from the Ru surface, the authors argued (Fig. 12.14, right). Electronic factors may also play a role.

Following the same synthetic procedure as that developed by Chaudret et al. [decomposition of Ru(COD)(COT) under H_2 in the presence of phosphine ligands] [45], Gomez and Masdeu-Bultó reported the synthesis of Ru NPs stabilized by a fluorinated phosphine tris(4-trifluoromethylphenyl)phosphine and tris(3,5-bistrifluoromethylphenyl)phosphine; Fig. 12.15) and their use as catalysts in supercritical carbon dioxide (sc-CO_2) [55]. More precisely, these Ru NPs were used in the hydrogenation of aromatic compounds in THF and sc-CO_2, showing a higher activity in the organic solvent. The lower catalytic activity of the Ru NPs in sc-CO_2 was explained by a lower solubility in spite of the presence of the fluorinated ligand, which was expected to facilitate the NP solubility in the supercritical medium.

In 2012, van Leeuwen et al. reported new phosphine ligands specifically designed to stabilize Ru NPs (Fig. 12.11) [56]. Nanoparticles of different shapes and sizes were obtained depending on the nature of the phosphine ligands (i.e., mono- or diphosphines, bite angle, alkyl- or aryl-phosphines, etc.). HRMAS NMR studies showed a partial ligand hydrogenation during the RuNP synthesis. Also, a marked ligand effect was evidenced in the catalytic hydrogenation of arene derivatives, such as benzene and 2-methylanisole. Ru NPs stabilized with partially hydrogenated phosphine **4** (dialkylaryl phosphine; Fig. 12.16) showed the highest activity in the hydrogenation of 2-methylanisole (TOF = 152 h^{-1}). This was explained by means of a substrate accessibility; the flexible coordination of monophosphines permits an easier access of the substrate to the active surface. Furthermore, triarylphosphine-stabilized NPs

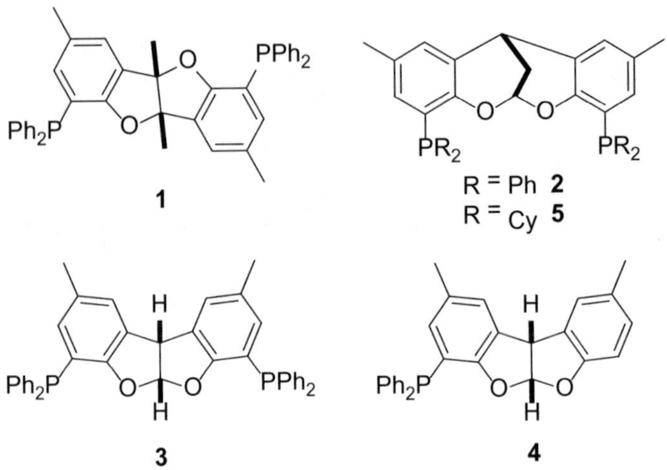

Fig. 12.16 Specific phosphine ligands designed by Leeuwen et al. to stabilize RuNPs

were inactive in this catalytic reaction since aromatic rings of the ligand block the NP faces necessaries for the anisole hydrogenation. It would seem that also electronic effects could explain the differences.

In 2012, the group of Philippot reported the synthesis of water-soluble Ru NPs with sulfonated diphosphines (Fig. 12.17) [57]. The resulting NPs were active in the biphasic hydrogenation of styrene in water. An interesting influence of the alkyl chain length of the diphosphine used as stabilizing ligand was observed. The longer the alkyl chain is, the higher the catalytic activity is. As NPs display similar sizes and shapes, the differences in catalytic activity were justified by the high flexibility of the longer alkyl chain, which facilitates the diffusion of the substrate to the active sites. The combined use of the sulfonated diphosphine (1,4-bis [(di-m-sulfonatophenyl)phosphino]butane; Fig. 12.17, $n = 4$) and a cyclodextrin (Me-β-CDs; Fig. 12.10) allowed the synthesis of water-soluble Ru NPs [58]. NMR studies evidenced the strong coordination of the phosphine ligand to the Ru surface as well as the presence of the cyclodextrin, which creates an inclusion complex between

Fig. 12.17 Sulfonated diphosphines (left) and 1,3,5-triaza-7-phosphaadamantane (PTA; left) used to stabilize water-soluble Ru NPs by Philippot et al.

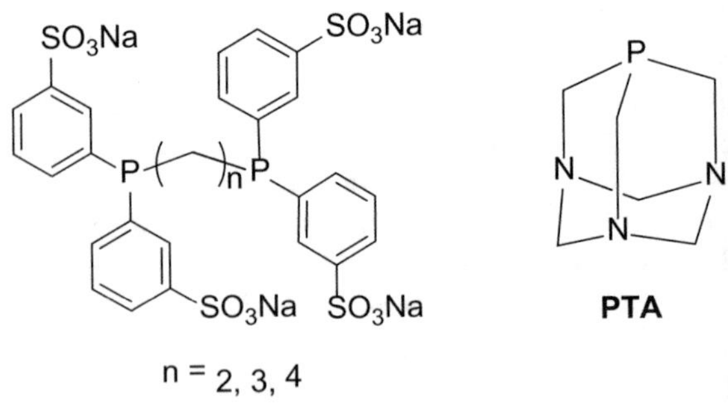

$n = 2, 3, 4$

the ligand and the supramolecular host. It was evidenced that cyclodextrin is able to modulate the catalytic reactivity of the sulfonated diphosphine-stabilized Ru NPs. Comparing the reactivity of this nanocatalyst with the diphosphine-stabilized Ru NPs (without cyclodextrin) in the biphasic hydrogenation of styrene, acetophenone and *m*-methylanisole, interesting differences were found. The presence of the cyclodextrin not only increases the catalytic activity, as it acts as phase-transfer promoter, but also influences the selectivity due to the formation of the inclusion complex between the cyclodextrin and the phosphine. Philippot and co-workers also reported another type of water-soluble phosphine-stabilized Ru NPs, but in this case, the NPs were stabilized by 1,3,5-triaza-7-phosphaadamantane (PTA) [59]. The resulting NPs showed interesting conversions and selectivities in the hydrogenation of olefins and arenes in biphasic (water/substrate) catalytic reactions without the necessity of a phase-transfer agent. The last results demonstrate the potential of stabilizing ligands not only in the control of selectivity and activity of Ru NPs, but also in the modification of their physical properties such as water solubility, which allow us to carry out catalysis in a more environmentally friendly way.

12.2.3.2   Secondary Phosphine Oxide Ligands (SPO)

In 2013, van Leeuwen and co-workers developed the use of secondary phosphine oxides (SPO) for Ru NP stabilization. The obtained SPO-stabilized Ru NPs were investigated in the hydrogenation of aromatic rings and carbonyl groups, turning out to be efficient nanocatalysts for the hydrogenation of arenes (TOFs up to 2700 mol.h^{-1} in the hydrogenation of benzene) [60]. The idea was to induce the heterolytic cleavage of hydrogen by using SPO ligands for Ru NPs, and in this way enhance the activity in hydrogenation reactions of polar substrates such as ketones. Hydrogenation of aromatics was strongly reduced by SPO ligands, but the activity for ketone hydrogenation remained similar. Later, for SPO-stabilized Au and Ir NPs, they found indeed that H_2 is split through a cooperative action between the SPO ligand and the neighboring metal center. Air-Stable Au NPs ligated by SPOs were employed as efficient catalysts in the hydrogenation of numerous substituted aldehydes, presenting high conversions and chemoselectivities [61]. In addition, a notable ligand effect was found by using Au NPs ligated by various SPO ligands with different electronic and steric properties (Fig. 12.18a) [62]. In particular, Au NPs stabilized with SPO ligands with aromatic groups (**1**, **2** and **3**) are highly selective in the hydrogenation of the carbonyl group, whereas Au NPs that contain alkyl groups (**4** and **5**) are less active and selective in the hydrogenation of aldehydes. This different catalytic behavior is explained through the different coordination forms of SPOs at the Au surface (Fig. 12.18b), which depend on the type of SPO ligand. While **1**, **2**, and **3** mainly coordinate as $R_2P = O$ anions, **3** and **4** also coordinates as $R_2P–OH$, changing the electronic effect of the catalytic systems and therefore their catalytic properties. Similar selectivity toward aldehyde hydrogenation was found for Ir NPs stabilized with SPO ligands **1**, **2**, and **5** [63, 64]. Lately, chiral Ir NPs stabilized by an asymmetric secondary phosphine oxide (Fig. 12.18c) was used as

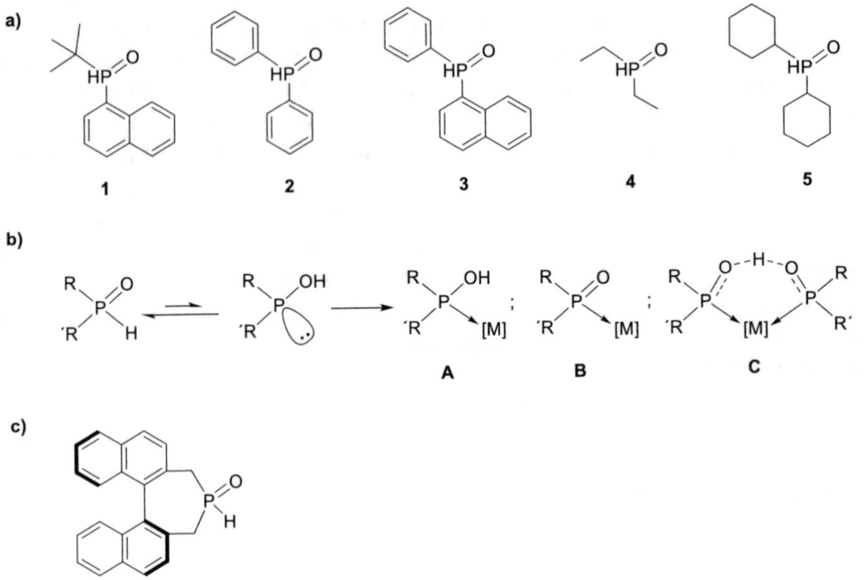

a)

1   2   3   4   5

b)

A   B   C

c)

Fig. 12.18 **a** Secondary phosphine oxides used by van Leeuwen et al. **b** Tautomeric forms of SPOs and possible coordination modes to a metal center. **c** Chiral secondary phosphine oxide 4,5-dihydro-3H-dinaphtho [2,1-c:1,2-e] phosphepine-4-oxide

catalyst in the hydrogenation of pro-chiral ketones, presenting the first example of asymmetric hydrogenation catalyzed by non-supported chiral Ir NPs [65]. Therefore, SPO has demonstrated to be an efficient bifunctional ligand, which not only stabilizes nanoparticles of different metals (Ru, Au or Ir), but is also actively involved in the reaction mechanism (as chirality inducer, and facilitating the cleavage of H_2).

12.2.3.3 Phosphite Ligands

In a multiple collaboration, Claver, Roucoux, and Philippot reported a series of Ru NPs stabilized with diphosphite ligands derived from carbohydrates (Fig. 12.19) [66]. Depending on the nature of the diphosphite ligand used as stabilizer, interesting differences were found in terms of nanoparticle size, dispersity, and catalytic activity. Specifically, the diphosphite bearing a long alkyl chain led to the formation of smaller and better dispersed Ru NPs, which showed the best activity in the asymmetric hydrogenation of pro-chiral arenes (*o*- and *m*-methyl anisoles). Although the enantiomeric excess obtained was very poor, these Ru NPs gave a high selectivity for the *cis* product. Comparable catalytic behavior was found for Ir and Rh NPs stabilized with the same ligands [67]. Rh nanoparticles displayed higher catalytic activity than their counterparts of Ru, Ir NPs being the least active.

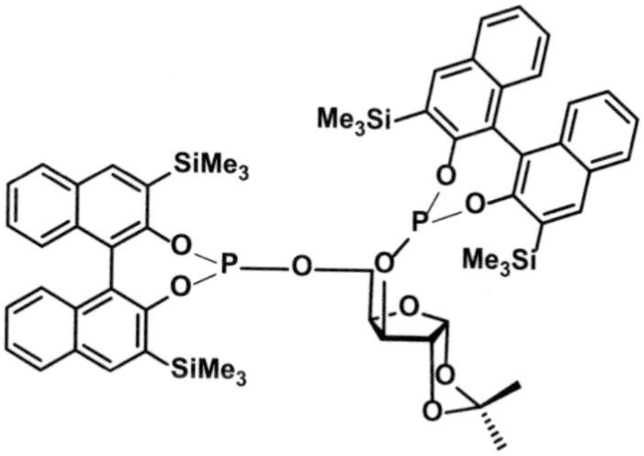

Fig. 12.19 a 1,3-Diphosphite ligand used to stabilize Ru, Rh, and Ir NPs by Claver et al.

12.2.4 N-Heterocyclic Carbene (NHC)-Stabilized Ru NPs

The first example of NHC-stabilized Ru NPs was described by Chaudret et al. in 2011. Following the organometallic approach developed years before, they decomposed Ru(COD)(COT) under H_2 pressure in the presence of two different free NHCs; N,N di(tert-butyl)imidazol-2-ylidene (ItBu) and 1,3-bis(2,6-diisopropylphenyl)imidazol-2-ylidene (IPr) (Fig. 12.20a) [68]. The catalytic activity of these Ru NPs was evaluated using the hydrogenation of styrene as model reaction. Both catalysts (Ru-ItBu and Ru-IPr) showed similar reaction rates in the stepwise hydrogenation of styrene, but the rate is lower than that observed for Ru NPs stabilized by PVP. This suggests that the coordination of NHCs partially blocks the Ru surface sites necessary for hydrogenation of both vinyl and phenyl groups. Using the same NHC-stabilized Ru NPs, the group of Chaudret in collaboration with van Leeuwen studied the hydrogenation of numerous arene derivatives finding an interesting ligand effect [69]. In general, they observed that Ru-IPr is more active than Ru-ItBu. The necessity of a larger amount of ligand to stabilize Ru-ItBu NPs (0.5 equiv.) leads to a high surface coverage, which may explain their lower activity. In any case, Ru-IPr (stabilized with 0.2 equiv of ligand) showed high activities in neat arene hydrogenation, with TOF up to 1300 h^{-1} in the hydrogenation of 2-methylanisole at 80 °C and a TOF of 421 h^{-1} for neat benzene at r. t.

This same group, but this time in collaboration of Glorius, developed a new methodology to stabilize Ru NPs with non-isolable NHCs [70]. This is based on the addition of the free carbene, formed *in situ* after the reaction of the imidazolium salt with KtBuO. The new route extended the use of NHCs for NP stabilization, overcoming the limitation of the previous synthetic route that required isolation of the free carbene [69]. Taking advantage of this novel approach, they prepared two different Ru NPs stabilized with non-isolable chiral NHCs (SIDPhNp and SIPhOH; Fig. 12.20b) with the intention to induce enantioselectivity in the hydrogenation of various substrates. Although no enantiomeric excess was observed in any case, interesting differences in the reactivity and the selectivity of the Ru NPs were observed in the hydrogenation of C=C as well as C=O double bonds. In general, both catalysts

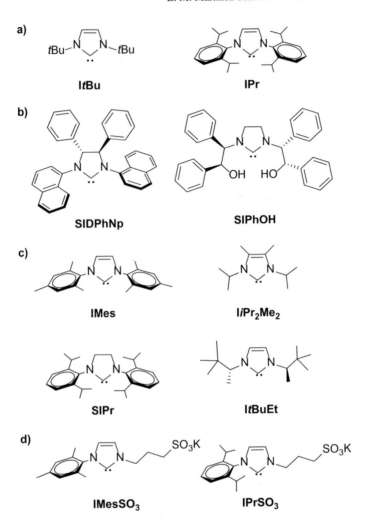

Fig. 12.20 **a** NHCs used by Chaudret et al. in Ref. [68]. **b** Non-isolable chiral N-heterocyclic carbenes used by Chaudret and Glorius in Ref. [70]. **c** Isolable N-heterocyclic carbenes used by Lara and Martínez-Prieto in Ref. [71]. **d** Sulfonated N-heterocyclic carbenes used by Chaudret and Martínez-Prieto in Ref. [76]

presented good levels of catalytic activity, but Ru NPs stabilized with the naphthalene derivative showed better activity than the other one. The lower catalytic activity of Ru-SIPhOH NPs was ascribed to the limited access of the substrate to the reactive sites of these NPs. It is principally due to the π-interactions between the aromatic rings and the metal surface and the strong interaction between hydroxyl groups of the ligand and the ruthenium. Shortly after, Lara, Martínez-Prieto and co-workers prepared a family of NHC-stabilized Ru NPs (Fig. 12.20c), including a chiral NHC [71]. They investigated their surface chemistry using the CO as probe molecule and tested the chiral one in a few asymmetric catalytic reactions. Despite the high activity and chemoselectivity observed, negligible enantiomeric excess was obtained. The lack of enantioinduction in asymmetric hydrogenation reactions with chiral ligand-stabilized Ru NPs reveals that it is not always possible to extrapolate the catalytic behavior of metal complexes to metal NPs.

In 2016, Scaiano and Fogg reported NHC-stabilized Ru NPs formed from the decomposition of second-generation Grubbs catalyst, $RuCl_2(H_2IMes)(PCy_3)(=CHPh)$ [H_2IMes = 1,3-bis(2,4,6-trimethylphenyl)-4,5-dihydroimidazol-2-ylidene] during olefin metathesis [72]. The resulting NPs demonstrated to be very active in isomerization of estragole, being the first example of olefin isomerization by Ru NPs. Also, by decomposition of NHC complexes, Stubbs and Huynh generated NHC-stabilized Ru NPs, which were selective catalysts in the hydrogenation of levulinic acid (LA) toward γ-valerolactone (GVL) [73]. Specifically, mono- and bidentate p-cymene Ru (II) NHC complexes were reduced by H_2 in water to form the catalytically active Ru NPs.

Recently, Chaudret et al. obtained water-soluble NHC-stabilized Pt [74] and Pd [75] NPs by decomposition of NHC complexes in water. After that, Martínez-Prieto and Chaudret employed for the first time the same water-soluble NHCs (WS-NHCs; Fig. 12.20d) to prepare Ru NPs characterized by their high activity, stability, and solubility in aqueous media [76]. Taking profit of their water-solubility, the catalyst/substrate interaction was investigated by solution NMR during the C–H deuteration of L-lysine. By using chemical shift perturbations (CSPs), a common NMR technique to investigate protein/substrate interactions, it was possible to study the coordination mode of L-lysine on the NP surface at different pHs, and therefore to understand the deuteration process catalyzed by these Ru/WS-NHCs. This C-H activation process was previously studied by DFT calculations and a four-membered dimetallacycle was found as the main intermediate [77]. The orientation of the substrate toward the Ru surface is highly dependent on the pH and crucial for the H/D exchange. The activity is maximum at high pH (~13), but it is almost suppressed at low pH (<3) due to the lack of substrate/nanoparticle interaction (Fig. 12.21). The same hydrosoluble Ru NPs were recently employed in the deuteration and tritiation of nucleobase pharmaceuticals and oligonucleotides showing an efficient hydrogen-isotope exchange [78]. Furthermore, Chaudret and Martínez-Prieto have demonstrated that the activity and selectivity of bimetallic WS-NHC-stabilized RuPt in the

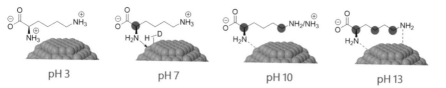

Fig. 12.21 Interaction of L-lysine with Ru NP surface as a function of pH. In red, C–H groups giving rise to H/D exchange. At low pH (3), the H/D exchange is blocked, while at higher pHs (7–13), not only the activity increases but also the selectivity changes. Adapted from Ref. [76] with permission from RSC

isotopic H/D exchange of L-lysine are highly dependent on the nanoparticle surface composition [79]. The interaction of L-lysine and the PtRu surface was also studied by chemical shift perturbation (CSP). It was shown that protons in epsilon position approach the surface of Ru and PtRu NPs and mainly interact with the Ru atoms. However, amines in alpha position do not interact in the same manner with Ru or PtRu surface. The L-lysine in the presence of both Ru and Pt atoms forms a chelate (amine-Ru/carboxylic acid-Pt), which is strongly coordinated to the metal surface and hinder the H/D exchange at this position (Fig. 12.22). All these works highlight the importance of the interaction substrate/nanoparticle for the activity and selectivity of MNP catalytic reactions.

In a multiple collaboration, Philippot, Chaudret, and Glorius evidenced the importance of the ligands for Ru NP stability. Employing NHCs with long aliphatic chains (LC-NHCs) in positions 4 and 5 of the imidazole ring (Fig. 12.23a), they obtained air-stable Ru NPs active in hydrogenation and oxidation reactions [80]. Moreover, these long-chain NHC-stabilized Ru NPs were presented as active catalysts for one-pot oxidation/hydrogenation processes. After the oxidation of the substrate was complete,

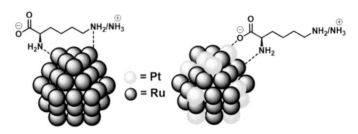

Fig. 12.22 Interaction of L-lysine with Ru and PtRu surfaces

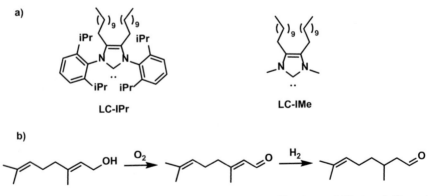

Fig. 12.23 a Long-chain N-heterocyclic carbenes used by Chaudret and Glorius. **b** One-pot oxidation/hydrogenation of geraniol catalyzed by Ru NPs stabilized with LC-NHCs

O$_2$ atmosphere was replaced by H$_2$ and the oxidation product was selective hydrogenated (Fig. 12.23b). Interesting differences in the activity were observed depending on the N-substituent of the LC-NHC ligand. The bulkier the N-substituents are, the larger the amount of open active sites is and the higher the activity. Therefore, LC-NHCs protect Ru NPs toward deep oxidation, which makes them suitable for oxidation-hydrogenation two-step processes. In addition, their catalytic reactivity can be modulated by modifying their N-substituents. The same LC-NHC ligands (Fig. 12.23) were used to stabilize Pt NPs, resulting in soluble Pt NPs active in the hydroboration of phenylacetylene [81]. Here, as well as in the previous example, a clear influence of the bulkiness of the N-substituents on the size, surface state, and catalytic activity was observed.

In 2018, the same multiple collaboration reported Ru NPs stabilized by NHCs derived from cholesterol and their use in the hydrogenation of aromatic substrates [82]. Two novel Ru NPs ligated by NHCs generated from their corresponding steroidal-based imidazolium salts (Fig. 12.24) were synthesized. The effect of these rigid ligands on the hydrogenation of arene derivatives was evaluated, which resulted in interesting differences in terms of activity and selectivity. For example, Ru NPs ligated by the NHC with the lipophilic part on the N-substituent (Ru/IMe-chol) normally present higher activity than the ones derived from the cholesterol moiety in the backbone of the imidazolium core (Ru/chol-IMe). Moreover, at low catalytic loading, Ru/IMe-chol totally hydrogenates styrene to ethylcyclohexane, while Ru/chol-IMe hydrogenates it selectively to ethylbenzene. The superior activity of Ru/IMe-chol is explained by the higher steric demand of IMe-chol, which results in a lower amount of ligands at the surface and more available faces necessary for the hydrogenation of aromatic rings. Moreover, the use of new biomimetic NHCs derived from cholesterol to stabilized Ru NPs presents a new type of catalysts with interesting possibilities for recognition of biological systems.

In conclusion, we have shown that ligand-stabilized Ru NPs are active nanocatalysts in many catalytic reactions, useful nano-objects for surface investigations, and easily tunable systems through the coordination of ancillary ligands (Tables 12.2, 12.3, and 12.4). Here, the ligands are not only able to modulate the catalytic proper-

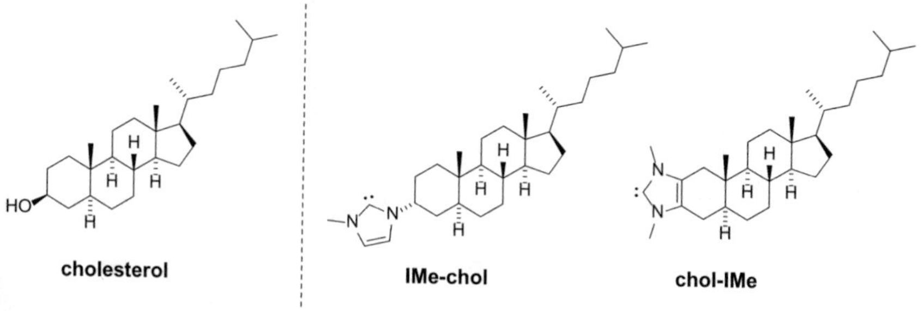

Fig. 12.24 Cholesterol (left) and cholesterol-based NHCs used for Ru NP stabilization. Reproduced from Ref. [82] with permission of RSC

Table 12.2 Nitrogen donor-stabilized Ru NPs synthesis and catalytic applications

Metal precursor	Stabilizing ligand	Synthetic approach	Catalysis	Comments	Refs.
Ru(acac)$_3$	Oleylamine	Thermal decomposition	Dehydrogenation of ammonia borane in water	Ru NPs supported in Al$_2$O$_3$	[16]
en–RuCl$_3$	Ethylenediamine	Reduction by NaBH$_4$	None	Fast exchange between coordinated and free ligand	[17]
Ru(COD)(COT)	Alkylamines (C8, C12 and C16)	H$_2$, −80 °C, 24h, THF	None	Dynamic studies	[18]
Ru(acac)$_2$	Dodecylamine	H$_2$, 140 °C, 70h, Mes	None	Ru hour-glass NPs	[19]
Ru$_3$(CO)$_{12}$	Hexadecylamine/hexadecanoic acid	H$_2$, 160 °C, 6h, toluene	C−C and C−O activation	Ru nanostars	[20]
Ru(COD)(COT)	Aminopropyltriethoxysilane	H$_2$, 25 °C, 12h, THF	Dehydrogenation of dimethylamine-borane	Used as building blocks to synthesize RuO$_2$@SiO$_2$	[21–23]
Ru(COD)(COT)	1-octylamine, 1-hexadecylamine	H$_2$, 30 °C, 12h, ILs	Hydrogenation of toluene and unsaturated compounds	Amines increased the stability of Ru NPs (recyclable up to 10 times)	[24, 25]
RuCl$_3$·3H$_2$O	Cyclodextrines + chiral aminoacids	(i) Reduction by H$_2$ (ii) Reduction by NaBH$_4$	Hydrogenation of pro-chiral compounds	No significant enantiomeric excess	[26]
Ru(COD)(COT)	4-(3-phenylpropyl)-pyridine	H$_2$, 25 °C, 1h, THF	None	Dual coordination mode	[27]
Ru(COD)(COT)	Phenylpyridine	H$_2$, 25 °C, THF	Electrocatalytic HER	In both acidic and basic media	[28]
Ru(COD)(COT)	Chiral N-donor ligands	H$_2$, −80 °C, 24h, THF-methanol	Reduction of organic prochiral unsaturated substrates	Poor enantioselectivity	[28]
Ru(COD)·(COT)	ICy(p-tol)NCN	H$_2$, 25 °C, 22h, THF	Styrene hydrogenation	Ultra-small Ru NPs (~1 nm)	[29]

(continued)

Table 12.2 (continued)

Metal precursor	Stabilizing ligand	Synthetic approach	Catalysis	Comments	Refs.
RuCl$_3$·3H$_2$O	PVPy	Reduction by NaBH$_4$	Hydrogenation of arenes and N-heteroaromatic compounds	Dual homolytic/heterolytic hydrogenation mechanism	[34]
Ru(COD)(COT)	Hexadecylamine, C60, PVP and IPr	H$_2$, −80 °C, 24h, THF	Ligand effect in the hydrogenation of nitrobenzene	More e$^-$ attractor ligand (C60) → higher activity/selectivity	[37]

Table 12.3 Oxygen and phosphorus donor-stabilized Ru NPs synthesis and catalytic applications

Metal precursor	Stabilizing ligand	Synthetic approach	Catalysis	Comments	Refs.
Ru(COD)(COT)	Alcohols (MeOH, nPrOH, iPrOH, pentanol, heptanol)	H_2, 25 °C, 24 h, neat alcohol; THF/MeOH mixtures	hydrogenation of benzene to cyclohexane	Polarity and alkyl chain length control the growth of the NPs.	[40, 41, 43]
$RuCl_3 \cdot 3H_2O$	Cyclodextrins (Me-CDs)	Reduction by $NaBH_4$	Hydrogenation of olefins, ketones and arenes	Me-CDs acts as stabilizer, phase-transfer agent and reactant host	[44]
Ru(COD)(COT)	Ethanoic acid	H_2, 25 °C, 24 h, pentane	None	Promising catalyst for HER	[46]
$RuCl_3 \cdot 3H_2O$	TPPTS	H_2 (50 bar), 120 °C, 4 h, ethanol	Asymmetric hydrogenation of aromatic ketones	$(1R, 2R)$-DPENDS and ILs as reaction medium	[47]
Ru(COD)(COT)	dppb	H_2, 25 °C, 24 h, THF	C–C activation/Hydrogenation/CO hydrogenation/isotopic exchange (H/D)	Selective poisoning of Ru-dppb increases their selectivity	[42, 48–52]
Ru(COD)(COT)	PPh_3	H_2, 25 °C, 24 h, THF	Hydrogenation of aromatic ketones	Their activity and selectivity was compared with Rh NPs	[53]

(continued)

Table 12.3 (continued)

Metal precursor	Stabilizing ligand	Synthetic approach	Catalysis	Comments	Refs.
Ru(COD)(COT)	Fluorinated phosphines	H_2, 25 °C, 18 h, THF	Hydrogenation of aromatic compounds	Catalysis performed in THF and sc-CO_2	[55]
Ru(COD)(COT)	Roof-shaped phosphines	H_2, 25 °C, 18 h, THF	Hydrogenation of arenes	Marked ligand effect	[56]
Ru(COD)(COT)	Sulphonated diphosphines	H_2, 25 °C, 20 h, THF	Hydrogenation of styrene/acetophenone/m-methylanisole	Biphasic hydrogenation in water combined or not with Me-CDs	[57, 58]
Ru(COD)(COT)	PTA	H_2, 25 °C, 20 h, THF	Hydrogenation of olefins and arenes	Biphasic catalytic reactions	[59]
Ru (COD)(COT)	Secondary Phosphine oxides (SPO)	H_2, 25 °C, 20 h, THF	Hydrogenation of aromatic rings and carbonyl groups	TOFs up to 2700 $mol.h^{-1}$ in hydrogenation of benzene	[60]
Ru (COD)(COT)	Phosphite ligands	H_2, 25 °C, 20 h, THF	Asymmetric hydrogenation of pro-chiral arenes (o- and m-methyl anisoles)	Poor *ee* but high selectivity for *cis* products	[66]

Table 12.4 Carbene-stabilized Ru NPs synthesis and catalytic applications

Metal precursor	Stabilizing ligand	Synthetic approach	Catalysis	Comments	Refs.
Ru (COD)(COT)	ItBu, IPr	H$_2$, 25 °C, 20 h, THF	Hydrogenation of arenes	High activity in neat arene hydrogenation (TOF up to 1300 h^{-1})	[68, 69]
Ru(COD)(COT)	SIDPhNp, SIPhOH	H$_2$, 25 °C, 20 h, THF (in situ formation of the free carbene)	Asymmetric hydrogenation of C–C and C–O double bonds	Use of non-isolable NHCs. No ee was observed	[70]
Ru(COD)(COT)	IMes, IiPr$_2$Me$_2$,SIPr, ItBuEt,	H$_2$, 25 °C, 20 h, pentane	Asymmetric hydrogenations	High activity and selectivity. Negligible ee	[71]
RuCl$_2$(H$_2$IMes) (PCy)(=CHPh)	2° generation Grubbs catalyst	Complex decomposition	Olefin isomerization of aromatic ketones	Ru NPs formed during olefin metathesis	[72]
p-cymene Ru(II) NHC complex	NHC ligands	H$_2$ atmosphere in water	Hydrogenation of levulinic acid to γ-valerolactone	NHC-complexes were reduced in water to form Ru NPs	[73]
Ru (COD)(COT)	Water-soluble NHCs (IMesSO$_3$, IPrSO$_3$)	H$_2$, 25 °C, 18 h, THF	C–H deuteration of L-lysine	Substrate/NP interactions studied by CSPs	[76]
Ru(COD)(COT)	Long aliphatic chains-NHCs (LC-IPr, LC-IMe)	H$_2$, 25 °C, 18 h, THF	One-pot oxidation/hydrogenation processes	Catalytic activity can be modulated by N-substituents	[79]
Ru (COD)(COT)	NHCs derived from cholesterol (IMe-chol, chol-IMe)	H$_2$, 25 °C, 18 h, THF	Hydrogenation of aromatic substrates	Higher activity of Ru/IMe-chol (higher steric demand)	[82]

ties of Ru NPs, but they can also be used as stabilizers for specific shapes and sizes by choosing the appropriate donor strength and quantity. For example, in the synthesis of Ru NPs by decomposition of Ru(COD)(COT) (organometallic approach), the stronger the donor ligand is, the smaller the size of the obtained NPs will be (Fig. 12.25) [4].

12.3 Supported Ligand-Functionalized Ru NPs

In the previous section, it has been discussed how ligands can modify the catalytic properties of colloidal Ru NPs, like they do in homogeneous catalysis. In the same way, ligands can be used to functionalize supported Ru NPs and thereby modify

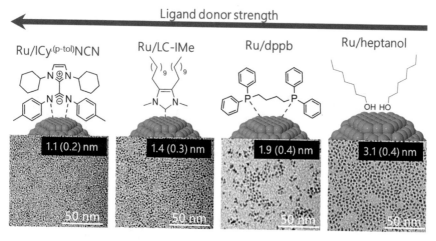

Fig. 12.25 Size-dependence with the ligand donor strength in RuNPs prepared by the organometallic approach. ICy.(p-tol)NCN = 1,3- dicyclohexylimidazolium-2-di-p-tolylcarbodiimide; LC-IMe = 1,3-dimethyl-4,5-diundecyl imidazol-2-ylidene; dppb = bis-(diphenylphosphino)butane. Reproduced from Ref. [4] with permission from ACS

their surface properties [83]. This is a novel approach to improve the selectivity of heterogeneous catalysts, which are the most important catalysts for industrial applications. In addition, immobilization will facilitate the separation of catalyst and products. Ligand modification of heterogeneous catalysts has a long tradition in asymmetric catalysis with the use of a chiral modifier, which is normally an organic molecule in an adsorption/desorption equilibrium that induces the enantioselectivity of the substrate (e.g., tartaric acid and cinchona alkaloids) [84]. However, in this section, we will focus on ligands fixed at the NP surface, and how they influence the catalytic properties of the supported Ru NPs. In principle, the coordination of the ligands partially blocks the active surface, eliminating the ability of active sites to adsorb and transform the substrate. Nevertheless, these ligands, like in ligand-stabilized Ru NPs, can modify electronically and sterically the NP surface, which may lead to more selective and sometimes even faster catalysts. In addition, the ligands can participate more intimately in the reaction via metal/ligand/substrate interactions. Because this novel approach has not yet been widely used, we will only mention a few recent examples, always centered on supported Ru NPs.

One of the more significant examples of the direct modification of supported Ru NPs by an ancillary ligand was reported by Tada and Glorius in 2016 [85]. Specifically, Ru NPs supported on potassium-doped alumina ($Ru/K-Al_2O_3$) was functionalized with two different NHCs (ICy and IMes). ^{13}C NMR studies corroborated that NHC ligands were mainly coordinated to the Ru surface. These NHC-modified $Ru/K-Al_2O_3$ catalysts were used in the hydrogenation of different substrates with the intention to study the effect of the surface functionalization on the catalytic activity. In the hydrogenation of acetophenone and *trans*-stilbene, they observed an influence

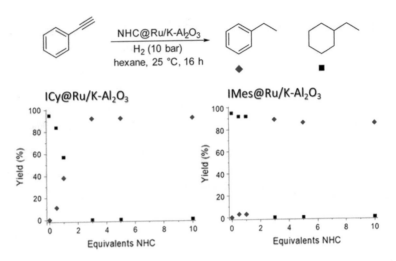

Fig. 12.26 Effect of NHC-loading on the chemoselectivity in the hydrogenation of phenylacetylene. Reproduced from Ref. [85] with permission from ACS

of the amount of ligand on the catalytic activity; the higher the amount of NHC ligand to functionalize Ru/K-Al$_2$O$_3$ is, the lower the catalytic activity is. Moreover, this tendency is more significant on the catalysts modified with the bulkier N-substituent (IMes). These results were explained through a blockage of surface atoms. Furthermore, an interesting ligand effect was found on the chemoselective hydrogenation of phenylacetylene. Increasing the amount of NHC ligand led to the selective hydrogenation of phenylacetylene to ethylbenzene with both catalysts (Fig. 12.26), this trend being more pronounced for Ru/K-Al$_2$O$_3$@IMes (higher steric demand of their N-substituents; addition of 0.5 eq. of PPh$_3$ led to the same effect). This is a nice example of tuning the activity and selectivity of supported Ru NPs by an a posteriori direct functionalization with organic ligands.

Pérez-Ramirez, in collaboration with López, reported the synthesis of Ru NPs stabilized with the surfactant HHDMA (hexadecyl(2-hydroxyethyl)dimethylammonium dihydrogen phosphate), which were supported on TiSi$_2$O$_6$ (Ru-HHDMA/TiSi$_2$O$_6$) and used in the continuous hydrogenation of levulinic acid to γ-valerolactone in water (Fig. 12.27a) [86]. The phosphate binds to Ru, which was found to be Ru(0), and the hydroxyethyl groups provide binding to TiSi$_2$O$_6$. Density functional theory (DFT) studies were done to understand the crucial role of HHDMA on this active and selective reaction (>95% selectivity at full conversion). These DTF analyses demonstrated that the acidic properties of the surfactant/metal interface enhance the hydrogenation of levulinic acid to γ-valerolactone (Fig. 12.26a). Additionally, this interfacial acidity helps to avoid surface oxidation of the catalyst, increasing the stability of the catalyst (Fig. 12.27b).

In 2013, van Leeuwen et al. registered a patent of a series of phosphine-functionalized Ru NPs supported on porous solid support (alumina or silica) or porous carbon (carbon nanotubes or carbides-derived carbons) for the hydrogenation

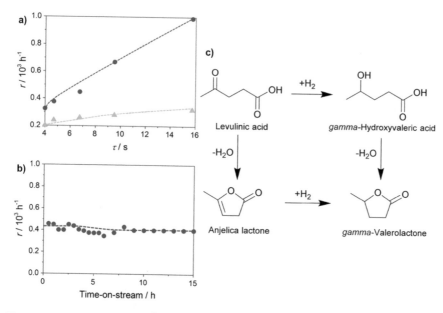

Fig. 12.27 Reaction rate (in 10^3 h^{-1}) as a function of the contact time (τ) in the hydrogenation of levulinic acid over Ru-HHDMA/TiSi$_2$O$_6$ (blue) and Ru/C (green) (**a**). Stability of Ru-HHDMA/TiSi$_2$O$_6$ in the hydrogenation of levulinic acid (**b**). Hydrogenation and dehydration of levulinic acid to γ-valerolactone (**c**). Adapted from Ref. [86] with permission from RSC

of aromatics compounds [87]. They presented more stable and robust Ru catalysts than those reported until the date, which besides do not require any activation pretreatment, notably they were ten times more active than the same Ru NPs in solution [56]. These catalysts did not show any sign of deactivation or decomposition, maintaining their activity after ten recycles. Ru NPs ligated by 4-(3-phenylpropyl)pyridine and immobilized on functionalized multi-walled carbon nanotubes (MWCNT) were reported by Gómez et al. in 2010 [88]. An interesting comparative hydrogenation study was done between supported and non-supported 4-(3-phenylpropyl)pyridine-stabilized Ru NPs. Overall, supported NPs (Ru-MWCNT) demonstrated to be more active than non-supported ones. This improvement of the catalytic activity was explained by a substrate/support interaction, which increases the conversion on the hydrogenation reactions. Finally, by comparison of the activity of the same Ru NPs but immobilized on different supports (silica, alumina, and activated carbon), they investigated the support effect. Ru-MWCNT showed the best activity due to an easier substrate/metal interaction as the Ru NPs remain on the surface of MWCNT.

Recently, Garcia-Antón, Philippot and Mas-Ballesté presented a series of Ru NPs supported on carbon microfibers (CFs) used for HER [89]. They observed that the nature of the support and the presence of organic ligand at the Ru surface are crucial factors for HER electroactivity, and the stability of these hybrid electrodes. Two different CFs [pristine (pCF) and oxidized (fCF)] were used to support "naked" and 4-phenylpyridine (PP)-functionalized Ru NPs (Fig. 12.28). In general, Ru NPs ligated

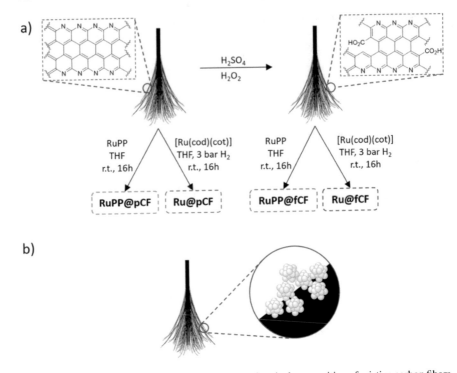

Fig. 12.28 **a** Schematic representation of the surface chemical composition of pristine carbon fibers (pCF; top-left) and functionalized carbon fibers (fCF; top-right) and experimental procedure for the deposition/synthesis of Ru NPs onto the surface of the carbon fibers. **b** Schematic representation of the surface of the carbon fibers once the Ru NPs have been deposited/synthesized. Reproduced from Ref. [89] with permission from Wiley

by PP showed better activities and stabilities than the ones without ligands. Between the two CFs studied, PP-functionalized Ru NPs supported on pCFs showed the higher stability. The great efficacy of RuPP@pCFs originates from π-interactions between the coordinated ligands and the support, which is hindered in RuPP@fCFs due to the presence of surface carboxylic groups. This work confirms the positive effect of the ancillary ligands on supported Ru NP catalysis, which in this case enhances the stability of the hybrid material thanks to π–π interaction between the ligand and the graphitic area of CFs.

Apart from the geometric and electronic effects induced by ligands on supported metal NPs, there is a third effect caused by a substrate/ligand interaction. Although to the best of our knowledge, there is no report on any example of ligand/reactant interaction of supported Ru NPs, it is worth mentioning a couple of examples centered on Pt and Pd. A clear example about controlling selectivity by substrate/ligand interactions was reported in 2014 by Medlin et al. [90] Pt NPs on Al_2O_3 were functionalized with self-assembled monolayers of various phenylthiols and used in the

Fig. 12.29 SAM-enhanced orientation of cinnamaldehyde with the catalyst surface. 3-Phenyl-1-propanethiol SAMs favoring aldehyde hydrogenation by creating an upright molecular orientation. Reproduced from Ref. [90] with permission from ACS

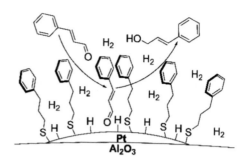

hydrogenation of cinnamaldehyde to cinnamyl alcohol. The chemoselective hydrogenation of the carbonyl group was induced by a non-covalent interaction between the substrate and the appropriate thiol as shown in Fig. 12.29. Interestingly, the maximum selectivity is achieved when the distance between the thiol and the phenyl group is the same as that in cinnamaldehyde (three carbons). The control of the selectivity through the ligand/substrate interaction shown by this system is reminiscent of biological catalysts, in which substrates interact with the enzymes by non-covalent bonds. In homogeneous supramolecular catalysis also, many examples can be found [91].

In 2010, Glorius et al. reported the induction of stereoselectivity by supported Pd NPs on magnetite (Fe_3O_4) functionalized with chiral NHCs (Fig. 12.30a) [92]. Here, also through a substrate/ligand interaction, asymmetric α-arylation of 2-methyl-1-tetralone using chloro- and bromobenzene was achieved (Fig. 12.30b). The magnetic character of the support, together with the heterogeneous nature of the catalysts allowed the magnetic recovery and the subsequent reuse of the catalyst without any loss of activity/selectivity. The possibility to control the chemoselectivity or stereoselectivity of supported nanoparticles through the substrate/ligand interaction is a promising method for nanoparticle catalysis that could be easily extended to supported ligand-functionalized Ru NPs.

12.4 Concluding Remarks

In this chapter, we have discussed the ability of ancillary ligands to modulate the catalytic properties of metal nanoparticles, focusing on Ru NP catalysis. The chapter is divided in two parts, colloidal solutions of Ru NPs (Sect. 12.2) and ligand modified Ru NPs on a support (Sect. 12.3). For the first systems, the presence of ligands, polymers, or detergents as stabilizers is a necessity as the particles would otherwise coalesce. Ligands bind to MNPs very much the same way as they bind to metals in metal complexes, although bridging of ligands is much more common in MNPs. Ideally, a metal complex catalyst contains only one "site," but MNPs will in most cases have a variety of sites, as is known from heterogeneous catalysis and surface studies. With the aid of ligands and reaction conditions, one can "control" size,

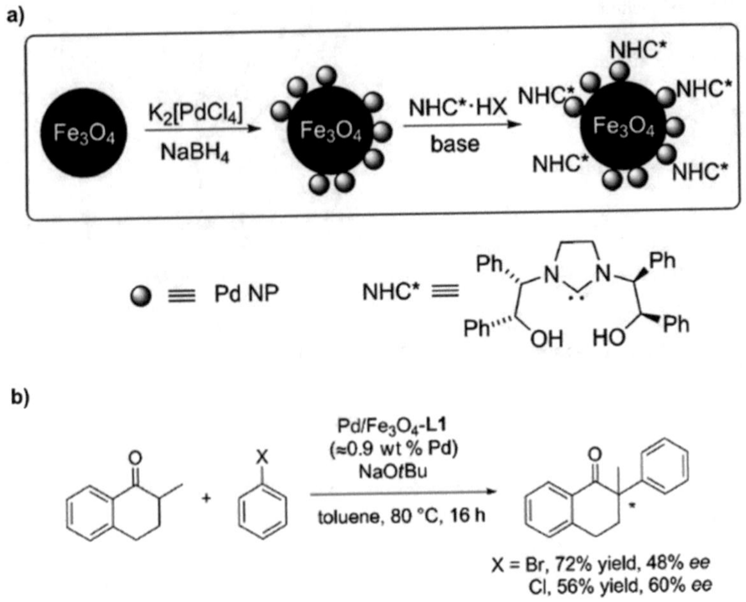

Fig. 12.30 **a** Preparation of Fe_3O_4/Pd NPs modified by chiral NHC; for clarity, the sizes are not represented proportionally. **b** The asymmetric a-arylation of 2-methyl-1-tetralone using chloro- and bromobenzene. Reproduced from [92] with permission from Wiley

exposed surfaces, electronic properties, etc., of the nanoparticles. As yet this is rather empirical, although on several occasions one has seen trends in sizes and catalytic activities due to electronic and steric effects of the ligands applied. The introduction of substrate/ligand interactions presents a further way to enhance rates and selectivity. As in homogeneous catalysts, ligands can be designed that foster heterolytic cleavage of dihydrogen and promote hydrogenation of polar bonds, while on classic metal catalysts the cleavage is always homolytic, apart from nearby support effects. The area of ligand-stabilized supported Ru NPs is considerably less studied and so far preformed ligated NPs have been reported. Post-ligand modification of immobilized Ru NPs is an interesting route to be developed, which has been neglected until now as ligands were mostly seen as inhibitors [93]. In the above, however, we have seen that in spite of dense ligand coverage many times fast catalysts were obtained, which could often be assigned to electronic effects of the ligands. In this chapter, we have shown the ability of surface ligands to modify the catalytic properties of Ru NPs, giving keys for future advances of this emerging field: ligand effects in metal NP catalysis.

References

1. (a) Heiz U, Landman U (2007) Nanocatalysis. Springer, Berlin; (b) Astruc D (2008) Nanoparticles and catalysis. Wiley-VCH, Weinheim; (c) Philippot K, Serp P (2013) Nanomaterials in catalysis. Wiley-VCH, Weinheim
2. Liu L, Corma A (2018) Metal catalysts for heterogeneous catalysis: from single atoms to nanoclusters and nanoparticles. Chem Rev 118:4981–5079; (b) Ni B, Wang X (2015) Face the edges: catalytic active sites of nanomaterials. Adv Sci 2:1500085; (c) Mostafa S, Behafarid F, Croy JR, Ono LK, Li L, Yang JC, Frenkel AI, Cuenya BR (2010) Shape-dependent catalytic properties of Pt nanoparticles. J Am Chem Soc 132:15714–15719
3. (a) McKenna F-M, Wells RPK, Anderson JA (2016) Enhanced selectivity in acetylene hydrogenation by ligand modified Pd/TiO$_2$ catalysts. Chem Commun 47:2351–2353; (b) McCue AJ, McKenna F-M, Anderson JA (2015) Triphenylphosphine: a ligand for heterogeneous catalysis too? Selectivity enhancement in acetylene hydrogenation over modified Pd/TiO$_2$ catalyst. Catal Sci Technol 5:2449–2459; (b) Marshall ST, O'Brien M, Oetter B, Corpuz A, Richards RM, Schwartz DK, Medlin JW (2010) Controlled selectivity for palladium catalysts using self-assembled monolayers. Nat Mater 9:853–858; (d) Liu P, Qin R, Fu G, Zheng N (2017) Surface coordination chemistry of metal nanomaterials. J Am Chem Soc 139:2122–2131; (e) Wang Y, Wan X-K, Ren L, Su H, Li G, Malola S, Lin S, Tang Z, Häkkinen H, Teo BK, Wang Q-M, Zheng N (2016) Atomically precise Alkynyl-protected metal nanoclusters as a model catalyst: observation of promoting effect of surface ligands on catalysis by metal nanoparticles. J Am Chem Soc 138:3278–3281
4. Martínez-Prieto LM, Chaudret B (2018) organometallic ruthenium nanoparticles: synthesis, surface chemistry, and insights into ligand coordination. Acc Chem Res 51:376–384
5. (a) Schrader I, Warneke J, Backenköhler J, Kunz S (2015) Functionalization of platinum nanoparticles with l-proline: simultaneous enhancements of catalytic activity and selectivity. J Am Chem Soc 137:905–912; (b) Makosch M, Lin W-I, Bumbálek V, Sá J, Medlin JW, Hungerbühler K, van Bokhoven JA (2012) Asymmetric heterogeneous catalysis: transfer of molecular principles to nanoparticles by ligand functionalization. ACS Catal 2:2079–2081
6. (a) Philippot K, Lignier P, Chaudret B (2014) organometallic ruthenium nanoparticles and catalysis, in ruthenium in catalysis, topics in organometallic chemistry. In: Dixneuf PH, Bruneau C (eds) Springer International Publishing; (b) Pla D, Gómez M (2016) ACS Catal 6:3537–3552
7. Lara P, Philippot K, Chaudret B (2013) ChemCatChem 5:28
8. (a) Rossi LM, Fiorio JL, Garcia MAS, Ferraz CP (2018) Dalton Trans 47:5889–5915; (b) An K, Somorjai GA (2012) Size and shape control of metal nanoparticles for reaction selectivity in catalysis. ChemCatChem 4:1512–1524; (c) Martinez-Prieto LM, Cano I, Marquez A, Baquero EA, Tricard S, Cusinato L, del Rosal I, Poteau R, Coppel Y, Philippot K, Chaudret B, Campora J, van Leeuwen PWNM (2017) Zwitterionic amidinates as effective ligands for platinum nanoparticle hydrogenation catalysts. Chem Sci 8:2931–2941
9. (a) Weare WW, Reed SM, Warner MG, Hutchison JE (2000) J Am Chem Soc 122:12890–12891; (b) Vignolle J, Tilley D (2009) Chem Commun 7230–7232; (c) Gomez S, Philippot K, Collière V, Chaudret B, Senocq F, Lecante P (2000) Chem Commun 1945–1946; (d) Baquero EA, Tricard S, Flores JC, de Jesús E, Chaudret B (2014) Angew Chem Int Ed 53:13220; (e) Asensio JM, Tricard S, Coppel Y, Andrés R, Chaudret B, de Jesús E (2017) Angew Chem Int Ed 56:865–869
10. (a) Richter C, Schaepe K, Glorius F, Ravoo BJ (2014) Chem Commun 50:3204–3207; (b) Hurst EC, Wilson K, Fairlamb IJS, Chechik V (2009) New J Chem 33:1837–1840
11. Ott LS, Finke RG (2007) Coord Chem Rev 251:1075–1100
12. Timonen JVI, Seppälä ET, Ikkala O, Ras RHA (2011) Angew Chem Int Ed 50:2080
13. (a) Philippot K, Chaudret B (2003) Organometallic approach to the synthesis and surface reactivity of noble metal nanoparticles. CR Chim 6:1019–1034; (b) Cormary B, Dumestre F, Liakakos N, Soulantica K, Chaudret B (2013) Dalton Trans 42:12546–12553
14. Amiens C, Chaudret B, Ciuculescu-Pradines D, Collière V, Fajerwerg K, Fau P, Kahn M, Maisonnat A, Soulantica K, Philippot K (2013) New J Chem 37:3374–3401

15. Mourdikoudis S, Liz-Marzán LM (2013) Nanoparticle synthesis. Chem Mat 25:1465–1476
16. Can H, Metin Ö (2012) A facile synthesis of nearly monodisperse ruthenium nanoparticles and their catalysis in the hydrolytic dehydrogenation of ammonia borane for chemical hydrogen storage. App Catal B Env 125:304–310
17. Lee JY, Yang J, Deivaraj TC, Too H-P (2003) A novel synthesis route for—protected ruthenium nanoparticles. J Colloid Interface Sci 268:77–80
18. Pan C, Pelzer K, Philippot K, Chaudret B, Dassenoy F, Lecante P, Casanove M-J (2001) Ligand-stabilized ruthenium nanoparticles: synthesis, organization, and dynamics. J Am Chem Soc 123:7584–7593
19. Watt J, Yu C, Chang SLY, Cheong S, Tilley RD (2013) Shape control from thermodynamic growth conditions: the case of hcp ruthenium hourglass nanocrystals. J Am Chem Soc 135:606–609
20. Lignier P, Bellabarba R, Tooze RP, Su Z, Landon P, Ménard H, Zhou W (2012) Facile synthesis of branched ruthenium nanocrystals and their use in catalysis. Cryst Growth Des 12:939–942
21. Zahmakıran M, Tristany M, Philippot K, Fajerwerg K, Özkar S, Chaudret B (2010) Amino-propyltriethoxysilane stabilized ruthenium(0) nanoclusters as an isolable and reusable heterogeneous catalyst for the dehydrogenation of dimethylamine–borane. Chem Commun 46:2938–2940
22. Zahmakıran M, Philippot K, Özkar S, Chaudret B (2012) Stabilized ruthenium(0) nanoparticles catalyst for the dehydrogenation of dimethylamine–borane at room temperature. Dalton Trans 41:590–598
23. Tristany M, Philippot K, Guari Y, Collière V, Lecante P, Chaudret B (2010) Synthesis of composite ruthenium-containing silica nanomaterials from amine-stabilized ruthenium nanoparticles as elemental bricks. J Mat Chem 20:9523–9530
24. Salas G, Santini CC, Philippot K, Colliere V, Chaudret B, Fenet B, Fazzini PF (2011) Influence of amines on the size control of in situ synthesized ruthenium nanoparticles in imidazolium ionic liquids. Dalton Trans 40:4660–4668
25. Salas G, Campbell PS, Santini CC, Philippot K, Costa GMF, Padua AAH (2012) Ligand effect on the catalytic activity of ruthenium nanoparticles. Dalton Trans 41:13919–13926
26. Chau NTT, Guegan J-P, Menuel S, Guerrero M, Hapiot F, Monflier E, Philippot K, Denicourt-Nowicki A, Roucoux A (2013) β-Cyclodextrins grafted with chiral amino acids: a promising supramolecular stabilizer of nanoparticles for? Appl Catal A 467:497–503
27. Favier I, Massou S, Teuma E, Philippot K, Chaudret B, Gomez M (2008) A new and specific mode of stabilization of metallic nanoparticles. Chem Commun: 3296–3298
28. Creus J, Drouet S, Surinach S, Lecante P, Colliere V, Poteau R, Philippot K, Garcia-Anton J, Sala X (2018) Ligand-capped Ru nanoparticles as efficient electrocatalyst for the hydrogen evolution reaction. ACS Catal 8:11094–11102
29. Jansat S, Picurelli D, Pelzer K, Philippot K, Gomez M, Muller G, Lecante P, Chaudret B (2006) Synthesis, characterization and catalytic reactivity of ruthenium nanoparticles stabilized by chiral N-donor ligands. New J Chem 30:115–122
30. Martinez-Prieto LM, Urbaneja C, Palma P, Campora J, Philippot K, Chaudret B (2015) A betaine adduct of N-heterocyclic carbene and carbodiimide, an efficient ligand to produce ultra-small ruthenium nanoparticles. Chem Commun 51:4647–4650
31. Novio F, Philippot K, Chaudret B (2010) Location and Dynamics of CO Co-ordination on Ru Nanoparticles: a study. Catal Lett 140:1–7
32. Martinez-Prieto LM, Cano I, Marquez A, Baquero EA, Tricard S, Cusinato L, del Rosal I, Poteau R, Coppel Y, Philippot K, Chaudret B, Campora J, van Leeuwen PWNM (2017) Zwitterionic amidinates as effective ligands for platinum nanoparticle hydrogenation catalysts. Chem Sci 8:2931–2941
33. López-Vinasco AM, Martínez-Prieto LM, Asensio JM, Lecante P, Chaudret B, Cámpora J, van Leeuwen PWNM (2020) Magnetic and recyclable Ni nanoparticles for selective semi-hydrogenation of alkynes to (Z)-alkenes. Catal Sci Tech 10:342–350
34. Fang M, Machalaba N, Sanchez-Delgado RA (2011) Hydrogenation of arenes and N-heteroaromatic compounds over ruthenium nanoparticles on poly(4-vinylpyridine): a versatile

catalyst operating by a substrate-dependent dual site mechanism. Dalton Trans 40:10621–10632

35. Fang M, Sánchez-Delgado RA (2014) Ruthenium nanoparticles supported on magnesium oxide: a versatile and recyclable dual-site catalyst for hydrogenation of mono- and poly-cyclic arenes, N-heteroaromatics, and S-heteroaromatics. J Catal 311:357–368
36. Martínez-Prieto LM, Puche M, Cerezo-Navarrete C, Chaudret B (2019) Uniform Ru nanoparticles on N-doped graphene for selective hydrogenation of fatty acids to alcohols. J Catal 77:429–437
37. Axet MR, Conejero S, Gerber IC (2018) Ligand effects on the selective hydrogenation of nitrobenzene to cyclohexylamine using ruthenium nanoparticles as catalysts. ACS Appl Nano Mater 1:5885–5894
38. Kurihara LK, Chow GM, Schoen PE (1995) Nanocrystalline metallic powders and films produced by the polyol method. Nanostruct Mater 5:607–613
39. Tran T-H, Nguyen T-D (2011) Controlled growth of uniform noble metal nanocrystals: aqueous-based synthesis and some applications in biomedicine. Coll Surf B 88:1–22
40. Vidoni O, Philippot K, Amiens C, Chaudret B, Balmes O, Malm J-O, Bovin J-O, Senoq F, Casanove M-J (1999) Novel, spongelike ruthenium particles of controllable size stabilized only by organic solvents. Angew Chem Int Ed 38:3736–3738
41. Pelzer K, Vidoni O, Philippot K, Chaudret B, Collière V (2003) Organometallic synthesis of size-controlled polycrystalline ruthenium nanoparticles in the presence of alcohols. Adv Funct Mat 13:118–126
42. Martínez-Prieto LM, Carenco S, Wu CH, Bonnefille E, Axnanda S, Liu Z, Fazzini PF, Philippot K, Salmeron M, Chaudret B (2014) Organometallic ruthenium nanoparticles as model catalysts for CO hydrogenation: a nuclear magnetic resonance and ambient-pressure x-ray photoelectron spectroscopy study. ACS Catal 4:3160–3168
43. Pelzer K, Philippot K, Chaudret B (2003) Z Phys Chem 217:1539–1548
44. Nowicki A, Zhang Y, Léger B, Rolland J-P, Bricout H, Monflier E, Roucoux A (2006) Supramolecular shuttle and protective agent: a multiple role of methylated in the chemoselective hydrogenation of benzene derivatives with ruthenium nanoparticles. Chem Commun: 296–298
45. Chau NTT, Handjani S, Guegan J-P, Guerrero M, Monflier E, Philippot K, Denicourt-Nowicki A, Roucoux A (2013) Methylated β-cyclodextrin-capped ruthenium nanoparticles: synthesis strategies, characterization, and application in hydrogenation reactions. ChemCatChem 5:1497–1503
46. González-Gómez R, Cusinato L, Bijani C, Coppel Y, Lecante P, Amiens C, del Rosal I, Philippot K, Poteau R (2019) Carboxylic acid-capped ruthenium nanoparticles: experimental and theoretical case study with ethanoic acid. Nanoscale 11:9392–9409
47. Wang J, Feng J, Qin R, Fu H, Yuan M, Chen H, Li X (2007) Asymmetric hydrogenation of aromatic ketones catalyzed by achiral monophosphine TPPTS-stabilized Ru. Tetrahedron Asymmetry 18:1643–1647
48. García-Antón J, Axet MR, Jansat S, Philippot K, Chaudret B, Pery T, Buntkowsky G, Limbach HH (2008) Reactions of olefins with ruthenium hydride nanoparticles: nmr characterization, hydride titration, and room-temperature C-C bond activation. Angew Chem Int Ed 47:2074–2078
49. Novio F, Monahan D, Coppel Y, Antorrena G, Lecante P, Philippot K, Chaudret B (2014) Surface chemistry on small ruthenium nanoparticles: evidence for site selective reactions and influence of ligands. Chem Eur J 20:1287–1297
50. Bonnefille E, Novio F, Gutmann T, Poteau R, Lecante P, Jumas J-C, Philippot K, Chaudret B (2014) a way to tune selectivity in hydrogenation reaction. Nanoscale 6:9806–9816
51. Rothermel N, Röther T, Ayvalı T, Martínez-Prieto LM, Philippot K, Limbach H-H, Chaudret B, Gutmann T, Buntkowsky G (2019) Reactions of D2 with 1,4-butane-stabilized metal nanoparticles-a combined gas-phase NMR, GC-MS and solid-state NMR study. ChemCatChem 11:1465–1471

52. Rothermel N, Bouzouita D, Roether T, de Rosal I, Tricard S, Poteau R, Gutmann T, Chaudret B, Limbach H-H, Buntkowsky G (2018) surprising differences of alkane catalyzed by ruthenium nanoparticles: complex surface-substrate recognition? ChemCatChem 10:4243–4247

53. Llop Castelbou J, Breso-Femenia E, Blondeau P, Chaudret B, Castillon S, Claver C, Godard C (2014) Tuning the selectivity in the hydrogenation of aromatic ketones catalyzed by similar ruthenium and rhodium nanoparticles. ChemCatChem 6:3160–3168

54. Breso-Femenia E, Castillon S, Godard C, Claver C, Chaudret B (2015) Selective catalytic deuteration of phosphorus ligands using ruthenium nanoparticles: a new approach to gain information on ligand coordination. Chem Commun 51:16342–16345

55. Escarcega-Bobadilla MV, Tortosa C, Teuma E, Pradel C, Orejon A, Gomez M, Masdeu-Bulto AM (2009) Ruthenium and rhodium nanoparticles as catalytic precursors in supercritical carbon dioxide. Catal Today 148:398–404

56. Gonzalez-Galvez D, Nolis P, Philippot K, Chaudret B, van Leeuwen PWNM (2012) Phosphine-stabilized ruthenium nanoparticles: the effect of the nature of the ligand in catalysis. ACS Catal 2:317–321

57. Guerrero M, Roucoux A, Denicourt-Nowicki A, Bricout H, Monflier E, Colliere V, Fajerwerg K, Philippot K (2012) Alkyl sulfonated diphosphines-stabilized ruthenium nanoparticles as efficient nanocatalysts in hydrogenation reactions in biphasic media. Catal Today 183:34–41

58. Guerrero M, Coppel Y, Chau NTT, Roucoux A, Denicourt-Nowicki A, Monflier E, Bricout H, Lecante P, Philippot K (2013) Efficient ruthenium nanocatalysts in liquid-liquid biphasic hydrogenation catalysis: towards a supramolecular control through a sulfonated smart combination. ChemCatChem 5:3802–3811

59. Debouttière P-J, Coppel Y, Denicourt-Nowicki A, Roucoux A, Chaudret B, Philippot K (2012) PTA-stabilized ruthenium and platinum nanoparticles: characterization and investigation in aqueous biphasic hydrogenation catalysis. Eur J Inorg Chem 2012:1229–1236

60. Rafter E, Gutmann T, Loew F, Buntkowsky G, Philippot K, Chaudret B, van Leeuwen PWNM (2013) Secondary phosphine oxides as pre-ligands for nanoparticle stabilization. Catal Sci Technol 3:595–599

61. Cano I, Chapman AM, Urakawa A, van Leeuwen PWNM (2014) Air-stable gold nanoparticles ligated for the chemoselective hydrogenation of aldehydes: crucial role of the ligand. J Am Chem Soc 136:2520–2528

62. Cano I, Huertos MA, Chapman AM, Buntkowsky G, Gutmann T, Groszewicz PB, van Leeuwen PWNM (2015) Air-stable gold nanoparticles ligated as catalyst for the chemoselective hydrogenation of substituted aldehydes: a remarkable ligand effect. J Am Chem Soc 137:7718–7727

63. Cano I, Martínez-Prieto LM, Chaudret B, van Leeuwen PWNM (2017) Iridium versus iridium: nanocluster and monometallic catalysts carrying the same ligand behave differently. Chem A Eur J 23:1444–1450

64. Cano I, Martinez-Prieto LM, Fazzini PF, Coppel Y, Chaudret B, van Leeuwen PWNM (2017) Characterization ligands on the surface of iridium nanoparticles. PCCP 19:21655–21662

65. Cano I, Tschan MJL, Martinez-Prieto LM, Philippot K, Chaudret B, van Leeuwen PWNM (2016) Enantioselective hydrogenation of ketones by iridium nanoparticles ligated with. Catal Sci Tech 6:3758–3766

66. Gual A, Axet MR, Philippot K, Chaudret B, Denicourt-Nowicki A, Roucoux A, Castillon S, Claver C (2008) Diphosphite ligands derived from carbohydrates as stabilizers for ruthenium nanoparticles: promising catalytic systems in arene hydrogenation. Chem Commun: 2759–2761

67. Gual A, Godard C, Philippot K, Chaudret B, Denicourt-Nowicki A, Roucoux A, Castillón S, Claver C (2009) Carbohydrate-Derived 1,3-Diphosphite Ligands as Chiral Nanoparticle Stabilizers: Promising Catalytic Systems for. ChemSusChem 2:769–779

68. Lara P, Rivada-Wheelaghan O, Conejero S, Poteau R, Philippot K, Chaudret B (2011) Ruthenium nanoparticles stabilized by N-heterocyclic carbenes: ligand location and influence on reactivity. Angew Chem Int Ed 50:12080–12084

69. Gonzalez-Galvez D, Lara P, Rivada-Wheelaghan O, Conejero S, Chaudret B, Philippot K, van Leeuwen PWNM (2013) NHC-stabilized ruthenium nanoparticles as new catalysts for the hydrogenation of aromatics. Catal Sci Tech 3:99–105
70. Martínez-Prieto LM, Ferry A, Lara P, Richter C, Philippot K, Glorius F, Chaudret B (2015) New Route to Stabilize Ruthenium Nanoparticles with Non-Isolable Chiral N-Heterocyclic Carbenes. Chem A Eur J 21:17495–17502
71. Lara P, Martínez-Prieto LM, Roselló-Merino M, Richter C, Glorius F, Conejero S, Philippot K, Chaudret B (2016) NHC-stabilized Ru nanoparticles: Synthesis and surface studies. NanoSO 6:39–45
72. Higman CS, Lanterna AE, Marin ML, Scaiano JC, Fogg DE (2016) Catalyst decomposition during yields isomerization-active ruthenium nanoparticles. ChemCatChem 8:2446–2449
73. Tay BY, Wang C, Phua PH, Stubbs LP, Huynh HV (2016) Selective hydrogenation of levulinic acid to γ-valerolactone using in situ generated ruthenium nanoparticles derived complexes. Dalton Trans 45:3558–3563
74. Baquero EA, Tricard S, Flores JC, de Jesús E, Chaudret B (2014) Highly stable water-soluble platinum nanoparticles stabilized by hydrophilic N-heterocyclic carbenes. Angew Chem Int Ed 126:13436–13440
75. Asensio JM, Tricard S, Coppel Y, Andrés R, Chaudret B, de Jesús E (2017) Knight shift in 13c nmr resonances confirms the coordination of N-heterocyclic carbene ligands to water-soluble palladium nanoparticles. Angew Chem Int Ed 129:883–887
76. Martinez-Prieto LM, Baquero EA, Pieters G, Flores JC, de Jesus E, Nayral C, Delpech F, van Leeuwen PWNM, Lippens G, Chaudret B (2017) Monitoring of nanoparticle reactivity in solution: interaction of l-lysine and Ru nanoparticles probed parallels regioselective H/. Chem Commun 53:5850–5853
77. Taglang C, Martínez-Prieto LM, del Rosal I, Maron L, Poteau R, Philippot K, Chaudret B, Perato S, Sam Lone A, Puente C, Dugave C, Rousseau B, Pieters G (2015) Enantiospecific using ruthenium nanocatalysts. Angew Chem Int Ed 54:10474–10477
78. Palazzolo A, Feuillastre S, Pfeifer V, Garcia-Argote S, Bouzouita D, Tricard S, Chollet C, Marcon E, Buisson D-A, Cholet S, Fenaille F, Lippens G, Chaudret B, Pieters G (2019) Efficient access to deuterated and tritiated nucleobase pharmaceuticals and oligonucleotides using hydrogen-isotope exchange. Angew Chem Int Ed 58:4891–4895
79. Bouzouita D, Lippens G, Baquero EA, Fazzini PF, Pieters G, Coppel Y, Lecante P, Tricard S, Martínez-Prieto LM, Chaudret B (2019) Tuning the catalytic activity and selectivity of water-soluble bimetallic by modifying their surface metal distribution. Nanoscale 11:16544–16552
80. Martinez-Prieto LM, Ferry A, Rakers L, Richter C, Lecante P, Philippot K, Chaudret B, Glorius F (2016) as versatile catalysts for one-pot/hydrogenation reactions. Chem Commun 52:4768–4771
81. Martínez-Prieto LM, Rakers L, López-Vinasco AM, Cano I, Coppel Y, Philippot K, Glorius F, Chaudret B, van Leeuwen PWNM (2017) Soluble platinum nanoparticles ligated by long-chain N-heterocyclic carbenes as catalysts. Chem A Eur J 23:12779–12786
82. Rakers L, Martínez-Prieto LM, López-Vinasco AM, Philippot K, van Leeuwen PWNM, Chaudret B, Glorius F (2018) Ruthenium nanoparticles ligated by and their application in the hydrogenation of arenes. Chem Commun 54:7070–7073
83. Kunz S (2016) Supported, ligand-functionalized nanoparticles: an attempt to rationalize the application and potential of ligands in heterogeneous catalysis. Top Catal 59:1671–1685
84. Mallat T, Orglmeister E, Baiker A (2007) Asymmetric catalysis at chiral metal surfaces. Chem Rev 107:4863–4890
85. Ernst JB, Muratsugu S, Wang F, Tada M, Glorius F (2016) Tunable heterogeneous catalysis: N-heterocyclic carbenes as ligands for supported heterogeneous Ru/ catalysts to tune reactivity and selectivity. J Am Chem Soc 138:10718–10721
86. Albani D, Li Q, Vile G, Mitchell S, Almora-Barrios N, Witte PT, Lopez N, Perez-Ramirez J (2017) Interfacial acidity in ligand-modified ruthenium nanoparticles boosts the hydrogenation of levulinic acid to gamma-valerolactone. Green Chem 19:2361–2370
87. van Leeuwen PWNM, González Gálvez D (2013) WO/2014/056889, 8 Oct 2013

88. Jahjah M, Kihn Y, Teuma E, Gomez M (2010) Ruthenium nanoparticles supported on multi-walled carbon nanotubes: highly effective catalytic system for hydrogenation processes. J Mol Catal A Chem 332:106–112

89. Creus J, Mallon L, Romero N, Bofill R, Moya A, Fierro JLG, Mas-Balleste R, Sala X, Philippot K, Garcia-Anton J (2019) Ruthenium nanoparticles supported for hydrogen evolution electrocatalysis. Eur J Inorg Chem: 2071–2077

90. Kahsar KR, Schwartz DK, Medlin JW (2014) Control of metal catalyst selectivity through specific noncovalent molecular interactions. J Am Chem Soc 136:520–526

91. Raynal M, Ballester P, Vidal-Ferran A, van Leeuwen PWNM (2014) Supramolecular catalysis. Part 1: non-covalent interactions as a tool for building and modifying homogeneous catalysts. Chem Soc Rev 43:1660–1733

92. Ranganath KVS, Kloesges J, Schäfer AH, Glorius F (2010) Asymmetric Nanocatalysis: N-heterocyclic carbenes as chiral modifiers of Fe_3O_4/Pd nanoparticles. Angew Chem Int Ed 49:7786–7789

93. Sonstrom P, Baumer M (2011) Supported colloidal nanoparticles in heterogeneous gas phase catalysis: on the way to tailored catalysts. Phys Chem Chem Phys 13:19270–19284

Index

© Springer Nature Switzerland AG 2020
P. W. N. M. van Leeuwen and C. Claver (eds.), *Recent Advances in Nanoparticle Catalysis*, Molecular Catalysis 1, https://doi.org/10.1007/978-3-030-45823-2

Printed in the United States
by Baker & Taylor Publisher Services